Zoological Monographs

Volume 9

The book series *Zoological Monographs* is devoted to the study of the animal kingdom and covers all aspects of animals, both living and extinct, including taxonomy, systematics and evolution, biodiversity and biogeography, ecology, morphology and anatomy, physiology and behavior. The volumes in the series either discuss the biology of a single animal group, approach a topic from an interspecies level, or present recent methodologies. The series is intended for graduate students, university teachers, scientists and professionals.

Louis H. du Preez • Willem J. Landman •
Olivier Verneau

Polystomatid Flatworms

State of Knowledge and Future Trends

Springer

Louis H. du Preez
Unit for Environment Science
and Management
North-West University
Potchefstroom, South Africa

Willem J. Landman
Unit for Environment Science
and Management
North-West University
Potchefstroom, South Africa

Olivier Verneau
Unit for Environment Science
and Management
North-West University
Potchefstroom, South Africa

ISSN 2523-3904 ISSN 2523-3912 (electronic)
Zoological Monographs
ISBN 978-3-031-35886-9 ISBN 978-3-031-35887-6 (eBook)
https://doi.org/10.1007/978-3-031-35887-6

This Springer imprint is published by the registered company Springer Nature Switzerland AG
The registered company address is: Gewerbestrasse 11, 6330 Cham, Switzerland

Authors LdP and OV would like to dedicate this book to Professor Claude Combes who introduced them to the world of polystomes

Acknowledgements

The authors wish to express their gratitude to the following people:

For assisting with the text and extracting information from the literature:

David Gibson; Amira Chaabane; Masters students at the North-West University (NWU): Michelle van Rooyen, Carina Coezer, Christa Morrisson and Joretha du Buisson.

For providing images:

Craig Adler, USA; Joe Jackson, UK; Jean-Lou Justine, Natural History Museum of Paris; Peter Olson, Natural History Museum of London; Anna Philips and Yolanda Villacampa and the: Scientific Imaging Lab, Smithsonian Institute Department of Invertebrate Zoology, National Museum of Natural History, Smithsonian Institution, Washington DC; Tom Platt, USA and Richard Tinsley, Bristol, UK.

For making museum specimens available:

Mal Bryant, Queensland Museum; Claude Combes, Perpignan University, France; Marcello Knoff, Institute Oswaldo Cruz, Brazil; Pat Pillit, US National Museum, USA and Richard Tinsley, Bristol University, UK.

For financial support:

National Research Foundation of South Africa and South African Institute for Aquatic Biodiversity.

Colleagues, friends and family for patience and support.

Contents

About the Authors

Louis H. du Preez is a professor of zoology at North-West University and heads the African Amphibian Conservation Research Group. He holds a Ph.D. in amphibian parasitology and his research interests include parasites, diseases and conservation of amphibians and freshwater turtles with a special focus on polystomatid flatworms.

Olivier holding a snapping turtle and Louis an alligator snapping turtle

Olivier Verneau is a professor of biology at the University of Perpignan. He holds a Ph.D. in host (fish)—parasite (tapeworms) coevolution. His research interests include biogeography, taxonomy, systematics and evolution of hosts and their parasites. He worked extensively on the herpetofauna of France with a special focus on the threatened Mediterranean turtle *Mauremys leprosa*.

About the Artist

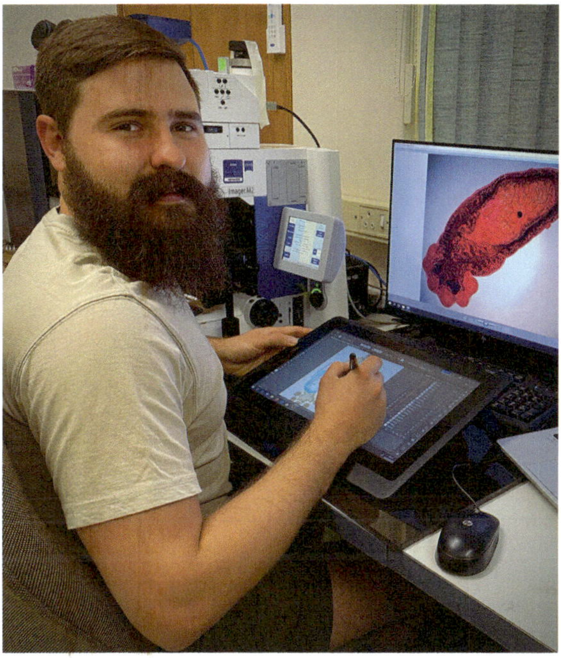

Willem J. Landman is a microscopy specialist and herpetology collections manager at the North-West University. He holds a M. Sc. in parasitology and his research interests include microscopy, parasitology with a particular focus on polystomatid flatworms, and taxonomic illustrations.

Chapter 1
Overview of the Polystomatidae: Systematics and Classification

Contents

Abstract Polystomatids (Polystomatidae) are platyhelminthes of the class Monogenea. While monogeneans are mainly fish parasites, all polystomatids, with a single exception, namely *Concinnocotyla australensis* (Reichenbach-Klinke, 1966) of the Australian lungfish, infect semi-aquatic tetrapods. The outlines of the Polystomatidae sensu Sinnappah et al. (Mol Phylogenet Evol 18:189–201, 2001) are given with the best modern classification, which now includes 30 genera of which 19 infect amphibians, 9 freshwater turtles, and one each infecting the Australian lungfish and the common hippopotamus, respectively. Diagnosis of subfamilies and genera is also presented, with type species and morphological characters. A family key to genus level is based on the most relevant morphological characters, host spectrum and distribution.

1.1 Systematics of the Polystomatidae

1.1.1 Outlines of the Family

The Neodermata comprises a clade of rhabditophoran flatworms that includes all parasitic platyhelminthes, namely the Cestoda (tapeworms), the Trematoda (digenetic flukes) and the Monogenea (monogenetic flukes). These parasites fascinated scientists from the early days, as depicted in one of the famous "Illustrations of Nature" by Ernst Haeckel (Fig. 1.1).

L. H. du Preez et al., *Polystomatid Flatworms*, Zoological Monographs 9,
https://doi.org/10.1007/978-3-031-35887-6_1

Fig. 1.1 Plate of flatworms as part of the series "Illustrations of Nature" by Ernst Haeckel. *Polystoma integerrimum* feature in the middle. © Wikipedia commons

In the 90s, the use of the terms Monogenoidea versus Monogenea was heavily debated (see Boeger and Kritsky 1993). Although there still exist two schools of thought, most scientists today accept Monogenea (see Wheeler and Chisholm 1995). Regardless of the characters used in systematics (morphological and/or molecular), the monogeneans are divided into two monophyletic groups, both infecting mainly marine and freshwater actinopterygian and chondrichthyan fishes and to a lesser extent semi-aquatic tetrapods (Boeger and Kritsky 1993, 1997, 2001; Mollaret et al. 1997; Campos et al. 1998; Littlewood et al. 1999a, b; Olson and Littlewood 2002; Lockyer et al. 2003). Regarding the classification of Boeger and Kritsky (2001), even if not being widely accepted by the international scientific community, the class Monogenoidea was further divided into two subclasses—Polyonchoinea and Heteronchoinea. The latter class is subdivided into two infrasubclasses—Oligonchoinea and Polystomatoinea. Irrespective of the nomenclature to be adopted for their classification, the monophyly and origins of monogeneans were still questioned (Littlewood et al. 2001; Perkins et al. 2010).

Within the Heteronchoinea, the Oligonchoinea encompasses worms which are strictly fish parasites, while the Polystomatoinea infests exclusively rhipidistians, such as the Australian lungfish *Neoceratodus forsteri* Krefft, 1870, amphibians, freshwater turtles and the common hippopotamus *Hippopotamus amphibius* Linnaeus, 1758. Polystomatoineans have long been divided into two families—the Sphyranuridae with three extant species after Price (1939), that were classified in a single genus *Sphyranura* Wright, 1879; and the Polystomatidae that was subdivided, until early 2001, into 19 genera. Based on morphological and biological structures, sphyranurids differ from polystomatids mainly by the number of suckers located on the opisthaptor, but also by their infestation site. Sphyranurids have a single pair of haptoral suckers and are gill or skin parasites of North American salamanders, while the remainder polystomatids have three pairs of cup-like suckers on a well-developed opisthaptor. They are mostly internal parasites of anurans, except for *Metapolystoma* and *Polystoma* that may be found either on the gills of frogs' tadpoles or in the bladder of adult hosts and internal or semi-external in freshwater turtles.

Regarding the ontogenetic development of the haptor in both sphyranurids and polystomatids, i.e., a single pair of haptoral suckers develop at the site of hooks III in sphyranurids (Llewellyn 1963) while three pairs of haptoral suckers develop successively at the sites of hooks III, IV and V in polystomatids (Gallien 1935; Paul 1938), Llewellyn (1963) considered a basal position of sphyranurids within polystomatoineans. This hypothesis was also defended by Prudhoe and Bray (1982), who suggested that polystomatids might have evolved from sphyranurids following similarities existing in the infestation site of sphyranurids and most of the fish monogenean parasites. From phylogenetic evidence based on the analysis of morphological characters, Boeger and Kritsky (1997) postulated that sphyranurids and amphibian polystomatids may have initially co-evolved within amphibians after initial divergence of the Batrachia, namely the clade grouping the Anura and the Urodela and secondarily switched to freshwater turtles. Regardless of the hypothesis that was advocated to explain the evolution of polystomatoineans, the systematic status of the Sphyranuridae has been disputed for a very long time. Wright (1879) originally placed representatives of *Sphyranura* in the Polystomatidae, while Monticelli (1903) transferred them to the Dicotylidae. Poche (1926), subsequently, established the definitive name Sphyranuridae. While Fuhrmann (1928) restored sphyranurids to the Polystomatidae, Alvey (1936) considered that sphyranurids should be reassigned in the Sphyranuridae. Finally, though Price (1939) also supported the Polystomatidae, Bychowsky (1957) concluded that sphyranurids should be regarded as representatives of the Sphyranuridae. New considerations about the larval development of polystomatoineans in the 90s contributed to the reconsideration of the systematic status of the Sphyranuridae. Kearn (1994) assumed that the two-suckers condition within sphyranurids could represent an adaptation to the lifestyle of these parasites that would have arisen by progenesis. Similarly, Williams (1995) considered that a single pair of haptoral suckers could indicate arrested development following neoteny. Regardless of the scenario leading to the possession of a unique pair of suckers, either by progenesis or neoteny, the presence of two suckers might have represented a derived character rather than a primitive one. Consequently, according to these authors, sphyranurids should have been assigned to the Polystomatidae instead of Sphyranuridae.

In order to clarify the taxonomy of the Sphyranuridae, Sinnappah et al. (2001) established the phylogeny of the Polystomatoinea by investigating the relationships of seven polystomatids and one sphyranurid from partial 18S rDNA sequences. They clearly showed that the Polystomatidae was not monophyletic since *Sphyranura oligorchis* Alvey, 1933 was nested within polystomatids of the Anura. Two sister clades were actually revealed within the Polystomatoinea—the first one grouping turtle polystomatids, the second one including all amphibian polystomatoineans, namely anuran polystomatids and the caudate polystomatid, *S. oligorchis*. Molecular evidence thus pointed to the need for a revision of the classification of polystomatoineans. Regarding the haptoral development of polystomatids and their phylogeny, one should consider that the ancestral state for all polystomatoineans might have been a haptor with three pairs of suckers. Therefore, the two-suckers condition which is observed within sphyranurids could have reflected an arrested development due to neoteny, as was previously suggested by Williams (1995) from comparative morphological studies conducted in *Polystoma integerrimum* (Fröhlich, 1791) by Williams and McKenzie (1995). Assuming *S. oligorchis* as a paedomorphic parasite led Sinnappah et al. (2001) to consider this species as a true polystome which, surprisingly, accomplishes his development on a permanently neotenic host, namely the American mudpuppy *Necturus maculosus* (Rafinesque, 1818). Therefore, *Sphyranura* should be restored to the family Polystomatidae, as proposed by Sinnappah et al. (2001). For the purpose of this document, we will refer to the Polystomatidae sensu Sinnappah et al. (2001).

1.1.2 Geographic and Host Distribution

The majority of polystomes are from the Ethiopian Realm (36%), followed by the Neotropical (18%), Oriental (15%), Nearctic (13%), Palaearctic (12%) and Australian (6%) Realms (Fig. 1.2).

They are known from the skin and gills of the Australian lungfish (Dipnoi); gills of tadpoles, kidneys, accessory bladders and urinary bladders of frogs (Anura); skin, gills and oral cavity of salamanders (Caudata); urinary bladder, cloaca and phalodeum of caecilians (Gymnophiona); surface of the eye, under eyelids, nasal cavities, mouth, pharyngeal pouches, cloaca, accessory bladders and urinary bladder of freshwater turtles (Testudines) and on the eye surface and under eyelids of the common hippopotamus (Cetartiodactyla). They are presently represented by 30 genera—one each infecting, respectively, the Australian lungfish and the common hippopotamus, 19 infecting amphibians and 9 freshwater turtles. Currently, slightly more than 200 polystomatid species are recognized, which are, in general, host and site-specific. Most polystomes (62%) are found in anuran hosts, followed by chelonian hosts (32.6%). Only a few species are known from the remaining hosts, namely salamanders (3%), caecilians (1.2%), the Australian lungfish (0.6%) and the common hippopotamus (0.6%) (Fig. 1.3).

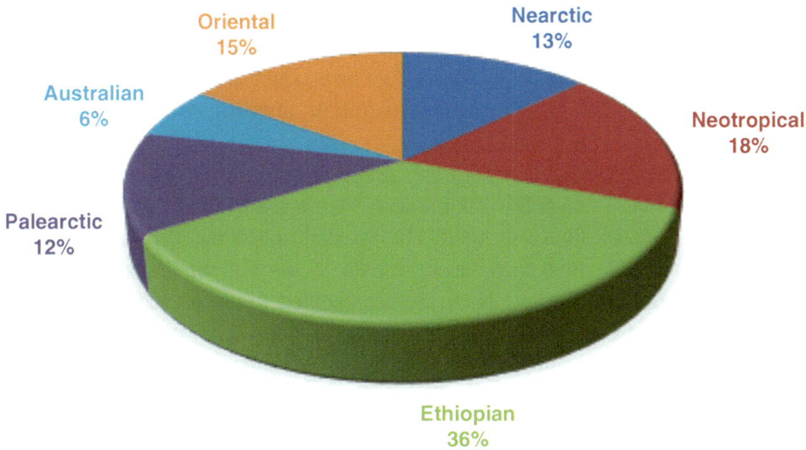

Fig. 1.2 Pie chart showing the respective proportions of polystomes from biogeographic realms

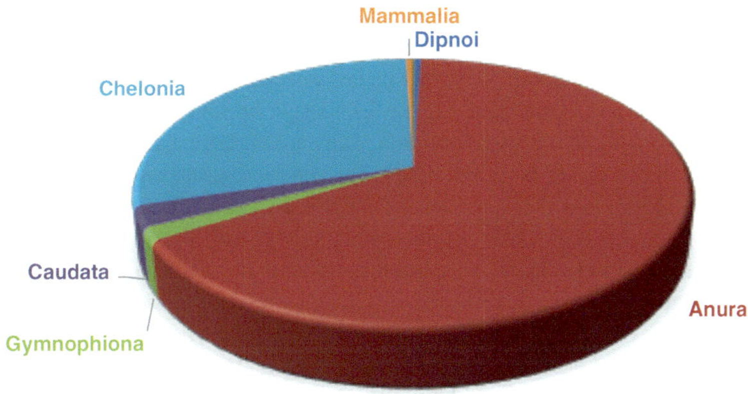

Fig. 1.3 Pie chart showing the respective proportions of polystomes parasitizing different host groups

Parasite transmission can only occur in an aquatic environment, involving an aquatic larva, the oncomiracidium. The exposure to the aquatic environment may be permanent for fully aquatic amphibians, like *Xenopus* Wagler, 1827, or cover a brief period of less than 1 h per year, for frogs adapted to arid conditions, like *Scaphiopus* Holbrook, 1836. As a result, the diversity of host species is limited to aquatic or semi-aquatic hosts, which include amphibians and freshwater turtles, but also, to a lesser extent, the Australian lungfish and the common hippopotamus. Suitable hosts are distributed worldwide except the Arctic and Antarctic regions.

The lungfish is at the crossroads between aquatic vertebrates, such as ray-finned fishes and coelacanths and tetrapods. It retains ancestral characteristics like the ability to breathe air and has lobed fins with a well-developed bony skeleton. It is represented nowadays by three genera and six species that occur in freshwaters of the South-American (*Lepidosiren* Fitzinger, 1837; 1 species), African (*Protopterus* Owen, 1839; 4 species) and Australian (*Neoceratodus* Castelnau, 1876; 1 species) continents. A single species, i.e. the Australian lungfish, is known to host a polystome.

Salamanders and newts are aquatic or semi-aquatic amphibians belonging to the Caudata. They have a lizard-like appearance with a long tail attached to a slender body, small eyes and a flattened head (Hickman et al. 2006). They are commonly found in temperate climates, particularly in habitats such as wetlands, rainforests or streams of the holarctic region. They are represented today by 820 species (Frost 2023). Three aquatic salamanders are known to host polystomes, namely the mudpuppy *Necturus maculosus*, the Oklahoma salamander *Eurycea tynerensis* Moore and Hughes, 1939 and the Japanese clawed salamander *Onychodactylus japonicus* (Houttuyn, 1782).

Frogs and toads, the most common amphibians, belong to the Anura. They are characterized by the absence of a tail in the adults and powerful hind limbs that enable them to jump. Other characteristics comprise a large mouth, two pairs of limbs, the fusion of their trunk and the presence of 6–10 vertebrae with an urostyle. They have a cosmopolitan distribution with a vast diversity comprising 7646 species (Frost 2023). There are about 140 frog species infected today by polystomes, which means that the Anura is the most successful host group for these parasites.

Caecilians belong to the Gymnophiona. They are elongated and limbless amphibians adapted to burrowing (soil-dwelling). The generic characteristics include a long, slender body, long ribs, a terminal anus, rudimentary eyes, a snout containing a pair of sensory tentacles, a terminal or subterminal mouth and a short tail when present (Hickman et al. 2006). They are commonly found in terrestrial, semi-aquatic or aquatic habitats in tropical environments of South America, central Africa, India and Southeast Asia. They are represented today by 221 species (Frost 2023). In addition, three caecilians are known to host polystomes, namely *Caecilia pachynema* Günther, 1859, *Caecilia gracilis* Shaw, 1802 and *Typhlonectes compressicauda* (Duméril and Bibron, 1841).

Freshwater turtles belong to the Testudines. They are oviparous reptilians characterized by a shell comprising a ventral plastron and a dorsal carapace, the absence of teeth (instead, the jaw contains keratinized plates) and limbs and limb girdles found inside the ribs (Hickman et al. 2006). Freshwater turtles are found everywhere except the Arctic. They are represented today by about 300 species (Rhodin et al. 2021). About 40 species of freshwater turtles are known to be infected by polystomes, some of which are infected by three distinct species.

The common hippopotamus is the third largest mammal commonly found in sub-Saharan Africa. It is semi-aquatic, has a barrel-shaped body and massive mouth with prominent and razor-sharp, canines. Hippopotami are territorial animals that reach maturity at approximately 7–8 years for the males and about 9–10 years for the females. Geographical distribution and populations have declined over the years due to anthropogenic activities, such as habitat fragmentation and poaching for ivory and meat. The common hippopotamus is the single mammal known to host a polystome.

1.2 Classification of the Polystomatidae

1.2.1 Diagnosis of the Family

Body elongate or somewhat pyriform, tapering anteriorly, with a haptor posteriorly; opisthaptor bilobed with two or six discoid suckers, with or without hamuli, 16 marginal hooklets; mouth terminal or subterminal, surrounded by false oral sucker, opening into a muscular pharynx; bucco-oesophageal canal; intestine bifurcate, crura with or without diverticula, united posteriorly or not; testis single, double or multiple, sometimes reticular; genital pore median, situated in region of intestinal bifurcation; cirrus with or without genital spines arranged in single or double ring; ovary compact, usually located in front of testes; vitellaria follicular diffuse throughout body, restricted in two lateral fields or densely compacted; vaginae, when present, in antero-lateral position; parasitic on skin and gills of the Australian lungfish, on gills of tadpoles and in kidneys and urinary bladders of anurans, on gills and skin of salamanders, in bladder and posterior lobes of caecilians, on eye or under eyelids, in nose, mouth, pharyngeal pouches or urinary bladder of chelonians and on eyes or under eyelids of the common hippopotamus.

In the classification of polystomes, several morphological features are being used, of which the most prominent are depicted in Fig. 1.4a & b. They include the size of the parasite; the position of the mouth, i.e. terminal or subterminal; the absence or presence of an uterus and its size and shape, i.e. small with few eggs, medium with 15–50 eggs, big with many eggs, up to a few hundreds, tubiform or sacciform; the absence or presence of vaginae and their size, i.e. small or big; the position and size of the ovary; the shape of the vitellarium or vitellaria, i.e. diffuse throughout body proper (pink stippling) or confined to two lateral fields, compact dense organs (dark pink); the position of the testis, number and shape, i.e. diffuse, single compact or multiple (dark blue); the shape of the intestine, i.e. diverticulated with or without pre-haptoral anastomoses or haptoral anastomoses, diverticulated not confluent posteriorly, not diverticulated with haptoral anastomose, not diverticulated and confluent posteriorly without haptoral anastomose, with two branches of equal length and not confluent posteriorly, with two branches of unequal length not confluent posteriorly; the number of haptoral suckers, i.e. 2 or 6, symmetrical or asymmetrical, with or without skeletal elements; the presence or not of hamuli, i.e. none, one or two pairs with their relative size to size of suckers.

1.2.2 Diagnosis of Subfamilies and Genera

1.2.2.1 The Concinnocotylinae Pichelin, Whittington and Pearson, 1991 (Fig. 1.5)

Body elongate; haptor with three pairs of suckers and one pair of very small hamuli; haptoral suckers asymmetrical, near triangular in shape, with sclerites in the shape of

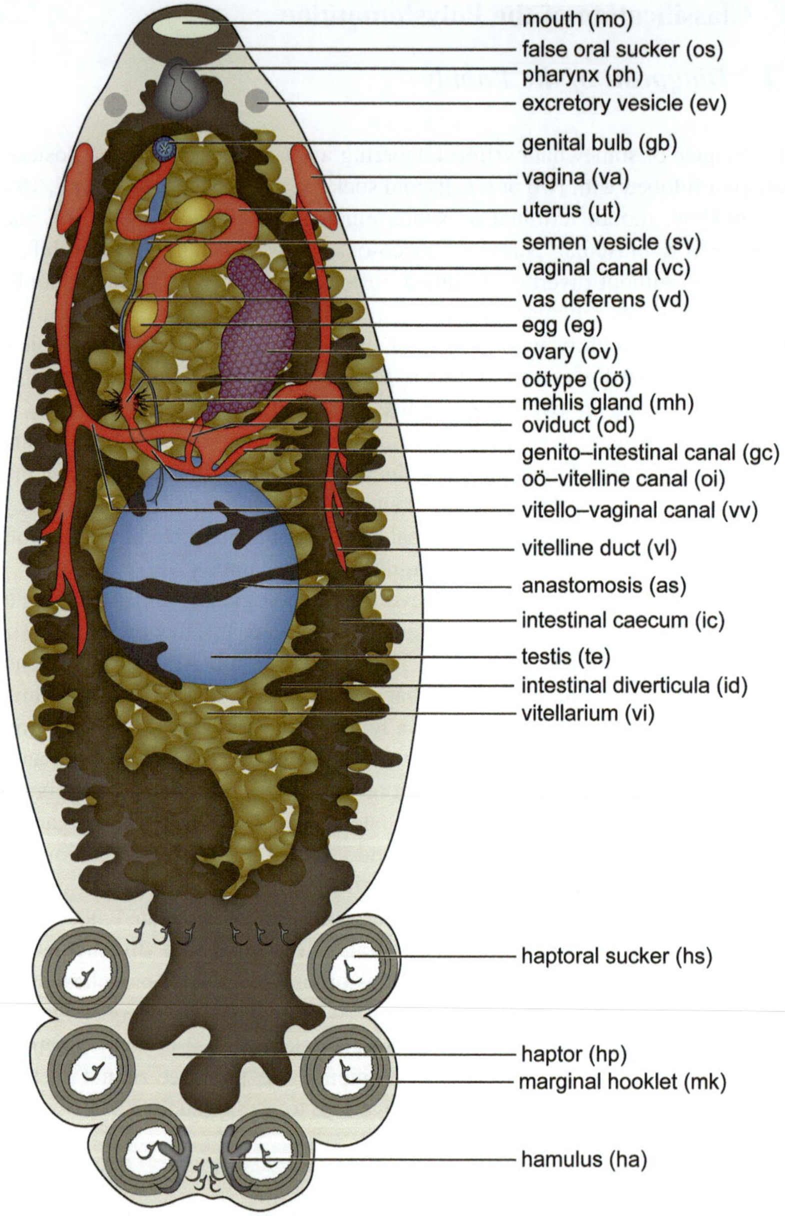

Fig. 1.4a Schematic drawing of a polystome with the placement and variation in shape and size of morphological features that are of taxonomic importance

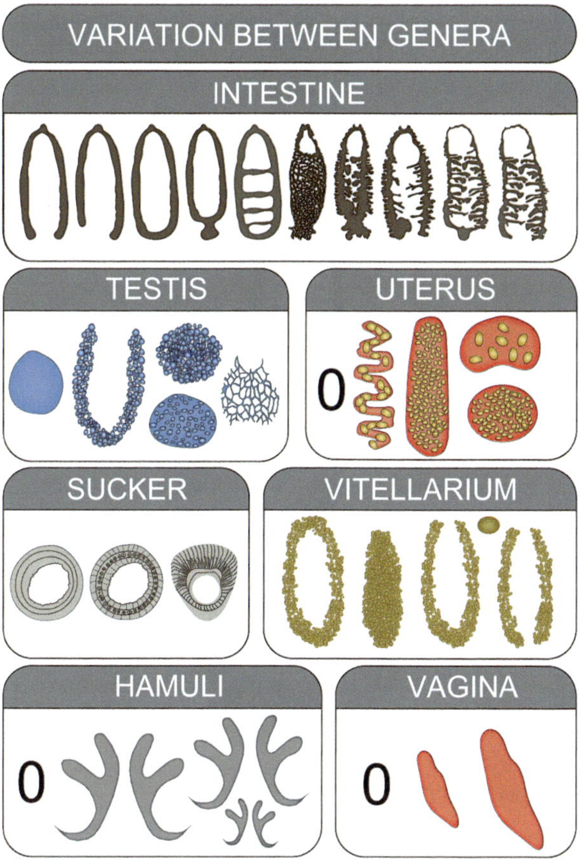

Fig. 1.4b Schematic variation of morphological features: Intestinum—confluent posterior or not, degree of diverticulae, Anastomoses or not, extending into haptor or not; Testis—follicular or dense, single or multiple, placement; Uterus—absent (0) or present, small or big, tubiform or sacciform; Ovary—size, placement; Haptoral suckers—one pair or three pairs, spherical or aspherical, with skeletal element or without; Vitellarium—follicular or dense, single or multiple, extent; Hamuli—absent (0) or present, one or two pairs; Vaginae—absent or present, size and placement

a fan; mouth subterminal; intestine bifurcate with caeca extending length of body, not entering haptor, confluent posteriorly; caeca unbranched but with single short anterior diverticulum near pharynx and conspicuous pocket; ovary pre-testicular; oötype usually containing a single egg, operculated, long ellipsoid, with abopercular appendage; uterus absent; vaginae absent; vitellarium follicular; testes numerous, intercaecal. One genus.

***Concinnocotyla* Pichelin, Whittington and Pearson, 1991**

Type species: *Concinnocotyla australensis* (Reichenbach-Klinke, 1966)

Body elongate; haptor with three pairs of suckers and one pair of hamuli; suckers bilateral, symmetric and cup-shaped, with skeleton of sclerites, sucker type 4;

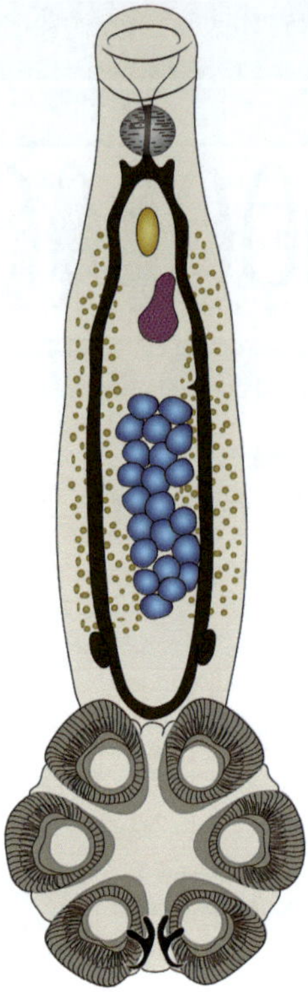

Fig. 1.5 Schematic drawing of *Concinnocotyla*

16 marginal hooklets; two pairs of eyes; mouth subterminal, surrounded by false sucker; pharynx large, muscular; bucco-oesophageal canal present; gut bifurcate, confluent posteriorly; caeca extending laterally length of body proper, unbranched but with single short anterior diverticulum on both sides near pharynx, connected posteriorly; two pockets, opposite, one connected to lateral face of each caecum post-testicularly, opening dorsally by short duct; sperm-filled sac between pockets; penis elongate, muscular, unarmed, lying in duct opening into genital atrium; seminal vesicle large, discrete, wall muscular; testes numerous, intercaecal; ovary pre-testicular, dorsal to seminal vesicle, amphitypic; genito-intestinal on ovarian side; ovo-vitelline duct short; öotype long, containing a single egg, opening into genital atrium; uterus absent; common genital pore median, ventral to pharynx; vaginae absent; vitellarium follicular, surrounding caeca, extending from penis bulb to posterior margin of testes, not confluent posteriorly; long egg ellipsoid, with abopercular appendage. Polystomes known only from gills and oral cavity of the Australian lungfish *N. forsteri*.

1.2.2.2 The Diplorchiinae Yamaguti, 1963 (Fig. 1.6)

Body oval to elongate; haptor discoid with three pairs of haptoral suckers; single pair of hamuli or absent; mouth subterminal; four eyespots present; intestine slightly to heavily diverticulated; pre-haptoral anastomoses in one genus; caeca not confluent posteriorly except in one genus; ovary pre-testicular; uterus tubiform or sacciform, extending full length of body; vaginae small, lateral, pre-ovarial; genital bulb unarmed or armed with 7–12 genital spines; vitellarium follicular in two lateral fields following gut caeca; two compact lateral pre-equatorial testes or multiple testes. Four genera.

***Diplorchis* Ozaki, 1931**
Type species: *Diplorchis ranae* Ozaki, 1931

Body elongate pyriform; haptor with three pairs of suckers and one pair of hamuli; eyespots present; gut bifurcate, not confluent posteriorly; caeca extending laterally length of body proper; two lateral pre-equatorial compact testes; ovary pre-testicular; uterus extensive, tubiform; genital bulb armed with less than ten genital spines; vaginae absent; vitellarium follicular with two lateral fields surrounding caeca; eggs operculated. Polystomes found in the urinary bladder of anurans in the Palaearctic and Oriental Realms.

***Neodiplorchis* Yamaguti, 1963**
Type species: *Neodiplorchis scaphiopodis* (Rodgers, 1941)

Body elongate pyriform; haptor with three pairs of suckers and one pair of hamuli; eyespots present; gut bifurcate; caeca diverticulated with no pre-haptoral anastomoses, confluent posteriorly; two lateral pre-equatorial compact testes; ovary pre-testicular; uterus extensive, tubiform; genital bulb armed with eight genital spines; vaginae present; vitellarium with two lateral pre-equatorial bodies; eggs operculated. Polystomes known only from the urinary bladder of the frog *Spea bombifrons* (Cope, 1863) in the Nearctic Realm.

***Parapolystoma* Ozaki, 1935**
Type species: *Parapolystoma bulliense* (Johnston, 1912)

Body elongate pyriform; haptor with three pairs of suckers and one pair of hamuli; gut bifurcate; caeca not confluent posteriorly, extending laterally length of body proper; two lateral pre-equatorial compact testes; ovary pre-testicular; uterus extensive, sacciform; genital bulb armed with less than ten genital spines; vaginae absent; vitellarium follicular, two lateral fields surrounding caeca; eggs operculated. Polystomes found in the urinary bladder of anurans in the Australian Realm.

***Pseudodiplorchis* Yamaguti, 1963**
Type species: *Pseudodiplorchis americanus* (Rogers and Kuntz, 1940)

Body elongate pyriform; haptor with three pairs of suckers; no hamuli; eyespots present; gut bifurcate; caeca confluent posteriorly, extending laterally length of body proper with some pre-haptoral anastomoses; two lateral pre-equatorial elongated compact testes; ovary pre-testicular; uterus extensive, tubiform; genital bulb armed with 10–12 genital spines; vaginae absent; eggs operculated; intra-uterine development. Polystomes known only from the urinary bladder of the frog *Scaphiopus couchii* Baird 1854 in the Nearctic Realm.

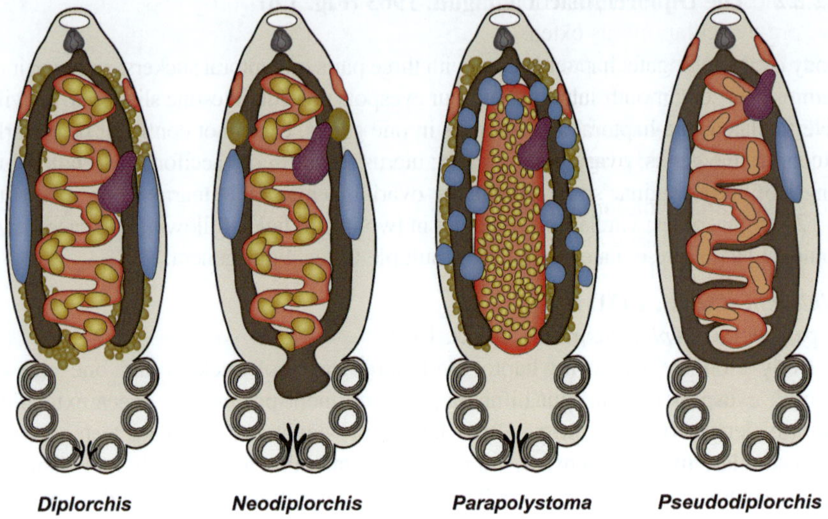

Fig. 1.6 Schematic drawings of Diplorchiinae genera

1.2.2.3 The Eupolystomatinae Yamaguti, 1963 (Fig. 1.7)

Body pyriform; haptor discoid with three pairs of haptoral suckers; hamuli absent or present; mouth subterminal; intestine bifurcate with small diverticula, confluent posteriorly and extending into haptor with no pre-haptoral anastomoses; ovary small, equatorial to posterior, pre-testicular; uterus sacciform, extending full length of body; eggs with soft membrane, not opercular; intra-uterine development; vaginae absent or small when present, lateral, pre-equatorial; genital bulb armed with 4–14 genital spines; vitellarium confined to two lateral fields; testis follicular, posterior, post-ovarian. Three genera.

***Eupolystoma* Kaw, 1950**
Type species: *Eupolystoma rajai* Kaw, 1950.

Body elongate pyriform; haptor with three pairs of suckers; hamuli absent; eyespots present; gut bifurcate, not diverticulated with no pre-haptoral anastomoses or rarely a single one, confluent posteriorly; testis follicular, post-equatorial to posterior; ovary post-equatorial, pre-testicular; uterus extensive, sacciform; genital bulb armed with 4–8 genital spines; vaginae present; vitellarium widely scattered or in two lateral fields; eggs not operculated, intra-uterine development. Polystomes found in the urinary bladder of anurans in Ethiopian and Oriental Realms.

***Madapolystoma* Du Preez, Verneau, Raharivololoniaina and Vences, 2010**
Type species: *Madapolystoma biritika* Du Preez, Verneau, Raharivololoniaina and Vences, 2010

Body small, elongate pyriform; haptor with three pairs of suckers and one pair of hamuli; eyespots absent; gut bifurcate with small diverticulae and no pre-haptoral anastomoses, confluent posteriorly; testis follicular, posterior in body; ovary posterior, pre-testicular; uterus extensive, tubiform; genital bulb armed with less than ten genital spines; vaginae present; vitellarium in two lateral fields, posterior; eggs thin membraned, not operculated; advanced intra-uterine development with embryos well developed having already hamuli and an F2 embryo. Polystomes found in the urinary bladder of Madagascan frogs.

***Kankana* Raharivololoniaina, Verneau, Berthier, Vences and Du Preez, 2011**
Type species: *Kankana manampoka* Raharivololoniaina, Verneau, Berthier, Vences and Du Preez, 2011

Body elongate pyriform; haptor with three pairs of suckers and one pair of hamuli; eyespots absent; gut bifurcate with small diverticulae and no pre-haptoral anastomoses, confluent posteriorly; testis follicular, post-equatorial; ovary equatorial, pre-testicular; uterus extensive, sacciform; genital bulb armed with eight genital spines; vaginae absent; vitellarium in two lateral fields, posterior; eggs thin membraned, not operculated; intra-uterine development. Polystomes known only from the urinary bladder of the frog *Cophyla pollicaris* (Boulenger, 1888) in Madagascar.

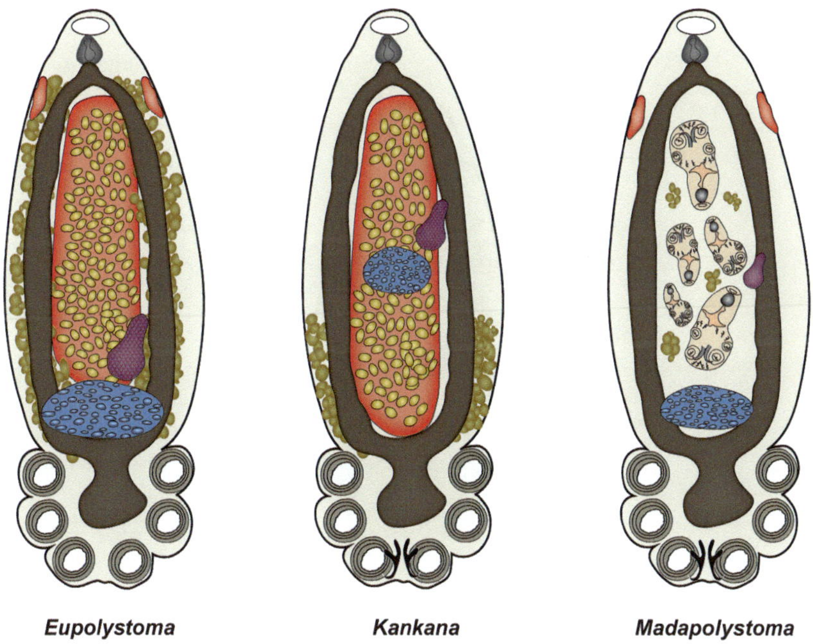

Fig. 1.7 Schematic drawings of Eupolystomatinae genera

1.2.2.4 The Nanopolystomatinae n. sub-fam. (Fig. 1.8)

Small ovoid body; no eyespots; haptor with three pairs of haptoral suckers each with a skeletal ring; one pair of hamuli with deep cut between narrow handle and guard; mouth subterminal; intestinal caeca not diverticulated, not extending full length of body proper and not joining posteriorly; ovary oval, equatorial; short uterus holding a single egg operculated; vagina lateral, at level of ovary; genital bulb armed with 10–12 spines; vitellarium follicular, dense, co-extensive with intestinal caeca, left and right fields confluent posteriorly; testis single, follicular, post-ovarian, median in midbody. One genus.

Nanopolystoma **Du Preez, Wilkinson and Huyse, 2008**

Type species: *Nanopolystoma lynchi* Du Preez, Wilkinson and Huyse, 2008

Body small, elongate; haptor with three pairs of suckers and one pair of hamuli; eyespots absent; gut bifurcate with no diverticulae and no pre-haptoral anastomoses, not confluent posteriorly; testis dense follicular, medial, post-equatorial; ovary equatorial, pre-testicular; uterus short, holding a single egg operculated; genital bulb armed with less than 20 genital spines; vaginae equatorial; vitellarium in two lateral fields, posterior. Polystomes found in the urinary bladder and phallodeum of caecilians in the Neotropical region.

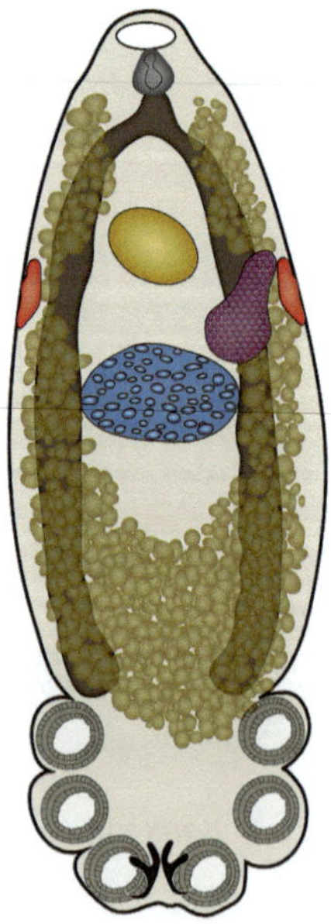

Fig. 1.8 Schematic drawing of *Nanopolystoma*

1.2.2.5 The Polystomatinae Gamble, 1896 (Fig. 1.9)

Medium to large; body pyriform to ovoid; haptor discoid with three pairs of haptoral suckers; one or two pairs of hamuli when present; mouth subterminal; eyespots absent; intestine bifurcate, caeca diverticulated with pre-haptoral anastomoses present or not; haptoral anastomosis usually present; ovary pyriform, medium to large, pre-equatorial, pre-testicular; uterus tubiform or sacciform, small to full length of body proper; eggs operculated, tan coloured; vaginae lateral, pre-ovarian; genital bulb armed with less than 12 genital spines; vitellarium throughout body proper; testis follicular, midbody equatorial to post-equatorial. Nine genera.

***Indopolystoma* Chaabane, Verneau and Du Preez, 2019**

Type species: *Indopolystoma viridi* Chaabane, Verneau and Du Preez, 2019

Body elongated; haptor small, discoid with three pairs of haptoral suckers; one pair of hamuli; mouth subterminal; eyespots absent; intestine bifurcate, with caeca diverticulated showing no pre-haptoral anastomoses; haptoral anastomosis barely entering haptor; ovary pyriform, anterior, pre-testicular; uterus tubiform, medium sized, holding as many as 40 eggs confined anteriorly; eggs operculated, tan coloured; vaginae lateral, pre-ovarian; genital bulb armed with 8–9 genital spines; vitellarium throughout body proper except area around ovary and reproductive ducts; testis follicular, medial, equatorial to post-equatorial. Polystomes found in the urinary bladder of anurans from Oriental and Palaearctic Realms.

***Mesopolystoma* Vaucher, 1981**

Type species: *Mesopolystoma samiriensis* Vaucher, 1981

Body pyriform; haptor discoid with three pairs of haptoral suckers; one pair of hamuli; mouth subterminal; intestine bifurcate, with caeca diverticulated showing three pre-haptoral anastomoses; haptoral anastomosis; ovary pyriform, equatorial, pre-testicular; uterus sacciform, medium sized, pre-ovarian; eggs operculated, tan coloured; vaginae lateral, pre-ovarian; genital bulb armed with eight genital spines; vitellarium throughout body proper except area around ovary and reproductive ducts; testis follicular, medial, equatorial to post-equatorial. Polystomes known only from the urinary bladder of the frog *Osteocephalus taurinus* Steindachner, 1862 in the Neotropical Realm.

***Metapolystoma* Combes, 1976**

Type species: *Metapolystoma cachani* (Gallien, 1957)

Body pyriform; haptor discoid with three pairs of haptoral suckers; one pair of hamuli; mouth subterminal; intestine bifurcate, confluent posteriorly, with caeca diverticulated showing no pre-haptoral anastomoses; ovary pyriform, post-equatorial to posterior, pre-testicular; uterus tubiform, extensive, pre-ovarian; eggs operculated, tan coloured; vaginae lateral, pre-ovarian, anterior third of body; genital bulb armed with 8–10 genital spines; vitellarium throughout body proper except area around ovary and reproductive ducts; testis follicular, medial, posterior. Polystomes found in the urinary bladder of frogs from the Ethiopian Realm.

***Neoriojatrema* Imkongwapang and Tandon, 2010**

Type species: *Neoriojatrema mocokchungensis* Imkongwapang and Tandon, 2010

Body pyriform; haptor discoid with three pairs of haptoral suckers; hamuli absent; mouth subterminal; intestine bifurcate, not confluent posteriorly, highly diverticulated with no anastomoses; ovary elongated, pre-equatorial in anterior third of body, pre-testicular; uterus tubiform, holding as many as 50 eggs, intra-uterine development; vaginae lateral, pre-ovarian, anterior third of body; genital bulb armed with 11–12 genital spines; vitellarium throughout body proper except area around ovary and reproductive ducts; testis follicular, medial, posterior. Polystomes known only from the urinary bladder of the frog *Xenophrys glandulosa* (Fei, Ye and Huang, 1990) of India.

***Polystoma* Zeder, 1800**

Type species: *Polystoma integerrimum* (Frölich, 1791)

Body pyriform; haptor discoid with three pairs of haptoral suckers; one pair of hamuli; mouth subterminal; intestine bifurcate, usually confluent posteriorly, with caeca diverticulated showing no pre-haptoral or extensive anastomoses that form a network; ovary pyriform, pre-equatorial, pre-testicular; uterus sacciform, pre-ovarian, holding a few eggs operculated and tan coloured; usually no intra-uterine development; vaginae antero-lateral, pre-ovarian, anterior third of body; genital bulb armed with genital spines; vitellarium throughout body proper except area around ovary and reproductive ducts; testis follicular, medial, posterior. Polystomes found in the urinary bladder of frogs in all continents except Australia.

***Protopolystoma* Bychowsky, 1857**

Type species: *Protopolystoma xenopodis* (Price, 1943)

Body pyriform; haptor discoid with three pairs of haptoral suckers; two pairs of hamuli; mouth subterminal; intestine bifurcate, not confluent posteriorly, with caeca diverticulated showing most of the times pre-haptoral anastomoses; ovary pyriform, pre-equatorial, pre-testicular; uterus small, pre-ovarian, holding a single fusiform egg operculated and tan coloured; usually no intra-uterine development; vaginae absent; genital bulb armed with 16–38 genital spines; vitellarium throughout body proper except area around ovary and reproductive ducts; testis follicular, medial, post-ovarian. Polystomes found in African frogs of the genus *Xenopus*.

***Riojatrema* Lamothe-Argumedo, 1963**

Type species: *Riojatrema bravoae* Lamothe-Argumedo, 1963

Body pyriform; haptor discoid with three pairs of haptoral suckers; no hamuli; mouth subterminal; intestine bifurcate, confluent posteriorly to form haptoral anastomosis; caeca diverticulated with pre-haptoral anastomoses; ovary pyriform, pre-equatorial, pre-testicular; uterus tubiform, pre-ovarian; eggs operculated, tan coloured; vaginae lateral, pre-ovarian, anterior third of body; genital bulb armed with 8–9 genital spines; vitellarium throughout body proper except area around ovary and reproductive ducts; testis follicular, medial, post-ovarian. Polystomes found in the urinary bladder of frogs of the Neotropical Realm.

Remark:

Parapseudopolystoma Nasir and Fuentes-Zambrano, 1983 nomen nudum

Type species: *Parapseudopolystoma cerrocoloradensis* Nasir and Fuentes-Zambrano, 1983.

This genus must be regarded as a junior synonym of *Riojatrema*. The generic features presented by Nasir and Fuentes-Zambrano (1983) are identical to the generic features for *Riojatrema*. While *Riojatrema* is known from Ecuador and Mexico, *Parapseudopolystoma* nomen nudum has been described from Venezuela, thus the same geographical Realm. Therefore, *P. cerrocoloradensis* is regarded as *Riojatrema cerrocoloradensis* (Nasir and Fuentes-Zambrano, 1983) n. comb.

***Sundapolystoma* Lim and Du Preez, 2001**

Type species: *Sundapolystoma chalconotae* Lim and Du Preez, 2001

Body ovoid, large; haptor discoid with three pairs of haptoral suckers; one pair of hamuli; mouth subterminal; intestine bifurcate, not confluent posteriorly, with caeca showing small diverticulae, pre-haptoral anastomoses often present; ovary pre-equatorial, pre-testicular; uterus extensive, pre-ovarian, tubiform; egg operculated and tan coloured; usually no intra-uterine development; vaginae antero-lateral; genital bulb armed with 7–13 genital spines; vitellarium throughout body proper except area around ovary and reproductive ducts; testis extensive, follicular, medial, post-ovarian. Polystomes found in frogs of the Oriental Realm.

***Wetapolystoma* Gray, 1993**

Type species: *Wetapolystoma almae* Gray, 1993

Body pyriform, large; haptor discoid, wider than body, with three pairs of haptoral suckers; one pair of hamuli; mouth subterminal; intestine bifurcate, confluent posteriorly, with caeca highly diverticulated showing pre-haptoral anastomoses often present; ovary large, ovoid, equatorial, pre-testicular; uterus, pre-ovarian, tubiform; eggs pale, thin membraned, not operculated; intra-uterine development; vaginae antero-lateral; genital bulb armed with eight genital spines; vitellarium throughout body proper except area around ovary and reproductive ducts; testis extensive, follicular, medial, post-ovarian. Polystomes known only from the urinary bladder of the frog *Rhinella margaritifera* (Laurenti, 1768) of the Neotropical Realm.

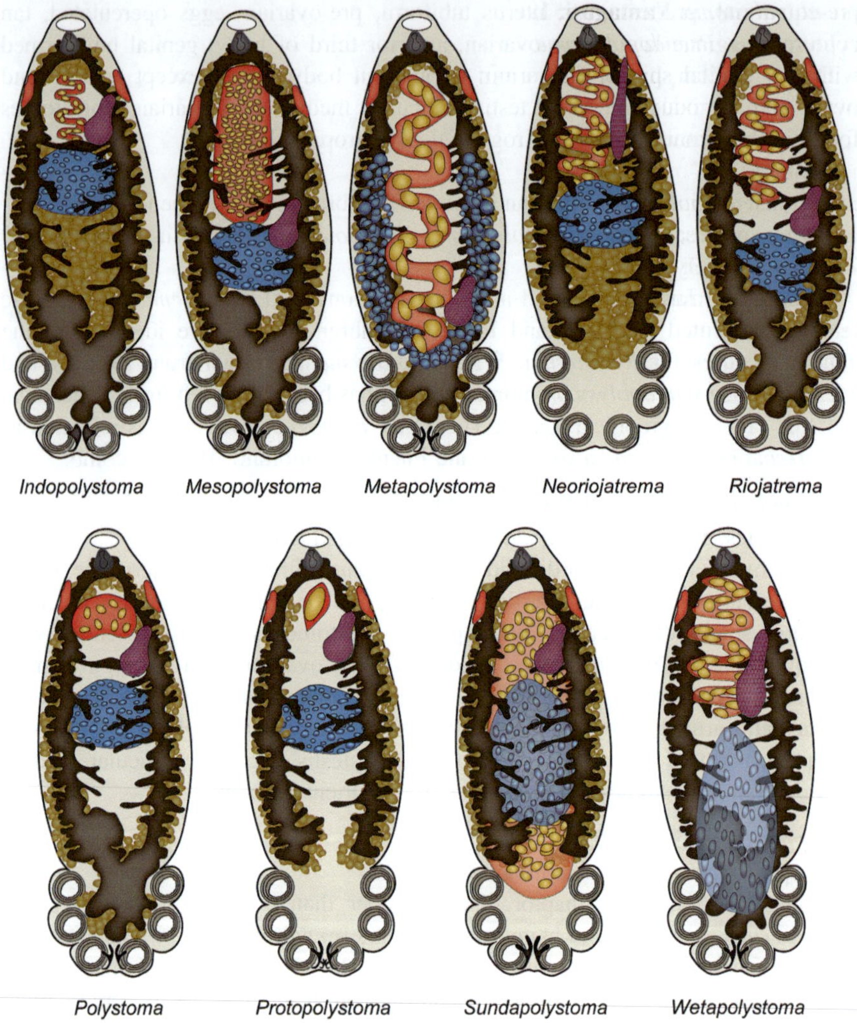

Fig. 1.9 Schematic drawings of Polystomatinae genera

1.2.2.6 The Pseudopolystomatinae Yamaguti, 1963 (Fig. 1.10)

Body small pyriform; haptor bilobed with three pairs of haptoral suckers; hamuli absent; mouth subterminal; eyespots absent; intestine bifurcate, with caeca diverticulated showing no pre-haptoral anastomoses; caeca not confluent posteriorly; ovary small, pre-equatorial; uterus small, holding one egg; vaginae absent; genital bulb armed with genital spines; vitellarium throughout body proper; testis follicular, midbody. One genus.

***Pseudopolystoma* Yamaguti, 1963**

Type species: *Pseudopolystoma dendriticum* Ozaki, 1948

Body pyriform; haptor discoid with three pairs of haptoral suckers; no hamuli; mouth subterminal; intestine bifurcate, not confluent posteriorly, with caeca diverticulated showing no pre-haptoral anastomoses; ovary, anterior, pre-testicular; uterus short, pre-ovarian, holding a single egg; no intra-uterine development; vaginae absent; genital bulb armed with genital spines; vitellarium throughout body proper except area around ovary and reproductive ducts; testis follicular, extensive, post-ovarian. Polystomes known only from the urinary bladder of the salamander *O. japonicus* of Japan.

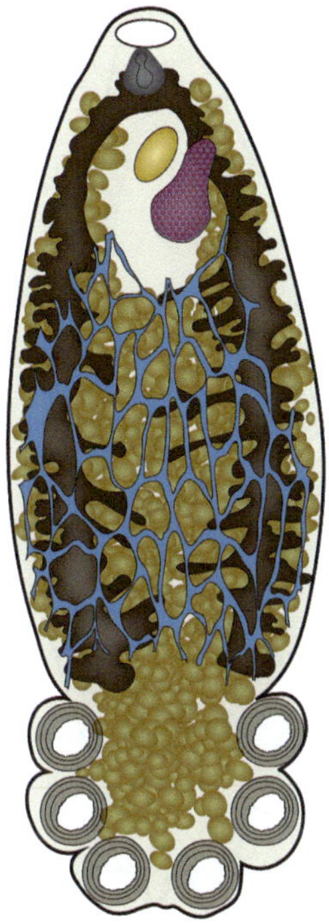

Fig. 1.10 Schematic drawing of *Pseudopolystoma*

1.2.2.7 The Sphyranurinae Price, 1939 (Fig. 1.11)

Body small; elongate; haptor bilobed with one pair of suckers; hamuli present; mouth prominent, subterminal; eyespots absent; intestine bifurcate, with caeca not diverticulated showing no pre-haptoral anastomoses; caeca join pre-haptoral; ovary small, pre-equatorial, pre-testicular; uterus holding a single big egg, pre-equatorial; vaginae absent; genital bulb armed with less than ten genital spines; vitellarium in two lateral fields along gut caecum joining posteriorly, pre-equatorial; testes multiple, post-equatorial, arranged linear along medial line. One genus.

***Sphyranura* Wright, 1879**

Type species: *Sphyranura osleri* Wright, 1879

Body small, elongate; haptor bilobed with one pair of suckers; one pair of hamuli; mouth prominent, subterminal; eyespots absent; intestine bifurcate, with caeca not diverticulated showing no pre-haptoral anastomoses; caeca confluent posteriorly; ovary small, pre-equatorial, pre-testicular; uterus holding a single big egg, pre-equatorial; vaginae absent; genital bulb armed with less than ten genital spines; vitellarium in two lateral fields along gut caecum joining posteriorly, pre-equatorial; testes multiple, post-equatorial, arranged linear along medial line. Polystomes found on gills and around base of limbs of salamanders of the Nearctic Realm.

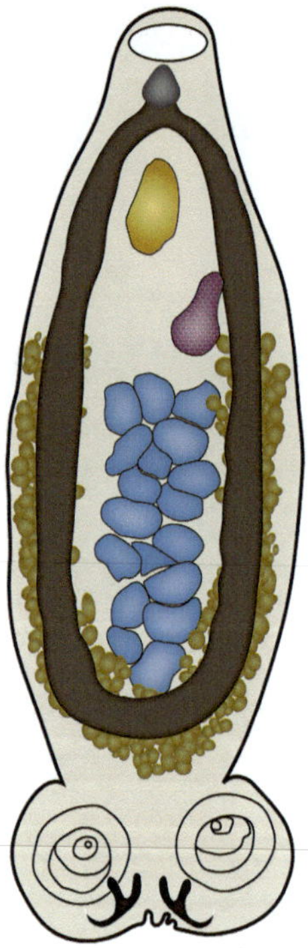

Fig. 1.11 Schematic drawing of *Sphyranura*

1.2.2.8 The Polystomoidinae Yamaguti, 1963 (Fig. 1.12)

Small to large ovoid body; haptor discoid with three pairs of haptoral suckers bearing skeletal elements; one or two pairs of hamuli when present; mouth subterminal; eyespots absent; intestine bifurcate with caeca not diverticulated, not joining posteriorly showing no anastomoses; ovary small, pre-equatorial; uterus present in a single genus, sacciform, pre-equatorial, holding up to 20 eggs operculated; vaginae antero-lateral, pre-ovarian; small to big, genital bulb armed with as many as

130 genital spines; vitellarium throughout body proper or in two lateral columns along gut caeca; testis compact, smooth to lobed, midbody, post-equatorial; eyes. Nine genera.

***Apaloneotrema* Du Preez and Verneau, 2020**

Type species: *Apaloneotrema moleri* (Du Preez and Morrison, 2012)

Body ovoid; haptor discoid with three pairs of haptoral suckers bearing skeletal elements; hamuli absent; mouth subterminal; eyespots absent; intestine bifurcate, with caeca not diverticulated, not joining posteriorly showing no anastomoses; ovary small, pre-equatorial; egg-cell-formation-chamber absent; uterus absent; egg ovoid to fusiform, tan coloured, operculated; vaginae antero-lateral, pre-ovarian; genital bulb armed with as many as 56 genital spines; vitellarium in two lateral columns along gut caeca, confluent posteriorly; testis compact, smooth to lobed, midbody, post-equatorial. Polystomes known only from the conjunctival sacs of *Apalone ferox* (Schneider, 1783) of the Nearctic Realm.

***Aussietrema* Du Preez and Verneau, 2020**

Type species: *Aussietrema tinsleyi* (Pichelin, 1995)

Body ovoid; haptor discoid with three pairs of haptoral suckers with skeletal elements; hamuli absent; mouth subterminal; eyespots absent; intestine bifurcate with caeca not diverticulated, not joining posteriorly showing no anastomoses; ovary small, pre-equatorial; egg-cell-formation-chamber absent; uterus absent; egg ovoid, tan coloured, operculated; vaginae antero-lateral, pre-ovarian; genital bulb armed with as many as 56 genital spines; vitellarium in two lateral columns along gut caeca, confluent posteriorly; testis compact, smooth to lobed, midbody, post-equatorial. Polystomes found in the conjunctival sacs of pleurodire turtles of the Australian Realm.

***Fornixtrema* Du Preez and Verneau, 2020**

Type species: *Fornixtrema palpebrae* (Strelkov, 1950)

Body pyriform to elongate; haptor discoid with three pairs of haptoral suckers bearing skeletal elements; hamuli absent; mouth subterminal; eyespots absent; intestine bifurcate, with caeca not diverticulated, not joining posteriorly, not extending beyond 80% of body proper showing no anastomoses; ovary small, pre-equatorial; egg-cell-formation-chamber present; uterus absent; egg fusiform, tan coloured, operculated; vaginae antero-lateral, pre-ovarian, small to big; genital bulb armed with as many as 56 genital spines; vitellarium in two lateral columns along gut caeca, confluent posteriorly; testis compact, smooth to lobed, midbody, post-equatorial. Polystomes found in the conjunctival sacs of cryptodire turtles in the Nearctic, Neotropic, Oriental and Palaearctic Realms.

***Manotrema* Du Preez, Domingues and Verneau, 2022**

Type species: *Manotrema brasiliensis* (Viera, Novelli, Sousa and SouzaLima, 2008)

Body ovoid; haptor discoid with three pairs of haptoral suckers; deep incisions between suckers appearing fingerlike; skeletal elements in suckers not arranged as a ring of blocks but rather thin spatulate outward bending needles; two small pairs of hamuli present with very deep cuts between handle and guard; mouth with false oral sucker subterminal; pharynx muscular; bifurcate intestinal caeca not confluent posteriorly, extending full length of body proper; testis single, compact equatorial; ovary pre-testicular, small; vaginae latero-ventral in line with anterior margin of testis; vitellaria throughout most of body proper, not extending into haptor; uterus absent; parasitic in urinary bladder of pleurodires of the Neotropical Realm.

***Pleurodirotrema* Du Preez, Domingues and Verneau, 2022**
Type species: *Pleurodirotrema chelodinae* (MacCallum, 1918)

Body ovoid; haptor discoid with three pairs of haptoral suckers; hamuli absent; mouth with false oral suckers subterminal; muscular pharynx; intestinal caeca bifurcate extending full length of body proper, not confluent posteriorly; testis single, compact equatorial; ovary pre-testicular, small; vitellaria throughout most of body proper, not extending into haptor; vaginae latero-ventral in line with anterior margin of testis; uterus absent; skeletal elements in suckers arranged as a ring of blocks; parasitic in urinary bladder and oral region of pleurodires of the Australian Realm.

***Polystomoidella* Price, 1939**
Type species: *Polystomoidella oblongum* (Wright, 1879)

Body ovoid; haptor discoid with three pairs of haptoral suckers bearing skeletal elements; one pair of hamuli; mouth subterminal; eyespots absent; intestine bifurcate, with caeca not diverticulated, not joining posteriorly showing no anastomoses; ovary small, pre-equatorial; uterus absent; egg ovoid, tan coloured, operculated; vaginae antero-lateral, pre-ovarian; genital bulb armed with as many as 16 genital spines; vitellarium throughout body proper; testis compact, smooth to lobed, midbody, post-equatorial. Polystomes found in the urinary bladder of cryptodire turtles of the Nearctic and Oriental Realms.

***Polystomoides* Ward, 1917 sensu Chaabane et al. 2022**
Type species: *Polystomoides coronatus* (Leidy, 1888)

Body ovoid; haptor discoid with three pairs of haptoral suckers (sensu Chaabane et al., 2022) with skeletal elements; two pairs of hamuli if any; hamuli when present smaller than haptoral suckers; mouth subterminal; eyespots absent; intestine bifurcate, with caeca not diverticulated, not joining posteriorly showing no anastomoses; ovary small, pre-equatorial; uterus absent; egg ovoid; vaginae antero-lateral, pre-ovarian; genital bulb armed with as many as 56 genital spines; vitellarium throughout body proper or in two lateral columns along gut caeca; testis compact, smooth to lobed, midbody, post-equatorial; polystomes without hamuli found only in the urinary bladder and cloaca while those with two pairs of hamuli found only in the oral region of their host. *Polystomoides* is widespread and occurs within cryptodires in all continents except Australia.

***Uropolystomoides* Tinsley and Tinsley, 2016**
Type species: *Uropolystomoides chabaudi* (Euzet and Combes, 1965)

Body ovoid; haptor discoid with three pairs of haptoral suckers bearing skeletal elements; two pairs of hamuli with one pair bigger than haptoral suckers; mouth subterminal; eyespots absent; intestine bifurcate, with caeca not diverticulated, not joining posteriorly showing no anastomoses; ovary small, pre-equatorial; uterus absent; egg ovoid, tan coloured, operculated; vaginae antero-lateral, pre-ovarian; genital bulb armed with as many as 95 genital spines; vitellarium in two lateral columns along gut caeca joining posteriorly; testis compact, smooth to lobed, midbody, post-equatorial. Polystomes found in the urinary bladder and cloaca of cryptodire and pleurodire turtles of the Australian, Ethiopian, Oriental and Palaearctic Realms.

***Uteropolystomoides* Tinsley, 2017**

Type species: *Uteropolystomoides multifalx* (Stunkard, 1924)

Body ovoid; haptor discoid with three pairs of haptoral suckers bearing skeletal elements; two pairs of hamuli smaller than haptoral suckers; mouth subterminal;

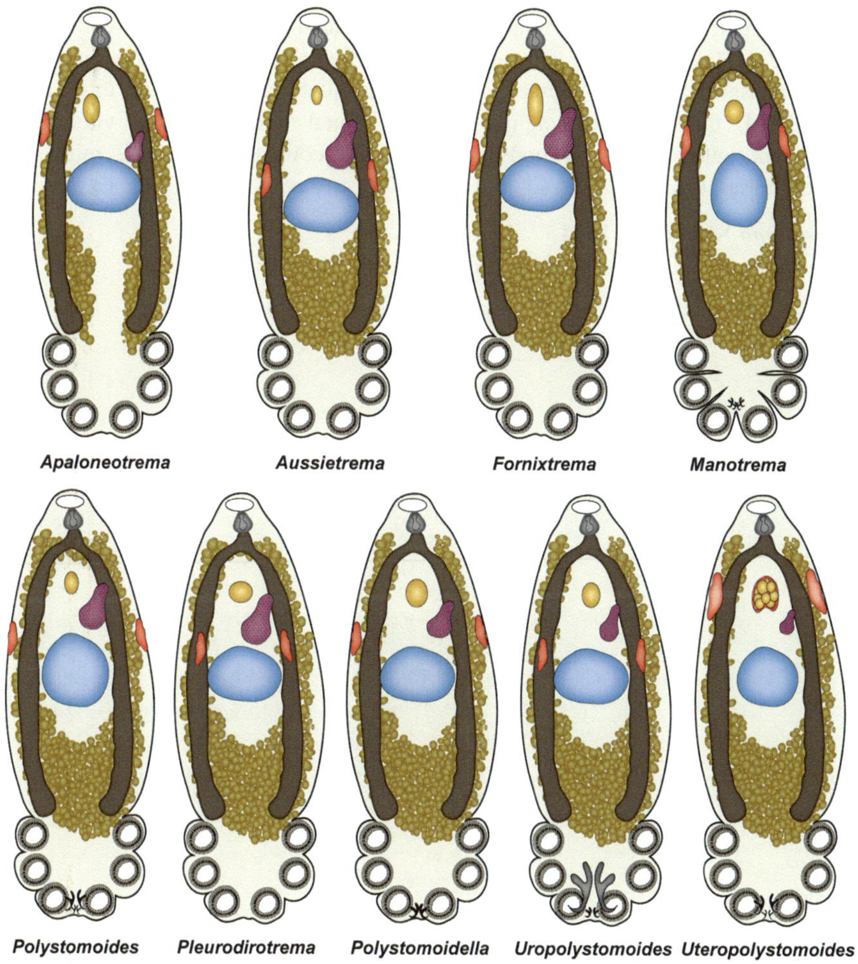

Fig. 1.12 Schematic drawings of Polystomoidinae genera

eyespots absent; intestine bifurcate, with caeca not diverticulated, not joining posteriorly showing no anastomoses; ovary small, pre-equatorial; uterus present, sacciform, holding as many as 20 ovoid eggs, tan coloured, operculated; vaginae very big, antero-lateral, pre-ovarian; genital bulb armed with 120–132 genital spines; vitellarium throughout body proper or in two lateral columns along gut caeca; testis compact, smooth to lobed, midbody, post-equatorial. Polystomes found in the oral region of *Pseudemys* Gray, 1856 in the Nearctic Realm.

1.2.2.9 The Oculotrematinae Yamaguti, 1963 (Fig. 1.13)

Body elongated with flexible region between haptor and anterior body, as long as 30 mm; haptor discoid with three pairs of haptoral suckers bearing skeletal elements; hamuli absent; mouth subterminal; eyespots absent; intestine bifurcate with caeca not diverticulated, of unequal length, not joining posteriorly; ovary small, pre-equatorial to anterior; uterus sacciform, pre-equatorial, holding up to 62 eggs operculated, thick walled; vaginae absent; genital bulb unarmed; vitellarium throughout anterior half of body proper; testis compact, spherical, midbody, pre-equatorial. One genus.

***Oculotrema* Stunkard, 1924**
Type species: *Oculotrema hippopotami* Stunkard, 1924

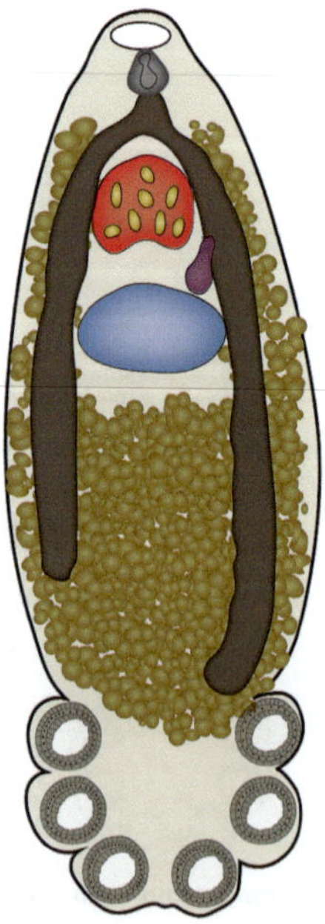

Fig. 1.13 Schematic drawing of *Oculotrema*

Body elongated with flexible region between haptor and anterior body, as long as 30 mm; haptor discoid with three pairs of haptoral suckers, suckers with skeletal elements; hamuli absent; mouth subterminal; eyespots absent; intestine bifurcate, with caeca not diverticulated, of unequal length, not joining posteriorly; ovary small, pre-equatorial to anterior; uterus sacciform, pre-equatorial, holding up to 62 operculated eggs, thick walled; vaginae absent; genital bulb unarmed; vitellarium throughout anterior half of body proper; testis compact, spherical, midbody, pre-equatorial. Polystomes known only from the eye and under eyelids of the common hippopotamus.

1.2.3 Key to the Polystomatidae

1.	Haptor with one pair of suckers, infecting aquatic salamanders	*Sphyranura*
	Haptor with three pairs of suckers	2
2.	Infecting the Australian lungfish	*Concinnocotyla*
	Infecting the common hippopotamus	*Oculotrema*
	Infecting amphibians	3
	Infecting chelonians	19
3.	Infecting caecilians; testis dense follicular and medial	*Nanopolystoma*
	Infecting the Japanese clawed salamander; testis forms a reticulate network	*Pseudopolystoma*
	Infecting anurans	4
4.	Two midbody compact testes or multiple testes	5
	Testis single and diffuse	8
5.	Opisthaptor without hamuli	*Pseudodiplorchis*
	Opisthaptor with one pair of hamuli	6
6.	Vitellaria as two compact pre-equatorial bodies	*Neodiplorchis*
	Vitellaria as two diffuse lateral fields	7
7.	Uterus tubiform	*Diplorchis*
	Uterus sacciform	*Parapolystoma*
8.	Opisthaptor without hamuli	9
	Opisthaptor with one or two pairs of hamuli	11
9.	Gut not diverticulate; uterus sacciform, pre and post-equatorial	*Eupolystoma*
	Gut diverticulate; uterus tubiform	10
10.	Gut caeca not confluent posteriorly; ovary pre-equatorial	*Neoriojatrema*
	Gut caeca confluent posteriorly; ovary equatorial	*Riojatrema*
11.	Gut caeca not confluent posteriorly; one or two pairs of hamuli	12
	Gut caeca confluent posteriorly; one pair of hamuli	13
12.	Opisthaptor with one pair of hamuli; uterus sacciform and extensive	*Sundapolystoma*
	Opisthaptor with two pairs of hamuli; uterus absent, known only from clawed frogs (Pipidae)	*Protopolystoma*
13.	Gut not diverticulate	14
	Gut diverticulate	15

(continued)

14.	Vitellaria posterior in two lateral fields; extended sacciform uterus; limited intra-uterine development of eggs	*Kankana*
	Vitellaria scattered in small clusters; advance intra-uterine development; oncomiracidia with well-developed hamuli	*Madapolystoma*
15.	Ovary equatorial to post-equatorial	16
	Ovary pre-equatorial	17
16.	Uterus tubiform, extensive, holding many eggs; testis U-shaped	*Metapolystoma*
	Uterus sacciform; testis medial	*Mesopolystoma*
17.	Uterus sacciform, holding <20 eggs	*Polystoma*
	Uterus tubiform, can hold >20 eggs	18
18.	Testis pre-equatorial; eggs yellow-tan and operculated	*Indopolystoma*
	Testis post-equatorial; eggs white-grey, not operculated	*Wetapolystoma*
19.	Uterus present with as many as 20 eggs; vaginae massive; > 120 genital spines	*Uteropolystomoides*
	No uterus	20
20.	Parasite on eye	21
	Parasite on oral region, urinary bladder or cloaca	22
21.	Egg fusiform, gut in anterior 70% length of body; egg-cell-formation-chamber present; infects cryptodires	*Fornixtrema*
	Egg spherical to oval, gut caeca full length of body proper; egg-cell-formation-chamber absent; infects Australian pleurodires	*Aussietrema*
	Egg big oval, gut caeca full length of body proper; egg-cell-formation-chamber absent; vaginae very big; > 120 genital spines; infects the Florida softshell turtle	*Apaloneotrema*
22.	Opisthaptor with one pair of hamuli	*Polystomoidella*
	Opisthaptor with no or two pairs of hamuli	23
23.	Opisthaptor with two pairs of hamuli, the first one being bigger than the haptoral suckers	*Uropolystomoides*
	Opisthaptor with no or two pairs of hamili: when present hamuli smaller than haptoral suckers	24
24.	Vaginae peripheral that extends well beyond the intestine; infecting cryptodires	*Polystomoides*
	Vaginae latero-ventral in line with anterior margin of testis; infecting pleurodires	25
25.	Opisthaptor with two pairs of hamuli; very deep cuts between handle and blade; found from the Neotropical Realm only	*Manotrema*
	Opisthaptor without hamuli; found from the Australian Realm only	*Pleurodirotrema*

References

Alvey CH (1936) The morphology and development of the monogenetic trematode *Sphyranura oligorchis* (Alvey 1933) and the description of *Sphyranura polyorchis* n. sp. Parasitology 28: 229–253

Boeger WA, Kritsky DC (1993) Phylogeny and a revised classification of the Monogenoidea Bychowsky, 1937 (Platyhelminthes). Syst Parasitol 26:1–32

Boeger WA, Kritsky DC (1997) Coevolution of the Monogenoidea (Platyhelminthes) based on a revised hypothesis of parasite phylogeny. Int J Parasitol 27:1495–1511. https://doi.org/10.1016/s0020-7519(97)00140-9

Boeger WA, Kritsky DC (2001) Phylogenetic relationships of the Monogenoidea. In: Littlewood DTJ, Bray RA (eds) Interrelationships of the Platyhelminthes. Taylor & Francis, New York, pp 92–100

Bychowsky BE (1957) Monogenetic Trematodes: their systematics and phylogeny. Izdatel'stvo Akademiya Nauk SSSR, Moscow-Leningrad

Campos A, Cummings MP, Reyes JL, Laclette JP (1998) Phylogenetic relationships of platyhelminthes based on 18S ribosomal gene sequences. Mol Phylogenet Evol 10:1–10. https://doi.org/10.1006/mpev.1997.0483

Frost DR. 2023. Amphibian Species of the World: an Online Reference. Version 6.2 (12 October 2023). Electronic Database accessible at https://amphibiansoftheworld.amnh.org/index.php. American Museum of Natural History, New York, USA. https://doi.org/10.5531/db.vz.0001

Fuhrmann O (1928) Trematoda: Kükenthal's Handbuch der Zoologie. W. de Gruyter, Berlin and Leipzig

Gallien L (1935) Recherches expérimentales sur le dimorphisme évolutif et la biologie de *Polystomum integerrimum* Fröl. Trav Stat Zool Wimereux 12:1–181

Hickman CP, Roberts LS, Larson A, l'Anson H, Eisenhour DJ (2006) Integrated principles of zoology. McGraw Hill, Boston

Kearn GC (1994) Evolutionary expansion of the Monogenea. Int J Parasitol 24:1227–1271. https://doi.org/10.1016/0020-7519(94)90193-7

Littlewood DTJ, Rohde K, Bray RA, Herniou EA (1999a) Phylogeny of the Platyhelminthes and the evolution of parasitism. Biol J Linn Soc 68:257–287. https://doi.org/10.1111/j.1095-8312.1999.tb01169.x

Littlewood DTJ, Rohde K, Clough KA (1999b) The interrelationships of all major groups of platyhelminthes: phylogenetic evidence from morphology and molecules. Biol J Linn Soc 66: 75–114. https://doi.org/10.1111/j.1095-8312.1999.tb01918.x

Littlewood DTJ, Cribb TH, Olson PD, Bray RA (2001) Platyhelminth phylogenetics – a key to understanding parasitism? Belg J Zool 131:35–46

Llewellyn J (1963) Larvae and larval development of monogeneans. Adv Parasitol 1:287–326

Lockyer AE, Olson PD, Littlewood DTJ (2003) Utility of complete large and small subunit rRNA genes in resolving the phylogeny of the Platyhelminthes: implications and a review of the cercomer theory. Biol J Linn Soc 78:155–171. https://doi.org/10.1046/j.1095-8312.2003.00141.x

Mollaret I, Jamieson BGM, Adlard RD, Hugall A, Lecointre G, Chombard C, Justine JL (1997) Phylogenetic analysis of the Monogenea and their relationships with Digenea and Eucestoda inferred from 28S rDNA sequences. Mol Biochem Parasitol 90:433–438. https://doi.org/10.1016/s0166-6851(97)00176-x

Monticelli FS (1903) Per una nouva classificatione delgi "Heterocotylea". Monit Zool Ital 14:334–337

Nasir P, Fuentes Zambrano JL (1983) Algunos trematodos monogeneticos venezolanos. Riv Parassitol 44:335–380

Olson PD, Littlewood DTJ (2002) Phylogenetics of the Monogenea – evidence from a medley of molecules. Int J Parasitol 32:233–244. https://doi.org/10.1016/S0020-7519(01)00328-9

Paul AA (1938) Life history studies of north American freshwater polystomes. J Parasitol 24:489–507

Perkins EM, Donnella SC, Bertozzi T, Whittington I (2010) Closing the mitochondrial circle on paraphyly of the Monogenea (Platyhelminthes) infers evolution in the diet of parasitic flatworms. Int J Parasitol 40:1237–1245. https://doi.org/10.1016/j.ijpara.2010.02.017

Poche F (1926) Das System der Platodaria. Arch Naturgesch 91:1–459

Price EW (1939) North American monogenetic trematodes. IV. The family Polystomatidae (Polystomatoidea). Proc Helminthol Soc Wash 6:80–92

Prudhoe S, Bray RA (1982) Platyhelminth parasites of the Amphibia. British Museum of Natural History, Oxford Univ. Press, London

Rhodin AGJ, Iverson JB, Bour R, Fritz U, Georges A, Shaffer HB, Dijk PV (2021) Turtles of the world: annotated checklist and atlas of taxonomy, synonymy, distribution, and conservation status. 9th edition. Chelonian Res Monogr 8:1–472

Sinnappah ND, Lim LHS, Rohde K, Tinsley R, Combes C, Verneau O (2001) A paedomorphic parasite associated with a neotenic amphibian host: phylogenetic evidence suggests a revised systematic position for Sphyranuridae within anuran and turtle polystomatoineans. Mol Phylogenet Evol 18:189–201. https://doi.org/10.1006/mpev.2000.0877

Wheeler TA, Chisholm LA (1995) Monogenea versus Monogenoidea: the case for stability in nomenclature. Syst Parasitol 30:159–164

Williams JB (1995) Phylogeny of the Polystomatidae (Platyhelminthes, Monogenea), with particular reference to *Polystoma integerrimum*. Int J Parasitol 25:437–441. https://doi.org/10.1016/0020-7519(94)00138-e

Williams JB, McKenzie J (1995) Scanning electron microscopy of *Polystoma integerrimum* (Monogenea, Polystomatidae). Int J Parasitol 25:335–342. https://doi.org/10.1016/0020-7519(94)00137-d

Wright RR (1879) Contributions to American helminthology. Proc Can Inst 1:54–75

Chapter 2
Contributions to the History of Polystomatid Flatworm Discoveries and Research

Contents

Abstract Polystomatid flatworms have been present in the scientific literature as far back as 1758, the year that Carl Linnaeus introduced the binomial classification system. At first discoveries came in at a snail's pace but dramatically increased to the more than 200 species currently known from 30 genera. This section reviews the history of discoveries over time and space. Its focus will instead be on scientists who made significant contributions, both in describing new taxa and adding to the knowledge of the biology, ecology, systematic and evolution of polystomes. The historical reviews will provide a background to the diverse polystome fauna, their uniqueness and their value as host–parasite models for testing proposed hypotheses.

2.1 Introduction

Over a span of more than two and a half centuries since the first polystome was discovered in the time of Carl Linnaeus, many scientists contributed to the knowledge of polystomatid flatworms. Historically type material was held in private collections or University collections. Later, scientific museums were used as repositories for type specimens. The first public national museum to be founded was the British Museum of London in 1753, followed by the Muséum National d'Histoire Naturelle of Paris in 1793. The US Natural Museum was founded much later, in 1846. Today museums exist in most countries around the globe but sadly many type specimens were lost over the years, especially some from private collections.

L. H. du Preez et al., *Polystomatid Flatworms*, Zoological Monographs 9,
https://doi.org/10.1007/978-3-031-35887-6_2

Over time as technology improved, scientists were able to study specimens using advanced microscopy techniques, including phase contrast, differential interference contrast (DIC), scanning electron microscopy (SEM), transmission electron microscopy (TEM), confocal laser scanning microscopy (CLSM) and more. Molecular tools that developed in the 1980s are nowadays regarded as essential in any taxonomic study. While some techniques advanced and changed intensively, others remained essentially unchanged. To catch a frog to inspect for the presence of polystomes, the bare hands remain the most functional and effective tool to do so, and Canada balsam remains the most reliable mounting medium for whole mounts. Rules, regulations and restrictions increased dramatically to the extent that some scientists no longer want to collect material and prefer to work on archived material in collections or focus their interest on predictive modelling and other computer-based applications. Many vertebrates are now listed as threatened and receive protection status ruling them off bounds. Permits are required for most collecting expeditions, and special ethical clearance from the host institution is needed before the onset of any study.

This chapter aims to reflect on the history of polystomatid flatworm discovery and acknowledge contributions made by many scientists around the globe. Where we were able to find suitable pictures of scientists and obtain permission to use them, they were included. This chapter is not a checklist of species, nor was the idea to mention every person that published on polystomes. We instead focus on the significant discoveries, trends and contributions.

2.2 The Early Discoveries

The first platyhelminth parasite reported from an amphibian host, was a trematode discovered by Jan Swammerdam as far back as 1737. In 1758, August Johann Rösel von Rosenhof (Fig. 2.1a) discovered the first monogenean from an anuran, the same year that the Swedish naturalist Carl Linnaeus (Fig. 2.1b) introduced the binomial classification system (Prudhoe and Bray 1982). Josef Aloys Frölich in 1791 described, from an unknown host, a flatworm (Frölich 1791) and named it *Linguatula integerrima* Frölich, 1791. Johann Georg Heinrich Zeder in 1800 (Fig. 2.1c) was of the opinion that this parasite was his newly described *Polystoma ranae* Zeder, 1800 (see Zeder 1800). Karl Rudolphi (1808) (Fig. 2.1d), however, considered *P. ranae* to conspecific with *L. integerrima* which is now *Polystoma integerrimum* (Frölich, 1791) (see Fig. 2.2). Pierre-Joseph van Beneden (Fig. 2.1e) proposed, as early as 1858, that trematodes be separated into two groups. Carl Gustav Carus in 1863 (Fig 2.1f) was given credit as the first person to use the term "Monogenea" (Yamaguti 1963). The Monogenea and Digenea were seen as orders of the Class Trematoda. At a "Round Table" meeting in Warsaw, Poland in 1978, it was decided to elevate the original epithet "Monogenea" to the level of Class (Boeger and Kritsky 1993).

Fig. 2.1 (**a**) August Johann Rösel von Rosenhof, (**b**) Carl Linnaeus, (**c**) Johann Georg Heinrich Zeder, (**d**) Karl Rudolphi, (**e**) Pierre-Joseph van Beneden, (**f**) Carl Gustav Carus, (**g**) Eduard Zeller, (**h**) Robert Ramsay Wright and (**i**) Joseph Leidy. Images (**a**), (**d**), (**f**), (**g**) and (**h**) © Creative Commons Attribution: No restrictions (public domain); image (**b**) oil on canvas by Alexander Roslin, Creative Commons Attribution: No restrictions (public domain); Image (**c**) Redrawn from a shadow sketch portrait of J.G.H. Zeder is in his family book, from 1787, in which the 23-year-old is shown with a wig and braid; Image (**e**) Portrait by Johann Carl Rößler, Creative Commons Attribution: No restrictions (public domain); image (**i**) by Gilbert Studios 1870, Creative Commons Attribution: No restrictions (public domain)

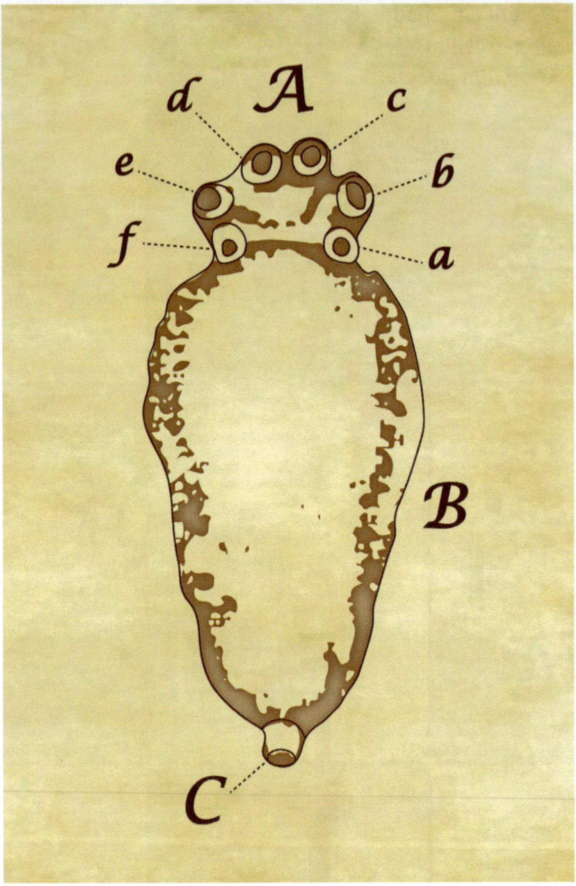

Fig. 2.2 First drawing of *Polystoma integerrimum* (Redrawn after Zeder 1800). Annotations: (**A**) anterior end; (**B**) body; (**C**) short tail with opening; a–f, six mouths

The first observations on monogenean development were by Eduard Zeller (Fig. 2.1g) with his studies on *P. integerrimum* (see Zeller 1872, 1876). He focussed on the morphology and development and noted that polystome eggs were deposited infrequently in the water where they developed and that the larva that emerged from the eggs had five incomplete rows of cilia, four eyespots, an intestinum, a well-developed excretory system and a haptoral disc with 16 marginal hooklets. Following the short free-swimming phase, the larva adheres to the gills of a tadpole and develops suckers. Zeller observed the existence of a neotenic phase for *P. integerrimum.*

Karl Rudolphi described the second polystome species from the Nearctic freshwater turtle *Emys orbicularis* (Linnaeus, 1758) and named it *Polystomum ocellatum* Rudolphi, 1819. It was later redescribed by Ozaki (1935a) and named *Polystomoides ocellatum* (Rudolphi, 1819). Robert Ramsay Wright (Fig. 2.1h) in 1879 described from North American mudpuppies *Necturus maculosus* (Rafinesque, 1818) *Sphyranura osleri* Wright, 1879, but also *Polystomoidella oblongum* (Wright, 1879) and *Polystomoidella whartoni* (Wright, 1879) from North American freshwater turtles *Sternotherus odoratus* (Latreille, 1802) and *Kinosternon* spp., respectively. Joseph Leidy (Fig. 2.1i) in 1888 also described *Polystomoides coronatum* (Leidy, 1888) from undetermined freshwater turtles from the Nearctic realm.

2.3 The American Era (1900–1960)

The early twentieth century discoveries in the USA were dominated by Horace Stunkard, George A MacCallum, Willian MacCallum (Fig. 2.3a), Emmet William Price (Fig. 2.3b), Henry Baldwin Ward, Paul Harwood (Fig. 2.3c), Clifford Harry Alvey (Fig. 2.3d) and Allard Anthony Paul. Stunkard (1916) described *Polystomoides orbiculare* (Stunkard, 1916) from the Painted turtle *Chrysemys picta* (Schneider,1783), and three other chelonian polystome species. Henry Ward published on the classification of North American parasitic worms (Ward 1917) and created the genus *Polystomoides* Ward, 1917. Following his retirement as a medical practitioner George MacCallum studied parasites of animals that died at the New York Zoological Garden. MacCallum (1918) thus described nine new polystome species from chelonians, several of whom were later synonymized.

Stunkard (1924a) described *Uteropolystomoides multifalx* (Stunkard, 1924) from the River Cooter *Pseudemys concinna* (LeConte, 1830). Stunkard (1924b) also described *Oculotrema hippopotami* Stunkard, 1924 from the eye of the common hippopotamus based on five specimens collected from the Giza Zoological Gardens in Cairo, Egypt. The scientific community at first rejected the description and people speculated that it must have been a mislabelled specimen as no monogenean had ever been described from a mammal. The validity of this species from the common hippopotamus was confirmed only in 1965 by June Thurston in a short note in the scientific journal Nature (Thurston and Laws 1965). Du Preez and Moeng (2004) provided a supplementary species description. Until today, this is the single polystome that has been recorded from a mammal.

Price conducted a comprehensive study and produced a series of four papers on North American trematodes, the fourth of which (Price 1939) dealt with the Polystomatidae. He proposed the genera *Neopolystoma* Price, 1939 and *Polystomoidella* Price, 1939, and provided formal names for *Polystoma gallieni* Price, 1939, a parasite discovered by Gallien in 1935 from *Hyla meridionalis* Boettger, 1874 but not formally named, and for *Polystoma ozakii* Price, 1939 a polystome that Ozaki (1935a) referred to as *Polystoma integerrimum* (Fröhlich, 1791) from *Rana ornativentris* Werner, 1903. Later Price described *Protopolystoma xenopodis* (Price, 1943) from *Xenopus laevis* (Daudin, 1802).

Paul Harwood described *Polystomoides terrapenis* (Harwood, 1932) from *Terrapene carolina triunguis* (Agassiz, 1857), and Allard Anthony Paul described the first anuran polystome from the USA, namely *Polystoma nearcticum* (Paul,

Fig. 2.3 (**a**) William MacCallum, (**b**) Emmet William Price, (**c**) Paul D Harwood, (**d**) Clifford Harry Alvey, (**e**) Yoshimasa Ozaki, (**f**) Satyu Yamaguti. Images: (**a**), (**b**) and (**f**) Creative Commons Attribution: No restrictions (public domain); (**c**) © U.S. National Animal Parasite Collection Records. Special Collections, USDA National Agricultural Library; (**d**) © Sherri Anderson (**e**) Repainted after Yamauchi (2013)

1935) from *Hyla versicolor* Le Conte, 1825. He also published on life-history strategies of North American polystomes (Paul 1935, 1938). Clifford Harry Alvey described two more species of *Sphyranura* Wright, 1879, also from *N. maculosus* of North America (Alvey 1933, 1936).

In Europe, Louis Gallien focussed largely on *P. integerrimum* and reported on the alternation of generations, namely how a bladder parasite can produce eggs that give rise to a neotenic form on tadpoles and how a branchial parasite can produce eggs that will infect older tadpoles to give rise to the bladder form on adult frogs (Gallien 1932a, b, 1933, 1934, 1935, 1938, 1947).

From Japan, Yoshimasa Ozaki (Fig. 2.3e) and Satyu Yamaguti (Fig. 2.3f) took the early lead. The first polystome to be described from the Oriental Realm was *Uropolystomoides kachugae* (Stewart, 1914) from the chelonian *Batagur kachuga* (Gray, 1831). This parasite was originally described as *Polystomum kachugae* Stewart, 1914 and renamed by Ozaki (1935a) as *Polystomoides kachugae* (Ozaki,

1935). Yoshimasa Ozaki focussed in the Oriental Realm and described *Diplorchis ranae* Ozaki, 1931 from the frog *Glandirana rugosa* (Temminck and Schlegel, 1838). He further worked extensively on the morphology, life history and the oncomiracidium of this species (Ozaki 1931a, b, 1932, 1935a, b, 1940). He also described three polystome species from chelonian hosts (Ozaki 1935a, 1936) and described *Pseudopolystoma dendriticum* (Ozaki, 1948) from the Japanese Clawed salamander *Onychodactylus japonicus* (Houttuyn, 1782) (Ozaki, 1948). He created the genera *Diplorchis* Ozaki, 1931 and *Parapolystoma* Ozaki, 1935. Although Yamaguti focussed largely on fish helminths, he also published on amphibian helminths (Yamaguti 1963) and his *Systema Helminthum* Vol IV dealt with the Monogenea and Aspidocotylea (Yamaguti 1963). He described one species (Yamaguti, 1936) and created three genera, *Neodiplorchis* Yamaguti, 1963, *Pseudodiplorchis* Yamaguti, 1963 and *Pseudopolystoma* Yamaguti, 1963.

In the Australian Realm, the first polystome species described was *Parapolystoma bulliense* (Johnston, 1912) from *Ranoidea phyllochroa* (Günther, 1863). In the Ethiopian Realm, it was *Eupolystoma alluaudi* (de Beauchamp, 1913), initially described as *Polystoma alluaudi* de Beauchamp, 1913 and later renamed by Euzet and Combes (1967) as *E. alluaudi*. In the Neotropical Realm, it was *Polystomoides domitilae* (Caballero, 1938) from *Trachemys ornata* (Gray, 1831).

2.4 The Polystome Revolution (1961–2022)

In the second half of the twentieth century, research on polystomes was mainly extended to the Ethiopian Realm where most of the new polystome discoveries and descriptions were made. However, besides Africa, other areas like Eurasia, Asia, Australia and South America were also investigated for their polystome diversity.

In Russia, Boris Bychowsky (Fig. 2.4a) published on the general morphology of *P. integerrimum* (see Bychowsky 1957, 1987). Joan Williams of the UK focused primarily on helminths of marine fish but also significantly contributed to the knowledge of polystomatid anatomy (Williams 1957, 1959a, b, 1960a, b, 1961). Likewise, June Thurston focused on polystomatid anatomy (Thurston 1964) and contributed to the knowledge on *Oculotrema* (see Thurston 1968a, b).

Louis Euzet (Fig 2.4b) from Montpellier (France) focused his research mainly on monogenean parasites. Over decades, he conducted outstanding research and supervised several postgraduate students. With his former student Claude Combes (Fig. 2.4c) and co-workers, he described three polystome species from Europe (Dollfus et al. 1965; Euzet and Combes 1966; Euzet et al. 1974b), one from Madagascar (Euzet and Combes 1964), and nine from Central and West Africa (Euzet et al. 1966, 1974a; Maeder et al. 1970). He also published in the field of polystome systematics, diversity and distribution (Euzet and Combes 1964, 1965, 1966, 1967, 1975, Euzet et al. 1969, 1975).

Claude Combes from Perpignan (France) continued investigations on polystomes from Europe, Asia and Africa and with co-workers described one species from

Fig. 2.4 (**a**) Boris Evseevitch Bychowsky, (**b**) Louis Euzet, (**c**) Claude Combes, (**d**) Raymond Laurent, (**e**) Robert Bourgat, (**f**) Jack Llewellyn, (**g**) Keshav Chandra Pandey, (**h**) Klaus Rohde, (**i**) Richard Tinsley. Images (**a**), (**b**), (**e**), (**g**) and (**h**) Creative Commons Attribution: No restrictions (public domain); Image (**d**) © Kraig Aldler, colour manipulated; Images (**c**) and (**f**) © Louis du Preez; Image (**i**) © Richard Tinsley

Malaysia and three from Africa. In collaboration with Raymond Laurent (Fig 2.4d) from Argentina, he described five species from the Neotropical Realm. His significant contributions on polystomatids were in the field of systematics, diversity and distribution (Combes 1964, 1966a, 1976; Combes and Channing 1978; Combes and Gerbeaux 1970; Combes and Justine 1982; Combes and Ktari 1976; Combes and Kulo 1978; Combes and Laurent 1974, 1978, 1979; Combes and Rohde 1978; Combes and Thiery 1983; Dollfus et al. 1965; Kohn et al. 1978); biology and ecology (Combes 1968, 1970, 1972a; Combes et al. 1976; Combes and Knoepffler 1977; Knoepffler and Combes 1977); morphology and behaviour of the oncomiracidium (Combes et al. 1978); chaetotaxy and sensory system (Combes and Lambert 1972; Combes et al. 1975); neotenic phase (Combes 1966a, 1967a, b Combes and Lambert 1970); host-specificity and host–parasite synchronization (Combes 1966b, 1967c, 1972b, 1981, 1983); internal cycles (Combes 1967b, 1973; Combes et al. 1973a, b); phylogeography and evolution (Sinnappah et al. 2001; Verneau et al. 2002; Bentz et al. 2006; Héritier et al. 2015). During the second International Symposium on Monogenea (ISM2) that was held in Montpellier in 1993, Claude Combes initiated the POLEE (Polystomatid Ecology and Evolution) collaborative programme with the objective to initiate a network to further the work on polystomatid parasites and to initiate molecular studies.

Annie Fournier, who was also based in Perpignan focused on ultrastructure of polystomes (Fournier and Combes 1978, 1979). Robert Bourgat (Fig. 2.4e), also from Perpignan was involved in the description of five species from Africa (Bourgat 1975, 1977; Euzet et al. 1974a; Bourgat and Murith 1980). Gueorgui Batchvarov was based in Bulgaria and described six species (Euzet et al. 1974b; Buchvarov 1980, 1984a, b; Batchvarov 1982; Biserkov et al. 2001). Jack Llewellyn (Fig. 2.4f) published on host–parasite associations (Llewellyn 1957, 1963, 1964, 1968) and the evolution of life-history patterns among monogeneans (Llewellyn 1970).

In the Oriental Realm, Keshav Chandra Pandey (Fig. 2.4g) described three species, one from amphibians, and two from turtles (Pandey 1969, 1973; Pandey and Agarwal 1978). Tandon also described two species from India (Diengdoh and Tandon 1991; Imkongwapang and Tandon 2010). Klaus Rohde (Fig 2.4h), who was based in Malaysia for several years, described five polystomes from freshwater turtles (Rohde 1963, 1965; Combes and Rohde 1978). Later, when he was based in Australia, he described a further four species from freshwater turtles of that country (Rohde 1984; Rohde and Pearson 1980). Rohde also published on the ultrastructure of the tegument (Rohde 1973a), protonephridial system and caecum morphology of turtle polystomes (Rohde 1973b). He further published on the ultrastructure of the pharynx and other structures (Rohde 1974, 1975) and the female genital system of *Polystomoides* (Rohde and Ebrahimzadeh 1969). Lixian Fan and co-workers described from China seven polystome species (Fan et al. 2004, 2006, 2007, 2008, 2020; Gao et al. 2012; Shen et al. 2013).

The most significant contribution to the current knowledge of polystomatid biology came from the work of Richard Tinsley (Fig 2.4i) and his team of co-workers and students. Since he completed his PhD in 1971, he has focused on polystomatids of Europe, Africa and North America. Over the following five decades, he conducted groundbreaking work on African polystomes, on various aspects of their morphology and biology. Co-workers who published extensively with Richard Tinsley include Helen Jackson, Joseph Jackson (Fig 2.5a), Jo Cable (Fig 2.5b) and Karen Tocque (Fig 2.5c). Tinsley covered a wide range of topics including systematics, diversity and distribution (Tinsley 1973a, c, 1974a, b, 1975, 1978a, b, 1981, 1982a; Tinsley and Jackson 1998a, 2002); biology, ecology, oviposition and hatching (Tinsley 1978c, 1980, 1983, 1984, 1989, 1990a, b; Tinsley and Owen 1975; Tocque and Tinsley 1991a, b, 1992, 1993a, b; Cable and Tinsley 1991a; Jackson and Tinsley 1988a, b, 1998a, b, c); co-speciation (Tinsley 1996; Tinsley and Jackson 1998b); population biology (Tinsley 1993); morphology and ultrastructure (Tinsley 1973a, 1974c, 1978b; Allen and Tinsley 1989; Cable and Tinsley 1991a, b, c, 1992a, b, 1993, 2001; Cable et al. 1997, 1998); attachment (Tinsley 1971); immune response (Armstrong et al. 1997; Tinsley 1995); biology and morphology of the oncomiracidium (Tinsley 1976, 1978a, 2013); life cycle and host–parasite synchronization (Tinsley 1973b, 1975, 1977, 1978c, 1982b, c, 1999a, b; Tinsley and Earle 1983; Tinsley and Jackson 1986, 1988); *Protopolystoma* (Tinsley and Jackson 1998b; Jackson and Tinsley 2001, 2007; Jackson et al. 2001); host–parasite interactions (Tinsley and Jackson 2002; Jackson and Tinsley 2003a, b, c, 2005; Jackson et al. 1998, 2006); pathology (Tinsley et al. 2002); reproduction (Tinsley 2004); survival and immunity defence systems (Tinsley et al. 2011, 2012); *Oculotrema* oncomiracidium (Tinsley 2013); systematics of amphibian and chelonian polystomes (Sinnappah et al. 2001; Tinsley 2017; Tinsley and Tinsley 2016).

From her thesis "Monogènes et Trématodes parasites d'Amphibiens en Côte d'Ivoire" Anne-Marie Maeder described six species (Maeder et al. 1970; Maeder 1973). Danièle Murith (Fig 2.5d), who was also based in the Ivory Coast as part of her studies published on the evolution of polystomes (Murith 1981a) and revealed the value of sclerites and marginal hooklets as taxonomic characters (Murith 1981a, b, 1982). She also published with co-workers on Cameroon's polystomes (Murith et al. 1978).

Rafael Lamothe-Argumedo (Fig. 2.5e) worked extensively on the anuran monogeneans of Mexico (Lamothe-Argumedo 1963, 1964, 1972, 1973a, b, 1976, 1985). He also published supplementary descriptions of *Neodiplorchis scaphiopodis* (Rodgers, 1941) and its oncomiracidium (see Lamothe-Argumedo 1973a, b, 1974), and described four Mexican turtle polystomes (Lamothe-Argumedo 1972). Claude Vaucher (Fig. 2.5f) from the Natural History Museum in Geneva described four species from the Neotropical Realm (Vaucher 1987, 1990). Thomas Platt from the USA (Fig. 2.5g) described two polystome species from the eyes of freshwater turtles, one species from Central America (Platt 2000a), the other one from North America (Platt 2000b), and redescribed one from Malaysia (Platt et al. 2011).

In Australia, Sylvie Pichelin (Fig. 2.5h), Ian Whittington and co-workers redescribed *Concinnocotyla australensis* (Reichenbach-Klinke, 1966) (Pichelin et al. 1991) and published on the attachment of *Concinnocotyla* eggs (Whittington and Pichelin 1991). Pichelin also described the second anuran polystome from

Fig. 2.5 (**a**) Joe Jackson, (**b**) Jo Cable, (**c**) Karen Tocque, (**d**) Daniele Murith, (**e**) Rafael Lamothe-Argumedo, (**f**) Claude Vaucher, (**g**) Thomas Platt, (**h**) Sylvie Pichelin, (**i**) Dawie Kok. Images (**a**), (**g**) and (**h**) received from authors; (**e**) © María Berenit Mendoza Garfias, Universidad Nacional Autónoma de México (UNAM), Mexico; (**f**) © Jean-Lou Justine; Images (**b**), (**c**) (**d**) and (**i**) © Louis du Preez

Australia (Pichelin 1995a) and conducted a detailed study on Australian turtle polystomes where five new species were described (Pichelin 1995b).

Dawie Kok (Fig. 2.5i) founded the study of polystomes in South Africa. With co-workers, he described seven polystome species (Du Preez and Kok 1992, 1993, 1995; Kok and Seaman 1987; Kok and Van Wyk 1986; Van Niekerk et al. 1993). Louis du Preez (Fig. 2.6a), student of Dawie Kok studied various aspects of the biology of *Polystoma australis* Kok and Van Wyk, 1986 (see Du Preez and Kok 1987; Kok and Du Preez 1987, 1989). With co-workers including Dawie Kok, Olivier Verneau (Fig 2.6b), Martins Aisien (Fig 2.6c), Lixian Fan (Fig 2.6d), Marcus Domingues (Fig 2.6e), and postgraduate students Lilian Raharivololoniaina (Fig. 2.6f), Mathieu Badets (Fig. 2.6g), Pauline Berthier, Adri Delport, Laurent Héritier (Fig. 2.6h), Leon Meyer, Christa Morrison, Maxine Theunissen, Michelle van Rooyen and Willie Landman (Fig 2.6i) studied various aspects of polystome biology including morphology (Du Preez and Kok 1987; Du Preez and Theunissen 2021; Du Preez et al. 2010, 2014, 2017; Héritier et al. 2018; Landman et al. 2018, 2021, 2023; Morrison and Du Preez 2011; Raharivololoniaina et al. 2011; Theunissen et al. 2014); life cycles and experimental infections (Du Preez and Kok 1987, 1992, 1997, 1998; Du Preez et al. 1997; Kok and Du Preez 1987, 1989); host-specificity (Du Preez and Kok 1997); systematics (Aisien and Du Preez 2009; Aisien et al. 2011; Berthier et al. 2014; Chaabane et al. 2019, 2022; Enabulele et al. 2012; Fan et al. 2020; Du Preez 2011, 2013, 2015; Du Preez and Delport 2015; Du Preez and Domingues 2019; Du Preez and Kok 1993, 1995; Du Preez and Lim 2000; Du Preez and Maritz 2006; Du Preez and Morrison 2012; Du Preez and Verneau, 2020; Du Preez et al. 2002, 2003, 2008, 2022; Lim and Du Preez 2001; Yildirimhan et al. 2012); phylogeny and evolution (Badets et al. 2011, 2013; Bentz et al. 2001; Héritier et al. 2015, 2017; Meyer et al. 2015; Verneau et al. 2002, 2009a, b, 2011, 2023).

Olivier Verneau from Perpignan focused on polystome systematics, phylogeography and evolution and published extensively on this topic (Badets et al. 2011, 2013; Bentz et al. 2001, 2006; Berthier et al. 2014; Chaabane et al. 2019, 2022; Du Preez and Verneau 2020; Du Preez et al. 2007, 2010, 2014, 2017; Héritier et al. 2015, 2017, 2018; Landman et al. 2018, 2021, 2023; Meyer et al. 2015; Raharivololoniaina et al. 2011; Verneau et al. 2002, 2009a, b, 2011, 2023). He and Louis Du Preez, in a fruitful collaboration with collaborators, described 35 polystome species and eight genera, among which three, i.e., *Indopolystoma* Chaabane, Verneau and Du Preez, 2019, *Kankana* Verneau, Berthier, Vences and Du Preez, 2011, and *Madapolystoma* Du Preez, Raharivololoniaina, Verneau and Vences, 2010 from amphibians and five, i.e., *Apaloneotrema* Du Preez and Verneau, 2020, *Aussietrema* Du Preez and Verneau, 2020, *Fornixtrema* Du Preez and Verneau, 2020, *Manotrema* Du Preez, Domingues and Verneau, 2022 and *Pleurodirotrema* Du Preez, Domingues and Verneau, 2022 from chelonians.

Fig. 2.6 (**a**) Louis du Preez, (**b**) Olivier Verneau, (**c**) Martins Aisien, (**d**) Lixian Fan, (**e**) Marcus Domingues, (**f**) Liliane Raharivololoniaina, (**g**) Mathieu Badets, (**h**) Laurent Héritier, (**i**) Willie Landman. Images (**a–d**; **f–i**) © authors; Image (**e**) © Jean-Lou Justine

2.5 Discussion

Since the first polystome was discovered, several spikes in the number of parasites described are noted. The discovery of new polystomes received a boost in the decade 1930–1940. In this period, equal numbers of amphibian and chelonian polystomes were described (Fig. 2.7) and mainly from the Nearctic and Oriental Realms, followed by the Palaearctic (Figs. 2.8 and 2.9). In the 1940s, the first polystomes from the Neotropical Realm were discovered, but interest in polystomes declined. In the 1950s and 1960s, the focus shifted to the Ethiopian Realm. From 1970 until 2000, a dramatic increase in the number of newly described species was observed. This renewed interest was the result of French researchers focusing on Central and West Africa, followed by the Oriental and Neotropical Realms, on the one hand, and of the international dynamics on polystomes following the emergence of the POLEE programme on the other (Fig. 2.10). Finally, with the advancement of the molecular tools from the 2000s, a large number of cryptic species were described, especially from amphibians of the Ethiopian Realm.

Currently, over 200 polystomes are known from 30 genera (Table 2.1). However, figuratively speaking this is only the "tip of the iceberg" or, as we say in Africa, "only the ears of the hippo". Vast areas are still unexplored and many fascinating discoveries remain, however, one thing is sure, polystomes will not fail us.

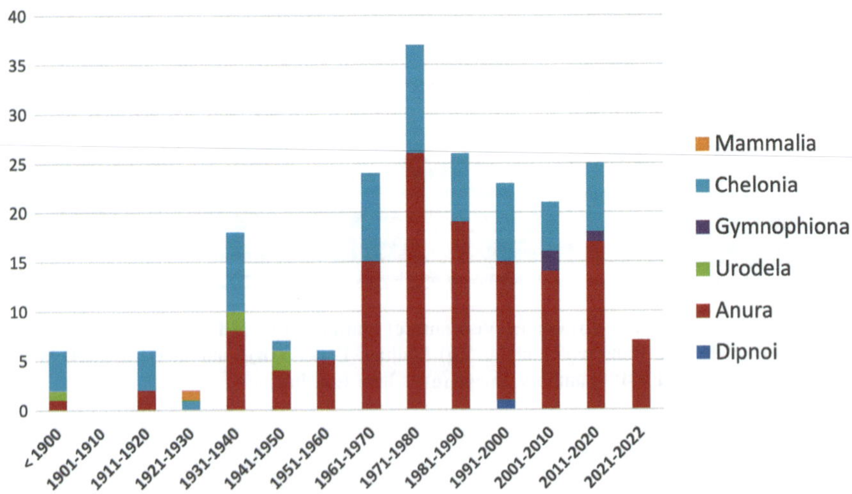

Fig. 2.7 Graphical representation of newly discovered polystome species from 1900 to 2022 in the various host species they parasitize

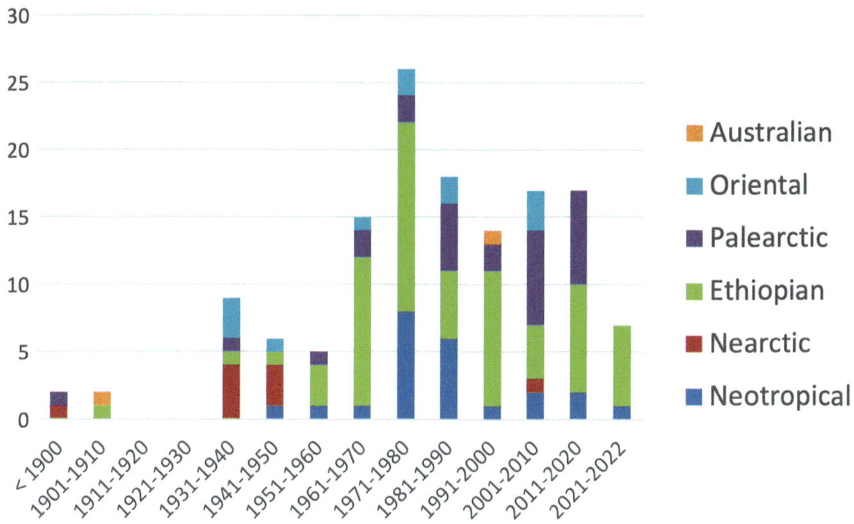

Fig. 2.8 Graphical representation of newly discovered amphibian polystome species from 1900 to 2022 in the various realms in which they are found to occur

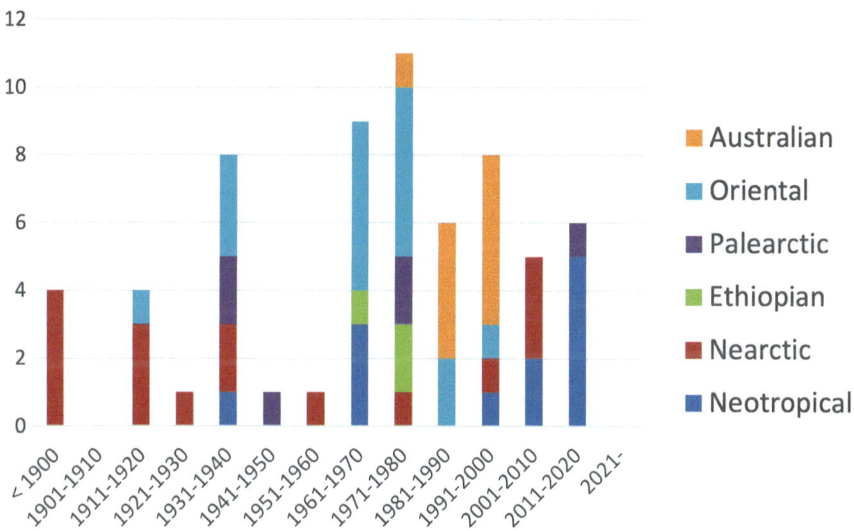

Fig. 2.9 Graphical representation of newly discovered chelonian polystome species from 1900 to 2022 in the various realms in which they are found to occur

Fig. 2.10 Polystomatid Ecology and Evolution (POLEE) group during ISM2 Symposium, Montpellier 1993. (First row: Claude Combes, Mary Beverly-Burton, Karen Tocque, Louis du Preez; Second row: Boris Batchvarov, Nirupama Agrawal, Annie Fournier, Daniele Murith, Richard Tinsley; Third row: Matthew Crosswaite, Ian Whittington)

Table 2.1 Recognized polystomatid genera with species number and host systematic

Polystome genus	No. of species	Host Class	Host Order
Concinnocotyla Pichelin, Whittington and Pearson, 1991	1	Sarcopterygii	Ceratodontiformes
Diplorchis Ozaki, 1931	7	Amphibia	Anura
Eupolystoma Kaw, 1950	6	Amphibia	Anura
Indopolystoma Chaabane, Verneau and Du Preez, 2019	11	Amphibia	Anura
Kankana Raharivololoniaina, Verneau, Berthier, Vences and Du Preez, 2011	1	Amphibia	Anura
Madapolystoma Du Preez, Raharivololoniaina,Verneau and Vences, 2010	5	Amphibia	Anura
Mesopolystoma Vaucher, 1981	1	Amphibia	Anura
Metapolystoma Combes, 1976	9	Amphibia	Anura
Neodiplorchis Yamaguti, 1963	1	Amphibia	Anura
Neoriojatrema Imkongwapang and Tandon, 2010	1	Amphibia	Anura
Parapolystoma Ozaki, 1935	2	Amphibia	Anura
Polystoma Zeder, 1800	70	Amphibia	Anura
Protopolystoma Bychowsky, 1957	6	Amphibia	Anura
Pseudodiplorchis Yamaguti, 1963	1	Amphibia	Anura
Riojatrema Lamothe-Argumedo, 1963	3	Amphibia	Anura
Sundapolystoma Lim and Du Preez, 2001	2	Amphibia	Anura
Wetapolystoma Gray, 1993	1	Amphibia	Anura
Nanopolystoma Du Preez, Wilkinson and Huyse, 2008	3	Amphibia	Gymnophiona
Pseudopolystoma Yamaguti, 1963	1	Amphibia	Urodela
Sphyranura Wright, 1879	4	Amphibia	Urodela
Apaloneotrema Du Preez and Verneau, 2020	1	Reptilia	Testudines
Aussietrema Du Preez and Verneau, 2020	4	Reptilia	Testudines
Fornixtrema Du Preez and Verneau, 2020	7	Reptilia	Testudines
Manotrema Du Preez, Domingues and Verneau, 2022	3	Reptilia	Testudines
Pleurodirotrema Du Preez, Domingues and Verneau, 2022	4	Reptilia	Testudines
Polystomoidella Price, 1939	3	Reptilia	Testudines
Polystomoides Ward, 1917, sensu Chaabane et al. (2022)	29	Reptilia	Testudines
Uropolystomoides Tinsley and Tinsley, 2016	13	Reptilia	Testudines
Uteropolystomoides Tinsley, 2017	1	Reptilia	Testudines
Oculotrema Stunkard, 1924	1	Mammalia	Cetartiodactyla

References

Aisien MS, Du Preez LH (2009) A redescription of *Polystoma africanum* Szidat, 1932 (Monogenea: Polystomatidae). Zootaxa 2095:37–46

Aisien MS, Du Preez LH, Imasuen AA (2011) *Polystoma okomuensis* n. sp. (Monogenea: Polystomatidae) from Boulenger's striped frog, *Phlyctimantis boulengeri* (Perret, 1986) in Nigeria. J Helminthol 85:153–159

Allen KM, Tinsley RC (1989) The diet and gastrodermal ultrastructure of polystomatid monogeneans infecting chelonians. Parasitology 98:265–273

Alvey CH (1933) *Sphyranura oligorchis* n. sp. from *Necturus maculosus*. J Parasitol 20:140

Alvey CH (1936) The morphology and development of the monogenetic trematode *Sphyranura oligorchis* (Alvey 1933) and the description of *Sphyranura polyorchis* n. sp. Parasitology 28: 229–253

Armstrong EP, Halton DW, Tinsley RC, Cable J, Johnston RN, Johnston CF, Shaw C (1997) Immunocytochemical evidence for the involvement of an FMRFamide-related peptide in egg production in the flatworm parasite *Polystoma nearcticum*. J Comp Neurol 377:41–48

Badets M, Whittington I, Lalubin F, Allienne JF, Maspimby JL, Bentz S, Du Preez LH, Barton D, Hasegawa H, Tandon V, Imkongwapang R, Ohler A, Combes C, Verneau O (2011) Correlating early evolution of parasitic platyhelminths to Gondwana breakup. Syst Biol 60:762–781

Badets M, Du Preez LH, Verneau O (2013) Alternative development in *Polystoma gallieni* (Platyhelminthes, Monogenea) and life cycle evolution. Exp Parasitol 135:283–286

Batchvarov GK (1982) *Polystoma combesi* n. sp. (Polystomatidae Gambl, 1896) a parasite on *Rana graeca* Boulanger, 1897, in Bulgaria. C R Acad Bulg Sci 35:249–251

Bentz S, Leroy S, Du Preez L, Mariaux J, Vaucher C, Verneau O (2001) Origin and evolution of African *Polystoma* (Monogenea: Polystomatidae) assessed by molecular methods. Int J Parasitol 31:697–705

Bentz S, Sinnappah-Kang ND, Lim L-HS, Lebedev B, Combes C, Verneau O (2006) Historical biogeography of amphibian parasites, genus *Polystoma* (Monogenea: Polystomatidae). J Biogeogr 33:742–749. https://doi.org/10.1111/j.1365-2699.2005.01402.x

Berthier P, Du Preez LH, Raharivololoniaina L, Vences M, Verneau O (2014) Two new species of polystomes (Monogenea: Polystomatidae) from the anuran host *Guibemantis liber*. Parasitology 63:108–119

Biserkov VY, Yildirimhan HS, Buchvarov G, Ugurtas JH (2001) *Polystoma macrocnemis* n. sp. (Monogenea: Polystomatidae) from the Iranian longlegged wood frog *Rana macrocnemis* (Ranidae) in Turkey. Syst Parasitol 48:61–66

Boeger WA, Kritsky DC (1993) Phylogeny and a revised classification of the Monogenoidea Bychowsky, 1937 (Platyhelminthes). Syst Parasitol 26:1–32

Bourgat R (1975) *Polystoma assoulinei* n. sp. (Monogenea, Polystomatidae) parasite vésical de *Ptychadaena huguettae* (Anura) au Togo. Vie Milieu C 25:197–202

Bourgat R (1977) Etude comparative des polystomes (Monogènes) de ranidés (Anoures) du Sud Togo. Description de *Polystoma togoensis* n. sp. Bull Mus Natl Hist Nat 312:447–463

Bourgat R, Murith D (1980) *Polystoma lamottei* n. sp. et *P. aeschlimanni* n. sp., deux polystomes (Monogènes) de la même espèce d'Amphibien: *Ptychadena pumilio* (Boulenger, 1920). Z Parasitenkd 62:293–301

Buchvarov GK (1980) *Polystoma mazurmovici* n. sp. (Polystomatidae Gambl, 1896), a parasite of *Rana dalmatina* Bonaparte, 1939 in Bulgaria. Trav Sci Univ Plovdiv – Biol 18:183–190

Buchvarov GK (1984a) *Polystoma skrjabini* n. sp. (Polystomatidae Gambl, 1896) – a parasite on *Hyla arborea* (Linné, 1758), in Bulgaria. C R Acad Bulgare Sci 37:697–699

Buchvarov GK (1984b) *Polystoma skuratovitchi* n. sp. (Polystomatidae Gambl, 1986) – a parasite on *Rana arvalis* Nicolsson, 1842 from West Poland. Trav Sci Univ Plovdiv – Biol 22:137–142

Bychowsky BE (1957) Monogenetic trematodes, their systematics and phylogeny. Akad Nauk USSR, p 509

Bychowsky BE (1987) Basic directions of phylogeny of monogenetic trematodes. In: Hargis WJ (ed) Monogenetic trematodes. Their systematics and phylogeny. Graphic Arts Press, Was Ch.1, pp 377–573

Cable J, Tinsley RC (1991a) Intra-uterine larval development of the polystomatid monogeneans, *Pseudodiplorchis americanus* and *Neodiplorchis scaphiopodis*. Parasitology 103:253–266

Cable J, Tinsley RC (1991b) Microsporidian hyperparasites and bacteria associated with *Pseudodiplorchis americanus* (Monogenea: Polystomatidae). Can J Zool 70:523–529

Cable J, Tinsley RC (1991c) The ultrastructure of photoreceptors in *Pseudodiplorchis americanus* and *Neodiplorchis scaphiopodis* (Monogenea: Polystomatidae). Int J Parasitol 21:81–90

Cable J, Tinsley RC (1992a) Tegumental ultrastructure of *Pseudodiplorchis americanus* larvae (Monogenea: Polystomatidae). Int J Parasitol 22:819–829

Cable J, Tinsley RC (1992b) Unique ultrastructural adaptations of *Pseudodiplorchis americanus* (Polystomatidae: Monogenea) to a sequence of hostile conditions following host infection. Parasitology 105:229–241

Cable J, Tinsley RC (1993) The ultrastructure of spermatogenesis and spermatozoa in *Pseudodiplorchis americanus* (Monogenea: Polystomatidae). Can J Zool 71:1609–1619

Cable J, Tinsley RC (2001) Ultrastructure of spermiogenesis and spermatozoa of *Discocotyle sagittata* (Monogenea: Polyopisthocotylea: Discocotylinea). Folia Parasitol 48:209–216

Cable J, Tocque K, Tinsley RC (1997) Histological analysis of the egg capsule of the ovoviviparous polystomatid monogenean, *Pseudodiplorchis americanus*. Int J Parasitol 27:1075–1080

Cable J, Harris PD, Tinsley RC (1998) Life history specializations of monogenean flatworms: a review of experimental and microscopical studies. Microsc Res Tech 42:186–199

Chaabane A, Verneau O, Du Preez LH (2019) *Indopolystoma* (n. gen.) Monogenea: Polystomatidae with the description of three new species and reassignment of eight known *Polystoma* species from Asian frogs (Anura, Rhacophoridae). Parasite 26:67

Chaabane A, Du Preez L, Johnston G, Verneau O (2022) Revision of the systematics of the Polystomoidinae (Platyhelminthes, Monogenea, Polystomatidae) with redefinition of *Polystomoides* Ward, 1917 and *Uteropolystomoides* Tinsley, 2017. Parasite 29:56. https://doi.org/10.1051/parasite/2022056

Combes C (1964) Trématodes parasites de *Rana temporaria* L. dans la haute-vallée de Carenca (P.-o.). Vie Milieu, 17:91–95

Combes C (1966a) Recherches sur les formes néoteniques de *Polystoma integerrimum pelobatis* Euzet et Combes, 1966, obtenus expérimentalement chez *Pelobates cultripes* (Cuvier). Bull Soc Neuchatel Sci Nat 89:5–17

Combes C (1966b) Recherches expérimentales sur la spécificite parasitaire des polystomes de *Rana temporaria* L. et de *Pelobates cultripes* (Cuv). Bull Soc Zool Fr 91:439–444

Combes C (1967a) (In French) Investigations on the neotenic form of Polystomatidae (Monogenea). Presence of a neotenic form in *Polystoma gallieni* (Price 1938). Bull Soc Neuchâtel Sci Nat 90:205–214

Combes C (1967b) (In French) Correlations between the sexual cycles of anuran amphibians and the Polystomatidae (Monogenea*), Hyla meridionales, Pelobates pelobatis, Pelobates gallieni, Pelobates cultripes*. CR Acad Sci Paris Ser D: Sci Natur 264:1051–1052

Combes C (1967c) Biologie des Polystomatidae (Monogenea): Existence et démonstration expérimentale des possibilités de cycle interne direct. Bull Soc Zool Fr 92:129–133

Combes C (1968) Biologie, écologie des cycles et biogéographie de digènes et monogènes d'amphibiens dans l'est des Pyrénées. Mém Mus Natl Hist Nat, Sér A Zool L1:150–155

Combes C (1970) Ecologie des monogènes parasites de vertébrés amphibiens. J Parasitol 56:60–61

Combes C (1972a) Écologie des Polystomatidae (Monogenea): facteurs influencant le volume et le rythme de la ponte. Int J Parasitol 2:233–238

Combes C (1972b) Influence of the behaviour of amphibians on helminth life-cycles. Zool J Linnean Soc 51:151–170

Combes C (1973) Biologie des Polystomatidae: Le cycle interne direct chez *Eupolystoma alluaudi* (de Beauchamp, 1913). Z Parasitenkd 42:69–75

Combes C (1976) World biogeography of Polystomatidae. Acad Sci USSR Proc 34:55–63

Combes C (1981) Invasion strategies in parasites of amphibious hosts. Parasitology 82:63–64

Combes C (1983) Application à l'écologie parasitaire des indices d'association fondés sur le caractère présence-absence. Vie Milieu 33:203–212

Combes C, Channing A (1978) Polystomatidae (Monogenea) d'amphibiens d'Afrique du Sud: *Polystoma natalensis* n. sp., parasite de *Strongylopus grayi* (Smith 1849). Vie Milieu 28–29:61–68

Combes C, Gerbeaux TM (1970) Recherches éco-parasitologiques sur l'helminthofaune de *Rana ridibunda perezi* (Amphibien Anoure) dans l'est des Pyrénées. Vie Milieu 21:121–156

Combes C, Justine JL (1982) Présence au Sénegal de *Polystomoides bourgati* Combes & Kulo, 1978 (Monogenea, Polystomatidae) chez la tortue *Pelusios adansoni* Schweigger. Bull Inst Fr Afr noire Sér A 44:323–325

Combes C, Knoepffler LP (1977) Parasitisme d'une population de *Pelobates cultripes* (Cuvier, 1829) à la sortie de l'eau par les postlarves de *Polystoma pelobatis* (Euzet & Combes, 1965). Vie Milieu 27:215–219

Combes C, Ktari MH (1976) *Neopolystoma euzeti* n. sp. (Monogenea, Polystomatidae) premier représentant du genre *Neopolystoma* Price, 1939 en Afrique. Ann Parasitol Hum Comp 51:221–225

Combes C, Kulo SD (1978) *Polystomoides bourgati* n. sp. (Monogenea, Polystomatidae) premier représentant du genre *Polystomoides* Ward, 1917 en Afrique occidentale. Rev Zool Afric 92: 622–626

Combes C, Lambert D (1970) Le processus de résorption ovocytaire chez les polystomes à ovaire double. Z Parasitenkd 33:262–266

Combes C, Lambert A (1972) Development of the sensory system during larval life in the Monogenea Polyopisthocotylea. CR Acad Sci Paris 274:402–404

Combes C, Laurent RF (1974) *Polystoma borellii* n. sp. (Monogenea, Polystomatidae) parasite de *Pleurodema borellii* Peracca (Anura, Leptodactylidae) en République Argentine. Acta Zool Lilloana 31:57–64

Combes C, Laurent RF (1978) Deux nouveaux Polystomatidae (Monogenea) de République Argentine. Acta Zool Lilloana 33:85–92

Combes C, Laurent RF (1979) Les monogènes Polystomatidae de République Argentine: description de deux nouvelles espèces et essai de synthèse. Rev Iber Parasitol 79:545–557

Combes C, Rohde K (1978) *Polystomoides platynotae* n. sp. (Monogenea, Polystomatidae), parasite du chelonien d'eau douce *Notochelys platynota* (Gray, 1834) en Malaisie. Vie Milieu 28–29:69–75

Combes C, Thiery A (1983) Sur un *Polystomoides* (Monogenea), parasite du chélonien *Clemys caspica* au Maroc. Bull Ins Sci 7:165–170

Combes C, Bourgat R, Salami-Cadoux M-L (1973a) Découverte d'un cycle interne direct intervenant comme mode habituel dans la reproduction d'un Plathelminthe. CR Acad Sci Paris 276:2005–2006

Combes C, Bourgat R, Salami-Cadoux M-L (1973b) (In French) Biology of the Polystomatidae: the direct internal cycle of *Eupolystoma alluaudi* (de Beauchamp, 1913). Z Parasitenkd (Berlin, Germany) 42:69–75

Combes C, Lambert A, Alméras MT (1975) La chétotaxie des larves nageantes de polystomes européens (Monogenea). Ann Parasitol Hum Comp (Paris) 50:25–37

Combes C, Bourgat R, Salami-Cadoux M-L (1976) Valeur adaptative du mode de transmission chez les Polystomatidae (Monogenea). Bull Ecol 4:207–214

Combes C, Lambert A, Bourgat R, Salami-Cadoux M-L (1978) L'oncomiracidium de *Eupolystoma alluaudi* (de Beauchamp, 1913): Monogène parasite de *Bufo regularis* Reuss en Afrique. Bull Mus Natl Hist Nat Zool 353:211–216

Diengdoh CR, Tandon V (1991) A new species of *Polystoma* (Monogenea) parasitic in rhacophorid amphibians in Meghalaya, India. Helminthologia 28:173–178

Dollfus RP, Euzet L, Combes C (1965) Trématodes de Batraciens (fin). Description d'un polystome. Ann Parasitol Hum Comp (Paris) 40:55–60

Du Preez LH (2011) *Polystoma vernoni* n. sp. (Monogenea: Polystomatidae) from the sharp nosed grass frog *Ptychadena oxyrhynchus* (Smith, 1849) in South Africa. J Helminthol 85:294–299

Du Preez LH (2013) Polystomatidae (Monogenea) of southern African Anura: *Polystoma channingi* n. sp. parasitic in two closely related *Cacosternum* species. Afr Zool 48:64–71

Du Preez LH (2015) *Eupolystoma namibiensis* n. sp. (Monogenea: Polystomatidae) parasitic in *Poyntonophrynus hoeschi* (Ahl, 1934) of Namibia. Afr Zool 50:141–145

Du Preez LH, Delport M (2015) A new polystomatid (Monogenea: Polystomatidae) from the mouth of the north American freshwater turtle *Pseudemys nelsoni*. ZooKeys 539:1–9

Du Preez LH, Domingues MV (2019) *Polystoma knoffi* n. sp. and *Polystoma travassosi* n. sp. (Monogenea: Polystomatidae): naming museum-archived specimens from Brazil. Syst Parasitol 96:755–765

Du Preez LH, Kok DJ (1987) *Polystoma australis* (Monogenea): loss of locomotory cilia associated with retarded hatching of oncomiracidia. Parasitol Res 74:50–54

Du Preez LH, Kok DJ (1992) Syntopic occurrence of new species of *Polystoma* and *Metapolystoma* (Monogenea: Polystomatidae) in *Ptychadena porosissima* in South Africa. Syst Parasitol 22: 141–150

Du Preez LH, Kok DJ (1993) Polystomatidae (Monogenea) of Anura in southern Africa: *Polystoma testimagna* n. sp. parasitic in *Strongylopus f. fasciatus* (Smith, 1849). Syst Parasitol 25:213–219

Du Preez LH, Kok DJ (1995) Polystomatidae (Monogenea) of southern African Anura: *Polystoma claudecombesi* n. sp. parasitic in *Rana angolensis* Bocage, 1866. Syst Parasitol 30:223–231

Du Preez LH, Kok DJ (1997) Supporting evidence of host-specificity among southern African polystomes (Polystomatidae: Monogenea). Parasitol Res 83:558–562

Du Preez LH, Kok DJ (1998) The relative importance of bladder versus neotenic stages of *Polystoma marmorati* and *P. umthakathi* in natural frog populations in South Africa. J Helminthol 72:117–121

Du Preez LH, Lim LHS (2000) *Neopolystoma liewi* sp. n. (Monogenea: Polystomatidae) from the eye of the Malayan box turtle (*Cuora amboinensis*). Folia Parasitol 47:11–16

Du Preez LH, Maritz MF (2006) Demonstrating morphometric protocols using polystome marginal hooklet measurements. Syst Parasitol 63:1–15

Du Preez LH, Moeng IA (2004) Additional morphological information on *Oculotrema hippopotami* Stunkard 1924 (Monogenea: Polystomatidae) parasitic on the African hippopotamus. Afr Zool 39:225–233

Du Preez LH, Morrison C (2012) Two new polystomes (Monogenea: Polystomatidae) from the eyes of north American freshwater turtles. Zootaxa 3392:47–59

Du Preez LH, Theunissen M (2021) A sucker for the job: morphology and functioning of suckers of polystomatid monogeneans. Folia Parasitol 68:006

Du Preez LH, Verneau O (2020) Eye to eye: classification of conjunctival sac polystomes (Monogenea: Polystomatidae) revisited with the description of three new genera *Apaloneotrema* n. g. *Aussietrema* n. g. and *Fornixtrema* n. g. Parasitol Res 119:4017–4031

Du Preez LH, Kok DJ, Seaman MT (1997) Host recognition of polystome oncomiracidia (Polystomatidae: Monogenea) in contact with natural and substitute anuran hosts. J Afr Zool 111:47–55

Du Preez LH, Vaucher C, Mariaux JP (2002) Polystomatidae (Monogenea) of African Anura: *Polystoma dawiekoki* n. sp. parasitic in *Ptychadena anchietae* (Bocage, 1867). Syst Parasitol 52: 35–41

Du Preez LH, Tinsley RC, De Sa R (2003) Polystomatidae (Monogenea) of southern African Anura: *Eupolystoma vanasi* n. sp. parasitic in *Schismaderma carens* (Smith, 1848). Syst Parasitol 54:71–79

Du Preez LH, Verneau O, Gross TS (2007) *Polystoma floridana* n. sp. (Monogenea: Polystomatidae) a parasite in the green tree frog, *Hyla cinerea* (Schneider), of North America. Zootaxa 1663:33–45

Du Preez LH, Wilkinson M, Huyse T (2008) The first record of polystomes (Monogenea: Polystomatidae) from caecilian hosts (Amphibia: Gymnophiona), with the description of a new genus and two new species. Syst Parasitol 69:201–209

Du Preez LH, Raharivololoniaina L, Verneau O, Vences M (2010) A new genus of polystomatid parasitic flatworm (Monogenea: Polystomatidae) without free-swimming life stage from the Malagasy poison frogs. Zootaxa 2722:54–68

Du Preez LH, Badets M, Verneau O (2014) Assessment of platyhelminth diversity within amphibians of French Guiana revealed a new species of *Nanopolystoma* (Monogenea: Polystomatidae) in the caecilian *Typhlonectes compressicauda*. Folia Parasitol 61:537–542

Du Preez LH, Badets M, Héritier L, Verneau O (2017) Tracking platyhelminth parasite diversity from freshwater turtles in French Guiana: first report of *Neopolystoma* Price, 1939 (Monogenea: Polystomatidae) with the description of three new species. Parasit Vectors 10:53

Du Preez L, Domingues MV, Verneau O (2022) Classification of pleurodire polystomes (Platyhelminthes, Monogenea, Polystomatidae) revisited with the description of two new genera from the Australian and Neotropical realms. Int J Parasitol Parasites Wildl 19:180–186

Enabulele EE, Aisien MSO, Du Preez LH (2012) *Polystomoides bourgati* (Monogenea: Polystomatidae) infecting *Pelusios castaneus* in southern Nigeria. Afr Zool 47:42–47

Euzet L, Combes C (1964) Sur un Polystomatidae (Monogenea) récolté à Madagascar chez *Rana mascareniensis* Duméril et Bibron. Bull Soc Zool France 89:392–401

Euzet L, Combes C (1965) Parasites des chéloniens malgaches. *Polystomoides chabaudi* n. sp. (Monogenea) chez la tortue d'eau douce *Pelomedusa subrufa* Lacépède 1788. Ann Parasitol Hum Comp 40:445–450

Euzet L, Combes C (1966) *Polystoma integerrimum pelobatis* n. subsp. (Monogenea) parasite de *Pelobates cultripes* (Cuvier, 1829). Ann Parasitol Hum Comp 41:109–118

Euzet L, Combes C (1967) Présence au Nord-Tchad de *Eupolystoma alluaudi* (de Beauchamp, 1913) (Monogenea, Polystomatidae). Ann Parasitol Hum Comp 42:403–406

Euzet L, Combes C (1975) Miscellanea helminthologica maroccana. XLI. *Polystoma gallieni* Price, 1939 (Monogenea) found in *Hyla meridionalis* Boettger, 1874 in Morocco. Bull Mus Natl Hist Nat Zool 212:655–657

Euzet L, Combes C, Knoepffler L-P (1966) Parasites d'amphibiens du Gabon Polystomatidae (Monogenea). Considérations sur la répartition géographique des Polystomatidae africains et malgaches actuellement connus. Biol Gabon 2:215–233

Euzet L, Combes C, Knoepffler LP (1969) Parasites d'amphibiens de Côte d'Ivoire et du Liberia: Polystomatidae (Monogenea). Biol Gabon 3:217–221

Euzet L, Bourgat R, Salami-Cadoux ML (1974a) *Polystoma galamensis* (Monogenea) parasite de *Rana galamensis* Duméril et Bibron, 1841, au Togo. Ann Parasitol Hum Comp 49:63–68

Euzet L, Combes C, Batchvarov G (1974b) Sur un nouveau Polystomatidae Européen, parasite de l'amphibien *Bufo viridis* Laur. Vie Milieu 24:129–140

Euzet L, Combes C, Knoepffler LP (1975) Parasites d'amphibiens de la République Centrafricaine. Polystomatidae (Monogenea). Vie Milieu 24:141–150

Fan LX, Wang ZL, Li JH (2004) A new species of *Polystoma* (Polystomatidae: Monogenea) parasitic in *Polypedates dugritei*. Anim Classific J 29:451–454

Fan XL, Pan H, Wang G (2006) A new species of the genus *Polystoma* (Polystomatidae: Monogenea) parasitic in *Rana pleuraden* Boulenger. Acta Zootaxon Sin 30:710–713

Fan LX, Wang ZL, Xu HJ (2007) A new species of the genus *Polystoma* (Polystomatidae: Monogenea) parasitic in *Rana grahami* Boulenger. Acta Zootaxon Sin 32:182–185

Fan LX, Li JH, He ZY (2008) A new species of the genus *Polystoma* (Polystomatidae: Monogenea) parasitic in the host *Rhacophorus carvirostris* Guenther. Acta Zootaxon Sin 33:340–343

Fan L, Xu W, Jia T, Netherlands EC, Du Preez LH (2020) *Polystoma luohetong* n. sp. (Monogenea: Polystomatidae) from *Rana chaochiaoensis* Liu (Amphibia: Ranidae) in China. Syst Parasitol 97:639–647

Fournier A, Combes C (1978) Structure of photoreceptors of *Polystoma integerrimum* (Platyhelminths, Monogenea). Zoomorphology 91:147–155

Fournier A, Combes C (1979) Demonstration of an embryonic developmental duality in *Eupolystoma alluaudi* (Monogenea, Polystomatidae) and its role in the genesis of the internal cycle. CR Acad Sci Paris 289:745–747

Frölich von JA (1791) Beiträge zur Naturgeschrichte der Eingeweidewürmer. Naturforscher, Halle 25:52–113

Gallien L (1932a) Sur l'évolution de la génération issue des formes néoténiques de *Polystomum integerrimum* Fröelich. CR Acad Sci Paris 195:77–79

Gallien L (1932b) Sur la reproduction néoténique chez *Polystomum integerrimum* Fröelich. CR Acad Sci 194:1852–1854

Gallien L (1933) Transformations histologiques corrélative du cycle sexuel chez *Polystomum integerrimum* Frölich. CR Acad Sci Paris 196:426–428

Gallien L (1934) Sur le déterminisme de la dualité d'évolution des larves chez *Polystomum integerrimum* Fröl. CR Acad Sci Paris 199:229–231

Gallien L (1935) Recherches expérimentales sur le dimorphisme évolutif et la biologie de *Polystomum integerrimum* Fröhl. Travaux Sta Zool Wimereux 12:1–181

Gallien L (1938) Sur la spécificité parasitaire de *Polystomum integerrimum* Fröelich. Bull Soc Zool Fr 63:249–251

Gallien L (1947) Description de *Polystomum gallieni* Price (Monogenea, Polystomatidae) parasite de *Hyla arborea* var. *meridionalis* Boettger. Ann Parasitol Hum Comp 22:24–29

Gao YY, Chen ZM, Fan LX (2012) A new species of the genus *Polystoma* (Monogenea: Polystomatidae) parasitic in *Hyla a. annectans*. Acta Zootaxon Sin 37:529–534

Héritier L, Badets M, Du Preez LH, Aisien MS, Lixian F, Combes C, Verneau O (2015) Evolutionary processes involved in the diversification of chelonian and mammal polystomatid parasites (Platyhelminthes, Monogenea, Polystomatidae) revealed by palaeoecology of their hosts. Mol Phylogenet Evol 92:1–10

Héritier L, Valdeón A, Sadaoui A, Gendre T, Ficheux S, Bouamer S, Kechemir-Issad N, Palacios C, Du Preez LH, Verneau O (2017) Introduction and invasion of the red-eared slider and its parasites in freshwater ecosystems of southern Europe: risk assessment for the European pond turtle in wild environments. Biodivers Conserv 26:1817–1843

Héritier L, Verneau O, Smith KG, Coetzer C, Du Preez LH (2018) Demonstrating the value and importance of combining DNA barcodes and discriminant morphological characters for polystome taxonomy (Platyhelminthes, Monogenea). Parasitology 67:38–46

Imkongwapang R, Tandon V (2010) *Neoriojatrema mokokchungensis* n. g., n. sp. (Monogenea: Polystomatidae: Polystomatinae) from the urinary bladder of a megophryid frog, *Xenophrys glandulosa*, in Nagaland, North-Eastern India. J Helminthol 84:1–6

Jackson HC, Tinsley RC (1988a) Environmental influences on egg production by the monogenean *Protopolystoma xenopodis*. Parasitology 97:115–128

Jackson HC, Tinsley RC (1988b) The capacity for viable egg production by the monogenean *Protopolystoma xenopodis* in single and multiple infections. Int J Parasitol 18:585–589

Jackson JA, Tinsley RC (1998a) Effects of temperature on oviposition rate in *Protopolystoma xenopodis* (monogenea: Polystomatidae). Int J Parasitol 28:309–315

Jackson JA, Tinsley RC (1998b) Incompatibility of *Protopolystoma xenopodis* (Monogenea: Polystomatidae) with an octoploid *Xenopus* species from southern Rwanda. Int J Parasitol 28: 1195–1199

Jackson JA, Tinsley RC (1998c) Reproductive interference in concurrent infections of two *Protopolystoma* species (Monogenea: Polystomatidae). Int J Parasitol 28:1201–1204

Jackson JA, Tinsley RC (2001) *Protopolystoma xenopodis* (Monogenea) primary and secondary infections in *Xenopus laevis*. Parasitology 123:455–463

Jackson JA, Tinsley RC (2003a) Parasite infectivity to hybridising host species: a link between hybrid resistance and allopolyploid speciation? Int J Parasitol 33:137–144

Jackson JA, Tinsley RC (2003b) Density dependence of postlarval survivorship in primary infections of *Protopolystoma xenopodis*. J Parasitol 89:958–960

Jackson JA, Tinsley TC (2003c) Postlarval *Protopolystoma* spp. kidney infections in incompatible *Xenopus* spp. induce weak resistance to heterospecifics. Parasitol Res 90:429–434

Jackson JA, Tinsley RC (2005) Geographic and within-population structure in variable resistance to parasite species and strains in a vertebrate host. Int J Parasitol 35:29–37

Jackson JA, Tinsley RC (2007) Evolutionary diversity in polystomatids infecting tetraploid and octoploid *Xenopus* in east African highlands: biological and molecular evidence. Parasitology 134:1223–1235

Jackson JA, Tinsley RC, Hinkel HH (1998) Mutual exclusion of congeneric monogenean species in a space-limited habitat. Parasitology 117:563–569

Jackson JA, Tinsley RC, Du Preez LH (2001) Differentiation of two locally sympatric *Protopolystoma* (Monogenea: Polystomatidae) species by temperature-dependent larval development and survival. Int J Parasitol 31:815–821

Jackson JA, Pleass RJ, Cable J, Bradley JE, Tinsley RC (2006) Heterogenous interspecific interactions in a host–parasite system. Int J Parasitol 36:1341–1349

Knoepffler LP, Combes C (1977) Présence en Corse de *Polystomoides ocellatum* (Rudolphi, 1819) chez *Emys orbicularis* (Chelonia, Emydidae). Considérations sur la répartition mondiale du genre *Polystomoides*. Vie Milieu 27:221–230

Kohn A, Combes C, Gomes DC (1978) Représentants du genre *Polystoma* Zeder (Monogenea) au Brésil. Bull Mus Natl Hist Nat Zool 514:227–229

Kok DJ, Du Preez LH (1987) *Polystoma australis* (Monogenea): life-cycle studies in experimental and natural infections of normal and substitute hosts. J Zool 212:235–243

Kok DJ, Du Preez LH (1989) *Polystoma australis* (Monogenea): development and reproduction in neotenic parasites. S Afr J Zool 24:225–230

Kok DJ, Seaman MT (1987) Polystomatidae (Monogenea) parasitic in the anuran genus *Natalobatrachus* in South Africa. S Afr J Zool 22:258–263

Kok DJ, Van Wyk JH (1986) Polystomatidae (Monogenea) parasitic in the anuran genus *Kassina* in South Africa. S Afr J Zool 21:189–195

Lamothe-Argumedo R (1963) Trematódos de los Anfibios de Mexico. I. Sobre un nuevo género de la familia Polystomatidae (Gamble, 1896), hallado en la vejiga urinaria de *Tomodactylus amulae* (Günther) y *Bufo simus* (Schmidt). Rev Soc Mex Hist Nat 24:73–88

Lamothe-Argumedo R (1964) Trematodos de los Anfibios de Mexico II. Una nueva subfamilia, Riojatrematinae (Monogenea: Polystomatidae) de Amphibia Anura. Rev Biol Trop 12:153–156

Lamothe-Argumedo R (1972) Monogeneos de reptiles I. Redescripción de cuatro especies de Monogenea (Polystomatidae) parasitos de la vejiga urinaria de tortugas de México. An Inst Biol Univ Nation Autón México Zool 43:1–16

Lamothe-Argumedo R (1973a) Monogéneos de los anfibios de México IV. Redescripción de *Neodiplorchis scaphiopi* (Rodgers, 1941) Yamaguti, 1963. An Inst Biol Univ Nation Autón México Zool 44:1–8

Lamothe-Argumedo R (1973b) Monogéneos de los anfibios de México V. Descripcion de la larva de *Neodiplorchis scaphiopi* (Rodgers, 1941) Yamaguti, 1963 (Monogenea: Polystomatidae). An Inst Biol Univ Nation Autón México Zool 44:9–14

Lamothe-Argumedo R (1974) Monogeneans of amphibians of Mexico.VI. Description of the larvae of *Neodiplorchis scaphiopi* (Rodgers, 1941) Yamaguti, 1963 (Monogenea: Polystomatidae). An Inst Biol Univ Nac Autón México Ser Zool 44:9–13

Lamothe-Argumedo R (1976) Monogéneos de los anfibios de México VI. Redescripción de *Polystoma naevius* Caballero y Zerecero, 1941. An Inst Biol Univ Nation Autón México Zool 47:1–8

Lamothe-Argumedo R (1985) Monogéneos de los anfibios de México. Hallazgo de *Pseudodiplorchis americanus* (Rodgers Y Kuntz, 1940) Yamaguti, 1963 en Baja California sur, México. An Inst Biol Univ Nation Autón México Zool 56:291–300

Landman W, Verneau O, Du Preez LH (2018) First record of viviparity in polystomatid flatworms (Monogenea: Polystomatidae) with the description of two new species of *Madapolystoma* from the Madagascan anuran hosts *Blommersia domerguei* and *Mantella expectata*. Int J Parasitol Parasites Wildl 7:343–354

Landman W, Verneau O, Raharivololoniaina L, Du Preez LH (2021) First record of *Metapolystoma* (Monogenea: Polystomatidae) from *Boophis* tree frogs in Madagascar, with the description of five new species. Int J Parasitol Parasites Wildl 14:161–178

Landman W, Verneau O, Vences M, Du Preez L (2023) *Metapolystoma ohlerianum* n. sp. (Monogenea: Polystomatidae) from *Aglyptodactylus madagascariensis* (Anura: Mantellidae). Acta Parasitol 68:1–15

Lim LHS, du Preez LH (2001) *Sundapolystoma chalconotae* n. g. n. sp. (Monogenea: Polystomatinae) from *Rana chalconota* (Schlegel) of peninsular Malaysia. Syst Parasitol 49: 223–231

Llewellyn J (1957) Host-specificity in monogenetic trematodes. Bull Soc Neuchâtel Sci Nat 32: 199–212

Llewellyn J (1963) Larvae and larval development of monogeneans. Adv Parasitol 1:287–326

Llewellyn J (1964) The effects of the host and its habits on the morphology and life-cycle of a monogenean parasite. In parasitic worms and aquatic conditions. Proceedings of symposium, Prague: Czech Acad Sci

Llewellyn J (1968) Larvae and larval development of monogeneans. Adv Parasitol 6:373–383

Llewellyn J (1970) The evolution of life history patterns among parasitic helminths: adaptive mechanisms in relation to the external environment. J Parasitol Sect II 56:211–212

MacCallum GA (1918) Studies on the Polystomatidae. Zoopathologica 1:103–120

Maeder A-M (1973) Monogènes et trématodes parasites d'amphibiens en Côte d'Ivoire. Rev Suisse Zool 80:267–322

Maeder AM, Euzet L, Combes C (1970) Espèces nouvelles du genre *Polystoma* (Monogenea) en Afrique Occidentale. Z Parasitenkd 35:140–155

Meyer L, Du Preez LH, Bonneau E, Héritier L, Quintana M-F, Valdeón A, Sadaoui A, Kechemir-Issad N, Palacios C, Verneau O (2015) Parasite host-switching from the invasive American red-eared slider, *Trachemys scripta elegans*, to the native Mediterranean pond turtle, *Mauremys leprosa*, in natural environments. Aquat Invasions 10:79–91

Morrison C, Du Preez LH (2011) Turtle polystomes of the world. *Neopolystoma*, *Polystomoidella* and *Polystomoides*. VDM Publishing, Germany

Murith D (1981a) Contribution à l'étude de la systématique des polystomes (Monogènes, Polystomatidae) parasites d'amphibiens anoures de basse Côte-d'Ivoire. Rev Suisse Zool 88: 475–533

Murith D (1981b) Contribution à l'étude de la biologie du développement des polystomes (Monogenea) parasites d'amphibiens anoures de basse Côte-d'Ivoire. Bull Soc Neuchâtel Sci Nat 10:5–33

Murith D (1982) In vivo-studies on the nature of host-parasite relationships in the amphibian-polystome (Monogenea)-complex. Rev Suisse Zool 89:957–965

Murith D, Miremad-Gassmann M, Vaucher C (1978) Contribution à l'étude des polystomes d'amphibiens du Cameroun. Rev Suisse Zool 853:681–698

Ozaki Y (1931a) A new polystome trematode, *Diplorchis ranae*, gen. et sp. n. Ann Mag Nat Hist 7: 181–184

Ozaki Y (1931b) On some polystome trematodes of Japan. Nihon Kiseichu Gakkai Kiji 3:3

Ozaki Y (1932) On the life-history of the polystome trematode, *Diplorchis ranae*. Dobuts Zasshi Tokyo 44:16–17

Ozaki Y (1935a) Studies on the frog-trematode, *Diplorchis ranae*. I. Morphology of the adult form with a review of the family Polystomatidae. J Sci Hiroshima Univ Ser B, Div, 1 3:193–223

Ozaki Y (1935b) Studies on the frog-trematode, *Diplorchis ranae*. II. Morphology and behaviour of the swimming larva. J Sci Hiroshima Univ Ser B, Div 1 4:23–34

Ozaki Y (1936) Two new trematodes from tortoise *Geomyda spengleri* (Gmelin). J Sci Hiroshima Univ Ser B (Zool) 4:81–90

Ozaki Y (1940) On *Polystomum integerrimum* from Japan. Dobuts Zasshi Tokyo 52:48

Ozaki Y (1948) A new trematode, *Polystoma dendriticum* from the urinary bladder of *Onychodactylus japonicus* in Shikoku. Biosphaera 2:33–37

Pandey KC (1969) Studies on monogenetic trematodes of India. I. On a new species of the genus *Eupolystoma* Kaw, 1950 from *Bufo* sp. Proc Indian Nat Sci Acad 39:191–193

Pandey KC (1973) Studies on monogenetic trematodes of India. II. On a new species of the rare genus *Polystomoides* Ward, 1917. Indian J Zootomy 14:143–145

Pandey KC, Agarwal M (1978) A new monogenean, *Polystomoides chauhani* n. sp., from *Hardella thurji* Gray. Indian J Helminthol 30:126–128

Paul AA (1935) *Polystoma integerrimum nearcticum* n. subsp. from the urinary bladder, genital ducts, kidneys and gills of *Hyla versicolor,* Le Conte. J Parasitol 21:442

Paul AA (1938) Life history studies of north American freshwater polystomes. J Parasitol 24:489–510

Pichelin S (1995a) *Parapolystoma johnstoni* n. sp. from *Litoria nyakalensis* (Amphibia) in Australia and taxonomic considerations on the Diplorchiinae (Monogenea). J Parasitol 81: 261–265

Pichelin S (1995b) The taxonomy and biology of the Polystomatidae (Monogenea) in Australian freshwater turtles (Chelidae, Pleurodira). J Nat Hist 29:1345–1381

Pichelin S, Whittington I, Pearson J (1991) *Concinnocotyla* (Monogenea: Polystomatidae), a new genus for the polystome from the Australian lungfish *Neoceratodus forsteri*. Syst Parasitol 18: 81–93

Platt TR (2000a) *Neopolystoma fentoni* n. sp. (Monogenea: Polystomatidae) a parasite of the conjunctival sac of freshwater turtles in Costa Rica. Mem Inst Ozwaldo Cruz Rio J 95:833–837

Platt TR (2000b) Helminth parasites of the western painted turtle, *Chrysemys picta belli* (Gray), including *Neopolystoma elizabethae* n. sp. (Monogenea: Polystomatidae), a parasite of the conjunctival sac. J Parasitol 86:815–818

Platt TR, Firth A, Sharma RS (2011) Redescription of *Neopolystoma liewi* Du Preez and Lim, 2000 (Monogenea: Polystomatidae), from *Cuora amboinensis* (Testudines: Geomydidae) with notes on specimen preparation. Comp Parasitol 78:286–290

Price EW (1939) North American monogenetic trematodes. IV. The family Polystomatidae (Polystomatoidea). Proc Helminthol Soc Wash 6:80–92

Prudhoe S, Bray RA (1982) Platyhelminth parasites of the Amphibia, 1st edn. Oxford, London, p 217

Raharivololoniaina L, Verneau O, Berthier P, Vences M, Du Preez LH (2011) First monogenean flatworm from a microhylid frog host: *Kankana*, a new polystome genus from Madagascar. Parasitol Int 60:465–473

Rohde K (1963) *Polystomoides malayi* n. sp. (Monogenea: Polystomatidae) aus der Harnblase von *Cyclemys amboinensis* in Malaya. Z Parasitenkd 22:278–282

Rohde K (1965) Studies on the genus *Polystomoides* Ward, 1917 (Monogenea). I. Description of 4 Malayan species, a key to the known species, and a comparison of the subcuticular layers in *Polystomoides* and some digenetic trematodes. Zool Jahrb Abt Syst Oekol Geogr Tiere 92:345–368

Rohde K (1973a) Light and electron microscopic studies of the tegument of *Polystomoides malayi* Rohde and *P. renschi* Rohde (Monogenea: Polystomatidae). Z Wiss Zool 186:161–172

Rohde K (1973b) Ultrastructure of the caecum of *Polystomoides malayi* Rohde and *P. renschi* Rohde (Monogenea: Polystomatidae). Int J Parasitol 3:461–466

Rohde K (1974) Light- and electron-microscopic studies of the pharynx and the anterior and posterior glands of *Polystomoides* (Monogenea: Polystomatidae). Zool Jahrb Abt Anat 92:1–17

Rohde K (1975) Fine structure of the Monogenea, especially *Polystomoides* Ward. Adv Parasitol 13:1–33

Rohde K (1984) Three new species of the genus *Neopolystoma* (Monogenea) from river tortoises in Australia. Syst Parasitol 6:99–105

Rohde K, Ebrahimzadeh A (1969) Das weibliche Geschlechtssystem der Gattung *Polystomoides* Ward, 1917 (Monogenea). Z Parasitenkde 33:110–134

Rohde K, Pearson JC (1980) Two polystomes (Monogenea) from Australian River tortoises (Pleurodira, Chelidae), *Polystomoides australiensis* sp. nov. *from Emydura krefftii*, and *Neopolystoma chelodinae* (MacCallum, 1919) from *Chelodina longicollis*. Zool Anz 204: 191–208

Rudolphi KA (1808). Entozoorum, sive vermium intestinalium: historia naturalis (Vol. 1). Sumtibus tabernae librariae et artium

Shen J, Wang G, Fan LX (2013) A new species of the genus *Polystoma* (Monogenea: Polystomatidae) parasitic in the tree frog *Philautus palpebralis* Smith. Acta Zootaxon Sin 38: 765–772

Sinnappah ND, Lim LHS, Rohde K, Tinsley R, Combes C, Verneau O (2001) A paedamorphic parasite associated with a neotenic amphibian host: phylogenetic evidence suggests a revised systematic position for Sphyranuridae within anuran and turtle Polystomatoineans. Mol Phylogenet Evol 18:189–201

Stunkard HW (1916) On the anatomy and relationships of some north American trematodes. J Parasitol 3:21–27

Stunkard HW (1924a) On some trematodes from Florida turtles. Trans Am Microsc Soc 43:97–117

Stunkard HW (1924b) A new trematode, *Oculotrema hippopotami* n. g., n. sp., from the eye of the hippopotamus. Parasitology 16:436–440

Theunissen M, Tiedt L, Du Preez LH (2014) The morphology and attachment of *Protopolystoma xenopodis* (Monogenea: Polystomatidae) infecting the African clawed frog *Xenopus laevis*. Parasite 21:20

Thurston JP (1964) The morphology and life cycle of *Protopolystoma xenopi* (Price) Bychovsky in Uganda. Parasitology 54:441–450

Thurston JP (1968a) The frequency distribution of *Oculotrema hippopotami* (Monogenea: Polystomatidae) on *Hippopotamus amphibius*. J Zool 154:481–485

Thurston JP (1968b) The larva of *Oculotrema hippopotami* (Monogenea: Polystomatidae). J Zool 154:475–480

Thurston JP, Laws RM (1965) *Oculotrema hippopotami* (Trematoda: Monogenea) in Uganda. Nature 205:1127

Tinsley RC (1971) The adaptation for attachment by the Polystomatidae (Monogenoidea). CR Mult Eur Parasitol 1:65–68

Tinsley RC (1973a) Ultrastructural studies on the form and function of the gastrodermis of *Protopolystoma xenopi* (Monogenoidea: Polyopisthocotylea). Biol Bull 144:541–555

Tinsley RC (1973b) Studies on the life-cycle, physiology and host relations of *Protopolystoma xenopi* (Price) Bychowsky and other Polystomatidea (Monogenoidea). PhD diss., at Univ. Leeds (1970–1971)

Tinsley RC (1973c) Observations on Polystomatidae (Monogenoidea) from East Africa with a description of *Polystoma makereri* n. sp. Z Parasitenkd 42:251–263

Tinsley RC (1974a) A redescription of *Polystoma pricei* Vercammen-Grandjean, 1960 (Monogenoidea; Polystomatidae). Rev Zool Afr 88:165–174

Tinsley RC (1974b) Observations on *Polystoma africanum* Szidat with a review of the inter-relationships of *Polystoma* species in Africa. J Nat Hist 8:355–367

Tinsley RC (1974c) Proceedings: the gut ultrastructure of the Polystomatidae (Monogenea). Parasitology 69:2–3

Tinsley RC (1975) The correlation between the life-cycle of *Eupolystoma alluaudi* (Monogenea) and the ecology of its host *Bufo regularis*. Parasitology 71:16–17

Tinsley RC (1976) Oncomiracidial morphology and evolutionary relationships within the Polystomatidae (Monogenoidea). Parasitology 73:25

Tinsley RC (1977) The influence of host sex hormones on levels of parasitization. Parasitology 75:5

Tinsley RC (1978a) Oviposition, hatching and the oncomiracidium of *Eupolystoma anterorchis* (Monogenoidea). Parasitology 77:121–132

Tinsley RC (1978b) The morphology and distribution of *Eupolystoma* species (Monogenoidea) in Africa, with a description of *E. anterorchis* sp. n. from *Bufo pardalis* at the cape. J Helminthol 52:291–302
Tinsley RC (1978c) The role of ovoviviparity in the transmission of polystomatid monogeneans. Parasitology 77:5–6
Tinsley RC (1980) Reproduction and transmission in an isolated parasite population. Parasitology 81:14
Tinsley RC (1981) The evidence from parasite relationships for the evolutionary status of *Xenopus* (Anura Pipidae). Monit Zool Ital 15:367–385
Tinsley RC (1982a) *Pseudodiplorchis americanus* (Monogenea): synchronisation of transmission with restricted host availability. In: Muller M, Gutteridge W, Kohler P (eds) Abstracts of the 5th Int Corg Parasitol, vol 7–1, Toronto, Canada, p 459
Tinsley RC (1982b) Relationships between host genotype and parasite specificity. Parasitology 84: 3–4
Tinsley RC (1982c) The reproductive strategy of a polystomatid monogenean in a desert environment. Parasitology 85:15
Tinsley RC (1983) Ovoviviparity in platyhelminth life-cycles. Parasitology 86:161–196
Tinsley RC (1984) Pulsed parasite transmission between desert-adapted amphibians. Parasitology 89:6
Tinsley RC (1989) The effect of host sex on transmission success. Parasitology 5:190–195
Tinsley RC (1990a) Host behaviour and opportunism in parasite life-cycles. In: Barnard CJ, Behmke JM (eds) Parasitism and host behaviour, pp 158–192
Tinsley RC (1990b) The influence of parasite infection on mating success in spadefoot toads, *Scaphiopus couchii*. Am Zool 30:313–324
Tinsley RC (1993) The population biology of polystomatid monogeneans. Bull Fr Pêche Piscic 328:120–136
Tinsley RC (1995) Parasitic disease in amphibians: control by the regulation of worm burdens. Parasitology 111:153–178
Tinsley RC (1996) Parasites of *Xenopus*. In: Tinsley RC, Kobel HR (eds) The Biology of *Xenopus*, 13, pp 233–259
Tinsley RC (1999a) Overview: extreme environments. Parasitology 119:1–6
Tinsley RC (1999b) Parasite adaptation to extreme conditions in a desert environment. Parasitology 119:31–56
Tinsley RC (2004) Platyhelminth parasite reproduction: some general principles derived from monogeneans. Can J Zool 82:270–291
Tinsley RC (2013) The oncomiracidium of *Oculotrema hippopotami* Stunkard, 1924 and relationships within the Polystomatidae (Monogenea). Syst Parasitol 84:123–135
Tinsley RC (2017) Reproductive innovation and the recognition of a new genus within the Polystomatidae (Monogenea) infecting chelonian vertebrates. Folia Parasitol 64:017
Tinsley RC, Earle CM (1983) Invasion of vertebrate lungs by the polystomatid monogeneans *Pseudodiplorchis americanus* and *Neodiplorchis scaphiopodis*. Parasitology 86:501–518
Tinsley RC, Jackson HC (1986) Intestinal migration in the life-cycle of *Pseudodiplorchis americanus* (Monogenea). Parasitology 93:451–469
Tinsley RC, Jackson HC (1988) Pulsed transmission of *Pseudodiplorchis americanus* (Monogenea) between desert hosts (*Scaphiopus couchii*). Parasitology 97:437–452
Tinsley RC, Jackson JA (1998a) Correlation of parasite speciation and specificity with host evolutionary relationships. Int J Parasitol 28:1573–1582
Tinsley RC, Jackson JA (1998b) Speciation of *Protopolystoma* Bychowsky, 1957 (Monogenea: Polystomatidae) in hosts of the genus *Xenopus* (Anura: Pipidae). Syst Parasitol 40:93–142
Tinsley RC, Jackson JA (2002) Host factors limiting monogenean infections: a case study. Int J Parasitol 32:353–365
Tinsley RC, Owen RW (1975) Studies on the biology of *Protopolystoma xenopodis* (Monogenoidea): the oncomiracidium and life-cycle. Parasitology 71:445–463

Tinsley RC, Tinsley MC (2016) Tracing ancient evolutionary divergence in parasites. Parasitology 143:1902–1916

Tinsley RC, Cable J, Porter R (2002) Pathological effects of *Pseudodiplorchis americanus* (Monogenea: Polystomatidae) on the lung epithelium of its host, *Scaphiopus couchii*. Parasitology 125:143–153

Tinsley RC, York JE, Stott LC, Everard AL, Chapple SJ, Tinsley MC (2011) Environmental constraints influencing survival of an African parasite in a north temperate habitat: effects of temperature on development within the host. Parasitology 138:1039–1052

Tinsley R, Stott L, York J, Everard A, Chapple S, Jackson J, Viney M, Tinsley MC (2012) Acquired immunity protects against helminth infection in a natural host population: long-term field and laboratory evidence. Int J Parasitol 42:931–938

Tocque K, Tinsley RC (1991a) Asymmetric reproductive output by the monogenean *Pseudodiplorchis americanus*. Parasitology 102:213–220

Tocque K, Tinsley RC (1991b) The influence of desert temperature cycles on the reproductive biology of *Pseudodiplorchis americanus* (Monogenea). Parasitology 103:111–120

Tocque K, Tinsley RC (1992) Ingestion of host blood by the monogenean *Pseudodiplorchis americanus*: a quantitative analysis. Parasitology 104:283–289

Tocque K, Tinsley RC (1993a) The effect of host age on infection by the monogenean *Pseudodiplorchis americanus* in the spadefoot toad, *Scaphiopus couchii*. 2nd international symposium on Monogenea Montpellier (France) 99–100

Tocque K, Tinsley RC (1993b) The relationship between *Pseudodiplorchis americanus* (Monogenea) density and host resources under controlled environmental conditions. Parasitology 108:175–183

Van Niekerk S, Kok DJ, Seaman MT (1993) A new species of *Polystoma* (Monogenea: Polystomatidae) parasitic in *Hyperolius marmoratus* (Anura: Hyperoliidae) in South Africa. Syst Parasitol 25:73–80

Vaucher C (1987) Polystomes d'Equateur, avec description de deux nouvelles espèces. Bull Soc Neuchâtel Sci Nat 110:45–56

Vaucher C (1990) *Polystoma cuvieri* n. sp. (Monogenea: Polystomatidae), a parasite of the urinary bladder of the leptodactylid frog *Physalaemus cuvieri* in Paraguay. J Parasitol 76:505–504

Verneau O, Bentz S, Sinnappah ND, Du Preez LH, Whittington I, Combes C (2002) A view of early vertebrate evolution inferred from the phylogeny of polystome parasites (Monogenea: Polystomatidae). Proc R Soc Lond Ser B Biol Sci 269:535–543

Verneau O, Du Preez LH, Badets M (2009a) Lessons from parasitic flatworms about evolution and historical biogeography of their vertebrate hosts. C R Biol 332:149–158

Verneau O, Du Preez LH, Laurent V, Raharivololoniaina L, Glaw F, Vences M (2009b) The double odyssey of Madagascan polystome flatworms leads to new insights on the origins of their amphibian hosts. Proc R Soc B Biol Sci 276:1575–1583

Verneau O, Palacios C, Platt T, Alday M, Billard E, Allienne JF, Basso C, Du Preez LH (2011) Invasive species threat: parasite phylogenetics reveals patterns and processes of host-switching between non-native and native freshwater turtles. Parasitology 138:1778–1792

Verneau O, Johnston GR, Du Preez L (2023) A quantum leap in the evolution of platyhelminths: host-switching from turtles to hippos illustrated from a phylogenetic meta-analysis of polystomes (Monogenea, Polystomatidae). Int J Parasitol 53(5–6):317–325

Ward HB (1917) On the structure and classification of north American parasitic worms. J Parasitol 4:1–2

Whittington ID, Pichelin S (1991) Attachment of eggs by *Concinnocotyla australensis* (Monogenea: Polystomatidae) to the tooth plates of the Australian lungfish, *Neoceratodus forsteri* (Dipnoi). Int J Parasitol 21:341–346

Williams JB (1957) Anatomy of *Polystoma integerrimum*. Nature 180:866

Williams JB (1959a) Abnormal migration in *Polystoma integerrimum*. J Helminthol 33:205–206

Williams JB (1959b) Preliminary notes on the anatomy of a polystome from the bladder of *Xenopus laevis* Daud. J Helminthol 33:207–208

Williams JB (1960a) The dimorphism of *Polystoma integerrimum* (Frölich) Rudolphi and its bearing on relationships within the Polystomatidae. J Helminthol 34:323–346

Williams JB (1960b) The dimorphism of *Polystoma integerrimum* (Frölich) Rudolphi and its bearing on relationships within the Polystomatidae: part I. J Helminthol 34:151–192

Williams JB (1961) The dimorphism of *Polystoma integerrimum* (Frölich) Rudolphi and its bearing on relationships within the Polystomatidae: part III. J Helminthol 35:181–202

Yamaguti S (1963) Systema helminthum. Vol IV. Monogenea and Aspidocotylea, vol 4

Yamauchi T (2013) Undergraduate theses in zoological laboratory, Hiroshima University (1932-1953). Bull Hiroshima Univ Mus 5:87–94

Yamaguti S (1936) Studies on the helminth fauna of Japan. Part 14. Amphibian trematodes. Jap J Zool 6(4): 551–576

Yildirimhan HS, Du Preez LH, Verneau O (2012) *Polystoma nacialtuneli* n. sp. (Monogenea: Polystomatidae) from the eastern spadefoot, *Pelobates syriacus* (Pelobatidae) in Turkey. J Helminthol 86:104–112

Zeder JGH (1800) Erster Nachtrag zur Naturgeschichte der Eingeweidewurmer, mit. Zufassen und Anmerkungen herausgegeben, Leipzig, p 320

Zeller E (1872) Untersuchungen über die Entwicklung und den Bau des *Polystomum integerrimum* Rud. Z Wiss Zool 22:1–28

Zeller E (1876) Weiterer Beitrag zur Kenntniss der Polystomen. Z Wiss Zool 27:238–274

Chapter 3
Collecting and Processing Polystomatid Flatworms

Contents

L. H. du Preez et al., *Polystomatid Flatworms*, Zoological Monographs 9,
https://doi.org/10.1007/978-3-031-35887-6_3

Abstract This chapter focus on specific techniques and procedures relevant to the study of polystomatid flatworms. We report from what is known in the literature and from personal experience. Information conveyed in this chapter should not be seen as a complete guide of all relevant techniques. Curiosity and innovation are equally important, and one should always try to develop new innovative techniques. Aspects that will be dealt with in this chapter include ethical clearances, collecting permits, safety, capturing potential hosts, screening of hosts for parasites, recovery of parasites and the processing of parasite material for various investigations.

3.1 Introduction

Over time, techniques have been developed for the study of specific taxonomic parasite groups and their associated hosts. These techniques range from the identification of potential hosts through the processing and studying of parasite material. Because parasite life cycles are interwoven and usually synchronized with that of their hosts, a basic knowledge and understanding of amphibians and freshwater turtles will, therefore, aid in the understanding of the polystome life strategies. It is of utmost importance to first read up on the hosts and their parasites before fieldwork activities for host and parasite investigations.

3.2 Ethical Clearance

Most academic institutions, as well as governmental or even non-governmental institutions, nowadays require ethical clearance for research studies. Many scientific journals also require proof of permits and ethical clearances when considering a paper for publication. As with permits, obtaining ethical clearance can be a lengthy and frustrating process and to complicate matters some institutions require proof of collecting permits before considering any ethical clearance application. On the other hand, some conservation authorities may require ethical clearance before they will issue a permit. Even working with material archived in collections may require ethical clearance. In most instances, this will require a simple 0-ethical application, which is merely a paper exercise. As time-consuming as ethical clearances may be, they are necessary as they ensure best practice and increasingly journals will refuse to publish contributions if clearances are not in place.

3.3 Permits

Most countries require collecting permits, even when collecting non-protected animals outside protected areas. Obtaining a permit is often a slow bureaucratic process and may take several months. This is why it is of the utmost importance to apply for permits well in advance and always for a longer period of time, as you never know when conditions will be suitable for collecting animals. The availability of amphibians is often determined by climatological conditions such as temperature and rainfall. It is also essential to keep copies of all permits as you might want to use collected data much later and then still need to cite permit numbers under which material has been collected.

3.4 Host Sampling

Hosts of polystomes include the Australian lungfish, the common hippopotamus, freshwater turtle species and species of the three extant orders of amphibians, namely salamanders, frogs and caecilians. Polystomes have direct life cycles without intermediate hosts. It cannot therefore be excluded that polystomes might in future be discovered in other aquatic or semi-aquatic hosts and it is thus important to consider all of them as potential hosts.

In preparation for fieldwork activities, there are several important aspects to be taken into consideration. It is thus necessary to:

- Conduct a proper locality investigation beforehand and identify sites where fieldwork will be conducted.
- Assess the safety hazards of such an operation. There might be potentially dangerous or venomous animals in the area. Landowners usually possess guns. From experience, it is a good idea to check whether the person with whom you made arrangements has informed other people of your presence on the property.
- Obtain written permission from the landowner. Develop a landowner consent form for your research group and keep records of signed agreements. This is usually a requirement set by ethics committees.
- Ensure that transfer permits are in place.
- Obtain ethical clearance for the whole methodology.
- Visit the study area in daytime prior to nightly collection. Orientate yourself and take GPS coordinates at potential sites.

- Never work alone. Take a friend or colleague with you.
- Make sure that concerned people know where you are going to work and when you should be expected back.
- Use only recommended and ethically approved catching techniques.
- Ensure that all traps used are properly marked with a tag providing information on the purpose of the trap and your identity and contact details. Mark the position of the trap with, for example, a marker attached to a tree. It is critically important not to miss a trap as they will continue to trap animals.
- Make sure to secure traps properly and set them correctly.
- Check traps regularly and remove unwanted by-catch organisms.
- Do not over-collect. Only collect what is necessary and adhere to permit restrictions.
- Work on the specimens as close to the collecting site as possible and take care not to expose the animals or containers to potential pathogens which might be introduced to the environment when animals are released.
- Take care not to expose animals to stressful conditions such as exposure to direct sunlight, extreme temperatures, desiccation, etc.
- Apply non-invasive screening parasite techniques where possible and release uninfected hosts and unwanted specimens where they were collected.
- Disinfect collecting equipment and footwear with a disinfectant such as weak hypochlorous acid or sodium hypochlorite (bleach) between sites.

Some museums allow archived amphibians and reptiles to be inspected for polystomes. In some instances, specimens are fixed remarkably well, especially those that are slightly flattened such as under the eyelids of turtles. Even if the specimens are not good enough to mount on slides, they are still valuable and may be used for histological sectioning and enzyme digestion for sclerites investigation, on the one hand, or as locality indicators of polystome infection on the other.

3.4.1 The Australian Lungfish

The Australian lungfish is fully protected and may not be captured without a special permit. Although it is not listed as threatened by the IUCN, it is listed in Appendix 2 of the Convention on International Trade in Endangered Species of Wild Fauna and Flora (CITES).

3.4.2 Amphibians

Amphibians are mostly captured by hand or with a dip net at night with the aid of a strong headlamp or handheld flashlight. Frogs are easy to locate by homing in on a calling male. Some aquatic frogs such as clawed frogs (*Xenopus* Wagler, 1827) or

Surinam toads (*Pipa* Laurenti, 1768) may be collected with baited net or bucket traps. A bucket trap can easily be constructed by mounting a plastic cone or the top half of a 2 l soda bottle in a hole made in the side of a plastic bucket (Fig. 3.1a). Lids with air holes in the lids must be secured. Traps must be baited with ox heart, ox liver, chicken liver or pieces of fish and set with at least 50 mm of the trap above the waterline (Fig. 3.1b). They must be covered with vegetation to prevent them from building up excessive heat when exposed to direct sunlight. It is very important to take care to retrieve them before excessive rain or when the water level is expected to rise, to prevent trapped animals from drowning. They have to be checked on a daily basis to remove trapped animals. They also need to be inspected before putting hands inside as non-target animals such as venomous snakes or electric eels may have entered the trap. When several traps are set out, it is strongly recommended that the position of traps be marked by attaching indicators to surrounding vegetation or by using GPS coordinates.

Fig. 3.1 (**a**) Bucket trap with a cone fitted; (**b**) Bucket trap set in a water body

Aquatic species may be placed in water in a bucket or other suitable container. For terrestrial frogs, we recommend specimens be placed in transparent plastic bags of about 200 X 400 mm with damp vegetation to keep the environment humid and provide shelter for the frogs to hide. Plastic bags prevent frogs from getting hurt when they jump around in distress. Frogs must be separated individually or according to species in distinct bags marked clearly according to the collecting site and closed off with a knot or an elastic band. It is also important to have a basic knowledge of amphibian diversity and its conservation status before setting off to collect specimens. Also, certain frog species can produce toxic skin secretions; wear protective clothing and gloves before handling them.

Caecilians are burrowing amphibians and are very difficult to find even after extensive digging through the soil. Furthermore, caecilians often occur in forested areas where tree roots may make digging almost impossible. However, they often surface after heavy rains have saturated the soil, and this increases the chances of capturing them above ground. Polystomes are only expected in aquatic or semi-aquatic caecilians, so it is more judicious to focus on these as they can be trapped with baited crayfish funnel traps (Fig. 3.2a) or baited bucket traps (Fig. 3.1). Ox liver, chicken liver or fish may be used as bait. Entrances of the traps can be modified to prevent larger non-target animals from entering.

3.4.3 Freshwater Turtles

Freshwater turtles can be collected by hand when snorkeling, provided the water is clear with good visibility. In muddy localities, turtles can be scooped up using a dip net or a seine net, however, the most effective way is to trap turtles with a collapsible crayfish trap baited with fish, chicken liver or ox liver (Fig. 3.2a, b, c). To prevent turtles from swallowing the bait, the bait can be placed inside any small perforated container (Fig. 3.2d) or a metal tea strainer (Fig. 3.2e), which still allows the scent to disperse. Add a closed plastic soda bottle inside the trap to serve as a float. Set the baited traps in such a way that part of the trap is above the waterline to allow turtles to breathe. Furthermore, it is important to secure traps with a line to a firm object to prevent them from washing away or becoming submerged due to movements from trapped animals. The position of traps must be marked with a marker and/or with accurate GPS coordinates and checked daily. Traps must be removed before heavy rains as the water level might rise rapidly submerging the traps. Using a basking trap (Fig. 3.2f) works well for some species, but it is a cumbersome process as turtles need to accept it as a basking site.

Fig. 3.2 (**a**) Collapsible crayfish net trap used to catch turtles, set in a river; (**b**) Trap set in a pond; (**c**) Turtles trapped overnight; (**d**) Perforated bait box; (**e**) Tea strainer to keep bait; (**f**) Basking trap

3.4.4 Crocodiles and Alligators

As of yet, no polystomes have been reported from crocodilians. We, however, believe it to be a matter of time before they are discovered. These animals can be captured in baited traps, with baited travel hooks or can be harpooned with special alligator harpoon tips connected to a cable. Crocodilians do not have urinary bladders but possible sites of polystome infection could include the cloaca, mouth, nostrils, eyes and ears.

3.4.5 The Common Hippopotamus

To obtain *Oculotrema* Stunkard, 1924 specimens from the eye of the hippopotamus is not an easy task. Hippos are sometimes culled when they cause problems in human settlements or die during prolonged periods of drought. We recommend that these opportunities be used to inspect carcasses and for the removal of parasites from the eyes and under the eyelids. Cut out the complete eye with eyelids intact from the eye sockets and then inspect them for the presence of parasites which often attach in a cluster on the surface of the eyeball or under the eyelids.

3.5 Procedure for Identifying Polystome Infection

Most polystomes release eggs. Searching the water in which a potential host has been kept for at least 12 h provides a non-invasive method of detecting the presence of polystome parasites. Two techniques are recommended to detect polystome eggs:

3.5.1 Procedure 1

Pour the water in which the amphibians or turtles have been kept overnight into a tall glass cylinder or beaker. Allow it to settle for 15 min. As the eggs are heavy, they will sink to the bottom. Decant 50% of the water and allow the remainder to stand for 10 min. Decant 50% of the remaining water. Continue until the water volume has been reduced to a volume that can be inspected in a Petri dish under a stereo microscope.

3.5.2 Procedure 2

This method requires water samples to be filtered. Water from containers in which potential hosts were housed is poured through two plankton sieves with respective mesh sizes of 500 μm and 100 μm (see Verneau et al. 2011). The first sieve removes coarse debris in the water, while the second retains finer debris and polystome eggs. Contents of both sieves are then washed into separate glass Petri dishes and examined under a dissecting microscope. Thereafter, both Petri dishes with their contents are screened separately. Contents from the course sieve are scanned for adult parasites which may have been dislodged and expelled, whereas contents from the fine sieve are scanned for polystome eggs. Use a centripetal force action (gently swirl the Petri dish in a circular motion) to concentrate heavier objects, including the polystome eggs in the middle of the Petri dish. Eggs can be separated from debris using a thin needle (Fig. 3.3a, b). If no eggs were obtained, screen the organism once more and release uninfected hosts where they were collected. Harvest eggs from infected hosts on a daily basis. It is very important to ensure that animals are well looked after and fed to reduce stress. As soon as the egg production reduces, the host can be dissected to recover its parasites. Polystome eggs are glossy yellow to golden in colour with no pattern or surface texture. On average, eggs measure between 150–250 microns long. The rule is "when not certain that you are looking at a polystome egg, it probably is not".

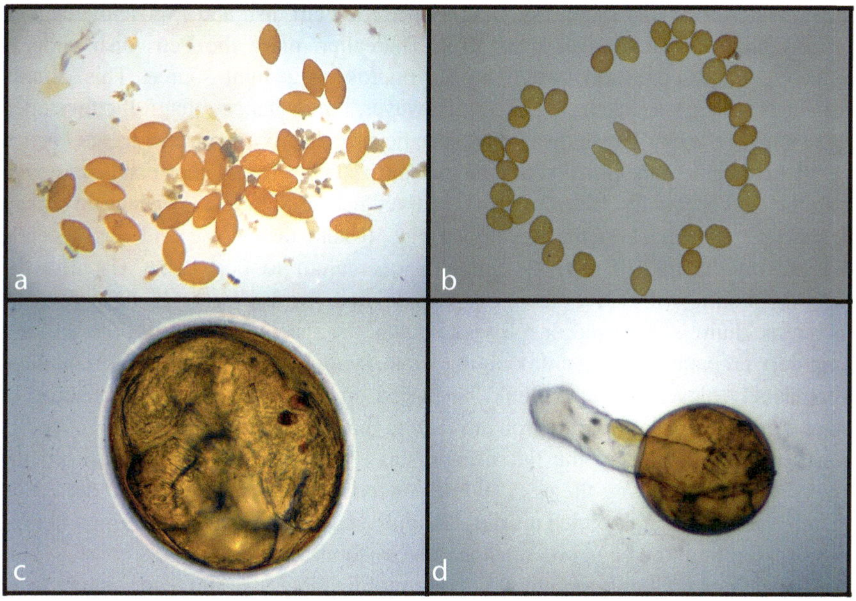

Fig. 3.3 (**a**) *Apaloneotrema moleri* eggs; (**b**) *Polystomoides* sp. (oval) and *Fornixtrema* sp. (fusiform) eggs; (**c**) Embryonated egg of *Polystoma australis*; (**d**) *Polystoma australis* oncomiracidium leaving the egg capsule. Note the operculum is hinged and stay attached to the egg

Take note that for some polystome genera like *Eupolystoma* Kaw, 1950, *Kankana* Raharivololoniaina, Verneau, Berthier, Vences and Du Preez, 2011, *Madapolystoma* Du Preez, Verneau, Raharivololoniaina and Vences, 2010, *Uropolystomoides* Tinsley and Tinsley, 2016, eggs are usually not released, and screening is thus not effective to determine whether a potential host is infected or not. To determine whether the host is infected in these genera, animals need to be euthanized and dissected.

3.6 Procedure for Polystome Eggs Development

Eggs should be carefully isolated from debris to limit bacterial and fungal growth on developing eggs. Transfer the eggs to a 40–50 mm Petri dish, containing bottled or filtered water. Avoid using distilled water. Eggs tend to develop better in glass Petri dishes. Cover the Petri dishes with their lids and place them in a dark area at room temperature. Leave the eggs undisturbed for at least a week before inspection. Top up the water level if necessary. Most polystome eggs require three to four weeks of incubation at 20–26 °C for the oncomiracidium to fully develop (Thurston 1968; Tinsley and Owen 1975). For some *Polystoma* Zeder, 1800 species, complete development of the oncomiracidium may be as fast as 10 days (Kok and Du Preez 1987) and 20–24 days for *Protopolystoma* Bychowsky, 1857 (see Tinsley and Owen 1975).

Movement of the oncomiracidium inside the egg indicates that the oncomiracidium is ready to emerge from the egg capsule. Four reflective eye spots are visible on the oncomiracidium at this stage (Fig. 3.3c). Open the Petri dish and expose the eggs to direct sunlight for no longer than 30 s. Thereafter, place the Petri dish under a dissecting microscope and switch on the microscope's light source. This usually stimulates the eggs to hatch (Fig. 3.3d). However, it may happen that fully developed eggs simply refuse to hatch, awaiting a specific stimulus. If oncomiracidia are needed merely to document the position of cilia tufts or to study sclerites, they can be removed from the eggs manually. Rest the tip of a curved needle in the Petri dish or on a glass microscope slide next to the developed egg. Rotate the needle and apply gentle pressure on the egg. This will force the operculum to pop open. Usually, the oncomiracidium will leave the egg capsule and swim freely. If however, the oncomiracidium is damaged or refuses to hatch, the specimen can still be used for a temporary preparation using ammonium picrate, lactophenol or Gray–Wess mountant. Alternatively, developed eggs could be placed on a microscope slide in a drop of water. Apply a small blob of petroleum jelly to the four corners of a 20x20 coverslip. Lower the coverslip onto the microscope slide. The petroleum jelly will keep the coverslip from squashing the eggs. While observing the eggs through the dissecting microscope use a sharp pointed needle and apply gentle pressure on the four corners of the coverslip until the eggs pop open and the oncomiracidia swim free.

A single egg contains enough DNA for amplification, it is, however, much better to have the eggs incubated for about a week before fixing them for molecular analyses. Keep eggs individually at −20 °C until used for molecular analyses.

3.7 Procedure for Amphibian Dissection

When an infected amphibian has been selected for dissection, it must be euthanized according to an internationally accepted protocol. We recommend the use of buffered Ethyl-3-aminobenzoate-methanesulfonate, also known as Tricaine-3-Methanesulfonate (MS-222) or Tricaine-4-Methanosulfonate (Benzocaine) for euthanasia. The dissection procedure for the specific groups of amphibians is hereafter described following a short description of the modes of preparation of chemicals and materials to be gathered for euthanasia and dissection.

3.7.1 Euthanasia

3.7.1.1 Ethyl 3-Aminobenzoate Methanesulfonate (MS-222)

- Prepare a 250 mg/L solution by adding 5 mg MS222 to 20 ml water. It will dissolve instantly.
- Add an equal weight of sodium bicarbonate ($NaHCO_3$) or titrate to pH = 7.0–7.5 to buffer MS222 which is acidic in solution.

Because MS222 is a schedule-4 substance, it is necessary to keep a drug dispensing catalogue. Ensure compliance with all regulations applicable in the country where the procedure is performed.

3.7.1.2 Tricaine-4-Methanosulfonate (Benzocaine)

- Prepare a 250 mg/L solution by adding 5 mg benzocaine to 10 ml hot water. Stir until benzocaine dissolves. Add 10 ml cold water. Alternatively, dissolve the benzocaine in 10 ml methanol and then dilute with water.
- Add an equal weight of sodium bicarbonate or titrate to pH = 7.0–7 to buffer Benzocaine which is acidic in solution.
- Pour some of the Benzocaine solution in the tub to just cover the surface of the base of the tub.
- Proceed as for MS222.

3.7.2 Dissecting Materials (Fig. 3.4)

Prior to the dissection, ensure the following items are in place and in working order:

- Full parasite account record sheet.
- Dissecting microscope.
- Dissecting scissors (Fig. 3.4a). While a pair of bigger scissors is needed for course dissections, a pair of small scissors is needed to cut open small animals or small structures such as the urinary bladder. A variety of very small scissors is available but could be expensive. Eye surgeons often dispose of retina scissors after one or two operations and are usually willing to pass these on. A pair of crochet scissors is fairly inexpensive and works very well for fine dissections.
- Bone scissors are important for dissecting terrapins. A pair of tin cutters works equally well.
- Dissecting needles. A curved sharp pointed needle is essential. Cheaper brands are available and work equally well. A very strong and functional needle can be made from a "root canal extractor" which can be obtained from a dentist who usually uses it once. The needle can be clamped in a mini vice (Fig. 3.4b).
- Camel-hair brush. A fine brush can be purchased at any shop specializing in fine arts.
- Dissecting tray. A dissecting tray can be prepared by melting beeswax and pouring a 10 mm deep layer into any small plastic tray. Beeswax is superior to candle wax as it will not crack and holds the dissecting needles securely.
- 0.06% saline solution (Ringer).
- Small glass Petri dishes.
- 1 mL insulin syringes with a fixed needle.
- Microscope slides.
- 20x20 mm coverslips.
- Dissecting equipment containing at least scissors, forceps and needles.
- 70% high-quality ethanol.
- 100% methanol.

- 10% neutral buffered formalin (NBF).
- Fixing tray.
- Collection jar.
- Specimen vials.

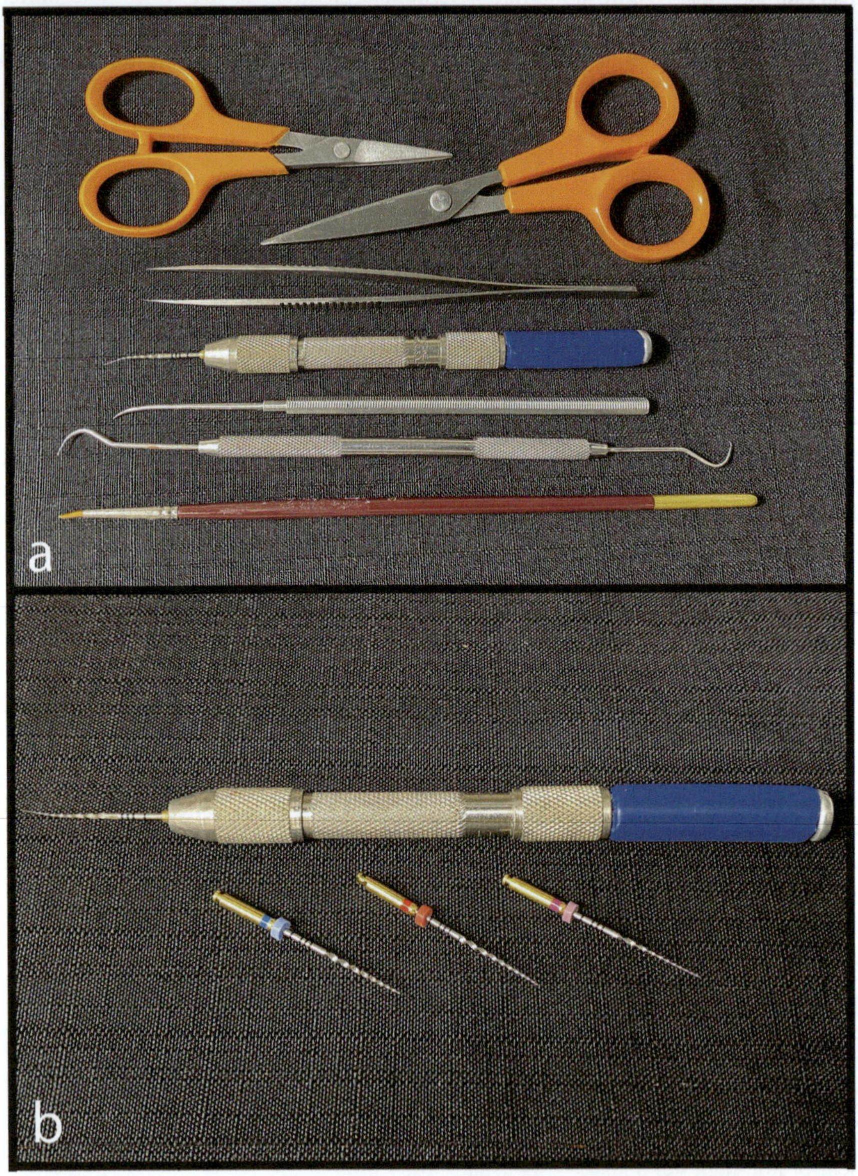

Fig. 3.4 (**a**) Basic dissecting equipment; (**b**) Root canal extractor needles and mini vice with a needle clamped

3.7.3 Frog Dissection Procedure

- Pour some of the prepared euthanasia solution (MS222 or Benzocaine) into the tub.
- Place the animal in the tub and cover it with a lid. Punch air holes in the lid.
- Place in an area where the animal will not be stressed.
- Allow for 10 min and test for vital signs. Gently squeeze its legs with blunt forceps just above the knee or touch the eye. If any movement is observed, wait a further 5–10 min. Because some countries may require double euthanasia, the brain must also be destroyed with a blunt needle (pithing) after the animal has been knocked out with MS222. The easiest way is to insert one blade of a pair of strong dissecting scissors and cut off the head behind the eyes. Then a blunt needle must be inserted into the brain and spinal cord in order to destroy the brain by rolling the needle forward and backwards. However, if the carcass is needed for a museum collection, this technique is not desirable.
- Cover the surface of the dissecting tray with either MS222 or Benzocaine solution to keep the animal sedated.
- Place the euthanized frog on its back in the dissecting tray.
- Pin the hands and feet in the dissecting tray.
- Cut through the skin and transverse across the body just behind the armpits.
- Cut backwards along the flanks and fold the skin flap in a posterior direction.
- Lift the abdominal muscles and cut carefully along the same line as for the skin taking care not to cut any of the underlying organs. Cut through the vena abdominalis close to the heart and fold the abdominal muscles which are now freed backwards and pin in the dissecting tray behind the frog.
- Locate the urinary bladder posteriorly in the abdominal cavity. Pull it gently with a pair of forceps to the side and inspect for the presence of any parasites (Fig. 3.5a, b).
- Loosen the urinary bladder with a pair of scissors and remove it intact after cutting it out as close to the cloaca as possible.
- Place the urinary bladder in a small Petri dish or cavity block in 0.6% amphibian saline solution (Ringer's solution).
- Cut open the bladder and carefully remove polystomes using a fine dissecting needle and fine camel-hair brush.
- Inspect the accessory bladders if present and also the kidneys for the presence of subadult parasites.
- Remove the kidneys, and in a Petri dish containing Ringers solution tease apart the kidneys with two fine dissecting needles. Take care to inspect every part of the kidneys and remove the parasites.
- When studying *Pseudodiplorchis* Yamaguti, 1963 or *Neodiplorchis* Yamaguti, 1963, also check the lungs.

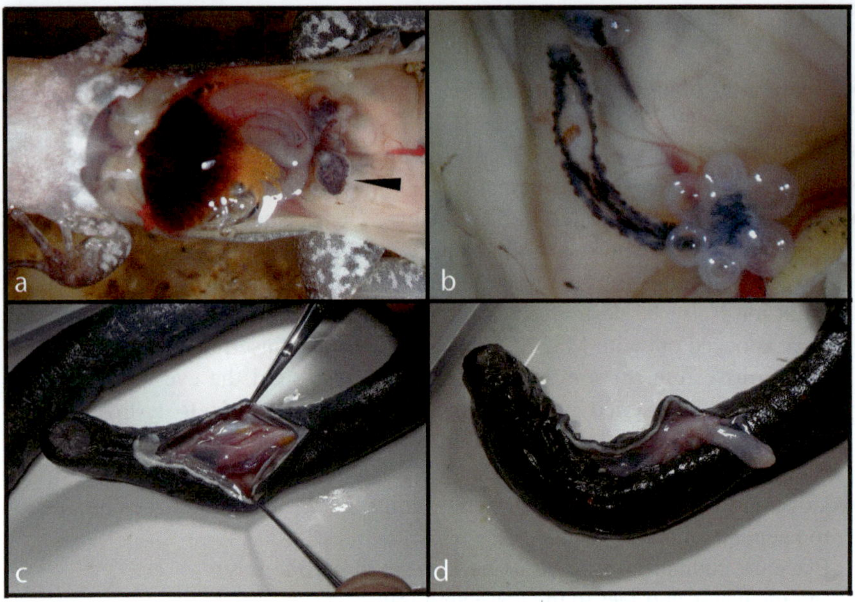

Fig. 3.5 (**a**) Frog pinned on a beeswax dissecting tray; arrow indicating polystomes in the urinary bladder (**b**) Polystome sp. attached to the bladder of a frog; (**c**) Caecilian dissection; (**d**) Caecilian bladder pulled out

3.7.4 Salamander Dissection Procedure

- Euthanize in the same manner as for frogs.
- Inspect the urinary bladder in a similar way as described for frogs.
- Also, inspect the area around the external gills (if present) and around the base of the salamander's legs and in the mouth.
- Pour hot 70% EtOH over the gills (if present) and in the mouth and check carefully for subadult parasites which will turn white after the hot alcohol treatment.

3.7.5 Caecilian Dissection Procedure

- Euthanize as for frogs.
- Place the euthanized animal on its back in a dissecting tray.
- Carefully cut through the skin and abdominal layers to expose the appendix shaped elongated urinary bladder (Fig. 3.5c, d). The fact that caecilians have no limbs makes it a bit more difficult to locate the urinary bladder which is situated about a fifth of the body length from the posterior end.
- Remove the bladder and transfer to a Petri dish containing amphibian saline.

- Also, inspect the remainder of the urinary tract including the phallodeum to the cloaca.
- After removing bigger parasites, pour hot EtOH on the bladder and other areas to make subadult parasites more noticeable.

3.8 Procedure for Freshwater Turtles Dissection

When an infected turtle has been selected for dissection, euthanize the animal according to an internationally accepted protocol. Several procedures can be used for euthanasia:

3.8.1 Means of Euthanasia

3.8.1.1 MS-222

An intracoelomic injection of 10 ml of a 250–500 mg/mL of neutralized MS222 is usually required. Consciousness is then lost rapidly (30 s – 4 min). A second intracoelomic injection of a 50% unbuffered solution of MS222 is subsequently administered to ensure death of the animal. The turtle is finally decapitated and its head with the neck is placed in a large Petri dish.

3.8.1.2 Sodium Soda-Pentabarbitone (Euthapent)

In some countries, the use of Euthapent is still allowed. This is a schedule 7 product and it may be necessary for a qualified veterinarian to administer the substance. It is considered as the most humane way of euthanasia for reptiles. Use 0.5 ml of Euthapent for small and 1 ml for big turtles, and dilute the used volume with 4 ml of lukewarm water. Fit a long needle to the syringe and inject alongside the neck aiming for the heart. This solution can alternatively be injected in the blood sinus above the neck under the carapace. Place a finger on the nose of the turtle and push the head back into the carapace. The needle is then gently inserted above the head close to the carapace. Pull gently on the plunger while inserting the needle. If blood flows freely into the syringe, it indicates that the sinus is located. The contents of the syringe is then gently injected. The turtle should be dead within three to 5 min. Note that this technique is not recommended for snapping turtles or other turtles that may bite.

3.8.1.3 Decapitation

According to ethical principles clubbing an animal is allowed. It is, however, difficult to stun a turtle by hitting it over the head with a blunt object. If this method is used, the animal should be decapitated following stunning.

3.8.1.4 Freezing

Most veterinarians and other experts agree that freezing is an unethical method with which to euthanize turtles. Freezing induces ice crystal formation in tissues as they freeze and may be painful for the animal. Placing a turtle in a refrigerator to cool it down prior to freezing is not a humane method of euthanasia either.

3.8.2 Dissection Procedure

Prior to the dissection ensure the following items listed in paragraph 3.7.1.3. are in place. Then, start the dissection following the procedure listed below:

- Place the animal in a dissecting tray.
- Decapitate by cutting the neck close to the body and place the head in a Petri dish.
- Using a hacksaw, cut through both sides of the joint between the carapace and plastron.
- Carefully cut through the remaining skin sections and loosen the plastron from the abdominal muscles. Remove the plastron.
- Open the abdomen to expose the viscera.
- Inspect the urinary bladder and remove the bladder with the accessory bladders intact and place it in a large Petri dish containing 0.06% Ringers saline.
- Cut the bladder open and carefully inspect for polystomes. Unlike amphibian polystomes, chelonian polystomes are flesh-coloured and more difficult to observe (Fig. 3.6a).
- After removing the parasites, pour hot 70% EtOH on the bladder to make subadult parasites more noticeable.
- Cut through the skin surrounding the tail and remove the tail and cloaca intact. Place in a Petri dish containing amphibian saline.
- Cut open the cloaca and carefully follow the reproductive canal while inspecting it for polystomes.
- After removing big parasites, pour hot EtOH on the opened reproductive canal to make subadult parasites more noticeable.
- Inspect the mouth and pharyngeal pouches.
- Open the mouth and inspect the anterior part of the oral cavity and mouth floor for parasites (Fig. 3.6b).

- Place the head upside down in a dissecting tray. Cut through the skin and underlying muscles following the inner-lower jawline along one side. Cut through the hyoid apparatus and further down to the oesophagus to expose the oral region and pharyngeal pouches. Inspect the inside frequently to ensure that you do not cut through a parasite.
- After removing big parasites, pour hot 70% EtOH on the mouth and pharyngeal pouches to make subadult parasites more noticeable.
- Inspect conjunctival sacs. Eye polystomes are usually small and often attach deep down under the nictitating membrane.
- Cut through the upper and lower eyelids in the middle of the eye to reduce tension on the eyelid and thus allowing you to pull the eyelid aside to be able to inspect deep down in the hidden ocular corners.
- It is quite difficult to observe the semitransparent flesh-coloured ocular polystomes. The presence of a golden fusiform egg in the uterus of the polystome often reveals the presence of a parasite (Fig. 3.6c, d).
- Remove the parasite with a camel-hair brush or a curved needle.

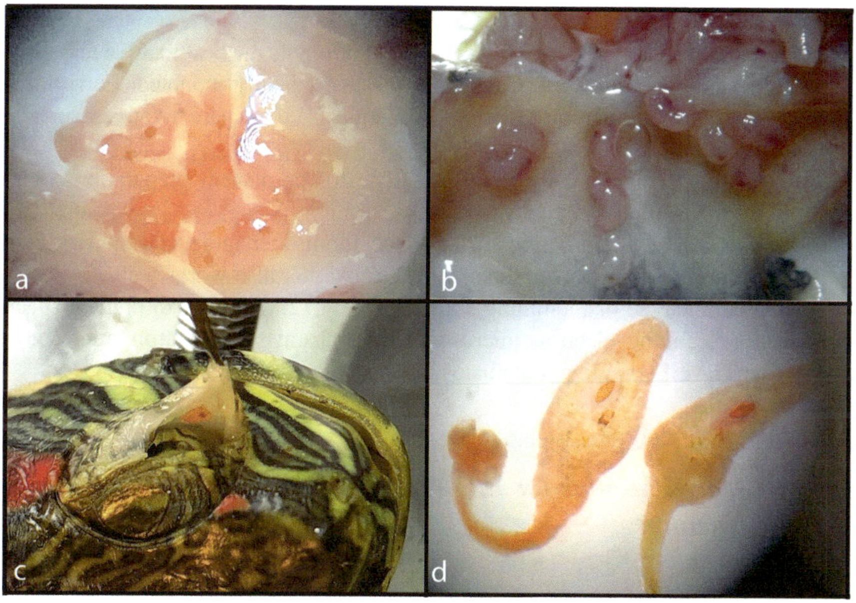

Fig. 3.6 (**a**) Polystomes in the bladder of a freshwater turtle; (**b**) Polystomes in the mouth of a freshwater turtle; (**c**) Polystomes on the eye of a freshwater turtle; (**d**) Fusiform yellow egg inside the parasite often reveals the presence of the flesh-coloured parasite

3.9 Processing Specimens

3.9.1 Temporary Mounts for Examination of General Structures

It is strongly recommended that a live polystome be placed in a drop of water under a coverslip. Focus especially on the reproductive ducts and gut structure to determine whether anastomoses are present. Take care that preparations do not dry out. The increased pressure from the coverslip may cause the parasites to burst. While some polystomes such as turtle polystomes and those of the hippopotamus are very tough, others such as *Protopolystoma* are very delicate.

3.9.2 Fixing Products for Morphological Structures or Molecular Studies

The envisaged use of the specimen determines the fixative to be used.

- Neutral buffered formalin as a 5% or a 10% solution is the best fixative and gives the best staining result for specimens to be permanently mounted. Formalin is often used in combination with other chemicals to prepare fixatives such as Bouin's and FAA.
- 70% ethanol is favoured by some scientists for permanent preparations or as fixative for scanning electron microscopy.
- 70% or 100% ethanol is used to fix tissue for further molecular studies. It is important to use analytical-grade alcohol. RNALater also works well.
- 5% Glutaraldehyde or Flemming solution (see 3.12.8) is used to prepare specimens for scanning electron microscopy or transmission electron microscopy SEM or TEM. Flemming contains osmium which counters shrinking.

3.9.3 Cleared Mounts for Examination of Sclerotized Structures

Sclerotized structures of all polystome life stages are often hard to see if the specimen is not cleared. Temporary mounts in lactophenol, ammonium picrate or Gray–Wess to clear the specimens are quite useful. Proceed as follows:

- Place the specimen in a small drop of water on a clean microscope slide.
- Place a drop of clearing agent (lactophenol or ammonium picrate) on a coverslip.
- Turn the coverslip over and lower it onto the slide in such a way that the droplet first makes contact with the specimen.

- For oncomiracidia it works better to place four small droplets of clearing agent near the corners of the coverslip. This prevents oncomiracidia from being washed out from under the coverslip.
- Blot excess fluid from the side of the coverslip. Optimally, the clearing agent should only fill the space beneath the coverslip. It is better to initially use less clearing agent and add a small droplet on the edge between the coverslip and the slide afterwards.
- Secure the four corners of the coverslip with clear nail varnish. Carefully clean the borders of the coverslip and seal the rest of the coverslip with nail varnish. When dry, seal the four sides of the coverslip with nail varnish all around the coverslip.

3.9.4 Permanent Preparations for Morphological Examination

When sufficient material is available we recommend that:

(i). Some of the specimens should be heat-killed by placing them in a drop of water on a microscope slide that is then heated from below with a butane lighter, only until the parasite stops moving. The parasite can be then fixed in 10% buffered formalin

(ii). The remaining specimens should be fixed in 10% buffered formalin under coverslip pressure. Body measurements and placement of organs should be studied from the unflattened specimens while sclerites should be measured in flattened specimens. However, in instances where the number of specimens is limited, we recommend that specimens be fixed flat under coverslip pressure. For describing polystomes it is of utmost importance to have the sclerites in a flat orientation

When selecting a stain, different scientists have their preferences. Some of the popular stains include acetocarmine, alum carmine, borax carmine, Ehrlich haematoxylin and Van Cleave's haematoxylin. Recipes for these stains are provided under 3.12.

To prepare a permanent preparation using acetocarmine, proceed as follows (Fig. 3.7):

- Hydrate fixed specimens earmarked for permanent mounts gradually from 70% EtOH (or formalin to water) to 30% EtOH.
- Stain overnight in a weak acetocarmine staining solution.
- Gradually dehydrate to absolute EtOH. If the specimens stain too dark, the excess stain can be removed by adding a drop of hydrochloric acid when the specimen is in 70% EtOH. The moment the stain starts oozing out of the specimen, continue with the dehydration to absolute alcohol.

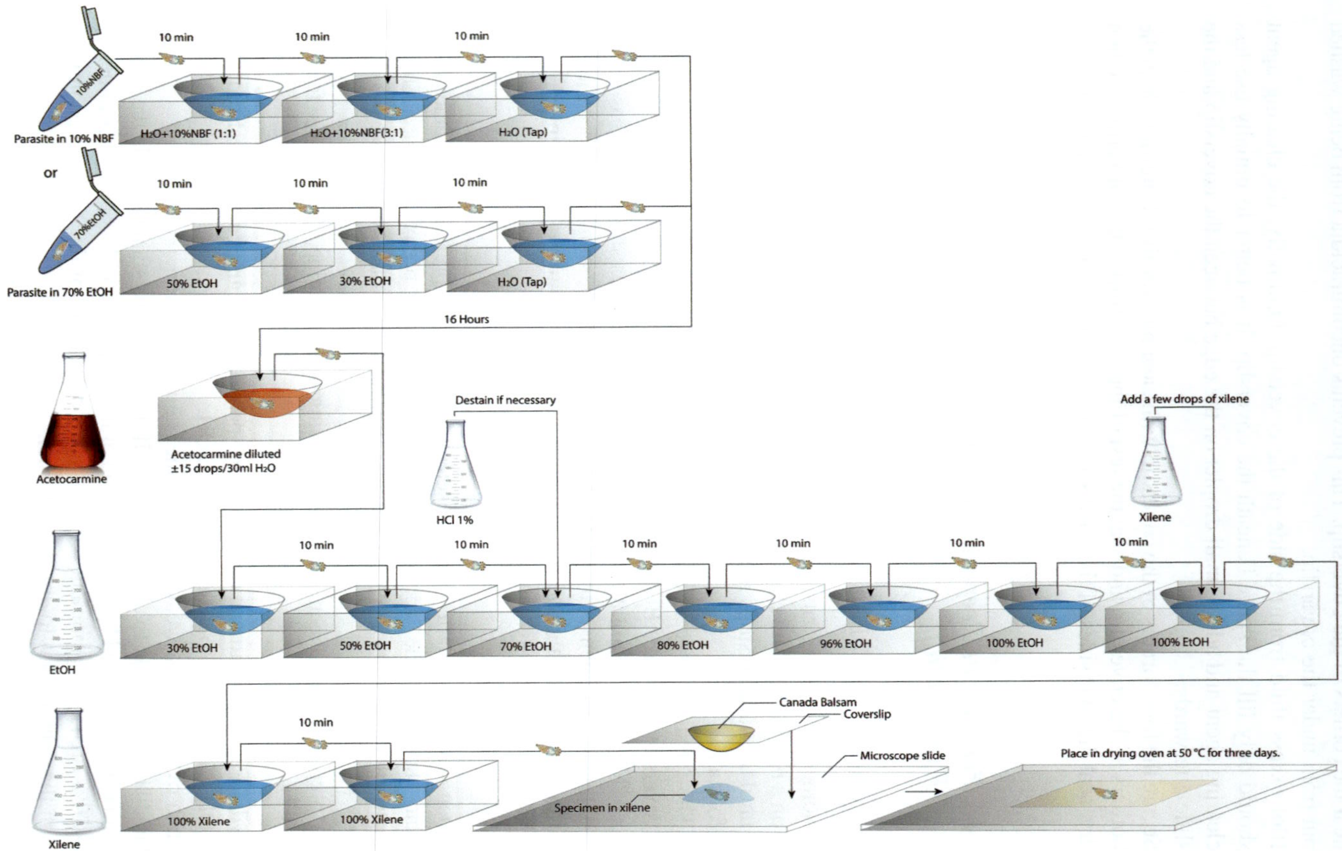

Fig. 3.7 Diagram illustrating the staining and permanently mounting of a polystome

- Transfer the specimen from absolute alcohol to a mixture of absolute ethanol and xylene (1:1) for 10 min. Other clearing agents, such as Histoclear or toluene, may be used instead of xylene.
- Transfer the specimen to pure xylene or alternative clearing agent for 10 min. Repeat this step.
- Mount the specimen in Canada balsam or Damar gum.
- Place the mounted specimen in a drying oven at 50 °C for 3 days.

To prepare a permanent preparation using haematoxylin, proceed as follows:

- Remove worm from vial and remove alcohol.
- Hydrate to tap water.
- Remove excess water.
- Add haematoxylin stain. Constantly check (5–7 min should be sufficient).
- Remove stain and add HCl (1%) to de-stain if necessary.
- If still overstained, remove HCl and add distilled water.
- Add ammonia (1%) to differentiate stain. Check constantly—Organs should be blue but other tissues semi-transparent.
- Remove ammonia.
- Add 70% EtOH (5 min). Keep covered.
- Add 80% EtOH (5 min).
- Add 90% EtOH (5 min).
- Add 96% EtOH (5 min).
- Add 100% EtOH (5 min).
- Add 100% EtOH (1 min).
- Remove excess EtOH but keep parasite covered with 100% EtOH.
- Add few drops of 100% methyl-salicylate.
- Transfer the specimen to 100% methyl-salicylate (5 min).
- Mount the specimen in Canada balsam or Damar gum.
- Place the mounted specimen in a drying oven at 50 °C for 3 days.

3.9.5 *Staining Oncomiracidia for Chaetotaxy*

Chaetotaxy is performed on oncomiracidia to study the placement of ciliated plates and sensilla. To prepare specimens, proceed as follows:

- Place the oncomiracidia in a minimal amount of artificial pond water (APW) cooled to 4 °C to minimize their swimming activity.
- Remove as much as possible of the APW and with a pipette add hot (60–65 °C) 0.5% silver nitrate.
- Place in the dark for 2.5 min.
- Wash the specimen vigorously 5–10 times in distilled water to remove any non-impregnated silver nitrate.

- Add a solution of one part glycerine and nine parts 80% alcohol in a glass cavity block.
- Place on a hot plate and allow alcohol to evaporate.
- Mount oncomiracidia in glycerine under coverslip.
- View the specimen under a light microscope and allow the stain to develop.
- Photograph.

3.9.6 Enzyme Digestion for Examination of Skeletal Structures

In order to study the harder embedded skeletal structures, soft tissue can be digested using an enzyme digestion technique adapted from Harris et al. (1999). Proceed as follows:

- Prepare a stock solution of 25 mg Proteinase K powder mixed with 10 mℓ ultra-distilled water (concentration of 2.5 mg/ml).
- Prepare a 50 mℓ digestion buffer (75 mM Tris-HCL pH 8.0, 10 mM EDTA, 5% SDS).
- Then prepare a 2 mℓ final concentration of 200 µg/mℓ Proteinase K by adding 160 µℓ stock solution Proteinase K to 1840 µℓ of digestion buffer.
- Dissect out the target tissue. For example, to study skeletal elements in the haptor only cut off the haptor of the parasite. Rehydrate in water.
- Cut discs with a diameter of about 5 mm from an acetate sheet (overhead transparency) with a hole punch and mount them on an aluminum SEM stub using double-sided carbon tape.
- Transfer the tissue pieces onto the disc. Harris et al. (1999) noted that varying degrees of dissociation could be achieved by varying the period of digestion and in effect improving the visualization of structures.
- Add a drop of enzyme digestion solution to the piece of tissue on the acetate disc mounted on the stub.
- Incubate at 50 °C for 10–15 min at a time.
- After one or two incubation sessions, add an equal volume of distilled water to the dried specimen and allow it to rehydrate for 2–5 min to remove excess salts.
- Remove the fluid using a small wedge cut from filter paper.
- Monitor the digestion progress microscopically until lysis occurs. The incubation process may have to be repeated several times to achieve the desired degree of digestion.
- Allow to air-dry for later observation under SEM.
- Sputter coat with gold palladium mixture and examine in the electron microscope.

3.9.7 *Scanning Electron Microscopy for Examination of Ultrastructure*

Material cleaning and fixation is of utmost importance for electron microscopy. Prior to parasite fixing, specimens should be thoroughly rinsed in saline solution and then briefly in water to remove any salts. 70% ethanol works well but better results are obtained when the specimens are fixed in 5% glutaraldehyde followed by post fixing in 1% osmium tetroxide. Alternatively, the Flemming solution which is a combination of glutaraldehyde and osmium tetroxide also works well.

3.10 Design of Primers for Amplification of Molecular Targets

Five genes or intergene are usually used for studying polystome systematics and evolution.

- ITS1 (internal transcribed spacer 1). It is amplified with forward L7, 5'--TGATTTGTCTGGTTTATTCCGAT-3′ and either reverse H7, 5'--GCTGCGTTCTTCATCGATACTCG-3′ or S1, 5'-ATTCCGATAACGAACGAGACT-3′ primers yielding a PCR fragment of about 1 kb depending of the size of ITS1 (see Verneau et al. 1997; Sinnappah et al. 2001). For all polystomes, sequencing is done with PCR primers, while for amphibian polystomes forward IF6, 5'-CCAAACTTGATCATTTAGAGG-3′ and IF4, 5'-GGGCAAGGCGTAAAGAAGCT-3′ primers and reverse IR7, 5'--ATGCAAAATGGTAGAGCTAAC-3′ and IR4, 5'--GGTACAGGAACCGGAATGAG-3′ primers can also be used (see Bentz et al. 2001).
- 18S rRNA gene: It is usually amplified in two overlapping fragments of approximately 1 kb each. The 5′ terminal end is amplified with forward F18, 5'--ACCTGGTTGATCCTGCCAGTAG-3′ and reverse 18RG, 5'-CTCTCTTAACCATTACTTCGG-3′ primers; the 3′ terminal end is amplified with forward 18F3, 5'-GGACGGCATGTTTACTTTGA-3′ and reverse IR5, 5'--TACGGAAACCTTGTTACGAC-3′ primers. The F18-18RG PCR portion is sequenced with reverse 18RC: 5'-TACGAGCTTTTTAACTGCAG-3′ and 18RG primers while the 18F3-IR5 portion is sequenced with forward 18F3 and S1: 5'-ATTCCGATAACGAACGAGACT-3′ primers (see Verneau et al. 2009).
- Partial 28S rRNA gene: It is usually amplified in two overlapping fragments of approximately 1 kb and 500 bp, respectively. The 5′ terminal end is amplified with forward LSU5', 5'-TAGGTCGACCCGCTGAAYTTAAGCA-3′ and reverse IR14, 5'-CATGTTAAACTCCTTGGTCCG-3′ primers; the 3′ terminal end is amplified with forward IF15, 5'-GTCTGTGGCGTAGTGGTAGAC-3′ and reverse LSU3', 5'-TAGAAGCTTCCTGAGGGAAACTTCGG-3′ primers.

The LSU5'-IR14 PCR fragment is sequenced with the reverse IR13: 5'--GTCGTGGCTTACACCCTGAGG-3′ and IR14 primers while the IF15-LSU3' fragment is sequenced with the forward IF15 primer (see Verneau et al. 2009). Reverse IR16, 5'-ATTCACACCCATTGACTCGCG-3′ primer may be also used instead of IR14 to amplify and sequence the 5′ terminal end of the 28S (see Héritier et al. 2015).

- Partial 12S rRNA gene: It is amplified with forward 12SpolF1, 5'--YVGTGMCAGCMRYCGCGGYYA-3′ and either reverse 12SpolR1, 5'--TACCRTGTTACGACTTRHCTC-3′ or 12SpolR9, 5'-TCGAAGATGACGGGCGATGTG-3′ primers. Sequencing is done with PCR primers (see Héritier et al. 2015).
- Partial COI gene: it is amplified with forward L-CO1p, 5'--TTTTTTGGGCATCCTGAGGTTTAT-3′ and reverse H-Cox1p2, 5'--TAAAGAAAGAACATAATGAAAATG-3′ primers yielding a PCR fragment of about 440 bp. Sequencing is done with PCR primers (see Littlewood et al. 1997). Reverse H-Cox1R, 5'-AACAACAAACCAAGAATCATG-3′ primer may be also used instead of H-Cox1p2 to amplify and sequence the partial COI yielding a PCR fragment of about 415 bp (see Héritier et al. 2015).

3.11 Morphological Features and Proposed Measurements of Taxonomic Importance for Polystomes

Descriptions for some of the known polystome species are inadequate, making comparisons challenging. To standardize species descriptions, we recommend that the following measurements (see Table 3.1) should be taken and reported in micrometers (μm). Measurements should include the range followed by the mean, ± standard deviation and sample size in parentheses, e.g. 2400–2750 (2600; 146; 4). Curved structures like ducts are measured as straight lines along the course of each duct.

3.12 Fixatives, Stains and Other Chemicals Commonly Used for Polystomes

3.12.1 Ammonium Picrate Solution

Like lactophenol, this solution is used to clear soft tissue, although it is much less aggressive than lactophenol. A strong advantage of this solution is that, after studying the ammonium picrate preparation, it can still be stained and permanently mounted. To prepare the solution, combine the following chemicals:

Table 3.1 Morphological and morphometric features to be reported for species descriptions

Source	Bibliographic references from which features were extracted
Sample size	Number of specimens studied
Body length	Total length of body including haptor. Measured along the midline of the parasite from the tip to a line between the posterior margins of the haptoral suckers.
Greatest width	Widest section of the body
Width at vagina	Body width at the level of the vaginae
Haptor length	Length of the haptor
Haptor width	Width of the haptor
False oral sucker width	Widest part of the false oral sucker
Mouth opening width	Width of the mouth
Pharynx length	Length of the pharynx
Pharynx width	Width of the pharynx
Ovary length	Total length of the ovary measured along the midline of the ovary
Ovary width	Widest part of the ovary
Egg length	Length of an egg
Egg width	Width of an egg
Egg incubation time	Time it takes for the eggs to hatch
Intra-uterine eggs number	Number of eggs in the uterus
Genital bulb width	Width of the genital bulb
Genital bulb position	Position of the genital bulb expressed as percentage of body length from the anterior end
Genital spines number	Number of genital spines present
Genital spine length	Average length of the genital spines in flat orientation
Vagina length	Length of the vagina
Vagina position	Position of the vagina expressed as percentage of body length from the anterior end
Haptoral sucker diameter	Diameter of the haptoral sucker. Only suckers that are spherical should be measured
Hamulus handle length X	Total length of the hamulus from the hook to the tip of the handle
Hamulus guard length Y	Length from the hook to the tip of the guard
Hamulus hook length	Length of the hamulus hook
Ovary position	Position of the ovary expressed as percentage of body length from the anterior end
Uterus length	Linear length of the uterus measured along the midline
Uterus width	Maximum width of the uterus
Marginal hooklet length	Length of the marginal hooklet
Haptoral L/body L	Ratio of the haptor length and body length
No of anastomoses	Number of anastomoses present

(continued)

Table 3.1 (continued)

Source	Bibliographic references from which features were extracted
Testis number	Number of testes length and/or testis width
Testis length (if measurable)	Length of the testis
Testis width (if measurable)	Width of the testis
Testis position	Position of the testis expressed as percentage of body length from the anterior end
Genito-intestinal canal length	Length of the genito-intestinal canal, measured following curves
Haptor length	Length of the haptor measured from the anterior margin of the haptor to the posterior margin
Haptor width	Widest part of the haptor

Formalin (10%)	1 ml
Glycerine	9 ml
Picric Acid	1 drop for every 10 ml formalin-glycerine mixture.

3.12.2 *Gray and Wess*

Gray and Wess' medium is used for temporary mounts. The mount will last longer if the edges of the coverslip is sealed with Canada balsam or any other permanent mounting medium. This medium is designed to view sclerites and eliminate any clouding musculature or tegument. Specimens mounted in Gray and Wess' medium will clear over a period of 2 weeks, depending on the size and thickness of specimens. To prepare the solution, mix the PVA and acetone. Mix half of the water with the glycerine and lactic acid. Stir to mix. Add the remaining water slowly using a pipette while stirring. Microwave the solution for 1 min, and prevent it from boiling over. Perform a final stir. The final product should be a homogenous, clear, gel. No white particles should be visible.

PVA (polyvinyl alcohol)	2 g 70%
Acetone	7 ml
Glycerin	5 ml
Lactic acid	5 ml
Distilled water	10 ml

3.12.3 *Lactophenol*

Lactophenol is used to clear soft tissue to reveal details of sclerotized parts. To prepare the solution, mix the following chemicals:

Phenol (carbolic acid) melted	3 ml
Lactic acid	1 ml
Glycerine	2 ml
Distilled water	1 ml

3.12.4 Ringer's Solution

This solution is isotonic for amphibian tissue. The heart of a frog will keep on beating for hours if submerged in Ringer's solution. To prepare the solution, the following chemicals must be dissolved in 1 litre of distilled water:

KCl	0.14 g
NaCl	6.5 g
CaC12	0.12 g
NaHCO3	1.2 g

3.12.5 Bouin's Fixative

Bouin's is a popular fixative for histological sectioning. This fixative penetrates quickly and causes hardly any shrinkage. Before staining it is necessary to remove the yellow stain caused by the picric acid. This is done by adding a few drops of saturated lithium carbonate to the 70% ethanol step when hydrating the sections before staining them. To prepare the solution, combine the following chemicals:

Picric acid (saturated solution)	75 ml
Formalin (100%)	25 ml
Glacial acetic acid	5 ml

3.12.6 Flemming's Fixative

Flemming's fixative works well for mitotic figures, but is not suitable for general work due to its poor penetrating qualities. A weak solution can be used for delicate objects. Tissue must be fixed for at least 24 h. The fixative is prepared by combining the following for a strong or a weak solution:

Strong solution:

Osmiumtetraoxide (OsO_4) (1%)	10 ml
Chromic acid (10%)	3 ml
Distilled water	19 ml
Glacial acetic acid	2 ml

Weak solution:

Osmiumtetraoxide (OsO_4) (2%)	5 ml
Chromic acid	25 ml
Distilled water	60 ml
Acetic acid (1%)	10 ml

3.12.7 Formalin (Neutral Buffered) (NBF)

For scanning electron-microscopical examination it is very important that the tissue is fixed in neutral buffered formalin to avoid distortion of tissue. Commercially available formalin is usually acidic, but can be buffered by adding the following chemicals:

Formalin (100%)	100 ml
NaH_2PO_4	3.5 g or (4 g NaH_2PO_4 H_2O)
Na_2HPO_4	6.5 g
Distilled water	900 ml

3.12.8 Glutaraldehyde (2,5%)

Glutaraldehyde (25%): 10 ml
Top up to 100 ml with phosphate buffer or cacodylate buffer
Store in refrigerator

3.12.9 Osmium Tetroxide (OsO_4)

Osmium tetroxide (OsO_4) is often used as a post-fixative in combination with glutaraldehyde when fixing material for electron-microscopical studies. It is commercially available in units of 0.5 g or 1 g, or in different volumes of 1% or 2% solutions. Because it is extremely dangerous, it must be handled with the greatest care in an extraction cabinet. The fumes alone act as a powerful fixative. OsO_4 fixes lipids but not carbohydrates. To prepare an OsO_4 solution, the neck of the glass ampule containing the osmium must be filed with a small iron file. However, the ampule should not be open. It must be put in an amber bottle with the dissolvent and shaken until the ampule breaks.

3.12.10 Acetocarmine Stain

Add 45 ml acetic acid to 55 ml distilled water in a glass beaker. Heat up, but do not boil. Add in small quantities of 2 g carmine powder. Stir after adding each quantity. Allow to cool. Chill in refrigerator and filter to remove undissolved carmine. Keep in an amber bottle at room temperature.

3.12.11 Alum Carmine

Add 7 g alum carmine (or 4 g alum +3 g carmine) to 100 ml distilled water and boil until dissolved (approximately 15 min). Leave to cool and add 100 ml 70% ethanol. Leave to stand for a few days and filter. Biological tissue can be transferred to the stain from 50% or 70% ethanol. If overstained, tissue can be de-stained by rinsing it briefly in acid alcohol.

3.12.12 Borax Carmine

Add 7 g borax carmine (or 4 g borax +3 g carmine) to 100 ml distilled water and boil until dissolved (+ 15 min). Allow to cool, then add 100 ml 70% ethanol. Leave to stand for a few days and filter. Biological tissue can be transferred to the stain from 50% or 70% ethanol. If overstained, tissue can be de-stained by rinsing it briefly in acid alcohol.

3.12.13 Ehrlich Haematoxylin

Haematoxylin	2 g
Ethyl alcohol, 95%	100 ml
Glycerin	100 ml
Distilled water	100 ml
Glacial acetic acid	10 ml
Potassium alum $KAl(SO_4)_2\ 12H_2O$	3 g

Ripen for several weeks or ripen artificially with addition of 0.2 g sodium iodate.

3.12.14 Van Cleave's Haematoxylin

Delafield's haematoxylin	1 ml
Ehrlich's haematoxylin	1 ml
Aluminum ammonium sulfate	6.5 g
Distilled water	100 ml

Usually takes several months to ripen.

References

Bentz S, Leroy S, Du Preez L, Mariaux J, Vaucher C, Verneau O (2001) Origin and evolution of African *Polystoma* (Monogenea: Polystomatidae) assessed by molecular methods. Int J Parasitol 31:697–705. https://doi.org/10.1016/S0020-7519(01)00179-5

Harris PD, Cable J, Tinsley RC, Lazarus CM (1999) Combined ribosomal DNA and morphological analysis of individual gyrodactylid monogeneans. J Parasitol 85:188–191

Héritier L, Badets M, Du Preez LH, Aisien MO, Lixian F, Combes C, Verneau O (2015) Evolutionary processes involved in the diversification of chelonian and mammal polystomatid parasites (Platyhelminthes, Monogenea, Polystomatidae) revealed by palaeoecology of their hosts. Mol Phylogenet Evol 92:1–10. https://doi.org/10.1016/j.ympev.2015.05.026

Kok DJ, Du Preez LH (1987) *Polystoma australis* (Monogenea): life-cycle studies in experimental and natural infections of normal and substitute hosts. J Zool 212:235–243

Littlewood DTJ, Rohde K, Clough KA (1997) Parasite speciation within or between host species? Phylogenetic evidence from site-specific polystome monogeneans. Int J Parasitol 27:1289–1297

Sinnappah ND, Lim LHS, Rohde K, Tinsley R, Combes C, Verneau O (2001) A paedomorphic parasite associated with a neotenic amphibian host: phylogenetic evidence suggests a revised systematic position for Sphyranuridae within anuran and turtle polystomatoineans. Mol Phylogenet Evol 18:189–201. https://doi.org/10.1006/mpev.2000.0877

Thurston JP (1968) The larva of *Oculotrema hippopotami* (Monogenea: Polystomatidae). J Zool 154:475–480

Tinsley RC, Owen RW (1975) Studies on the biology of *Protopolystoma xenopodis* (Monogenoidea): the oncomiracidium and life-cycle. Parasitology 71:445–463

Verneau O, Renaud F, Catzeflis FM (1997) Evolutionary relationships of sibling tapeworm species (Cestoda) parasitizing teleost fishes. Mol Biol Evol 14:630–636

Verneau O, Du Preez LH, Laurent VR, Raharivololoniaina L, Glaw F, Vences M (2009) The double odyssey of Madagascan polystome flatworms leads to new insights on the origins of their amphibian hosts. Proc Roy Soc B Biol Sci 276:1575–1583. https://doi.org/10.1098/rspb.2008.1530

Verneau O, Palacios C, Platt T, Alday M, Billard E, Allienne J-F, Basso C, Du Preez LH (2011) Invasive parasite threat: parasite phylogenetics reveals patterns and processes of host-switching between non-native and native captive freshwater turtles. Parasitology 138:1778–1792. https://doi.org/10.1017/S0031182011000333

Chapter 4
Polystome Species of Amphibians

Contents

Abstract This chapter brings together all known polystomes of amphibians, namely 135 spp. from all three extant orders Anura (frogs and toads), Urodela (salamanders) and Apoda (caecilians). For each of the species, we provide information on type material, reference to the description, etymology, type locality, host

L. H. du Preez et al., *Polystomatid Flatworms*, Zoological Monographs 9,
https://doi.org/10.1007/978-3-031-35887-6_4

identity, geographical distribution, site of infection, morphometrics and a species drawing based on the original description.

4.1 Morphological Features of Taxonomic Importance in Amphibian Polystomes

Although most polystomatid species have a similar general body plan, they do have characteristics that distinguish them from one another. These characteristics include size, body ratios, shape and extent of the intestine, position and number of ovaries, testes, eggs, suckers and hooks among others. Because they are soft-bodied organisms that can vary and can distort depending on the degree of flattening during fixation, a high premium is placed on the number, shape and size of sclerotized parts like hamuli and marginal hooklets for description (Du Preez and Maritz 2006). Each of the 19 genera has distinguishing characteristics that helped with their classification. Because of the high degree of host-specificity, the identity of the host was for a long time an important characteristic for species delimitation, however, a polystome should never be identified based on the identity of the host alone.

Molecular studies have been widely applied to infer phylogenetic relationships of polystomes as well as to understand their evolutionary history (Bentz et al. 2001; Verneau et al. 2002; Bentz et al. 2006; Verneau et al. 2009; Badets et al. 2011; Héritier et al. 2015). Molecular tools also played a very important and often determining role in defining the identity of both host and parasite. In the past 10 years, the DNA taxonomy or DNA barcoding approach has indeed greatly increased the rate of polystome species description, from both amphibian and chelonian hosts (Berthier et al. 2014; Chaabane et al. 2019, 2022; Du Preez et al. 2007, 2010, 2014, 2017; Fan et al. 2020; Héritier et al. 2018; Landman et al. 2018, 2021, 2023; Sales et al. 2023; Raharivololoniaina et al. 2011; Yildirimhan et al. 2012). Molecular data are more accurate and less subjective in taxonomic identification than morphological evidence, especially when cryptic species occur in similar environments. Therefore, the DNA barcoding method greatly improves our classifications that are based on morphology only. However, there is still a great need to acquire DNA from species discovered prior to the availability of this approach to make substantial improvements in systematics.

Literature on polystomes is scattered and many papers are difficult to access. Furthermore, many of the descriptions were originally published in the mother tongue of the author rendering their study sometimes very difficult. In this chapter, we compile an extensive list of all known polystomes. For each species, the original paper has been translated where necessary, studied and summarized to address the most relevant information concerning the species. Supplementary information, when available, was also extracted from the literature. For each species we provide:

- *Collection:* The institution(s) where the type material is housed.
- *Holotype:* Museum code or number of the holotype specimen.
- *Paratypes:* Museum codes or numbers of the paratype specimens.
- *Syntypes:* Museum codes or numbers of the syntypes.
- *Original description:* The reference to the article in which the species was first described.
- *Other taxonomic contributions:* Authors who contributed to the taxonomy of the species in the same or distinct geographical areas.
- *Etymology:* Origin and explanation of the species name.
- *Type locality:* Location where the type specimen was collected.
- *Other localities:* Other locations where the species was reported.
- *Type host:* Scientific and common name of type host.
- *Additional hosts:* When additional host species were reported, they are mentioned.
- *Geographical distribution:* Distribution of the host species in its natural habitat.
- *Host conservation status:* The IUCN conservation status of the host species.
- *Site on host:* Site location on the host where the parasite was found.
- *DNA sequence:* GenBank accession numbers are provided for sequences.
- *Morphology and morphometrics (μm):* Standard measurements are summarized in a table and supplemented by data reported by other authors. They are given in micrometers. Where no count or measurements are provided, it indicates that this information is not available. Apart from measurements other information is given when accessible.
- *Remarks:* Additional information.

4.2 Taxonomic Corrections

In this section, we sink a genus that we believe not to be valid. Furthermore, during the course of history, several polystomes were incorrectly named. Here we correct these species names to comply with the International Code of Zoological Nomenclature.

4.2.1 Parapseudopolystoma cerrocoloradensis Nasir and Fuentes Zambrano, 1983

Nasir and Fuentes Zambrano (1983) described a new polystome they found in the urinary bladder of *Rhinella marina* (Linnaeus, 1758) from Venezuela and created a new genus *Parapseudopolystoma* Nasir and Fuentes Zambrano, 1983 and a new subfamily Parapseudopolystomatinae. They named the species *Parapseudopolystoma cerrocoloradensis* Nasir and Fuentes Zambrano, 1983. In their original description, they did not compare this species with the species already existing in the genus *Riojatrema* Lamothe-Argumedo, 1964, known from the same continent. The morphological features of *P. cerrocoloradensis* perfectly fit the generic characteristics of *Riojatrema* Lamothe-Argumedo, 1963, which takes seniority. We thus regard *Parapseudopolystoma* as well as Parapseudopolystomatinae as *nomen nudum* and propose the new species name *Riojatrema cerrocoloradense* (Nasir and Fuentes Zambrano, 1983) n. comb.

4.2.2 *Polystoma ozakii* Uchida, Machida, Uchida and Itagaki, 1988

Uchida et al. (1988) described a new polystome from *Bufo japonicus* (Temminck and Schlegel, 1838) from Japan and named it *Polystoma ozakii* Uchida, Machida, Uchida and Itagaki, 1988. It was observed that this species is a primary junior homonym of *Polystoma ozakii* Price, 1939. According to the International Code of Zoological Nomenclature (ICZN) article 53.3, two taxa erected within a genus with the same epithet constitutes primary homonymy. Following ICZN articles 60.1 and 60.3, the junior homonym is hereby rejected and replaced by *Polystoma uchidai* nom. nov.

4.2.3 *Polystoma floridanum* Du Preez, Verneau and Gross, 2007

At this stage *Neopolystoma floridanum* (Stunkard, 1924), a junior synonym of *Polystomoides orbiculare*, from the bladder of *Pseudemys loridana* (Le Conte, 1830) was originally named *Polystoma floridanum.* Du Preez et al. (2007) described a new polystome from the urinary bladder of *Hyla cinerea* Schneider, 1799, now *Dryophytes cinereus* (Schneider, 1799), and named it *Polystoma floridana* Du Preez, Verneau and Gross, 2007. As *P. floridanum* was previously occupied by we herein propose to change it to *Polystoma cinereum* nom. nov.

4.2.4 *Correction for Gender*

Generic names ending in -soma, -stoma, -trema and -nema are neuter and Greek in origin. They have been Latinized and look feminine, but they are not. There was a practice in the nineteenth century of changing them into a true Latin format. e.g. -stomum for -stoma, but the Greek forms have priority if they are older (Gibson pers. comm.). The correct form is thus *Polystoma* Zeder, 1800 (family

Polystomatidae) and not *Polystomum* Zeder, 1800 (family Polystomatidae). All other genera based on *Polystoma* are also neuter. A number of species were incorrectly published according to the rules for gender. Therefore, we provide a list with the corrected names (Table 4.1).

Table 4.1 Polystome taxa of amphibians corrected for gender

Current taxon	New taxon
Eupolystoma namibiensis	*Eupolystoma namibiense*
Indopolystoma pingbianensis	*Indopolystoma pingbianense*
Madapolystoma isaloensis	*Madapolystoma isaloense*
Mesopolystoma samiriensis	*Mesopolystoma samiriense*
Neoriojatrema mokokchungensis	*Neoriojatrema mokokchungense*
Polystoma australis	*Polystoma australe*
Polystoma dorsalis	*Polystoma dorsale*
Polystoma ebriensis	*Polystoma ebriense*
Polystoma ezoensis	*Polystoma ezoense*
Polystoma fuscus	*Polystoma fuscum*
Polystoma gabonensis	*Polystoma gabonense*
Polystoma galamensis	*Polystoma galamense*
Polystoma napoensis	*Polystoma napoense*
Polystoma sodwanensis	*Polystoma sodwanense*
Polystoma testimagna	*Polystoma testimagnum*
Polystoma togoensis	*Polystoma togoense*
Protopolystoma fissilis	*Protopolystoma fissile*
Protopolystoma occidentalis	*Protopolystoma occidentale*
Protopolystoma orientalis	*Protopolystoma orientale*
Protopolystoma ramulosus	*Protopolystoma ramulosum*
Parapseudopolystoma cerrocoloradensis	*Riojatrema cerrocoloradense*
Riojatrema ecuadorensis	*Riojatrema ecuadorense*

4.3 Polystomes of Salamanders (Fig. 4.1)

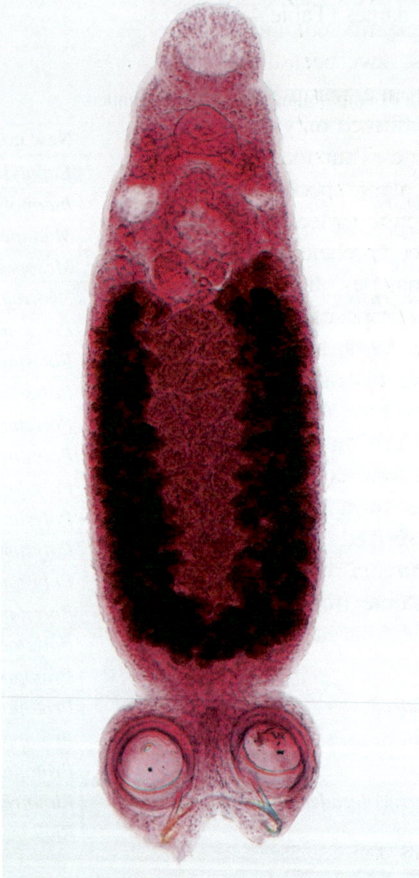

Fig. 4.1 Micrograph of *Shyranura osleri*

4.3.1 Pseudopolystoma Yamaguti, 1963

4.3.1.1 *Pseudopolystoma dendriticum* (Ozaki, 1948) (Fig. 4.2)

Collection Unknown.

Holotype Unknown.

Paratypes Unknown.

Original description Ozaki (1948).

Other taxonomic contributions Uchida and Itagaki (1979).

Etymology Unknown

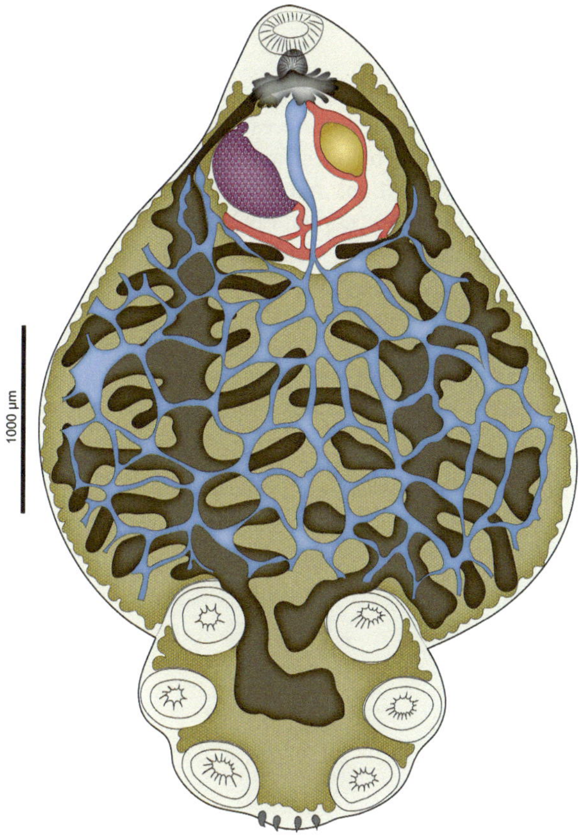

Fig. 4.2 *Pseudopolystoma dendriticum*. Redrawn from Ozaki (1948)

Type locality Mt. Isizuti, Shikoku Island, Japan.

Other localities Hakone, Kanagawa, Japan.

Type host *Onychodactylus japonicus* (Houttuyn, 1782), Japanese Clawed Salamander.

Additional hosts None.

Host geographical distribution *Onychodactylus japonicus* is native and endemic to Japan, where it can be found in mountainous regions in Honshu and Shikoku. It occurs in freshwater aquatic (small streams) as well as terrestrial (forested habitats adjacent to streams) systems (Kaneko and Matsui 2004a; Yoshikawa et al. 2008).

Host conservation status Because of its wide distribution, occurring in several protected regions with a presumably relatively large population, the Japanese Clawed Salamander is listed as Least Concern by The IUCN Red List of Threatened Species 2021: e.T79101232A79100404. https://doi.org/10.2305/IUCN.UK.2021-1.RLTS.T79101232A79100404.en. Accessed on 30 January 2022.

Site on host Urinary bladder.

DNA sequence

- 18S rRNA gene, 1993 bp. GenBank Accession: FM992700
- 28S rRNA gene, 1404 bp. GenBank Accession: FM992707
- Cytochrome c Oxidase I, 377 bp. GenBank Accession: KR856180
- 12S rRNA gene, 491 bp. GenBank Accession: KR856122
- Xlox gene, 75 bp. GenBank Accession: FN298172
- Abd-A/Lox4 gene, allele A, 75 bp. GenBank Accession: FN298166
- Abd-A/Lox4 gene, allele B, 75 bp. GenBank Accession: FN298167
- Dfd/Hox4 gene, 75 bp. GenBank Accession: FN298159
- Hox2/3 gene, 75 bp. GenBank Accession: FN298154
- Lab/Hox1 gene, 75 bp. GenBank Accession: FN298149

Morphology and morphometrics (μm):

Source	Ozaki (1948)	Uchida and Itagaki (1979)
Sample size		
Body length	4360–5750	3100
Greatest width	3280–4100	1800
Width at vagina		
Haptor length		1100
Haptor width	1600–2200	1500
False oral sucker width		
Pharynx length		
Pharynx width		
Testis		Multilobe, extensive
Ovary length	520	520
Ovary width	290	250
Egg length	281–300	
Egg width	171–195	
Egg incubation time		
Number of intra-uterine eggs	1	
Genital bulb width		
Number of genital spines		20 (10 small; 10 large)
Genital spine length		
Haptoral sucker diameter	350	
Hamulus handle length X		
Hamulus guard length Y		
Hamulus hook length		
Marginal hooklet length		
Haptoral L/Body L		
No of anastomoses		0

Remarks The species was originally described as *Polystoma dendriticum* (Ozaki, 1948).

4.3.2 Sphyranura Wright, 1879

4.3.2.1 *Sphyranura euryceae* Hughes and Moore, 1943 (Fig. 4.3)

Collection United States National Museum (USNM).

Syntype USNM 1337573.

Original description Hughes and Moore (1943).

Other taxonomic contributions McAllister et al. (1991).

Etymology Named after the host.

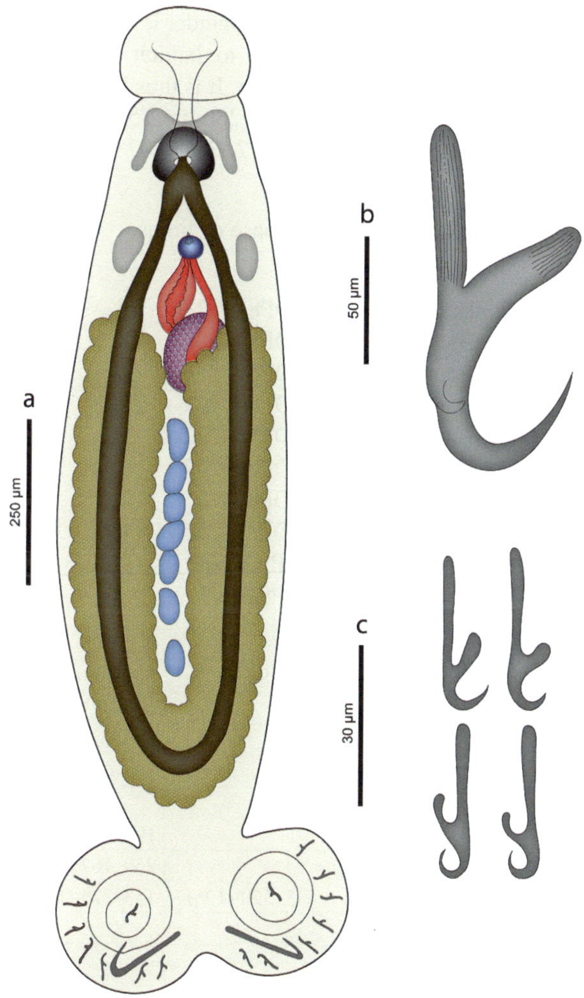

Fig. 4.3 (**a**) *Sphyranura euryceae*, (**b**) hamulus, (**c**) marginal hooklets. Redrawn from Hughes and Moore (1943)

Type locality Peavine Creek, Oklahoma, USA.

Other localities None.

Type host *Eurycea tynerensis* Moore and Hughes, 1939, Oklahoma Salamander.

Additional hosts *Eurycea lucifuga* Rafinesque, 1822 and *Eurycea multiplicata* (Cope, 1869).

Host geographical distribution *Eurycea tynerensis* occurs in terrestrial as well as in freshwater aquatic systems, where it is found in shallow, slow running water. This species is native to the United States where it is present in eastern Oklahoma, north-western Arkansas and south-western Missouri (Hammerson 2004a; Hughes and Moore 1943).

Host conservation status The Oklahoma Salamander is not widely distributed and its distribution range may be declining due to habitat destruction, agriculture, pollution of the aquatic habitat and urbanization. It is thus listed as Near Threatened by The IUCN Red List of Threatened Species 2004: e.T59277A11908828. https://doi.org/10.2305/IUCN.UK.2004.RLTS.T59277A11908828.en. Accessed on 30 January 2022.

Site on host Gills and skin, particularly at base of limbs.

DNA sequence
- 18S rRNA gene. GenBank Accession: OP879228
- 18S rRNA gene. GenBank Accession: OP879229
- 28S rRNA gene. GenBank Accession: OP879230
- 28S rRNA gene. GenBank Accession: OP879231
- 28S rRNA gene. GenBank Accession: OP879232
- 28S rRNA gene. GenBank Accession: OP879233
- 12S rRNA gene. GenBank Accession: OP879225
- 12S rRNA gene. GenBank Accession: OP879226
- Mitochondrion. GenBank Accession: OP920606
- Mitochondrion. GenBank Accession: OP920607

Morphology and morphometrics (μm):

Source	Hughes and Moore (1943)	McAllister et al. (1991)
Sample size	15–30	20
Body length	760–2700 (1329)	800–2400 (1620)
Greatest width	200–667 (393)	300–600 (420)
Width at vagina		
Haptor length	141–314 (227)	191–355 (259)
Haptor width	246–633 (399)	269–767 (463)
False oral sucker width	82–220 (128) × 135–320 (196)	153–225 (186) × 155–284 (203)
Pharynx length	53–153 (93)	
Pharynx width	60–153 (117)	
Number of testes	7 (3–10)	

(continued)

Morphology and morphometrics (µm):		
Source	Hughes and Moore (1943)	McAllister et al. (1991)
Testis length	37–98	53–98 (77)
Testis width	30–105	78–120 (102)
Ovary length		
Ovary width		
Egg length	240–373 (308)	254–282 (268)
Egg width	180–240 (199)	145–217 (190)
Egg incubation time		
Intra-uterine eggs (n)	1	1
Genital bulb width	30–61 (46)	
Number of genital spines	7–9 (8)	
Genital spine length	22–29 (25)	
Haptoral sucker diameter	96–220 (141)	
Hamulus handle length X		
Hamulus guard length Y		
Hamulus hook length	119–170 (131)	
Marginal hooklet length	18–31 (29)	
Haptoral L/Body L	0.17	0.16
No of anastomoses	0	0

Remarks None.

4.3.2.2 *Sphyranura oligorchis* Alvey, 1933 (Fig. 4.4)

Collection American Museum of Natural History.

Holotype AMNH1432.1.

Paratypes AMNH 1432.1; AMNH 1432.2; AMNH 1432.3.

Original description Alvey (1933).

Other taxonomic contributions Alvey (1936); Meyer and Olsen (1972).

Etymology Refers to the small testes.

Type locality Pennsylvania, USA.

Other localities None.

Type host *Necturus maculosus* (Rafinesque, 1818), Mudpuppy.

Additional hosts None.

Host geographical distribution *Necturus maculosus* is a bottom dweller, occurring in freshwater systems such as lakes and rivers. The species is native to Canada, but can also be found in various regions throughout North America including southern Manitoba, northern Mississippi, northern Louisiana, northern Georgia, northern Alabama, southern Quebec and south to Oklahoma (Hammerson 2004b).

Host conservation status Because of its wide distribution and relatively high tolerance of habitat modification, the Mudpuppy is listed as Least Concern by The IUCN Red List of Threatened Species2015: e.T59433A64731610. https://doi.org/10.2305/IUCN.UK.2015-4.RLTS.T59433A64731610.en. Accessed on 30 January 2022.

Site on host Gills and skin, particularly at base of limbs.

DNA sequence
- 18S rRNA gene, 2008 bp. GenBank Accession: FM992701
- 18S rRNA gene, 532 bp. GenBank Accession: AJ287994
- 28S rRNA gene, 1411 bp. GenBank Accession: FM992708
- Cytochrome c Oxidase I, 395 bp. GenBank Accession: KR856174
- 12S rRNA gene, 489 bp. GenBank Accession: KR856098
- Abd-A/Lox4 gene, 75 bp. GenBank Accession: FN298165
- Hox2/3 gene, 75 bp. GenBank Accession: FN298153

Morphology and morphometrics (μm):			
Source	Alvey (1933)	Alvey (1936)	Meyer and Olsen (1972)
Sample size			
Body length	2500–3500	2500–3500	4000
Greatest width	300–400	300–400	700
Width at vagina			
Haptor length	350–470		
Haptor width	700–800		
False oral sucker width			
Pharynx length			
Pharynx width			
Number of testes	5–7	5–7, usually 6	
Testis length	80–100	80–100	
Testis width	80–160	80–100	
Ovary length		80	
Ovary width	65–130	100	
Egg length	280–410	280–410	280
Egg width	140–180	220–260	140
Egg incubation time		28–32	
Number of intra-uterine eggs		1	1
Genital bulb width			
Number of genital spines		8	
Genital spine length			
Haptoral sucker diameter			
Hamulus handle length X		260	
Hamulus guard length Y			
Hamulus hook length			

(continued)

Morphology and morphometrics (μm):			
Source	Alvey (1933)	Alvey (1936)	Meyer and Olsen (1972)
Marginal hooklet length		25	
Haptoral L/Body L			
No of anastomoses			

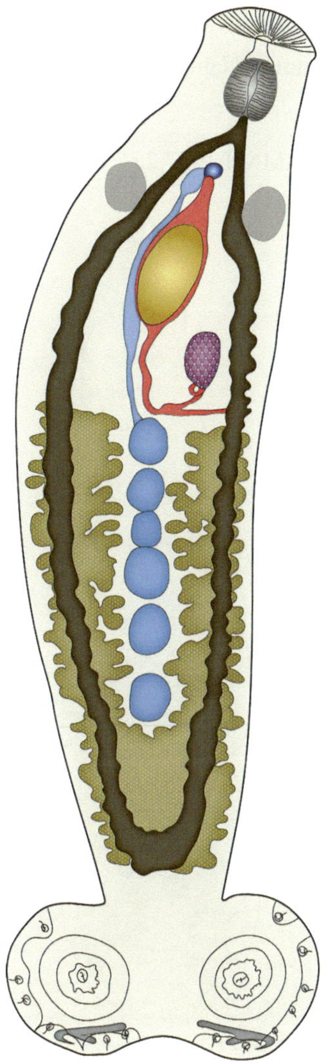

Fig. 4.4 *Sphyranura oligorchis*. Redrawn from Alvey (1933)

Remarks Alvey (1936) reported an ovary length of 80 in its manuscript and elsewhere gave the range of 65–75. This species has been reported from the same host than *Sphyranura osleri* Wright, 1879 in Canada and *Sphyranura polyorchis* Alvey, 1936 in the USA. Molecular phylogenetic studies showed that *S. oligorchis* was nested within the Polystomatidae. This led Sinnappah et al. (2001) to consider *Sphyranura* as a true genus of the Polystomatidae.

4.3.2.3 *Sphyranura osleri* Wright, 1879 (Fig. 4.5)

Collection American Museum of Natural History (AMNH).

Holotype Unknown.

Paratypes AMNH 1427.1; AMNH 1427.2.

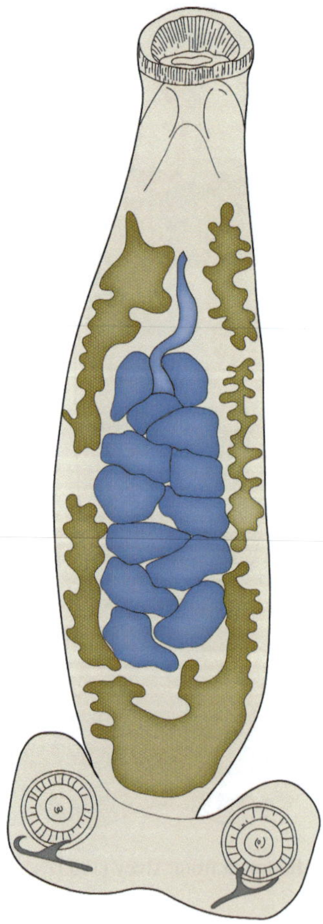

Fig. 4.5 *Sphyranura osleri*. Redrawn from Wright (1879)

Original description Wright (1879).

Other taxonomic contributions Wright and Macallum (1887); Coggins and Sajdak (1982); Alvey (1936).

Etymology Named after Prof Osler, Montreal, Canada.

Type locality Wisconsin, Canada.

Other localities None.

Type host *Necturus maculosus*, Mudpuppy.

Additional hosts None.

Host geographical distribution *Necturus maculosus* is a bottom dweller occurring in freshwater systems such as lakes and rivers. The species is native to Canada, but can also be found in various regions throughout North America including southern Manitoba, northern Mississippi, northern Louisiana, northern Georgia, northern Alabama, southern Quebec and south to Oklahoma (Hammerson 2004b).

Host conservation status Because of its wide distribution and its relatively high tolerance for habitat modification, the Mudpuppy is listed as Least Concern by The IUCN Red List of Threatened Species 2015: e.T59433A64731610. https://doi.org/10.2305/IUCN.UK.2015-4.RLTS.T59433A64731610.en. Accessed on 30 January 2022.

Site on host Gills and mouth cavity.

DNA sequence None.

Morphology and morphometrics (μm):		
Source	Wright (1879)	Alvey (1936)
Sample size	2	
Body length	2600	
Greatest width	700	
Width at vagina		
Haptor length	450	
Haptor width	1000	
False oral sucker width		
Pharynx length		
Pharynx width		
Number of testes		12–16
Testis length		
Testis width		
Ovary length		100
Ovary width		160
Egg length	364	
Egg width	247	
Egg incubation time		
Number of intra-uterine eggs		1
Genital bulb width		

(continued)

Morphology and morphometrics (μm):		
Source	Wright (1879)	Alvey (1936)
Number of genital spines	< 8	
Genital spine length		
Haptoral sucker diameter	270	
Hamulus handle length X	200	
Hamulus guard length Y		
Hamulus hook length		
Marginal hooklet length	25	
Haptoral L/Body L		
No of anastomoses		0

Remarks This species has been reported from the same host than *S. oligorchis* and *S. polyorchis* in the USA.

4.3.2.4 *Sphyranura polyorchis* Alvey, 1936 (Fig. 4.6)

Collection American Museum of Natural History (AMNH).

Holotype AMNH 1431.

Paratypes None.

Original description Alvey (1936).

Other taxonomic contributions None.

Etymology Refers to the multiple testes reported for the species.

Type locality Pennsylvania, USA.

Other localities None.

Type host *Necturus maculosus*, Mudpuppy.

Additional hosts None.

Host geographical distribution *Necturus maculosus* is a bottom dweller occurring in freshwater systems such as lakes and rivers. The species is native to Canada but can also be found in various regions throughout North America, including southern Manitoba, northern Mississippi, northern Louisiana, northern Georgia, northern Alabama, southern Quebec and south to Oklahoma (Hammerson 2004b).

Host conservation status Because of its wide distribution and a relatively high tolerance of habitat modification, the mudpuppy is listed as Least Concern by The IUCN Red List of Threatened Species2015: e.T59433A64731610. https://doi.org/10.2305/IUCN.UK.2015-4.RLTS.T59433A64731610.en. Accessed on 30 January 2022.

Site on host Gills and skin, particularly at base of limbs.

DNA sequence None.

Morphology and morphometrics (µm) of *S. polyorchis*	
Source	Alvey (1936)
Mature specimens (n)	
Body length	2400–3000
Greatest width	410–770
Width at vagina	
Haptor length	
Haptor width	
False oral sucker width	300
Pharynx length	
Pharynx width	150
Number of testes	20–23
Testis length	
Testis width	
Ovary length	85
Ovary width	65
Egg length	
Egg width	
Egg incubation time	
Number of intra-uterine eggs	
Genital bulb width	
Number of genital spines	
Genital spine length	
Haptoral sucker diameter	
Hamulus handle length X	
Hamulus guard length Y	
Hamulus hook length	
Marginal hooklet length	
Haptoral L/Body L	
No of anastomoses	0

Remarks This species has been reported from the same host than *S. oligorchis* in the USA and *S. osleri* in the Canada. Although Price (1939) and McAllister et al. (1991) questioned the validity of this species, we keep it now based on reported morphological differences (Figs. 4.4 and 4.5).

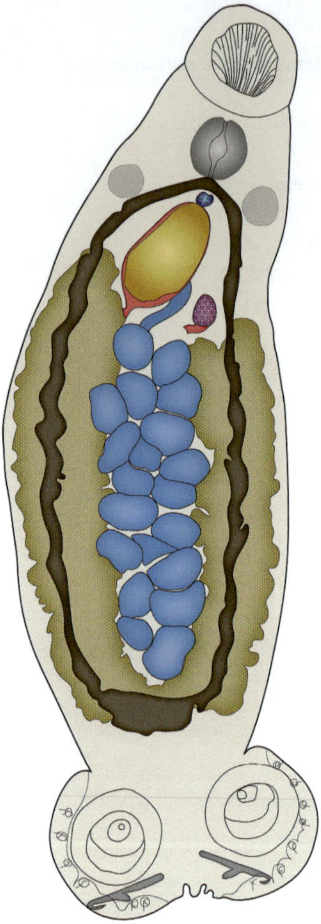

Fig. 4.6 *Sphyranura polyorchis*. Redrawn from Alvey (1936)

4.4 Polystomes of Anurans (Fig. 4.7)

Fig. 4.7 Micrographs of (**a**) *Eupolystoma vanasi*, (**b**) *Madapolystoma biritika*, (**c**) *Polystoma channingi*, (**d**) *Protopolystoma xenopodis*

4.4.1 Diplorchis Ozaki, 1931

4.4.1.1 *Diplorchis grahami* Fan, Wang and Xu, 2007 (Fig. 4.8)

Collection Yunnan College of Life Sciences.

Holotype 200505G112.

Paratypes 200505G241-2 & 200505G115.

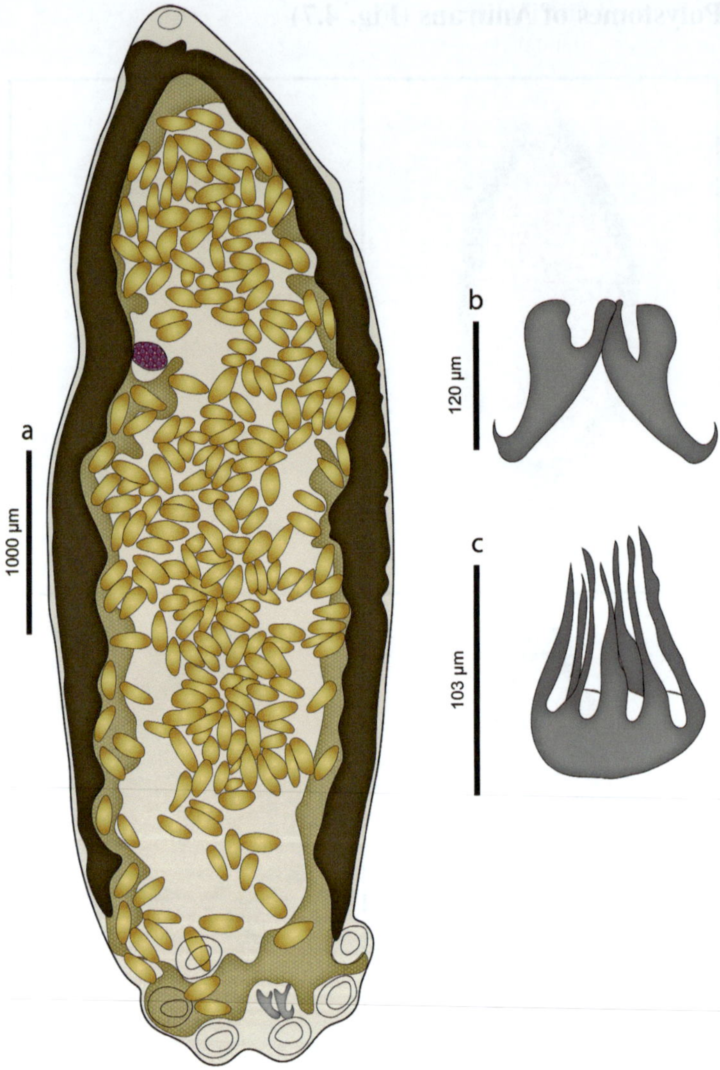

Fig. 4.8 (**a**) *Diplorchis grahami*, (**b**) hamuli, (**c**) genital spines. Redrawn from Fan et al. (2007)

Original description Fan et al. (2007).

Other taxonomic contributions None.

Etymology Named after the host

Type locality Yunnan Province, China.

Other localities None.

Type host *Odorrana grahami* (Boulenger, 1917), Yunnanfu Frog. In the original description of the polystome, the host was known as *Rana grahami* Boulenger, 1917.

Additional hosts None.

Host geographical distribution *Odorrana grahami* is native to China being distributed in Yunnan, Sichuan, Guizhou and perhaps in Hunan Provinces and northern Vietnam where it can be found in the Lien Son National Park. It inhabits hill streams in montane forest and occasionally water pools. It may leave the water to inhabit grassland, forest and shrubland habitats (Datong et al. 2004).

Host conservation status The Yunnanfu Frog population is declining due to overharvesting for food and habitat loss and degradation caused by smallholder farming and shifting agriculture (Datong et al. 2004). It is listed as Near Threatened by The IUCN Red List of Threatened Species 2004:e.T58607A11808625. https://doi.org/10.2305/IUCN.UK.2004.RLTS.T58607A11808625.en. Accessed on 27 January 2022.

Site on host Urinary bladder.

DNA sequence None.

Morphology and morphometrics (μm):	
Source	Fan et al. (2007)
Sample size	32
Body length	3139–6306 (4627)
Greatest width	861–2083 (1611)
Width at vagina	(1244)
Haptor length	472–1028 (681)
Haptor width	889–1694 (1178)
False oral sucker width	
Pharynx length	
Pharynx width	
Testis number	2
Testis length	
Testis width	
Ovary length	
Ovary width	
Egg length	208–312 (248.43)
Egg width	77–122 (96.90)
Egg incubation time	
Number of intra-uterine eggs	230
Genital bulb width	
Number of genital spines	7
Genital spine length	
Haptoral sucker diameter	204–408 (284.22)
Hamulus handle length X	295.87
Hamulus guard length Y	
Hamulus hook length	
Hamulus X/Y ratio	1.255
Marginal hooklet length	
Haptoral L/Body L	0.147
No of anastomoses	0

Remarks None.

4.4.1.2 *Diplorchis hangzhouensis* Shu-Yi and Suo, 1987 (Fig. 4.9)

Collection Parasitology Research Laboratory, Biological Department, East China Normal University.

Holotype Unknown.

Paratypes Unknown.

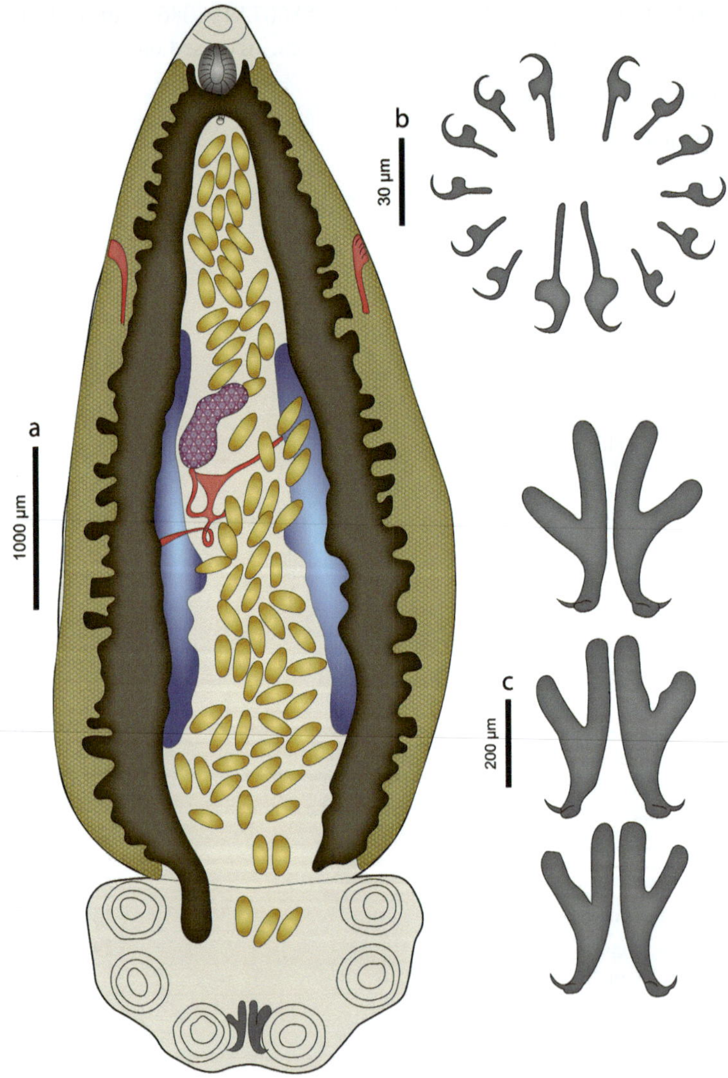

Fig. 4.9 (**a**) *Diplorchis hangzhouensis*, (**b**) marginal hooklets, (**c**) hamuli. Redrawn from Zhang and Long (1987)

Original description Shu-Yi and Suo (1987).

Other taxonomic contributions None.

Etymology Named after type locality.

Type locality Hangzhou, Zhejiang Province, China. 30°17'N, 120°09′E.

Other localities None.

Type host *Sylvirana guentheri* (Boulenger, 1882), Guenther's Frog. In the original description of the polystome, the host was known as *Rana guentheri* Boulenger, 1882.

Additional hosts None.

Host geographical distribution *Sylvirana guentheri* can be found in Vietnam, Taiwan, Province of China, China, Macao, Hong Kong as well as being introduced into Guam. This species inhabits marshes, pools, paddy fields, slow-moving streams and ditches (Kuangyang et al. 2009).

Host conservation status The Guenther's Frog has a widespread distribution, is abundant and is found in a variety of habitats. It is listed as Least Concern by The IUCN Red List of Threatened Species 2004: e.T58612A86479584. https://doi.org/10.2305/IUCN.UK.2004.RLTS.T58612A11810324.en. Accessed on 27 January 2022.

Site on host Urinary bladder.

DNA sequence

- Mitochondrion, 13,918 bp. GenBank Accession: JQ038227

Morphology and morphometrics (μm):	
Source	Shu-Yi and Suo (1987)
Sample size	16
Body length	3225–9137 (5617)
Greatest width	1321–2950 (2101)
Width at vagina	
Haptor length	516–1550 (977)
Haptor width	1109–2279 (1683)
False oral sucker width	134–246 × 218–355 (164 × 281)
Pharynx length	123–0.232 (186)
Pharynx width	107–0.205 (168)
Testis number	2
Testis length	1851–2069 (1914)
Testis width	731–820 (783)
Ovary length	328–714 (516)

(continued)

Morphology and morphometrics (μm):	
Source	Shu-Yi and Suo (1987)
Ovary width	137–344 (212)
Egg length	191–246 (214)
Egg width	116–144 (125)
Egg incubation time	
Number of intra-uterine eggs	
Genital bulb width	61–76
Number of genital spines	8–9
Genital spine length	22–30 (26)
Haptoral sucker diameter	205–506 (368)
Hamulus handle length X	271–480 (433)
Hamulus guard length Y	
Hamulus hook length	
Marginal hooklet length	19–44
Haptoral L/Body L	0.17
No of anastomoses	0

Remarks None.

4.4.1.3 *Diplorchis latouchii* Shu-Yi and Suo, 1987 (Fig. 4.10)

Collection Parasitology Research Laboratory, Biological Department, East China Normal University.

Holotype Unknown.

Paratypes Unknown.

Original description Shu-Yi and Suo (1987).

Other taxonomic contributions None.

Etymology Named after the host.

Type locality Hangzhou, Zhejiang Province, China. 30°17'N, 120°09′E.

Other localities None.

Type host *Hylarana latouchii* (Boulenger, 1899), Kuatun Frog. In the original description of the polystome, the host was known as *Rana latouchii* Boulenger, 1899.

Additional hosts None.

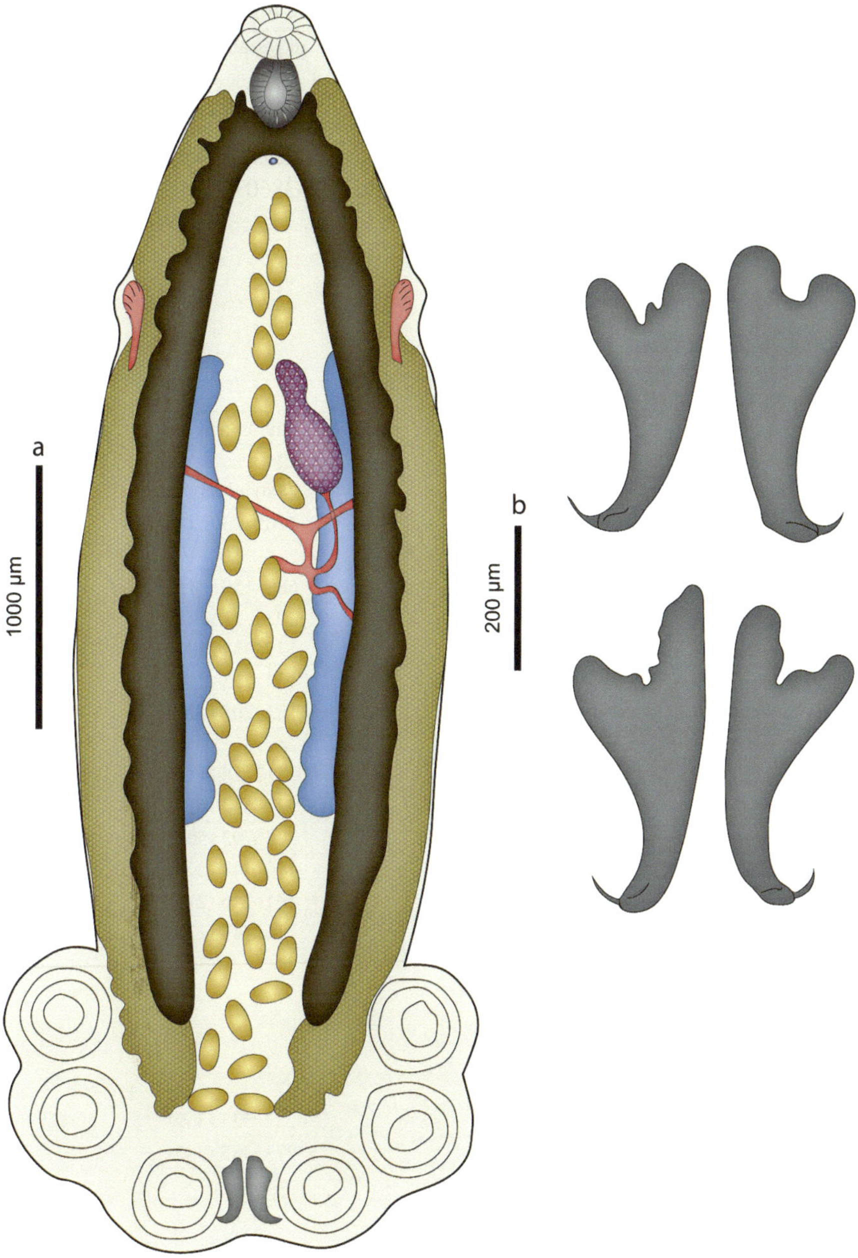

Fig. 4.10 (**a**) *Diplorchis latouchii*, (**b**) hamuli. Redrawn from Zhang and Long (1987)

Host geographical distribution *Hylarana latouchii* is distributed in the Province of China in China, Taiwan and Hong Kong, where it inhabits forests (Lau et al. 2004a).

Host conservation status The Kuatun Frog has a large population, a wide distribution and is tolerant to various habitats. It is listed as Least Concern by The IUCN Red List of Threatened Species 2020: e.T58640A63860302. https://doi.org/10.2305/IUCN.UK.2020-1.RLTS.T58640A63860302.en. Accessed on 27 January 2022.

Site on host Urinary bladder.

DNA sequence None.

Morphology and morphometrics (μm):	
Source	Shu-Yi and Suo (1987)
Sample size	12
Body length	2580–6858 (4423)
Greatest width	1105–2302 (1570)
Width at vagina	
Haptor length	688–1591 (1193)
Haptor width	1036–2352 (1780)
False oral sucker width	179–314 × 246–383 (228 × 324)
Pharynx length	174–253 (211)
Pharynx width	122–205 (173)
Testis number	2
Testis length	1517–1843 (1650)
Testis width	535–642 (615)
Ovary length	328–793 (567)
Ovary width	150–348 (228)
Egg length	191–246 (214)
Egg width	116–144 (125)
Egg incubation time	
Number of intra-uterine eggs	
Genital bulb width	
Number of genital spines	7–8
Genital spine length	27
Haptoral sucker diameter	232–523 (369)
Hamulus handle length X	273–485 (385)
Hamulus guard length Y	
Hamulus hook length	
Marginal hooklet length	
Haptoral L/Body L	0.269
No of anastomoses	0

Remarks None.

4.4.1.4 *Diplorchis lividae* Song, Xiao and Ding, 2008 (Fig. 4.11)

Collection Lab of Fish Parasitology, College of Life Sciences, South China Normal University (HNWZS).

Holotype HNWZS 20060716-1.

Paratypes HNWZS 20060716-2-9.

Original description Song et al. (2008).

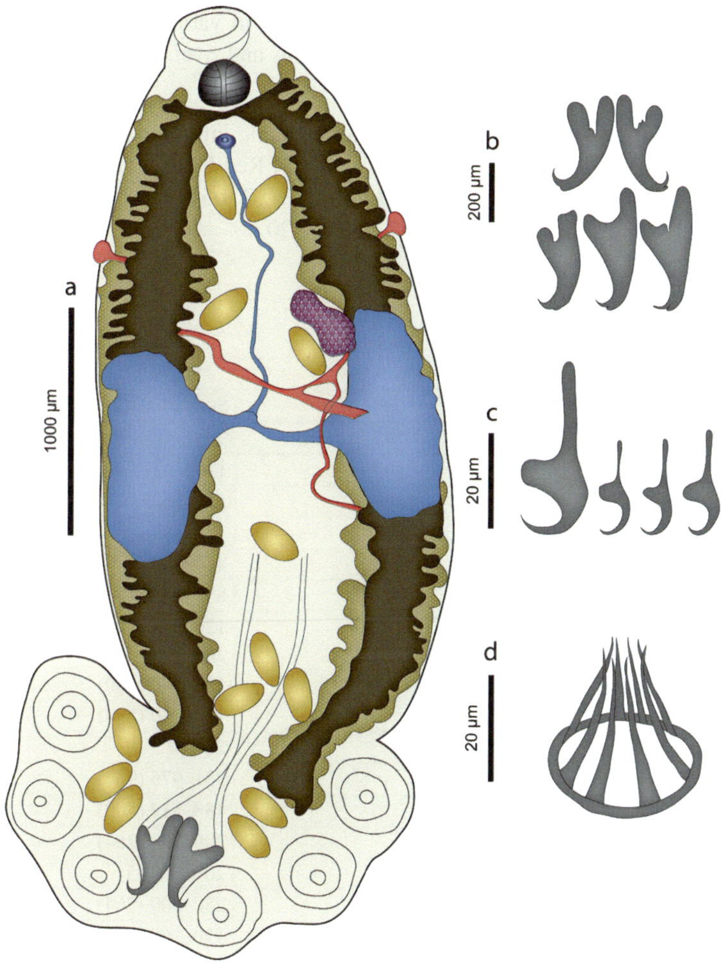

Fig. 4.11 (**a**) *Diplorchis lividae*, (**b**) hamuli, (**c**) marginal hooklets, (**d**) genital spines. Redrawn from Song et al. (2008)

Other taxonomic contributions None.

Etymology Named after the host.

Type locality Hainan Island, China.

Other localities None.

Type host *Odorrana livida* (Blyth, 1856), Tenasserim Odorfrog. In the original description of the polystome, the host was known as *Rana livida* Boulenger, 1887.

Additional hosts None.

Host geographical distribution *Odorrana livida* is native to Myanmar, however its distribution is unclear as the species may occur in various regions. The large-eared rock frog likely inhabits fast-flowing streams and rivers in tropical (montane) forest habitats (Bain 2004).

Host conservation status Due to a lack of available information, the Tenasserim Odorfrog is listed as Data Deficient by The IUCN Red List of Threatened Species 2004: e.T58645A11819303. https://doi.org/10.2305/IUCN.UK.2004.RLTS.T58645A11819303.en. Accessed on 27 January 2022.

Site on host Urinary bladder.

DNA sequence None.

Morphology and morphometrics (μm):	
Source	Song et al. (2008)
Sample size	9
Body length	1700–4175 (3294)
Greatest width	575–1625 (1122)
Width at vagina	
Haptor length	475–950 (647)
Haptor width	850–1850 (1364)
False oral sucker width	120–250 (173) × 150–380 (282) 200–360
Pharynx length	130–200 (160)
Pharynx width	130–320 (199)
Testis number	2
Testis length	951–976 (964)
Testis width	488–512 (500)
Ovary length	200–350 (264)
Ovary width	20–170 (95)
Egg length	250–300 (264)
Egg width	110–160 (136)
Egg incubation time	

(continued)

Morphology and morphometrics (μm):	
Source	Song et al. (2008)
Number of intra-uterine eggs	1–3
Genital bulb width	20–23 (22)
Number of genital spines	8–9
Genital spine length	22–25 (24)
Haptoral sucker diameter	200–360 (284)
Hamulus handle length X	250–400 (308)
Hamulus guard length Y	200–310 (251)
Hamulus hook length	
Marginal hooklet length	20–38 (29)
Haptoral L/Body L	0.196
No of anastomoses	0

Remarks None.

4.4.1.5 *Diplorchis nigromaculatus* Lee, 1936 (Fig. 4.12)

Collection Department of Biology, National University of Peking, Peiping, China.

Holotype Unknown.

Paratypes Unknown.

Original description Lee (1936).

Other taxonomic contributions None.

Etymology Named after the host.

Type locality Pekin, China.

Other localities None.

Type host *Pelophylax nigromaculatus* (Hallowell, 1861), Dark-Spotted Frog. In the original description of the polystome, the host was known as *Rana nigromaculata* Hallowel, 1861.

Additional hosts None.

Host geographical distribution *Pelophylax nigromaculatus* is present in China, Japan, Republic of Russian Federation, Korea and Democratic People's Republic of Korea where it inhabits meadows, forests, deserts and bush land as well as lakes, swamps, paddy fields and ponds (Kuzmin et al. 2004a).

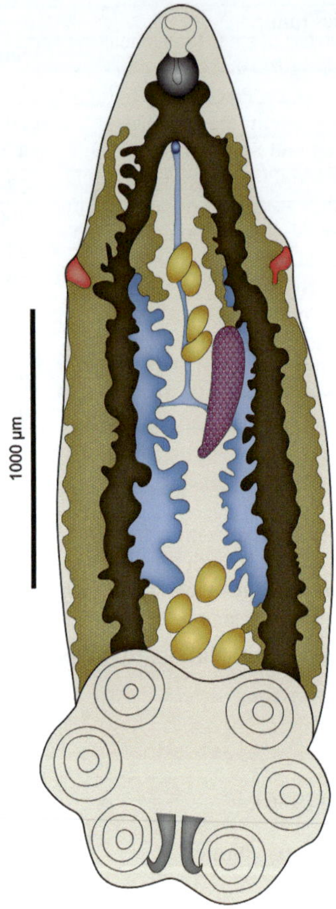

Fig. 4.12 *Diplorchis nigromaculatus*. Redrawn from Lee (1936)

Host conservation status Due to overexploitation for human consumption, the Dark-Spotted Frog is listed as Near Threatened by The IUCN Red List of Threatened Species 2004: e.T58679A11809026. https://doi.org/10.2305/IUCN.UK.2004.RLTS.T58679A11809026.en. Accessed on 27 January 2022.

Site on host Urinary bladder.

DNA sequence None.

Morphology and morphometrics (µm):	
Source	Lee (1936)
Sample size	174
Body length	3100
Greatest width	1000
Width at vagina	

(continued)

Morphology and morphometrics (μm):	
Source	Lee (1936)
Haptor length	1000
Haptor width	
False oral sucker width	
Pharynx length	
Pharynx width	
Testis number	2
Testis length	1200
Testis width	
Ovary length	300–430
Ovary width	
Egg length	
Egg width	16 × 90
Egg incubation time	
Intra-uterine eggs (n)	
Genital bulb width	
Number of genital spines	7–8
Genital spine length	
Haptoral sucker diameter	
Hamulus handle length X	
Hamulus guard length Y	
Hamulus hook length	
Marginal hooklet length	
Haptoral L/Body L	
No of anastomoses	0

Remarks None.

4.4.1.6 *Diplorchis ranae* Ozaki, 1931 (Fig. 4.13)

Collection Unknown.

Holotype Unknown.

Paratypes Unknown.

Original description Ozaki (1931).

Other taxonomic contributions Ozaki (1935).

Etymology Named after the host.

Type locality Japan.

Other localities None.

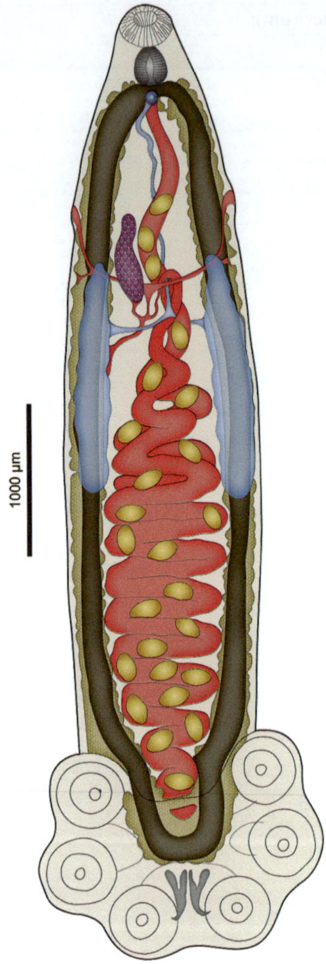

Fig. 4.13 *Diplorchis ranae*. Redrawn from Ozaki (1931)

Type host *Glandirana rugosa* (Temminck and Schlegel, 1838), Wrinkled Frog. In the original description of the polystome, the host was known as *Rana rugosa* Temminck and Schlegel, 1838.

Additional hosts None.

Host geographical distribution *Glandirana rugosa* is native to Japan. It is present in artificial ponds, rivers, wetlands and streams in several regions including Honshu, Sado Island, Kyushu and Hokkaido. This species has been introduced into Hawaii, USA (Frost 2021).

Host conservation status The Wrinkled Frog is quite common. It has a wide distribution and as a result is listed as Least Concern by The IUCN Red List of Threatened Species 2021: e.T58706A118983374. https://doi.org/10.2305/IUCN.UK.2021-1.RLTS.T58706A118983374.en. Accessed on 30 January 2022.

Site on host Urinary bladder.

DNA sequence
- 18S rRNA gene, 2100 bp. GenBank Accession: AM157184
- 28S rRNA gene, 1449 bp. GenBank Accession: AM157198
- 12S rRNA gene, 474 bp. GenBank Accession: KR856070
- Cytochrome c Oxidase I, 388 bp. GenBank Accession: JF699304

Morphology and morphometrics (μm):	
Source	Ozaki (1931)
Sample size	
Body length	4500–5400
Greatest width	750–1150
Width at vagina	
Haptor length	1500
Haptor width	1750
False oral sucker width	250–300
Pharynx length	160–200
Pharynx width	140–170
Testis number	2
Testis length	1200–1500
Testis width	170–230
Ovary length	470–500
Ovary width	130–160
Egg length	170–190
Egg width	80–100
Egg incubation time	
Intra-uterine eggs (n)	
Genital bulb width	50–80
Number of genital spines	10–11
Genital spine length	
Haptoral sucker diameter	300–390
Hamulus handle length X	200–320
Hamulus guard length Y	
Hamulus hook length	
Marginal hooklet length	25–35
Haptoral L/Body L	
No of anastomoses	0

Remarks None.

4.4.1.7 *Diplorchis shilinense* Fan, Pan and Wang, 2006 (Fig. 4.14)

Collection Unknown.

Holotype Unknown.

Paratypes Unknown.

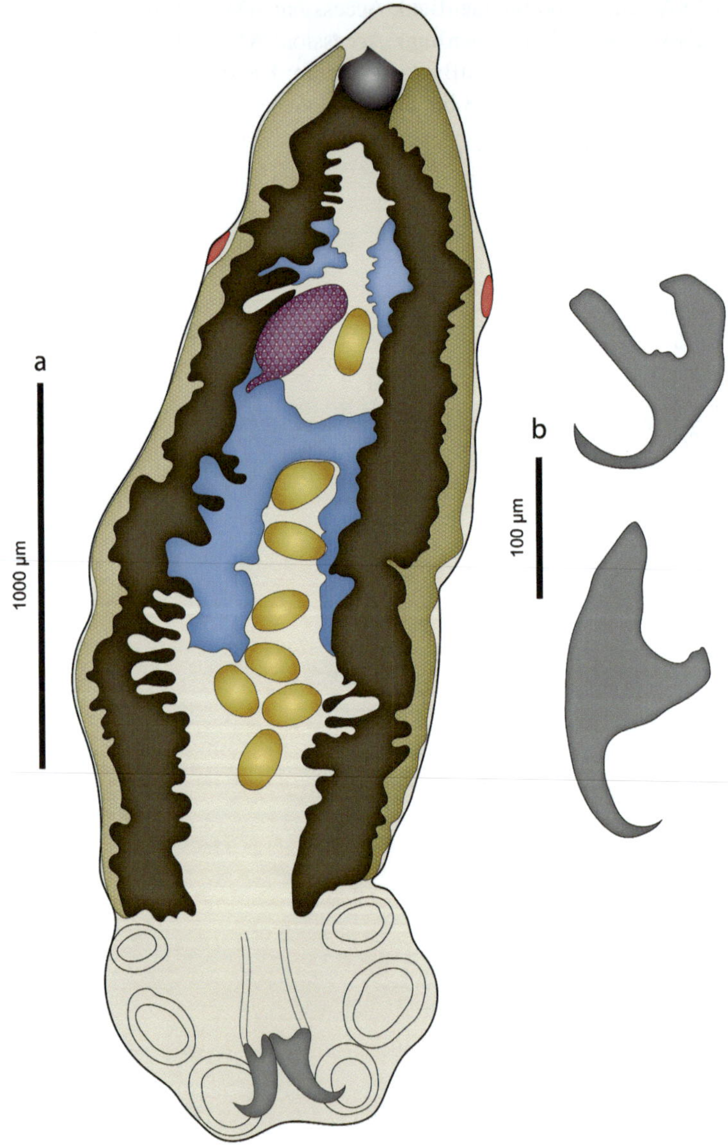

Fig. 4.14 (**a**) *Diplorchis shilinense*, (**b**) hamuli. Redrawn from Fan et al. (2006)

Original description Fan et al. (2006).

Taxonomic note The original name Dilporchis shilinensis was corrected for gender.

Other taxonomic contributions None.

Etymology Named after the type locality.

Type locality Shilin County, Yunnan Province, China.

Other localities None.

Type host *Nidirana pleuraden* (Boulenger, 1904), Yunnan Pond Frog. In the original description of the polystome, the host was known as *Rana pleuraden* Boulenger, 1904.

Additional hosts None.

Host geographical distribution *Nidirana pleuraden* is native to China and can be found in regions such as Yunnan provinces, southwestern Sichuan and western Guizhou. It inhabits ponds, paddy fields and ditches (Datong and Shunqing 2004a).

Host conservation status The Yunnan Pond Frog has a wide distribution range, a large population and is tolerant for a broad range of habitats. However, this species may decline in the future due to overexploitation as food source, habitat destruction caused by urbanization and water pollution. Currently it is still listed as Least Concern by The IUCN Red List of Threatened Species 2020: e. T58698A63867379. https://doi.org/10.2305/IUCN.UK.2020-1.RLTS. T58698A63867379.en. Accessed on 30 January 2022.

Site on host Urinary bladder.

DNA sequence

- 18S rRNA gene, 2067 bp. GenBank Accession: KR856123
- 18S rRNA gene, 1517 bp. GenBank Accession: EU979387
- 28S rRNA gene, 1449 bp. GenBank Accession: KR856141
- 12S rRNA gene, 493 bp. GenBank Accession: KR856071
- Cytochrome c Oxidase I, 395 bp. GenBank Accession: KR856162

Morphology and morphometrics (μm):	
Source	Fan et al. (2006)
Sample size	
Body length	2314–2971 (2643)
Greatest width	914–743 (943)
Width at vagina	714–743 (729)
Haptor length	486–629 (557)
Haptor width	800–886 (843)
False oral sucker width	
Pharynx length	143–159 (152)
Pharynx width	127 (127)
Testis number	2

(continued)

Morphology and morphometrics (μm):	
Source	Fan et al. (2006)
Ovary length	276–318 (297)
Ovary width	
Egg length	138–180 (165)
Egg width	74–106 (90)
Egg incubation time	
Number of intra-uterine eggs	
Genital bulb width	
Number of genital spines	
Genital spine length	
Haptoral sucker diameter	156–191 (173)
Hamulus handle length X	
Hamulus guard length Y	
Hamulus hook length	
Marginal hooklet length	
Haptoral L/Body L	0.298–0.35 (0.32)
No of anastomoses	0

Remarks *Diplorchis shilinense was originally described as Diplorchis shilinensis.*

4.4.2 *Eupolystoma* Kaw, 1950

4.4.2.1 *Eupolystoma alluaudi* (de Beauchamp, 1913) (Fig. 4.15)

Collection Unknown.

Holotype Unknown.

Paratypes Unknown.

Original description De Beauchamp (1913).

Other taxonomic contributions Beverley-Burton (1962); Euzet and Combes (1967); Combes et al. (1973); Bourgat and Salami (1973); Salami-Cadoux (1975); Bwathondi (1977); Combes et al. (1978); Tinsley (1978a); Bourgat et al. (1983); Du Preez et al. (2003); Aisien et al. (2004).

Etymology Named after Charles A. Alluaud (1861–1949), French explorer and entomologist.

Type locality Aberdare, Kenya.

Other localities Cameroon, Ethiopia, Senegal, Nigeria, Tanzania, Tchad, Togo and Zimbabwe.

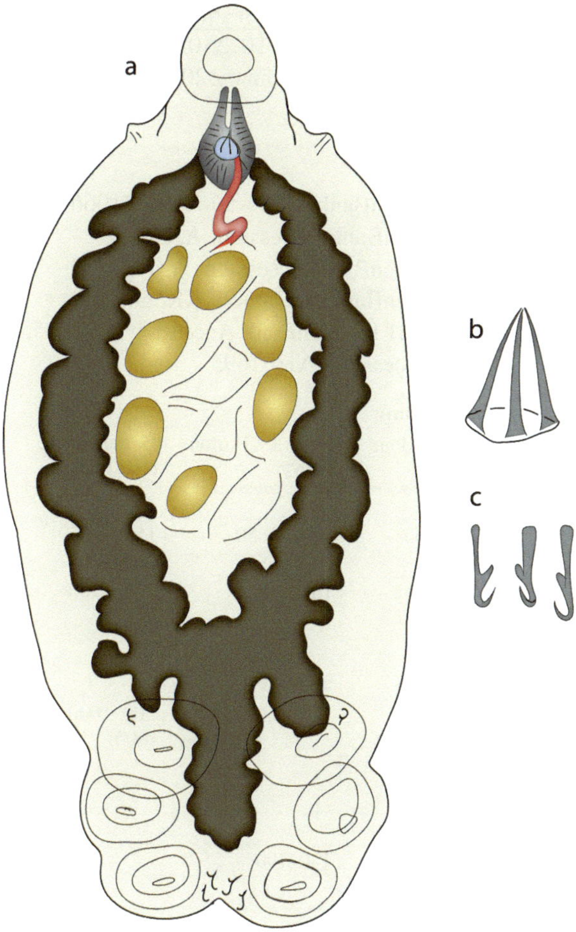

Fig. 4.15 (**a**) *Eupolystoma alluaudi*, (**b**) genital spines, (**c**) marginal hooklets. Redrawn from De Beauchamp (1913)

Type host *Sclerophrys regularis* (Reuss, 1833), Common African Toad. In the original description of the polystome, the host was known as *Bufo regularis* Reuss, 1833.

Additional hosts *Altiphrynoides malcolmi* (Grandison, 1978), *Pyxicephalus adspersus* (Tschudi, 1838), *Bufo* sp., *Sclerophrys pentoni* (Anderson, 1893), *Sclerophrys xeros* (Tandy, Tandy, Keith and Duff-MacKay, 1976), *Sclerophrys maculata* (Hallowell, 1854) and *Sclerophrys gutturalis* (Power, 1927).

Host geographical distribution *Sclerophrys regularis* is widely distributed. It occurs in various countries including among others Angola, Uganda, Egypt, Ethiopia, Kenya, Nigeria, Congo, Ghana and Sudan (Tandy et al. 2004).

Host conservation status The Common African Toad is widely distributed, has a tolerance for a wide range of habitats and has a presumed large population. It is listed as Least Concern by The IUCN Red List of Threatened Species 2016: e. T54747A107349840. https://doi.org/10.2305/IUCN.UK.2016-3.RLTS.T54747A107349840.en. Accessed on 30 January 2022.

Site on host Urinary bladder.

DNA sequence

- 18S rRNA gene, 2096 bp. GenBank Accession: AM051066
- 18S rRNA gene, 615 bp. GenBank Accession: AJ287991
- 28S rRNA gene, 1450 bp. GenBank Accession: AM157199
- 12S rRNA gene, 500 bp. GenBank Accession: KR856072
- Cytochrome c Oxidase I, 361 bp. GenBank Accession: FR667558
- ITS1, 279 bp. GenBank Accession: AJ301695

Morphology and morphometrics (μm):				
Source	Salami-Cadoux (1975)	Beverley-Burton (1962)	Bourgat et al. (1983)	Bourgat et al. (1983)
Sample size	200	3	21	121
Body length	1530–3100	2700–3850	1500–3500 (2900)	4000–4730
Greatest width	610–1520	1030–1850	700–1700 (1300)	
Width at vagina				
Haptor length	310–460	560–650		
Haptor width	460–1040	720–920	500–1000 (800)	1200–1400
False oral sucker width				
Pharynx length				
Pharynx width		140–190		
Ovary length				
Ovary width				
Egg length	125–160		140	121–162
Egg width	48–80		70	70–101
Egg incubation time				
Number of intra-uterine eggs				
Genital bulb width				
Number of genital spines	8–9			
Genital spine length	20			
Haptoral sucker diameter	140–210		140–239 (216)	330–381
Hamulus handle length X				
Hamulus guard length Y				
Hamulus hook length				

(continued)

Morphology and morphometrics (μm):				
Source	Salami-Cadoux (1975)	Beverley-Burton (1962)	Bourgat et al. (1983)	Bourgat et al. (1983)
Marginal hooklet length	17–23			
Haptoral L/Body L	0.15–0.20	0.17–0.21		
No of anastomoses	0			

Remarks *Eupolystoma alluaudi* was originally described as *Polystomum alluaudi* De Beauchamp, 1913. The original description by de Beauchamp (1913) did not provide body measurements except for a statement that the parasite was 2000 × 800 microns. Measurements reported by Bourgat et al. (1983) that are presented in columns 3 and 4 of the table are for two different host species, *S. xeros* and *S. pentoni*, respectively. The large spectrum of host species for *E. alluaudi* may be questioned and the polystome species could be a complex of cryptic species. This species has been reported from the same host as that of *Polystoma aethiopiense* Meskal, 1970 in Ethiopia and *Polystoma africanum* Szidat, 1932 in Togo, Liberia and Sierra Leone. *Sclerophrys gutturalis* also hosts *Polystoma mashoni* Beverley-Burton, 1962 in Zimbabwe.

4.4.2.2 *Eupolystoma anterorchis* Tinsley, 1978 (Fig. 4.16)

Collection Natural History Museum of London (NHNL).

Holotype NHML 1977.9.13.

Paratypes NHML 1977.9.1.4–11.

Original description Tinsley (1978a).

Other taxonomic contributions Tinsley (1978b).

Etymology Refers to the more forward position of the testis compared to other species in the genus.

Type locality Cape, South Africa.

Other localities None.

Type host *Sclerophrys pantherina* (Smith, 1828), Western Leopard Toad. In the original description of the polystome, the host was known as *Bufo pardalis* Poynton, 1964. However, according to Frost (2021), this species from the Cape is now considered as *Sclerophrys pantherina* instead of *Sclerophrys pardalis* (Hewitt, 1935).

Additional hosts None.

Host geographical distribution *Sclerophrys pantherina* only occupies a small geographical distribution range in the Western Cape Province of South Africa It

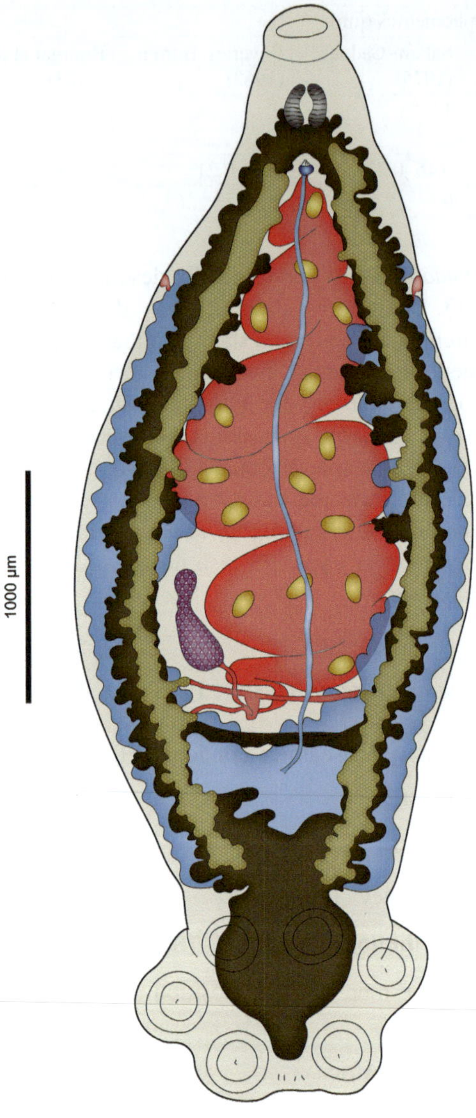

Fig. 4.16 *Eupolystoma anterorchis*. Redrawn from Tinsley (1978a)

inhabits gardens, farmlands and fynbos heathland habitats that are close to freshwater (Du Preez and Carruthers 2017).

Host conservation status The Western Leopard Toad has a limited distribution due to habitat fragmentation, urbanization, agricultural expansion and invasive species, such as *Sclerophrys gutturalis* which increases competition for resources and may result in hybridization. It is listed as Endangered by The IUCN Red List of

Threatened Species 2016: e.T54723A77159333. https://doi.org/10.2305/IUCN.UK. 2016-3.RLTS.T54723A77159333.en. Accessed on 30 January 2022.

Site on host Urinary bladder, kidney and kidney ducts.

DNA sequence None.

Morphology and morphometrics (μm):		
Source	Tinsley (1978a)	Tinsley (1978b)
Sample size	20	
Body length	3390–4760 (4320)	
Greatest width	1450–1860 (1720)	
Width at vagina	730–1210 (1110)	
Haptor length	550–830 (790)	
Haptor width	1070–1330 (1310)	
False oral sucker width	360–480 (440)	
Pharynx length	200–240 (230)	
Pharynx width	170–210 (210)	
Ovary length	360–560 (470)	
Ovary width	90–160 (120)	
Egg length	135–185	
Egg width	80–105	
Egg incubation time		
Number of intra-uterine eggs	300	
Genital bulb width		
Number of genital spines	8	
Genital spine length	24	
Haptoral sucker diameter	220–270 (260)	
Hamulus handle length X		
Hamulus guard length Y		
Hamulus hook length		
Marginal hooklet length	26	
Haptoral L/Body L		0.18
No of anastomoses	1	

Remarks None.

4.4.2.3 *Eupolystoma chauhani* Pandey, 1969 (Fig. 4.17)

Collection Unknown.

Holotype Unknown.

Paratypes Unknown.

Original description Pandey (1969).

Other taxonomic contributions None.

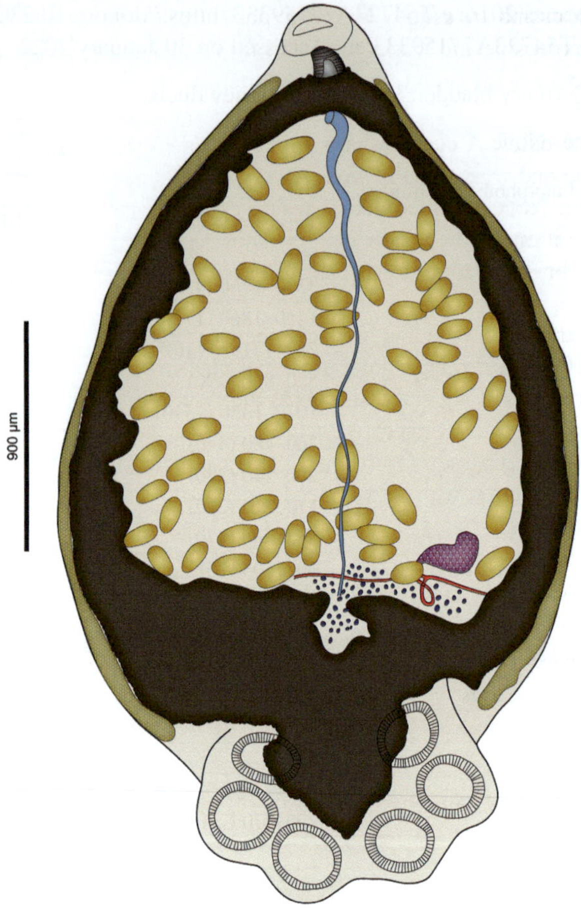

Fig. 4.17 *Eupolystoma chauhani*. Redrawn from Pandey (1969)

Etymology Named after BS Chauhan.

Type locality Gazipur, Lucknow, India.

Other localities None.

Type host *Bufo* sp.

Additional hosts None.

Host geographical distribution Unknown.

Host conservation status Unknown.

Site on host Urinary bladder.

DNA sequence None.

Morphology and morphometrics (μm):	
Source	Pandey (1969)
Sample size	2
Body length	3240–4250
Greatest width	1200–1750
Width at vagina	
Haptor length	1210–1350
Haptor width	1320–1450
False oral sucker width	210 × 300
Pharynx length	80–90
Pharynx width	60–70
Ovary length	
Ovary width	
Egg length	70–80
Egg width	20–30
Egg incubation time	
Number of intra-uterine eggs	
Genital bulb width	
Number of genital spines	6
Genital spine length	
Haptoral sucker diameter	120–150
Hamulus handle length X	
Hamulus guard length Y	
Hamulus hook length	
Marginal hooklet length	
Haptoral L/Body L	0.34
No of anastomoses	0

Remarks None.

4.4.2.4 *Eupolystoma namibiense* Du Preez, 2015 (Fig. 4.18)

Collection Parasitic Worm Collection, National Museum, Bloemfontein, South Africa (NMBP); Parasitic Worms Collection at the Natural History Museum, London (NHML).

Holotype NMBP 372.

Paratypes NMBP 373–379; NHML 2015.4.16.1–2.

Original description Du Preez (2015).

Taxonomic note The original name *Eupolystoma namibiensis* was corrected for gender.

Other taxonomic contributions None.

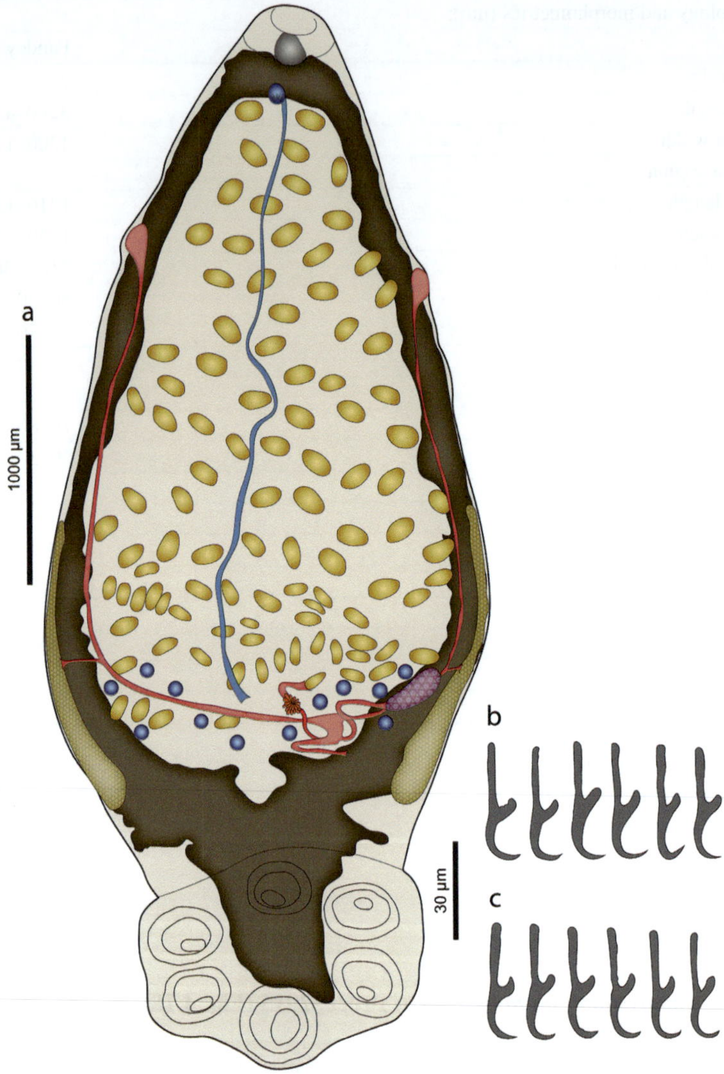

Fig. 4.18 (**a**) *Eupolystoma namibiense*, (**b**) marginal hooklets (1), (**c**) marginal hooklets (7–8). Redrawn from Du Preez (2015)

Etymology Named after the country of origin, Namibia.

Type locality Rocky pools below the dam wall of the Hardap dam near the town of Mariental, Namibia (24,49694S; 17,87194E).

Other localities None.

Type host *Poyntonophrynus hoeschi* (Ahl, 1934), Hoesch's Toad.

Additional hosts None.

Host geographical distribution *Poyntonophrynus hoeschi* is restricted to Namibia where it occurs in the central and western parts in association with rock outcrops (Frost 2021).

Host conservation status The Hoesch's Toad is listed as Least Concern by The IUCN Red List of Threatened Species 2013: e.T54663A18374358. https://doi.org/10.2305/IUCN.UK.2013-2.RLTS.T54663A18374358.en. Accessed on 30 January 2022.

Site on host Urinary bladder.

DNA sequence None.

Morphology and morphometrics (μm):	
Source	Du Preez (2015)
Sample size	
Body length	3477–6522 (4401)
Greatest width	826–2496 (1438)
Width at vagina	
Haptor length	575–1267 (876)
Haptor width	873–1572 (1091)
False oral sucker width	206–401 (315)
Pharynx length	147–228 (177)
Pharynx width	109–214 (143)
Ovary length	202–278 (244)
Ovary width	66–126 (95)
Egg length	128–183 (156)
Egg width	70–113 (96)
Egg incubation time	
Number of intra-uterine eggs	118
Genital bulb width	53–81 (68)
Number of genital spines	5
Genital spine length	32–38 (35)
Haptoral sucker diameter	223–436 (299)
Hamulus handle length X	
Hamulus guard length Y	
Hamulus hook length	
Marginal hooklet length	31–34 (32)
Haptoral L/Body L	0.165–0.194 (0.199)
No of anastomoses	0

Remarks *Eupolystoma namibiense* was originally described as *Eupolystoma namibiensis.*

4.4.2.5 *Eupolystoma rajai* Kaw, 1950 (Fig. 4.19)

Collection Dr. Thapar's Helminthological Collections, Lucknow University, India.

Holotype Unknown.

Paratypes Unknown.

Original description Kaw (1950).

Other taxonomic contributions None.

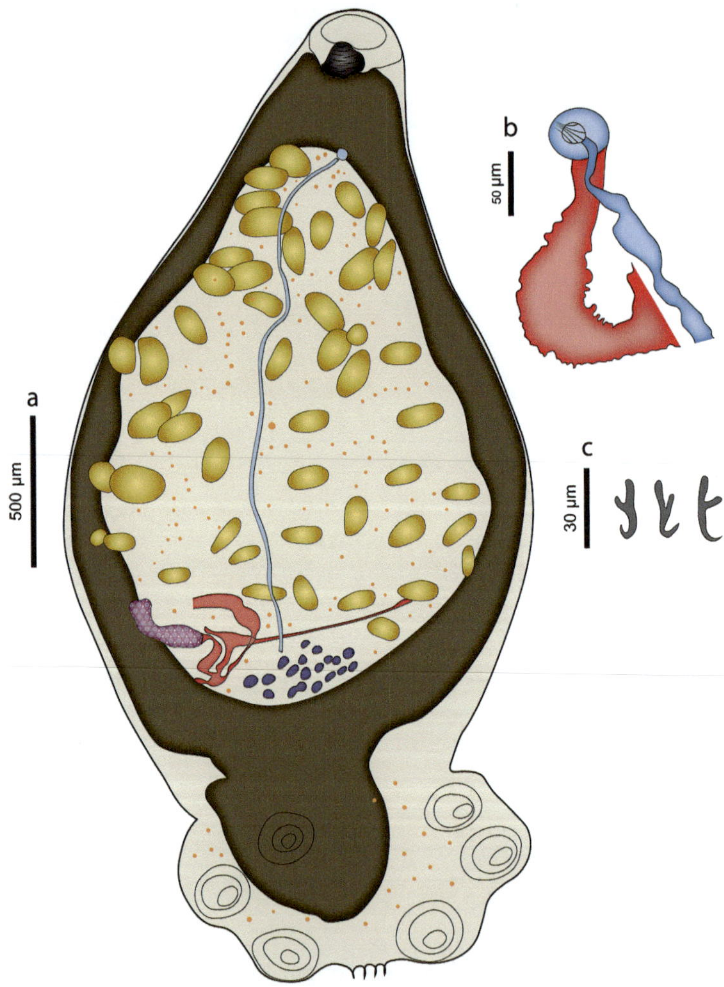

Fig. 4.19 (**a**) *Eupolystoma rajai*, (**b**) reproductive system, (**c**) marginal hooklets. Redrawn from Kaw (1950)

Etymology Named after Mr Rajai.

Type locality Punch, Kashmir, India.

Other localities None.

Type host *Rana* sp.

Additional hosts None.

Host geographical distribution Unknown.

Host conservation status Unknown.

Site on host Urinary bladder.

DNA sequence None.

Morphology and morphometrics (μm):	
Source	Kaw (1950)
Sample size	6
Body length	2400–3220
Greatest width	890–1400
Width at vagina	
Haptor length	580–660
Haptor width	940–1240
False oral sucker width	180–220 × 280–350
Pharynx length	120–170
Pharynx width	150
Ovary length	280
Ovary width	80
Egg length	157–170
Egg width	81–99
Egg incubation time	
Intra-uterine eggs (n)	30
Genital bulb width	44–54
Number of genital spines	8
Genital spine length	27
Haptoral sucker diameter	150–250
Hamulus handle length X	
Hamulus guard length Y	
Hamulus hook length	
Marginal hooklet length	22–27
Haptoral L/Body L	0.204–0.241
No of anastomoses	0

Remarks None.

4.4.2.6 *Eupolystoma vanasi* Du Preez, Tinsley and De Sá, 2003 (Fig. 4.20)

Collection Parasitic Worm Collection, National Museum, Bloemfontein, South Africa (NMBP); Parasitic Worm Collection, National History Museum, London (NHML).

Holotype NMBP 262.

Paratypes NMBP 263–270; NHML 2002.5.7.3–6.

Original description Du Preez et al. (2003).

Other taxonomic contributions None.

Etymology Named after Prof Jo van As, Free State University, South Africa.

Type locality Northern Province, South Africa.

Other localities None.

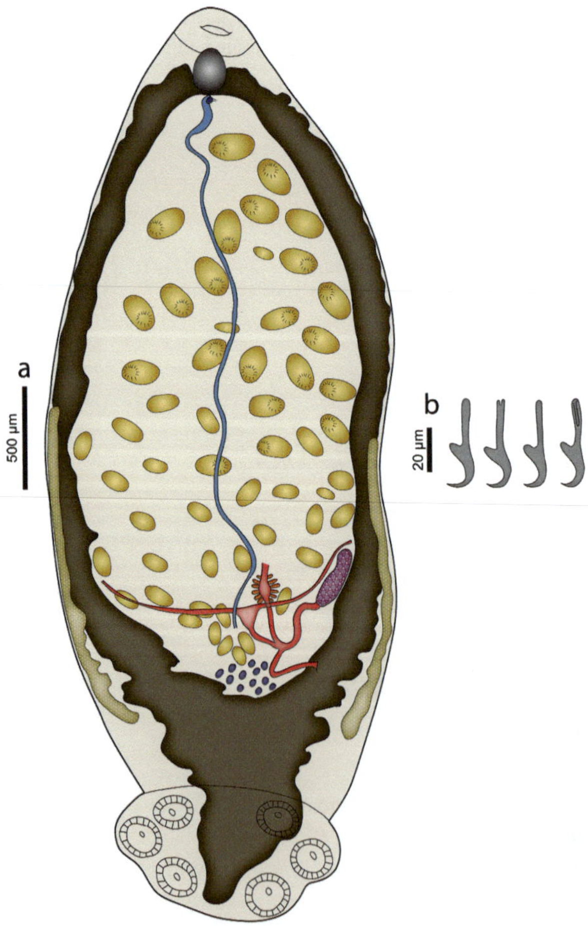

Fig. 4.20 (**a**) *Eupolystoma vanasi*, (**b**) marginal hooklets. Redrawn from Du Preez et al. (2003)

Type host *Schismaderma carens* (Smith, 1848), Red Toad.

Additional hosts None.

Host geographical distribution *Schismaderma carens* can be found in South Africa, Tanzania, Congo, Swaziland, Angola, Zimbabwe, Malawi, Botswana, Zambia, Mozambique, Kenya and Namibia (Frost 2021). It usually lives in grassland, around human settlements, livestock ranches, agricultural land and savannah habitats.

Host conservation status The Red Toad has a large population, a wide distribution and is tolerant for various habitats. It is listed as Least Concern by The IUCN Red List of Threatened Species 2013: e.T54885A3021025. https://doi.org/10.2305/IUCN.UK.20ti13-2.RLTS.T54885A3021025.en. Accessed on 30 January 2022.

Site on host Urinary bladder.

DNA sequence
- 18S rRNA gene, 2100 bp. GenBank Accession: AM157185
- 28S rRNA gene, 1445 bp. GenBank Accession: AM157200
- 12S rRNA gene, 495 bp. GenBank Accession: KR856073
- Cytochrome c Oxidase I, 379 bp. GenBank Accession: FR667559

Morphology and morphometrics (μm):	
Source	Du Preez et al. (2003)
Sample size	38
Body length	3775–7733 (5977)
Greatest width	1505–3275 (2356)
Width at vagina	
Haptor length	564–1128 (836)
Haptor width	1076–1924 (1447)
False oral sucker width	160–290 (222)
Pharynx length	165–270 (206)
Pharynx width	131–230 (166)
Ovary length	184–470 (250)
Ovary width	100–150 (121)
Egg length	136–150 (129)
Egg width	79–89 (83)
Egg incubation time	
Number of intra-uterine eggs	81
Genital bulb width	29.5–32.5 (31.8)
Number of genital spines	4
Genital spine length	29.5–32.5 (31.8)
Haptoral sucker diameter	160–285 (226)
Hamulus handle length X	
Hamulus guard length Y	
Hamulus hook length	
Marginal hooklet length	33.5–36.0 (34.9)

(continued)

Morphology and morphometrics (µm):	
Source	Du Preez et al. (2003)
Haptoral L/Body L	0.12–0.18 (0.14)
No of anastomoses	0

Remarks None.

4.4.3 Indopolystoma Chaabane, Verneau and Du Preez, 2019

4.4.3.1 *Indopolystoma carvirostris* (Fan, Li and He, 2008) (Fig. 4.21)

Collection School of Life Science of Yunnan Normal University, Kunming.

Holotype 200505C0801.

Paratypes 200505C0400, 200505C0701 & 200505C0802.

Original description Fan et al. (2008).

Other taxonomic contributions Shen et al. (2013).

Etymology Named after the host *Rhacophorus carvirostris* which was later pointed out to be a misidentified host (Chaabane et al. 2019).

Type locality Yunnan Province, China.

Other localities None.

Type host *Kurixalus bisacculus* (Taylor, 1962), Taylor's Rough-armed Treefrog or *Kurixalus verrucosus* (Boulenger, 1893), Boulenger's Bushfrog. In the original description of the polystome, the host was known as *Rhacophorus carvirostis* Boulenger 1982.

Additional hosts None.

Host geographical distribution *Kurixalus bisacculus* is known with certainty from north-central to south-central Thailand and higher elevations of the Cardamom Mountains in Cambodia. The form known from southeastern Yunnan and likely western Guangxi, China, south along the highlands through Vietnam to southern Laos may be an unnamed cryptic species. *Kurixalus verrucosus* is known from the highlands of central and northern Vietnam and southern China, through western and northern peninsular Thailand (Frost 2021).

Host conservation status The two host species are abundant species with a wide distribution. They are both listed as Least Concern by The IUCN Red List of Threatened Species 2004: e.T58982A11853973. https://doi.org/10.2305/IUCN.UK.2004.RLTS.T58982A11853973.en. Accessed on 30 January 2022.

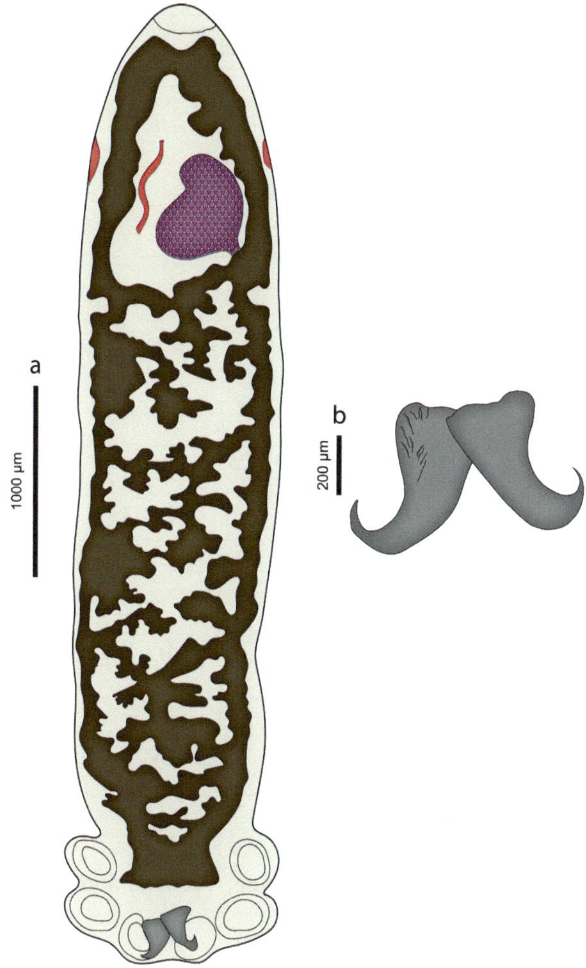

Fig. 4.21 (**a**) *Indopolystoma carvirostris*, (**b**) hamuli. Redrawn from Fan et al. (2008)

Site on host Urinary bladder.

DNA sequence

- 18S rRNA gene, 1517 bp. GenBank Accession: EU734835

Morphology and morphometrics (µm):		
Source	Fan et al. (2008)	Shen et al. (2013)
Sample size	10	
Body length	4030–7790 (5070)	
Greatest width	810–1400 (1130)	
Width at vagina	830	
Haptor length	500–750 (680)	
Haptor width	640–1500 (890)	

(continued)

Morphology and morphometrics (μm):		
Source	Fan et al. (2008)	Shen et al. (2013)
False oral sucker width		
Pharynx length		
Pharynx width		
Ovary length	497	
Ovary width		
Egg length		
Egg width		
Egg incubation time		
Number of intra-uterine eggs		
Genital bulb width		
Number of genital spines		8
Genital spine length		
Haptoral sucker diameter	212–339 (259)	
Hamulus handle length X		
Hamulus guard length Y		
Hamulus X/Y ratio	1.17	
Hamulus hook length		
Marginal hooklet length		
Haptoral L/Body L	0.096–0.124 (0.134)	
No of anastomoses	>1	

Remarks *Indopolystoma carvirostris* was originally described as *Polystoma carvirostris* Fan, Li and He, 2008.

4.4.3.2 *Indopolystoma elongatum* Chaabane, Verneau and Du Preez, 2019 (Fig. 4.22)

Collection Parasite Collection, National Museum of Natural History, Paris, France (MNHN).

Holotype MNHN 1184.

Paratypes MNHN 1185–1186.

Original description Chaabane et al. (2019).

Other taxonomic contributions None.

Etymology Refers to the elongated body of the species.

Type locality Upstream of Kunigami-gun, city of Nago, Okinawa Prefecture, Japan.

Other localities Sado Island, Niigata Prefecture, Japan.

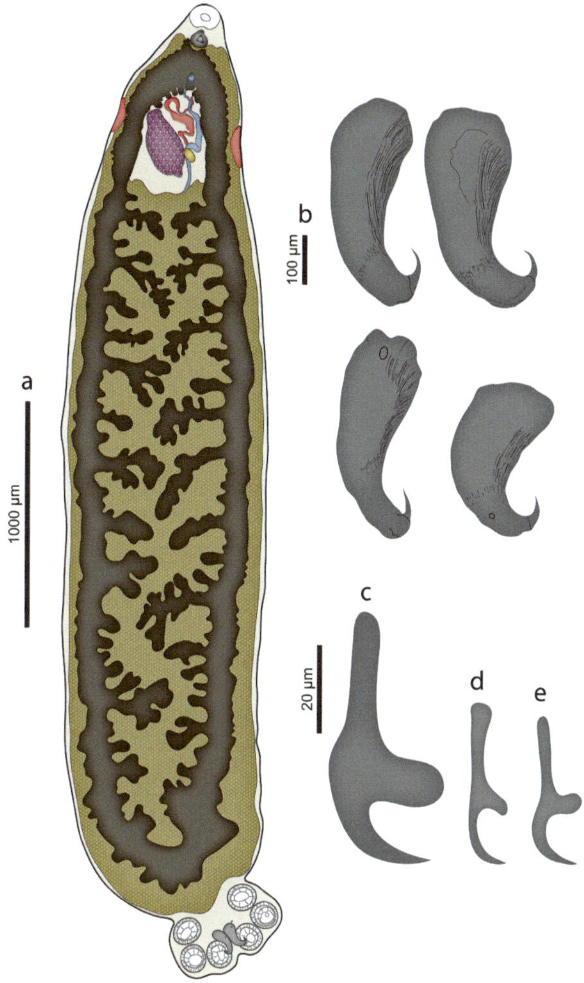

Fig. 4.22 (**a**) *Indopolystoma elongatum*, (**b**) hamuli, (**c**) marginal hooklets (1), (**d**) marginal hooklets (3), (**e**) marginal hooklets (8). Redrawn from Chaabane et al. (2019)

Type host *Zhangixalus arboreus* (Okada and Kawano, 1924), Kinugasa Flying Frog. In the original description of the polystome, the host was known as *Rhacophorus arboreus* Dubois, 1987.

Additional hosts None.

Host geographical distribution *Zhangixalus arboreus* is known from Honshu Island, Japan (Frost 2021).

Host conservation status The Kinugasa Flying Frog is widespread and listed as Least Concern by The IUCN Red List of Threatened Species 2021: e.

T58973A177226592. https://doi.org/10.2305/IUCN.UK.2021-1.RLTS.T58973A177226592.en. Accessed on 30 January 2022.

Site on host Urinary bladder.

DNA sequence

- 18S rRNA gene, 2098 bp. GenBank Accession: AM157190
- 28S rRNA gene, 1452 bp. GenBank Accession: AM157213
- 12S rRNA gene, 483 bp. GenBank Accession: KR856094
- Cytochrome c Oxidase I, 395 bp. GenBank Accession: KR856170

Morphology and morphometrics (μm):	
Source	Chaabane et al. (2019)
Sample size	3
Body length	12,847–14,878 (14,791)
Greatest width	23,171–3270 (3182)
Width at vagina	1902–2031 (1967)
Haptor length	734–1098 (916)
Haptor width	1512–1810 (1661)
False oral sucker width	528–573 (551)
Pharynx length	227–247 (237)
Pharynx width	256–268 (259)
Ovary length	1062–1108 (1072)
Ovary width	514–563 (520)
Egg length	224–256 (240)
Egg width	106–130 (118)
Egg incubation time	
Number of intra-uterine eggs	1
Genital bulb width	129–211 (139)
Number of genital spines	8
Genital spine length	38–44 (41)
Haptoral sucker diameter	380–459 (420)
Hamulus handle length X	303–419 (407)
Hamulus guard length Y	
Hamulus hook length	72–85 (78.5)
Marginal hooklet length	36
Haptoral L/Body L	0.05–0.07 (0.06)
No of anastomoses	>1

Remarks None.

4.4.3.3 *Indopolystoma hakgalense* (Crusz and Ching, 1975) (Fig. 4.23)

Collection Unknown.

Holotype RTS 39.

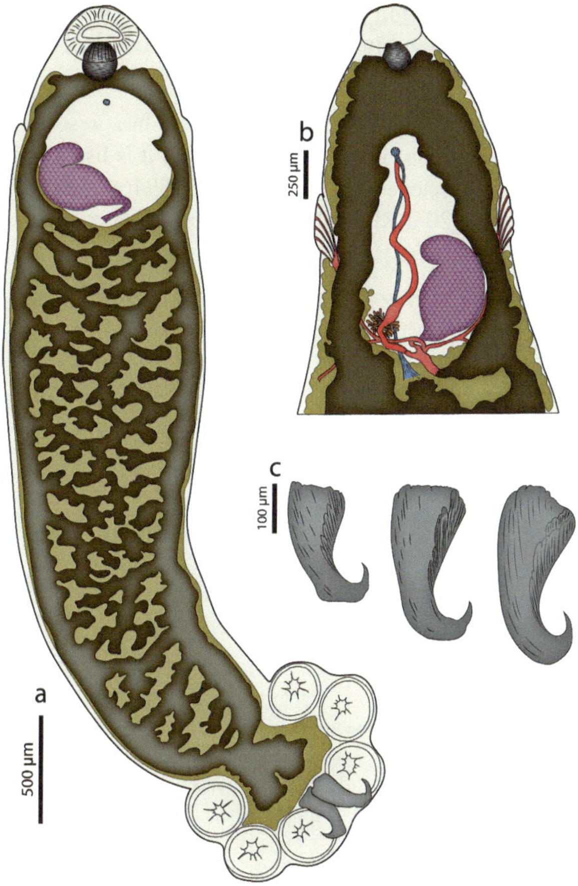

Fig. 4.23 (**a**) *Indopolystoma hakgalense*, (**b**) reproductive system, (**c**) hamuli. Redrawn from Crusz and Ching (1975)

Paratypes Unknown.

Original description Crusz and Ching (1975).

Other taxonomic contributions None.

Etymology Named after the type locality.

Type locality Hakgala, Ceylon.

Other localities None.

Type host *Taruga eques* (Günther, 1858), Saddled Tree Frog. In the original description of the polystome, the host was known as *Rhacophorus cruciger eques* Kirtisinghe, 1957.

Additional hosts None.

Host geographical distribution *Taruga eques* occurs in Sri Lanka where it inhabits tree canopies and tree trunks in montane tropical, moist forests and grass on the edge of ponds (Frost 2021).

Host conservation status The Saddled Tree Frog has a severely fragmented distribution with degradation of its habitat. As a result it is listed as Endangered by The IUCN Red List of Threatened Species 2020: e.T58946A156586528. https://doi.org/10.2305/IUCN.UK.2020-3.RLTS.T58946A156586528.en. Accessed on 30 January 2022.

Site on host Urinary bladder.

DNA sequence None.

Morphology and morphometrics (μm):	
Source	Crusz and Ching (1975)
Sample size	12
Body length	3870–9230
Greatest width	890–1790
Width at vagina	
Haptor length	635–794
Haptor width	1016–1509
False oral sucker width	189–402
Pharynx length	
Pharynx width	135–197
Ovary length	492–874
Ovary width	207–349
Egg length	
Egg width	
Egg incubation time	
Number of intra-uterine eggs	
Genital bulb width	
Number of genital spines	
Genital spine length	
Haptoral sucker diameter	250–320
Hamulus handle length X	380–440
Hamulus guard length Y	
Hamulus hook length	
Marginal hooklet length	20.5–36.9
Haptoral L/Body L	0.09
No of anastomoses	Network

Remarks *Indopolystoma hakgalense* was originally described as *Polystoma hakgalense* Crusz and Ching, 1975.

4.4.3.4 *Indopolystoma indicum* (Diengdoh and Tandon, 1991) (Fig. 4.24)

Collection Eastern Regional Station, Zoological Survey of India (IV/ERS/ZSI) Helminthological collection of Department of Zoology, NEHU, Shillong, India (NEHU).

Holotype IV/ERS/ZSI 304.

Paratypes IV/ERS/ZSI 305; NEHU/Z-MA/1.

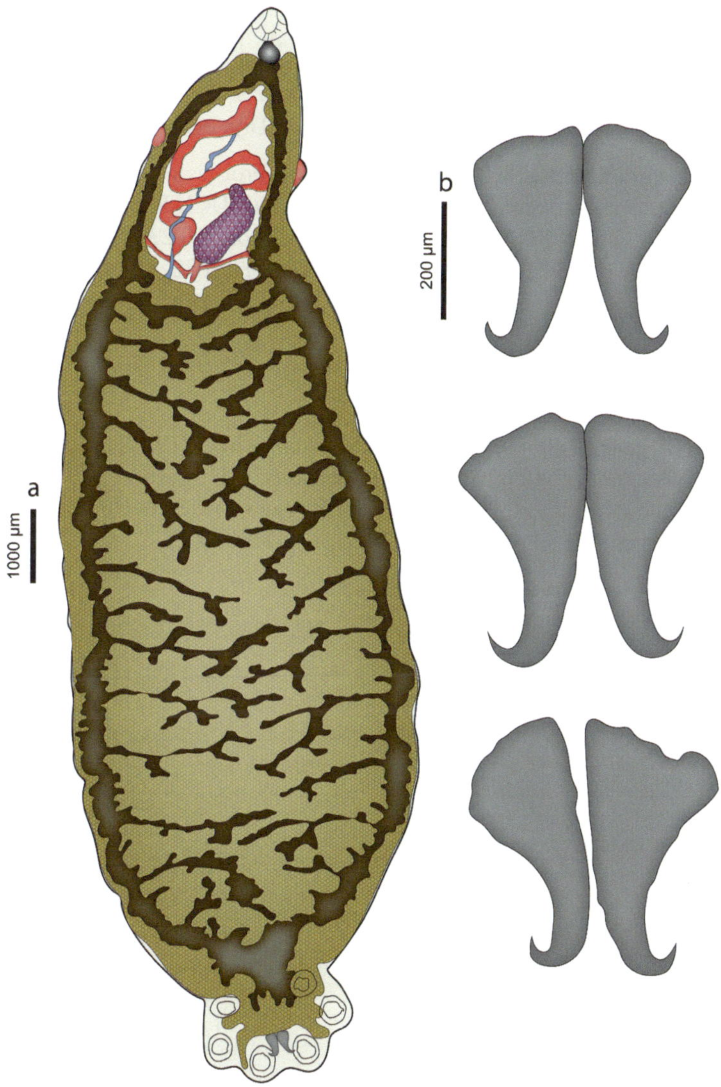

Fig. 4.24 (**a**) *Indopolystoma indicum*, (**b**) hamuli. Redrawn from Diengdoh and Tandon (1991)

Original description Diengdoh and Tandon (1991).

Other taxonomic contributions None.

Etymology Named after the type locality country.

Type locality Cherrapunji (25°18' N: 91°46′ E), East Khasi, Hills District, Meghalaya state, India.

Other localities None.

Type host *Rhacophorus nigropalmatus* Boulenger, 1895, Wallace's Flying Frog.

Additional hosts *Zhangixalus smaragdinus* (Blyth, 1850).

Host geographical distribution *Rhacophorus nigropalmatus* occurs in Thailand, Indonesia, Brunei, Darussalam and Malaysia (Van Dijk et al. 2004a).

Host conservation status The Wallace's Flying Frog as a large population and wide distribution and is listed as Least Concern by The IUCN Red List of Threatened Species 2004: e.T59008A11866922. https://doi.org/10.2305/IUCN.UK.2004.RLTS.T59008A11866922.en. Accessed on 30 January 2022.

Site on host Urinary bladder.

DNA sequence The sequences below were reported from *I. indicum* infecting *Z. smaragdinus* in India.

- 18S rRNA gene, 2088 bp. GenBank Accession: AM157193
- 28S rRNA gene, 1454 bp. GenBank Accession: AM157216
- Cytochrome c Oxidase I, 396 bp. GenBank Accession: JF699303
- 12S rRNA gene, 483 bp. GenBank Accession: KR856085

Morphology and morphometrics (μm):	
Source	Diengdoh and Tandon (1991)
Sample size	7
Body length	6732–16,614 (12,556)
Greatest width	2222–5641 (4164)
Width at vagina	1197–2622 (2057)
Haptor length	977–1710 (1298)
Haptor width	1584–2466 (2104)
False oral sucker width	186–532 (390) × 236–574 (466)
Pharynx length	164–395 (291)
Pharynx width	164–349 (249)
Ovary length	504–1368 (1049)
Ovary width	288–630 (499)
Egg length	190–214 (205)
Egg width	125–140 (132)
Egg incubation time	
Number of intra-uterine eggs	40

(continued)

Morphology and morphometrics (μm):	
Source	Diengdoh and Tandon (1991)
Genital bulb width	
Number of genital spines	8
Genital spine length	38–40 (39)
Haptoral sucker diameter	264–465 (353)
Hamulus handle length X	178–414 (320)
Hamulus guard length Y	
Hamulus hook length	
Marginal hooklet length	
Haptoral L/Body L	0.1
No of anastomoses	1–2

Remarks *Indopolystoma indicum* was originally described as *Polystoma indicum* Diengdoh and Tandon, 1991.

4.4.3.5 *Indopolystoma leucomystax* (Shu-Yi and Suo, 1987) (Fig. 4.25)

Collection Parasitology Research Laboratory, Biological Department, East China Normal University.

Holotype Unknown.

Paratypes Unknown.

Original description Shu-Yi and Suo L (1987).

Other taxonomic contributions None.

Etymology Named after the host.

Type locality Hangzhou, Zhejiang Province, China.

Other localities None.

Type host *Polypedates leucomystax* (Gravenhorst, 1829), Common Tree Frog. In the original description of the polystome, the host was known as *Rhacophorus leucomystax* Boulenger, 1889.

Additional hosts None.

Host geographical distribution *Polypedates leucomystax* occurs in Bangladesh, Indonesia, China, India, Vietnam, Philippines, Thailand, Nepal, Malaysia and Singapore. It has also been introduced to Japan. This species usually inhabits human habitats including gardens, agricultural areas, artificial ponds, lakes and ditches (Diesmos et al. 2004).

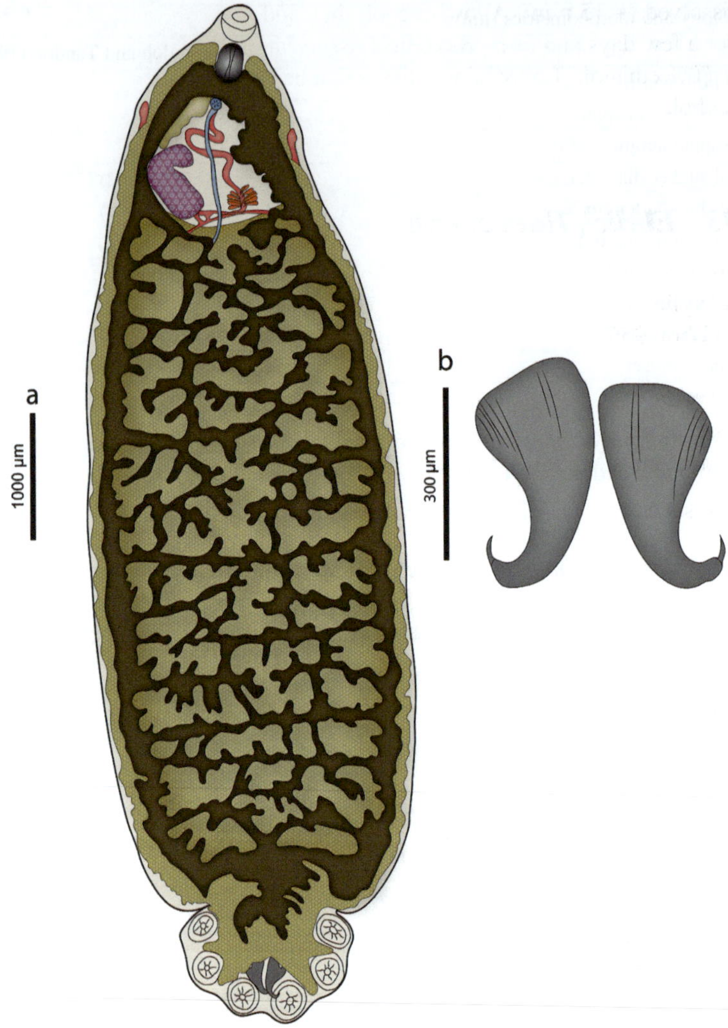

Fig. 4.25 (**a**) *Indopolystoma leucomystax*, (**b**) hamuli. Redrawn from Zhang and Long (1987)

Host conservation status The Common Tree Frog has a large population, a wide distribution and is tolerant for various habitats. It is listed as Least Concern by The IUCN Red List of Threatened Species 2004: e.T58953A86477485. https://doi.org/10.2305/IUCN.UK.2004.RLTS.T58953A11861409.en. Accessed on 30 January 2022.

Site on host Urinary bladder.

DNA sequence None.

Morphology and morphometrics (μm):	
Source	Shu-Yi and Suo (1987)
Sample size	5
Body length	6665–8514 (7568)
Greatest width	2542–3311 (2732)
Width at vagina	
Haptor length	753–989 (824)
Haptor width	1355–2040 (1693)
False oral sucker width	164–364 × 194–424 (261 × 392)
Pharynx length	0.219–0.286 (252)
Pharynx width	194–246 (216)
Ovary length	451–683 (608)
Ovary width	191–294 (259)
Egg length	
Egg width	
Egg incubation time	
Number of intra-uterine eggs	
Genital bulb width	75–89 × 69–76
Number of genital spines	8
Genital spine length	
Haptoral sucker diameter	207–356 (289)
Hamulus handle length X	260–410 (332)
Hamulus guard length Y	
Hamulus hook length	
Marginal hooklet length	
Haptoral L/Body L	0.11
No of anastomoses	Network

Remarks *Indopolystoma leucomystax* was originally described as *Polystoma leucomystax* Shu-Yi and Suo, 1987.

4.4.3.6 *Indopolystoma mutus* (Meng, Song and Ding, 2010) (Fig. 4.26)

Collection Lab of Fish Parasitology, College of Life Sciences, South China Normal University (HNJFL).

Holotype HNJFL 20070401–1.

Paratypes HNJFL 20070401–2.

Original description Meng et al. (2010).

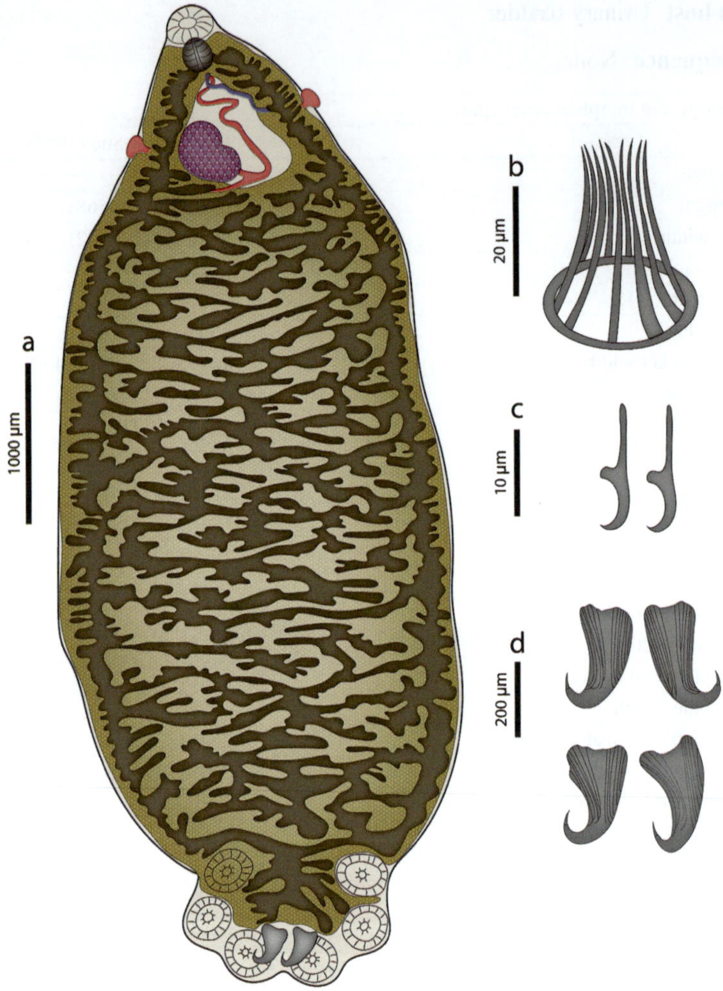

Fig. 4.26 (**a**) *Indopolystoma mutus*, (**b**) genital spines, (**c**) marginal hooklets, (**d**) hamuli. Redrawn from Meng et al. (2010)

Other taxonomic contributions Shen et al. (2013).

Etymology Named after the host.

Type locality Jianfengling, Hainan Island, China.

Other localities None.

Type host *Polypedates mutus* (Smith, 1940), Burmese Whipping Frog.

Additional hosts None.

Taxonomy None.

Host geographical distribution *Polypedates mutus* occurs in northern Vietnam, northern Myanmar as well as in southern and southwestern China where it can be found in Guangdong, Yunnan, Guangxi, Hainan and Guizhou Provinces. This species is also suspected to occur in northern Thailand. It inhabits forest habitats (van Dijk et al. 2004b).

Host conservation status The Burmese Whipping Frog has a large population size, wide distribution range and tolerance of habitat modification. This species may be threatened by deforestation and degradation (van Dijk et al. 2004b). It is thus listed as Least Concern by The IUCN Red List of Threatened Species 2004: e. T58958A11862355. https://doi.org/10.2305/IUCN.UK.2004.RLTS. T58958A11862355.en. Accessed on 30 January 2022.

Site on host Urinary bladder.

DNA sequence None.

Morphology and morphometrics (μm):		
Source	Meng et al. (2010)	Shen et al. (2013)
Sample size		
Body length	6875–7800 (7338)	7338
Greatest width	2550–3575 (3063)	3063
Width at vagina		
Haptor length	875–950 (913)	913
Haptor width	1425–1875 (1650)	1650
False oral sucker width	175–300 (238) × 275–500 (388)	310
Pharynx length	210–280 (245)	245
Pharynx width	200–270 (235)	235
Ovary length	550–625 (588)	588
Ovary width	182–333 (256)	
Egg length		
Egg width		
Egg incubation time		
Number of intra-uterine eggs		
Genital bulb width		
Number of genital spines	9	
Genital spine length	38–40 (39)	
Haptoral sucker diameter	270–350 (310)	
Hamulus handle length X	330–350 (340)	
Hamulus guard length Y	280–300 (290)	
Hamulus hook length		
Marginal hooklet length	14–16 (15)	
Haptoral L/Body L	0.127–0.121 (0.124)	
No of anastomoses	Network	

Remarks *Indopolystoma mutus* was originally described as *Polystoma mutus* Meng, Song and Ding, 2010.

4.4.3.7 *Indopolystoma parvum* Chaabane, Verneau and Du Preez, 2019 (Fig. 4.27)

Collection Parasite Collection, National Museum of Natural History, Paris, France (MNHN).

Holotype MNHN 1187.

Paratypes MNHN 1188.

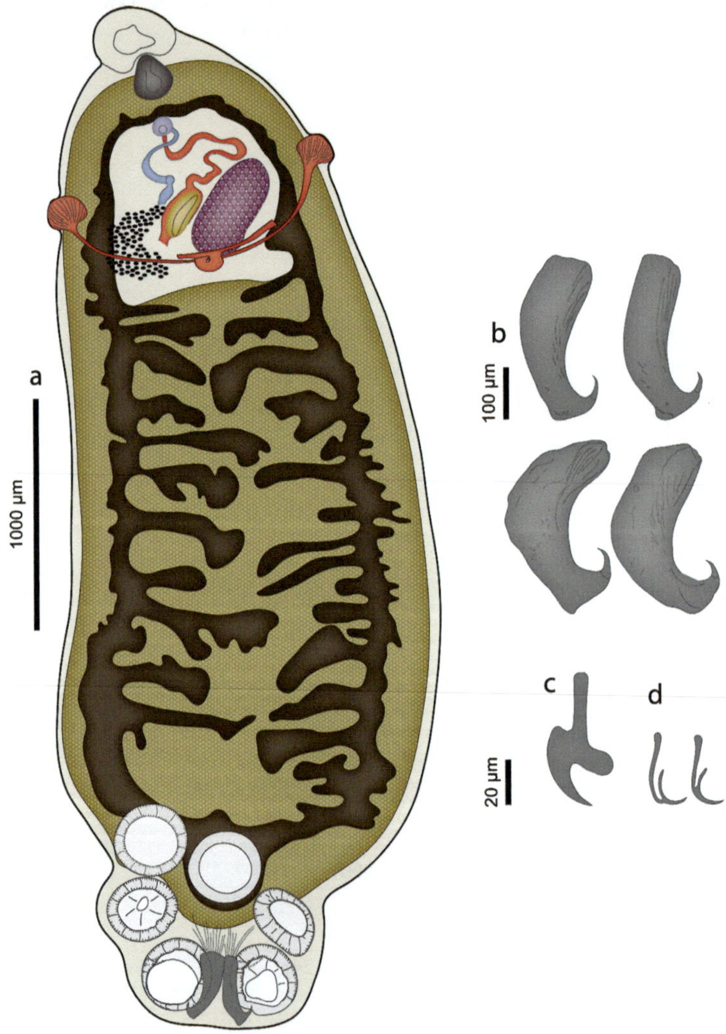

Fig. 4.27 (**a**) *Indopolystoma parvum*, (**b**) hamuli, (**c**) marginal hooklets (1), (**d**) marginal hooklets (2–7). Redrawn from Chaabane et al. (2019)

Original description Chaabane et al. (2019).

Other taxonomic contributions None.

Etymology Refers to the small body size.

Type locality Wawu Shan, Hongya Xian, Sichuan province, China.

Other localities None.

Type host *Zhangixalus omeimontis* (Stejneger, 1924), Omei Treefrog. In the original description of the polystome, the host was known as *Rhacophorus omeimontis* Liu, 1950.

Additional hosts None.

Host geographical distribution *Zhangixalus omeimontis* is found from Sichuan and southeastern Yunnan eastward to Hubei and Guangxi, China (Frost 2021).

Host conservation status The Omei Treefrog is listed as Least Concern by The IUCN Red List of Threatened Species 2020: e.T58961A63884080. https://doi.org/10.2305/IUCN.UK.2020-1.RLTS.T58961A63884080.en. Accessed on 30 January 2022.

Site on host Urinary bladder.

DNA sequence

- 18S rRNA gene, 2088 bp. GenBank Accession: AM157189
- 28S rRNA gene, 1454 bp. GenBank Accession: AM157212
- Cytochrome c Oxidase I, 396 bp. GenBank Accession: KR856093
- 12S rRNA gene, 483 bp. GenBank Accession: KR856169

Morphology and morphometrics (μm):	
Source	Chaabane et al. (2019)
Sample size	2
Body length	4536–4891 (4714)
Greatest width	1701–2130 (1916)
Width at vagina	1279–1525 (1402)
Haptor length	584–749 (667)
Haptor width	1023–1494 (1258)
False oral sucker width	358–431 (395)
Pharynx length	187–194 (191)
Pharynx width	174–205 (190)
Ovary length	483–606 (545)
Ovary width	231–306 (269)
Egg length	219–222 (221)
Egg width	93–113 |(103)
Egg incubation time	
Number of intra-uterine eggs	1
Genital bulb width	75–90 (83)

(continued)

Morphology and morphometrics (μm):	
Source	Chaabane et al. (2019)
Number of genital spines	8
Genital spine length	16–18 (17)
Haptoral sucker diameter	281–389 (335)
Hamulus handle length X	311–340 (326)
Hamulus guard length Y	
Hamulus hook length	39–64 (52)
Marginal hooklet length	32
Haptoral L/Body L	0.12–0.17 (0.14
No of anastomoses	1

Remarks None.

4.4.3.8 *Indopolystoma pingbianense* (Fan, Wang and Li, 2004) (Fig. 4.28)

Collection Unknown.

Holotype Unknown.

Paratypes Unknown.

Original description Fan et al. (2004).

Taxonomic note The name *Indopolystoma pingbianensis* was corrected for gender.

Other taxonomic contributions None.

Etymology Named after the type locality.

Type locality Pingxian County, Yunnan Province, China.

Other localities None.

Type host *Zhangixalus dugritei* (David, 1872), Baoxing Tree Frog. In the original description of the polystome, the host was known as *Polypedates dugritei* David, 1872.

Additional hosts None.

Host geographical distribution *Zhangixalus dugritei* occurs in southwestern and central China in Yunnan and Sichuan Province where it inhabits montane forest habitats (Lau et al. 2004b).

Host conservation status The Baoxing Tree Frog has a large population, a wide distribution range and is tolerant for habitat modification. It is listed as Least Concern by The IUCN Red List of Threatened Species 2004: e. T58945A11860286. https://doi.org/10.2305/IUCN.UK.2004.RLTS.T58945A11860286.en. Accessed on 30 January 2022.

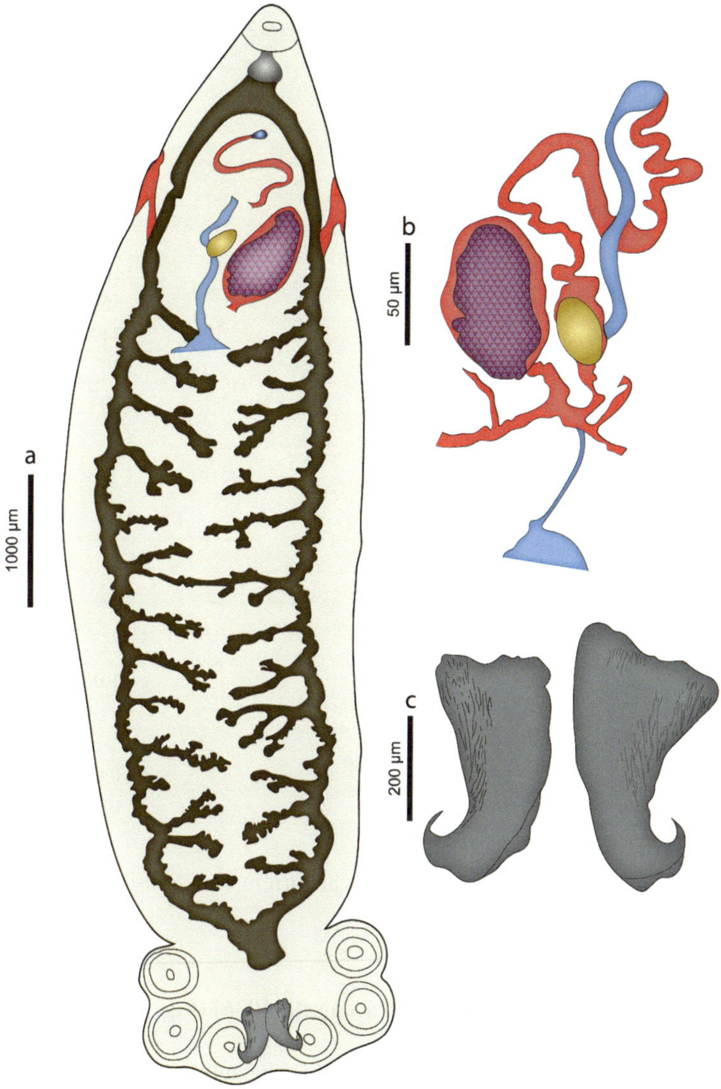

Fig. 4.28 (**a**) *Indopolystoma pingbianense*, (**b**) reproductive system, (**c**) hamuli. Redrawn from Fan et al. (2004)

Site on host Urinary bladder.

DNA sequence None.

Morphology and morphometrics (µm):	
Source	Fan et al. (2004)
Sample size	
Body length	6000–12,000 (9428)

(continued)

Morphology and morphometrics (μm):	
Source	Fan et al. (2004)
Greatest width	1640–2390 (1990)
Width at vagina	1572
Haptor length	424–1253 (869)
Haptor width	415–1374 (784)
False oral sucker width	
Pharynx length	218–328 (253)
Pharynx width	189–252 (229)
Ovary length	453–927 (712)
Ovary width	349–491 (434)
Egg length	307
Egg width	165
Egg incubation time	
Number of intra-uterine eggs	
Genital bulb width	
Number of genital spines	
Genital spine length	
Haptoral sucker diameter	434
Hamulus handle length X	
Hamulus guard length Y	
Hamulus X/Y ratio	1.06
Hamulus hook length	
Marginal hooklet length	
Haptoral L/Body L	0.07–0.104 (0.092)
No of anastomoses	0–1

Remarks *Indopolystoma pingbianense* was originally described as *Polystoma pingbianensis* Fan, Wang and Li, 2004.

4.4.3.9 *Indopolystoma rhacophori* (Yamaguti, 1936) (Fig. 4.29)

Collection The type and paratypes are stored in the author's private collection.

Holotype Unknown.

Paratypes Unknown.

Original description Yamaguti (1936).

Other taxonomic contributions None.

Etymology Named after the host.

Type locality Kyoto, Japan.

Other localities None.

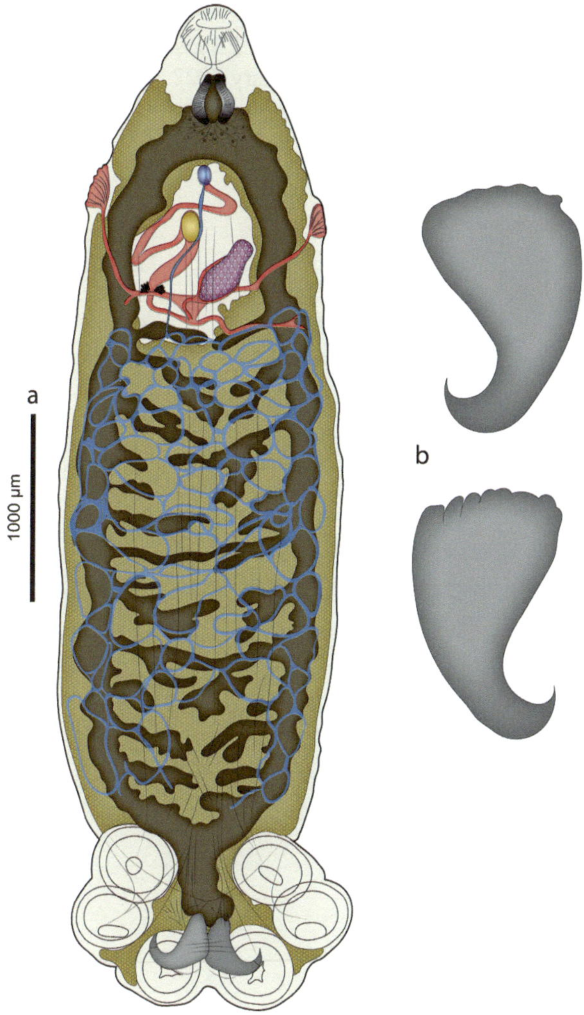

Fig. 4.29 (**a**) *Indopolystoma rhacophori*, (**b**) hamuli. Redrawn from Yamaguti (1936)

Type host *Zhangixalus schlegelii* (Günther, 1858), Schlegel's Green Tree Frog. In the original description of the polystome, the host was known as *Rhacophorus schlegelii* Boulenger, 1882.

Additional hosts None.

Host geographical distribution *Zhangixalus schlegelii* is native to Japan where it is found in Kyushu, Honshu and Shikoku. This species inhabits wetlands and is particularly common in paddy fields (Kaneko and Matsui 2004b).

Host conservation status The Schlegel's Green Tree Frog has a certain tolerance for habitat modification, has a presumed large population and is widely distributed. It is listed as Least Concern by The IUCN Red List of Threatened Species 2020: e. T58961A63884080. https://doi.org/10.2305/IUCN.UK.2020-1.RLTS.T58961A63884080.en. Accessed on 30 January 2022.

Site on host Urinary bladder.

DNA sequence None.

Morphology and morphometrics (μm):	
Source	Yamaguti (1936)
Sample size	
Body length	4500–6000
Greatest width	1000–1900
Width at vagina	
Haptor length	
Haptor width	
False oral sucker width	800–1100 × 1100–1600
Pharynx length	188–240
Pharynx width	188–220
Ovary length	420–630
Ovary width	210–340
Egg length	238–300
Egg width	140–163
Egg incubation time	
Number of intra-uterine eggs	1–10
Genital bulb width	
Number of genital spines	8
Genital spine length	42
Haptoral sucker diameter	320–400
Hamulus handle length X	350–420
Hamulus guard length Y	
Hamulus hook length	
Marginal hooklet length	24–42
Haptoral L/Body L	
No of anastomoses	1–2

Remarks *Indopolystoma rhacophori* was originally described as *Polystoma rhacophori* Yamaguti, 1936.

4.4.3.10 *Indopolystoma viridi* Chaabane, Verneau and Du Preez, 2019 (Fig. 4.30)

Collection Parasite Collection, National Museum of Natural History, Paris (MNHN); Parasitic Worm Collection, National Museum, Bloemfontein, South Africa (NMBP).

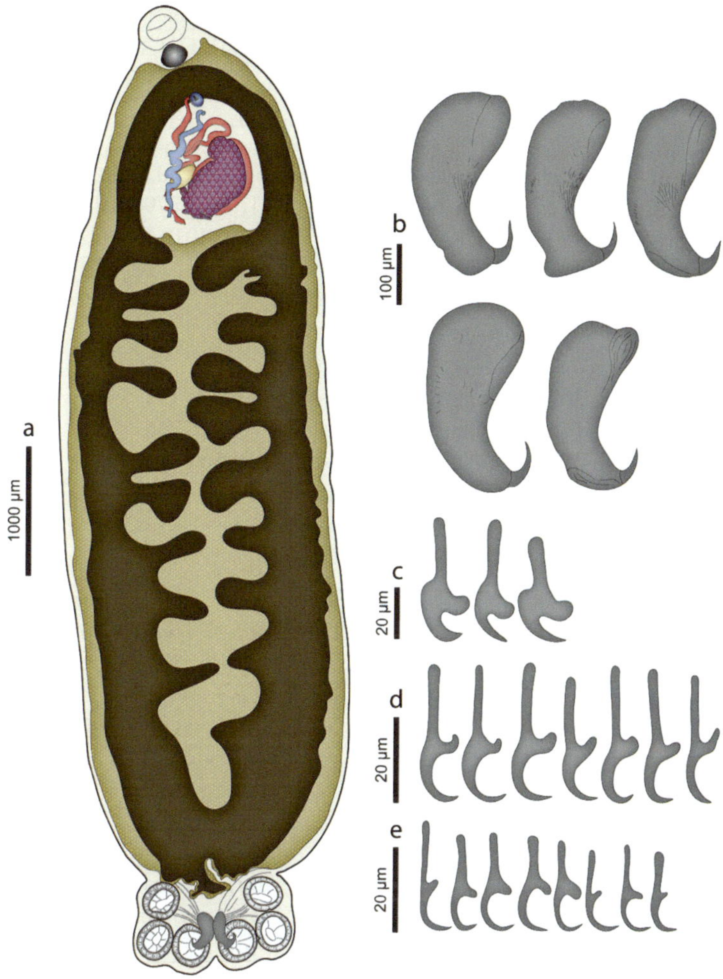

Fig. 4.30 (**a**) *Indopolystoma viridi*, (**b**) hamuli, (**c**) marginal hooklets (1), (**d**) marginal hooklets (8), (**e**) marginal hooklets(2–7). Redrawn from Chaabane et al. (2019)

Holotype NHN HEL1173.

Paratypes MNHN HEL1174–HEL1183, NMBP 512–P515.

Original description Chaabane et al. (2019).

Other taxonomic contributions None.

Etymology Named after the host.

Type locality Tokunoshima Island, Kagoshima Prefecture, Japan.

Other localities None.

Type host *Zhangixalus viridis* (Hallowell, 1861), Okinawa Green Tree Frog.

Additional hosts None.

Host geographical distribution *Zhangixalus viridis* is known from the Okinawajima, Iheyajima and Kumejima Islands of the Okinawa Island Group, Japan (Frost 2021).

Conservation status The Okinawa Green Tree Frog is listed as Least Concern by The IUCN Red List of Threatened Species 2021: e.T187984019A177225773. https://doi.org/10.2305/IUCN.UK.2021-1.RLTS.T187984019A177225773.en. Accessed on 30 January 2022.

Site on host Urinary bladder.

DNA sequence

- 18S rRNA gene, 2098 bp. GenBank Accession: AM157191
- 28S rRNA gene, 1452 bp. GenBank Accession: AM157214
- 12S rRNA gene, 491 bp. GenBank Accession: KR856095
- Cytochrome c Oxidase I, 371 bp. GenBank Accession: MN564839

Morphology and morphometrics (μm):	
Source	Chaabane et al. (2019)
Sample size	15
Body length	5532–11,907 (85,500
Greatest width	1534–2859 (2112
Width at vagina	1175–1800 (1481)
Haptor length	526–1354 (773)
Haptor width	588–1592 (1184)
False oral sucker width	266–465 (394)
Pharynx length	124–268 (210)
Pharynx width	164–244 (209)
Ovary length	577–925 (775)
Ovary width	286–536 (401)
Egg length	191–268 (214)
Egg width	74–165 (125)
Egg incubation time	
Number of intra-uterine eggs	1
Genital bulb width	67–124 (105)
Number of genital spines	8–9
Genital spine length	27–49 (40)
Haptoral sucker diameter	242–423 (333)
Hamulus handle length X	276–373 (313)
Hamulus guard length Y	
Hamulus hook length	48–74 (66)
Marginal hooklet length	C1: 31–44 (40); C2–8: 16–31 (21.5)
Haptoral L/Body L	0.05–0.17 (0.09)
No of anastomoses	0

Remarks None.

4.4.3.11 *Indopolystoma zuoi* (Shen, Wang and Fan, 2013) (Fig. 4.31)

Collection Life Sciences Department, Yunnan Normal University, Department of Zoology (Pal).

Holotype Pal2011052601.

Paratypes Pal2011052602–03.

Original description Shen et al. (2013).

Other taxonomic contributions None.

Etymology Named after Prof Zuo Yangxian, Yunnan University, China.

Type locality Pingbian County, Yunnan Province, China.

Other localities None.

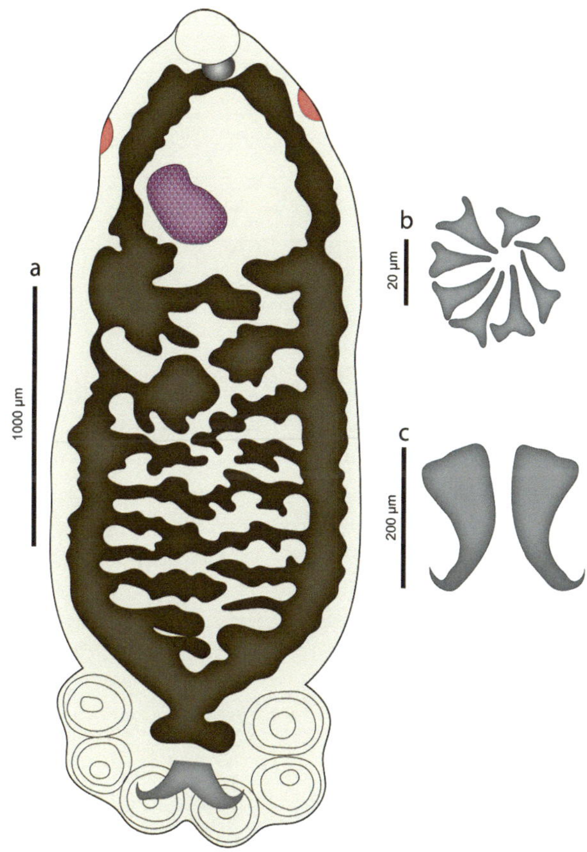

Fig. 4.31 (**a**) *Indopolystoma zuoi*, (**b**) genital spines, (**c**) hamuli. Redrawn from Shen et al. (2013)

Type host *Feihyla palpebralis* (Smith, 1924), Vietnamese Bubble-Nest Frog. In the original description of the polystome, the host was known as *Philautus palpebralis* Smith, 1924.

Additional hosts None.

Host geographical distribution *Feihyla palpebralis* is native to Vietnam (Tam Dao and Lang Bian) and China (Yunnan) where it inhabits forests (van Dijk et al. 2004c).

Host conservation status The Vietnamese Bubble-Nest Frog has a fragmented distribution due to water pollution, agricultural activities and forest degradation. It is listed as Near Threatened and close to being listed as Vulnerable by The IUCN Red List of Threatened Species 2004: e.T58792A11830377. https://doi.org/10.2305/IUCN.UK.2004.RLTS.T58792A11830377.en. Accessed on 30 January 2022.

Site on host Urinary bladder.

DNA sequence

- 18S rRNA gene, 1096 bp. GenBank Accession: KF850147

Morphology and morphometrics (μm):	
Source	Shen et al. (2013)
Sample size	
Body length	1624–3533 (2716)
Greatest width	1202–1904 (1280)
Width at vagina	
Haptor length	632–924 (701)
Haptor width	831–1317 (1063)
False oral sucker width	
Pharynx length	164
Pharynx width	151
Ovary length	329–400 (369)
Ovary width	
Egg length	
Egg width	
Egg incubation time	
Number of intra-uterine eggs	
Genital bulb width	
Number of genital spines	8
Genital spine length	
Haptoral sucker diameter	209–311 (251)
Hamulus handle length X	185–307 (240)
Hamulus guard length Y	173–265 (205)
Hamulus hook length	
Marginal hooklet length	
Haptoral L/Body L	0.26
No of anastomoses	2

Remarks *Indopolystoma zuoi* was originally described as *Polystoma zuoi* Shen, Wang and Fan, 2013.

4.4.4 *Kankana* Raharivololoniaina, Verneau, Berthier, Vences and Du Preez, 2011

4.4.4.1 *Kankana manampoka* Raharivololoniaina, Verneau, Berthier, Vences and Du Preez, 2011 (Fig. 4.32)

Collection Parasitic Worm Collection, National Museum, Bloemfontein 9300, South Africa (NMBP); Parasitic Worms Collection, Natural History Museum, London (NHML).

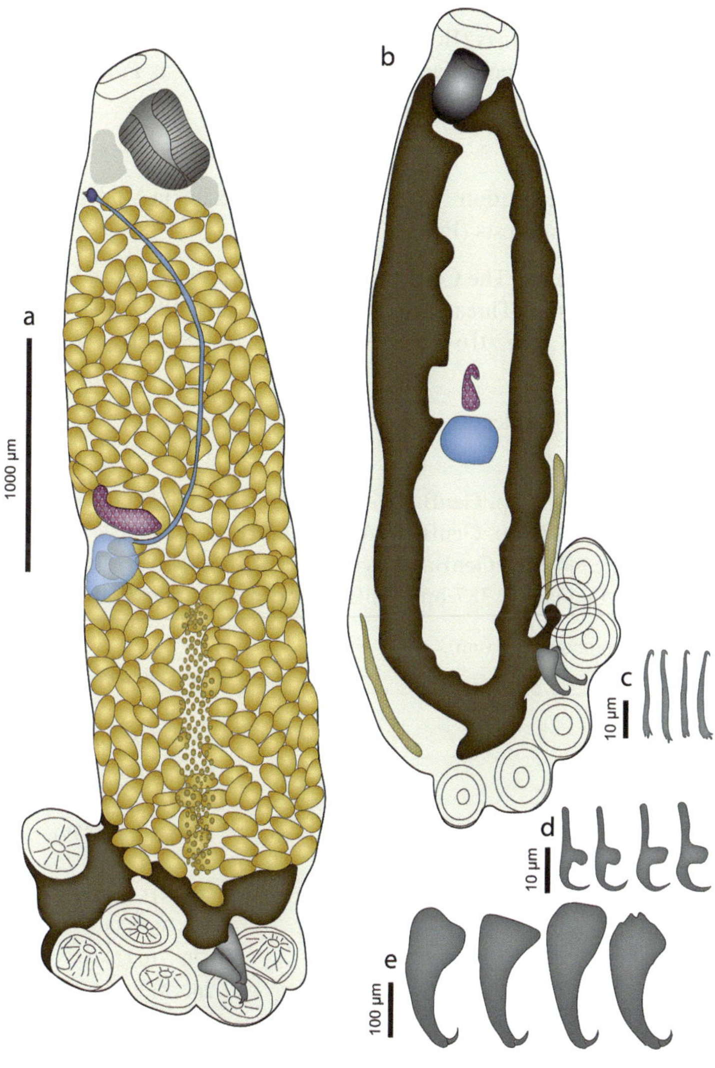

Fig. 4.32 (**a**) *Kankana manampoka* (Holoype), (**b**) Paratype, (**c**) genital spines, (**d**) marginal hooklets, (**e**) hamuli. Redrawn from Raharivololoniaina et al. (2011)

Holotype NMBP 323.

Paratypes NMBP 324–326; NHML 2010.8.27.3.

Original description Raharivololoniaina et al. (2011).

Other taxonomic contributions None.

Etymology Named after the words kankana that means "worm" in Malagasy and manampoka that means "unexpected".

Type locality Ranomafana National Park, Madagascar.

Other localities None.

Type host *Cophyla pollicaris* (Boulenger, 1888), Common Giant Treefrog. In the original description of the polystome, the host was known as *Platypelis pollicaris* Boulenger, 1888.

Additional hosts None.

Host geographical distribution *Cophyla pollicaris* occurs in Madagascar where it is found in arboreal rainforests (Frost 2021).

Host conservation status The Common Giant Treefrog is listed as Least Concern by The IUCN Red List of Threatened Species 2016: e.T57961A67477873. https://doi.org/10.2305/IUCN.UK.2016-1.RLTS.T57961A67477873.en. Accessed on 30 January 2022.

Site on host Urinary bladder.

DNA sequence

- 18S rRNA gene, 2087 bp. GenBank Accession: HM854292
- 28S rRNA gene, 1421 bp. GenBank Accession: HM854293
- 12S rRNA gene, 440 bp. GenBank Accession: KR856074
- Cytochrome c Oxidase I, 387 bp. GenBank Accession: JF699307

Morphology and morphometrics (μm):	
Source	Raharivololoniaina et al. (2011)
Sample size	3
Body length	3672–4488 (4092)
Greatest width	1002–1074 (1035)
Width at vagina	
Haptor length	1342–1670 (1479)
Haptor width	735–999 (883)
False oral sucker width	305–315 (310)
Pharynx length	372–398 (389)
Pharynx width	270–310 (294)
Ovary length	260–313 (286)
Ovary width	95–128 (112)
Egg length	146–164 (157)

(continued)

Morphology and morphometrics (μm):	
Source	Raharivololoniaina et al. (2011)
Egg width	81–86 (83)
Egg incubation time	
Number of intra-uterine eggs	231
Genital bulb width	46–48 (47)
Number of genital spines	8
Genital spine length	20–23 (22)
Haptoral sucker diameter	247–338 (289)
Hamulus handle length X	251–268 (260)
Hamulus guard length Y	
Hamulus hook length	31–36 (34)
Marginal hooklet length	16–19 (18)
Haptoral L/Body L	0.372 (0.365)
No of anastomoses	0

Remarks None.

4.4.5 *Madapolystoma* Du Preez, Raharivololoniaina, Verneau and Vences, 2010

4.4.5.1 *Madapolystoma biritika* Du Preez, Raharivololoniaina, Verneau and Vences, 2010 (Fig. 4.33)

Collection Parasitic Worm Collection, National Museum, Bloemfontein, South Africa (NMBP).

Holotype NMBP 318.

Paratypes NMBP 319–22.

Original description Du Preez et al. (2010).

Other taxonomic contributions None.

Etymology Named after the Malagasy word "biritika" meaning extremely small.

Type locality Madagascar.

Other localities None.

Type host *Mantella madagascariensis* (Grandidier, 1872), Madagascar Golden Frog.

Additional hosts *Mantella baroni* Boulenger 1888 and *Mantella milotympanum* Staniszewski 1996.

Fig. 4.33 (**a**) *Madapolystoma biritika* (Holoype), (**b**) Paratype, (**c**) marginal hooklets (2–7), (**d**) marginal hooklets (2–7), (**e**) marginal hooklets(8), (**f**) hamuli. Redrawn from Du Preez et al. (2010)

Host geographical distribution *Mantella madagascariensis* can be found in forested areas of East-central Madagascar in upland locations (700–1050 m elevation) (Frost 2021).

Host conservation status The Madagascar Golden Frog has a very limited extent of occurrence and is known from less than 10 locations. It is therefore listed as Vulnerable by The IUCN Red List of Threatened Species 2017: e. T57446A84167257. https://doi.org/10.2305/IUCN.UK.2017-2.RLTS. T57446A84167257.en. Accessed on 30 January 2022.

Site on host Urinary bladder.

DNA sequence

- 18S rRNA gene, 2077 bp. GenBank Accession: FM897295 (*M. baroni*)
- 28S rRNA gene, 1470 bp. GenBank Accession: FM897278 (*M. baroni*)
- Cytochrome c Oxidase I, 396 bp. GenBank Accessions: FR667564 to FR667567, JN015510 (*M. baroni*)
- Cytochrome c Oxidase I, 331 bp. GenBank Accession: FR667568 (*M. baroni*)
- Cytochrome c Oxidase I, 331 bp. GenBank Accessions: FR667569 and FR667570 (*M. madagascareniensis*)
- Cytochrome c Oxidase I, 397 bp. GenBank Accession: FR667571 (*M. madagascareniensis*)
- Cytochrome c Oxidase I, 397 bp. GenBank Accession: FR667572 (*M. mylotympanum*)

Morphology and morphometrics (μm):

Source	Du Preez et al. (2010)
Sample size	5
Body length	1304–3041 (2041)
Greatest width	420–597 (512)
Width at vagina	369–548 (481)
Haptor length	594–788 (679)
Haptor width	707–1046 (882)
False oral sucker width	95–206 (160)
Pharynx length	117–164 (136)
Pharynx width	105–125 (119)
Ovary length	36–85 (61)
Ovary width	18–34 (26)
Egg length	
Egg width	
Egg incubation time	
Number of intra-uterine eggs	
Genital bulb width	18–24 (20)
Number of genital spines	8
Genital spine length	10.5–10.9 (10.7)
Haptoral sucker diameter	160–255 (200)
Hamulus handle length X	190–238 (215)
Hamulus guard length Y	175–223 (199)
Hamulus hook length	34–50 (45)
Marginal hooklet length	21–26 (24)
Haptoral L/Body L	0.33
No of anastomoses	0

Remarks None.

4.4.5.2 *Madapolystoma cryptica* Berthier, Du Preez, Raharivololoniaina, Vences and Verneau, 2014 (Fig. 4.34)

Collection Parasitic Worm Collection, National Museum, Bloemfontein, South Africa (NMBP).

Holotype NMBP 327.

Paratypes NMBP 328–329.

Original description Berthier et al. (2014).

Other taxonomic contributions None.

Etymology Referring to the cryptic morphology of the species.

Type locality Tsaratanana, Madagascar.

Other localities None.

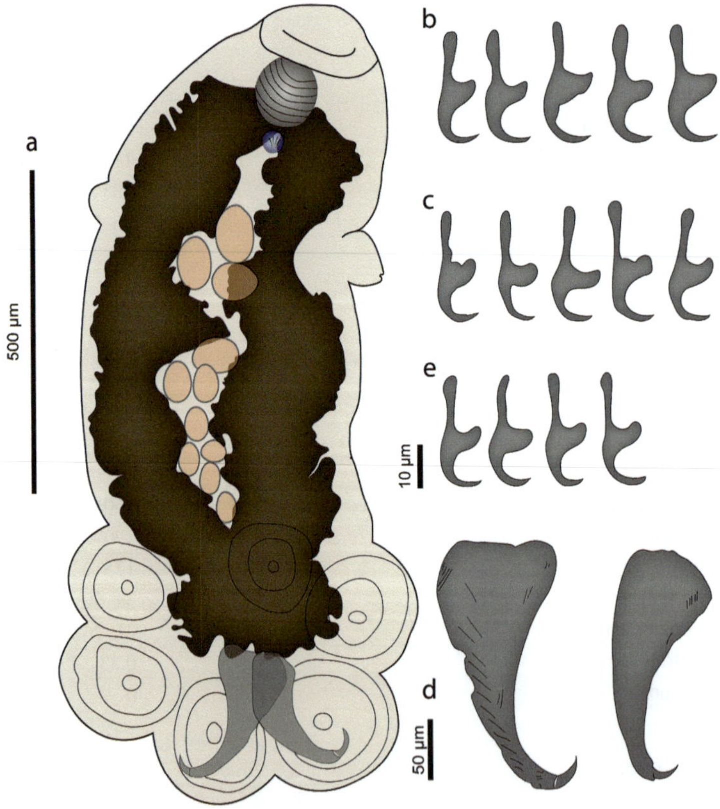

Fig. 4.34 (**a**) *Madapolystoma cryptica*, (**b**) marginal hooklets (1), (**c**) marginal hooklets (2–7), (**d**) marginal hooklets (8), (**e**) hamuli. Redrawn from Berthier et al. (2014)

Type host *Guibemantis liber* (Peracca, 1893), Free Madagascar Frog.

Additional hosts None.

Host geographical distribution *Guibemantis liber* can be found in Madagascar (Nussbaum et al. 2008).

Host conservation status The Free Madagascar Frog has a large population, a wide distribution and a certain tolerance for habitat modification. However, this species is threatened by habitat modification following deforestation, habitat fragmentation, habitat loss due to activities such as grazing and agriculture. It inhabits rainforests, pools and swamps where it can be found on plants. It is currently listed as Least Concern by The IUCN Red List of Threatened Species 2016: e.T57495A84173286. https://doi.org/10.2305/IUCN.UK.2016-1.RLTS.T57495A84173286.en. Accessed on 30 January 2022.

Site on host Urinary bladder.

DNA sequence
- 28S rRNA gene, 1466 bp. GenBank Accession: JN800275
- 28S rRNA gene, 1465 bp. GenBank Accessions: JN800276, JN800277
- 28S rRNA gene, 1464 bp. GenBank Accession: JN800278
- Cytochrome c Oxidase I, 396 bp. GenBank Accessions: JN015511, JN015518, JN015519, JN015522
- Cytochrome c Oxidase I, 377 bp. GenBank Accession: JN015515
- Cytochrome c Oxidase I, 387 bp. GenBank Accession: JN015520

Morphology and morphometrics (μm):	
Source	Berthier et al. (2014)
Sample size	3
Body length	1027–1239 (1151)
Greatest width	411–439 (429)
Width at vagina	
Haptor length	337–424 (387)
Haptor width	431–550 (475)
False oral sucker width	58–83 (71)
Pharynx length	94
Pharynx width	58
Ovary length	
Ovary width	
Egg length	
Egg width	
Egg incubation time	
Number of intra-uterine eggs	
Genital bulb width	31
Number of genital spines	7
Genital spine length	14

(continued)

Morphology and morphometrics (μm):	
Source	Berthier et al. (2014)
Haptoral sucker diameter	122–186 (160)
Hamulus handle length X	215–239 (227)
Hamulus guard length Y	197–219 (208)
Hamulus hook length	31–39 (36)
Marginal hooklet length	
Haptoral L/Body L	0.33–0.34 (0.34)
No of anastomoses	0

Remarks None.

4.4.5.3 *Madapolystoma isaloense* Landman, Verneau and Du Preez, 2018 (Fig. 4.35)

Collection Parasitic Worm Collection, National Museum, Bloemfontein, South Africa (NMBP).

Holotype NMBP 482.

Paratypes NMBP 483–488.

Original description Landman et al. (2018).

Taxonomic note The original name *Madapolystoma isaloensis* was corrected for gender.

Other taxonomic contributions None.

Etymology Named after the type locality.

Type locality Isalo, Madagascar.

Other localities None.

Type host *Mantella expectata* Busse and Böhme, 1992; Blue-legged Mantella.

Additional hosts None.

Host geographical distribution *Mantella expectata* is only known from a few localities around the Isalo Massif (Frost 2021).

Host conservation status The Blue-legged Mantella has a very limited distribution and is known from fewer than five threat-defined localities. Because of its limited extent of occurrence, it is listed as Endangered by The IUCN Red List of Threatened Species 2017: e.T57443A84166737. https://doi.org/10.2305/IUCN.UK.2017-2.RLTS.

Site on host One mature parasite was found in the urinary bladder while immature forms were found in both urinary and accessory bladders.

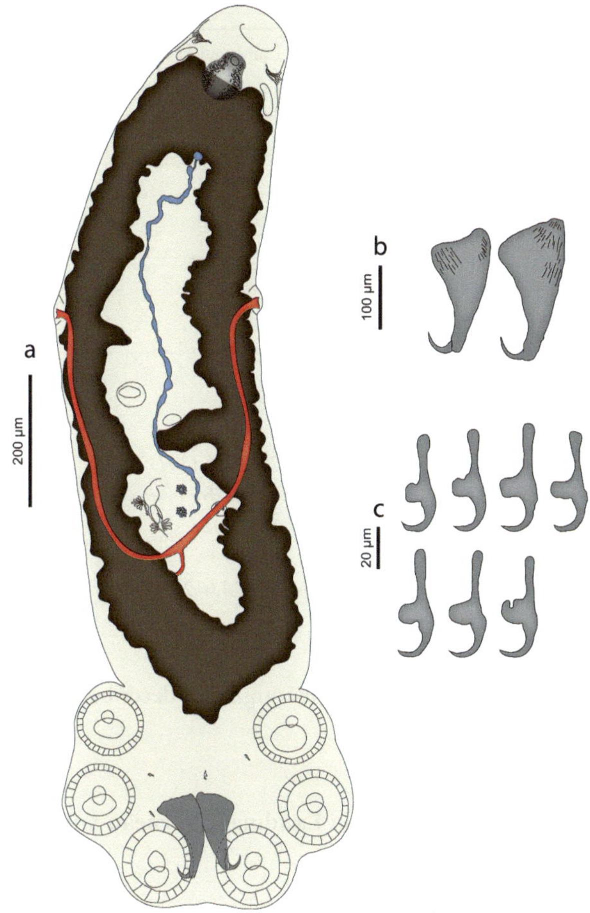

Fig. 4.35 (**a**) *Madapolystoma isaloense*, (**b**) hamuli, (**c**) marginal hooklets. Redrawn from Landman et al. (2018)

DNA sequence

- 28S rRNA gene, 1496 bp. GenBank Accession: FM897279

Morphology and morphometrics (µm):	
Source	Landman et al. (2018)
Sample size	One mature and 19 immature
Body length	2672
Greatest width	871
Width at vagina	683
Haptor length	618
Haptor width	887
Haptor length to body length ratio	0.23

(continued)

Morphology and morphometrics (μm):	
Source	Landman et al. (2018)
False oral sucker width	
Pharynx length	149
Pharynx width	130
Ovary length	
Ovary width	
Egg length	
Egg width	
Egg incubation time	
Intra-uterine eggs (n)	
Genital bulb width	27
Number of genital spines	7
Genital spine length	7.4–8.4 (7.9)
Haptoral sucker diameter	201–216 (207)
Hamulus handle length X	173; 228
Hamulus guard length Y	183; 202
Hamulus hook length	46; 49
Marginal hooklet length	C1: 24.6–26.6 (25.5); C2–8: 26.4–26.7 (25.8)
No of anastomoses	0

Remarks *Madapolystoma isaloense* was originally described as *Madapolystoma isaloensis.*

4.4.5.4 *Madapolystoma magnahami* Landman, Verneau and Du Preez, 2018 (Fig. 4.36)

Collection Parasitic Worm Collection, National Museum, Bloemfontein, South Africa (NMBP).

Holotype NMBP 474.

Paratypes NMBP 475–NMBP 481.

Original description Landman et al. (2018)

Other taxonomic contributions None.

Etymology Refers to the exceptionally large hamuli.

Type locality Ambohitantely Special Reserve, Madagascar (18,166667S; 47,273333E).

Other localities None.

Type host *Blommersia domerguei* (Guibé, 1973), Domergue's Madagascar Frog.

Additional hosts None.

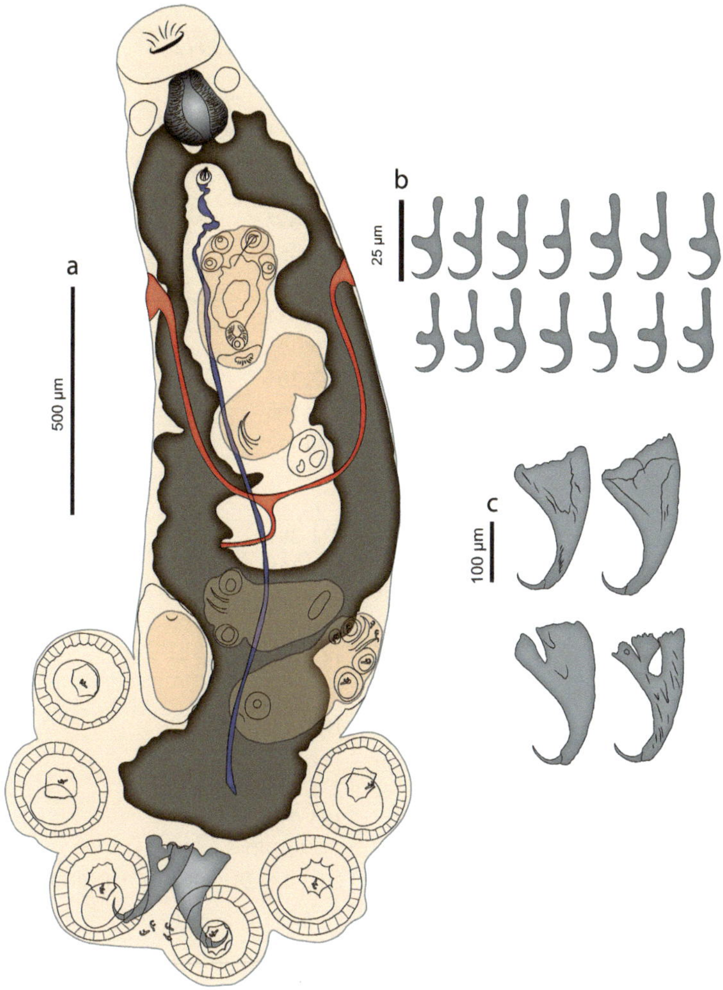

Fig. 4.36 (**a**) *Madapolystoma magnahami*, (**b**) marginal hooklets, (**c**) hamuli. Redrawn from Landman et al. (2018)

Host geographical distribution *Blommersia domerguei* occurs on the plateau of central Madagascar and has a wide distribution (Frost 2021).

Host conservation status Because of its wide distribution and abundance the Domergue's Madagascar Frog is listed as Least Concern by The IUCN Red List of Threatened Species 2016: e.T57477A84171195. https://doi.org/10.2305/IUCN.UK.2016-1.R.

Site on host Mature parasites were found in the urinary bladder while immature stages were found in both urinary and accessory bladders.

DNA sequence

- 28S rRNA gene, 1470 bp. GenBank Accession: FM897278

Morphology and morphometrics (μm):	
Source	Landman et al. (2018)
Sample size	2 mature and 27 immature
Body length	2179; 2340
Greatest width	507; 597
Width at vagina	414; 457
Haptor length	550; 648
Haptor width	769; 934
Haptor length to body length ratio	0.29
False oral sucker width	157, 161
Pharynx length	156; 166
Pharynx width	140; 153
Ovary length	
Ovary width	
Embryo length	224–391
Embryo width	152–168
Egg incubation time	
Intra-uterine embryos (n)	5–8
Genital bulb width	15.1
Number of genital spines	6
Genital spine length	16.1–16.8 (16.5)
Haptoral sucker diameter	193–221 (209)
Hamulus handle length X	231–244 (239)
Hamulus guard length Y	193–212 (204)
Hamulus hook length	47–49 (48)
Marginal hooklet length	C1: 26.6–28.4 (27.6); C2–8: 26.3–27.4 (27.0)
Haptoral L/Body L	
No of anastomoses	0

Remarks None.

4.4.5.5 *Madapolystoma ramilijaonae* Berthier, Du Preez, Raharivololoniaina, Vences and Verneau, 2014 (Fig. 4.37)

Collection Parasitic Worm Collection, National Museum, Bloemfontein, South Africa (NMBP).

Holotype NMBP 330.

Paratypes NMBP 331–333.

Original description Berthier et al. (2014).

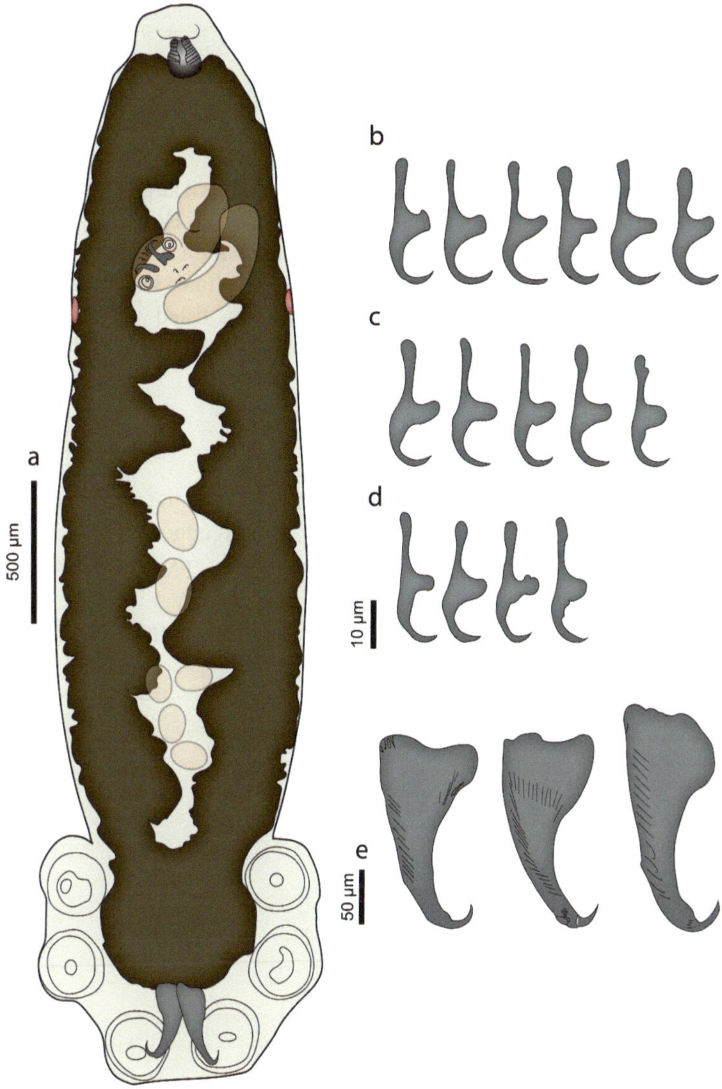

Fig. 4.37 (**a**) *Madapolystoma ramilijaonae*, (**b**) marginal hooklets (1), (**c**) marginal hooklets (2–7), (**d**) marginal hooklets (8), (**e**) hamuli. Redrawn from Berthier et al. (2014)

Other taxonomic contributions Landman et al. (2018).

Etymology Named after Prof Olga Ramilijaona Ravoahangimalala, Madagascar.

Type locality An'Ala, Ranomeda, Ranomafana, Andasibe and Ranomafanakely, Madagascar.

Other localities None.

Type host *Guibemantis liber*, Free Madagascar Frog.

Additional hosts None.

Host geographical distribution *Guibemantis liber* can be found in Madagascar (Nussbaum et al. 2008).

Host conservation status The Free Madagascar Frog has a large population, a wide distribution and a certain tolerance for habitat modification. However, this species is threatened by habitat modification following deforestation, habitat fragmentation, habitat loss due to activities such as grazing and agriculture. It inhabits rainforests, pools and swamps where it can be found on plants. It is listed as Least Concern by The IUCN Red List of Threatened Species 2016: e.T57495A84173286. https://doi.org/10.2305/IUCN.UK.2016-1.RLTS.T574.

Site on host Urinary bladder.

DNA sequence
- 18S rRNA gene, 2131 bp. GenBank Accession: FM897293
- 18S rRNA gene, 2093 bp. GenBank Accession: FM897294
- 28S rRNA gene, 1465 bp. GenBank Accessions: FM897276, FM897277, JN800271 to JN800274
- Cytochrome c Oxidase I, 396 bp. GenBank Accessions: JN015512, JN015514, JN015516, JN015517, JN015523 to JN015525

Morphology and morphometrics (μm):		
Source	Berthier et al. (2014)	Landman et al. (2018)
Sample size	4	
Body length	1493–3481 (2948)	
Greatest width	602–857 (705)	
Width at vagina		
Haptor length	567–771 (661)	
Haptor width	765–1013 (922)	
False oral sucker width	90–156 (123)	
Pharynx length	164–184 (175)	
Pharynx width	151–156 (154)	
Ovary length		
Ovary width		
Egg length		
Egg width		
Egg incubation time		
Number of intra-uterine eggs		21
Genital bulb width	30–38 (33)	
Number of genital spines	5–8	
Genital spine length	14–16 (15)	
Haptoral sucker diameter	188–244 (216)	
Hamulus handle length X	163–195 (179)	

(continued)

Morphology and morphometrics (µm):		
Source	Berthier et al. (2014)	Landman et al. (2018)
Hamulus guard length Y	154–193 (178)	
Hamulus hook length	31–47 (39)	
Marginal hooklet length		20.6–26.2 (23.0)
Haptoral L/Body L	0.221–0.379 (0.224)	
No of anastomoses	0	

Remarks None.

4.4.6 *Mesopolystoma* Vaucher, 1981

4.4.6.1 *Mesopolystoma samiriense* Vaucher, 1981 (Fig. 4.38)

Collection Natural History Museum, Geneva, Switzerland (MHNG).

Holotype MHNG 0014636.

Paratypes Unknown.

Original description Vaucher (1981).

Taxonomic note The original name *Mesopolystoma samiriensis* was corrected for gender.

Other taxonomic contributions Vaucher (1987).

Etymology Named after the type locality.

Type locality Pithecia Research station, Rio Samiria, Peru.

Other localities None.

Type host *Osteocephalus taurinus* Steindachner, 1862; Manaus Slender-Legged Treefrog.

Additional hosts None.

Taxonomy None.

Host geographical distribution *Osteocephalus taurinus* is present in Colombia, Peru, Venezuela, Bolivia, Guyana, French Guiana, Brazil, Suriname and Ecuador where it can be found on tree trunks or on the banks of ponds in rainforests and gallery forests (La Marca et al. 2010a).

Host conservation status The Manaus Slender-Legged Treefrog has a large population, a wide distribution and is tolerant to a certain degree of habitat modification. It is listed as Least Concern by The IUCN Red List of Threatened Species 2010: e.

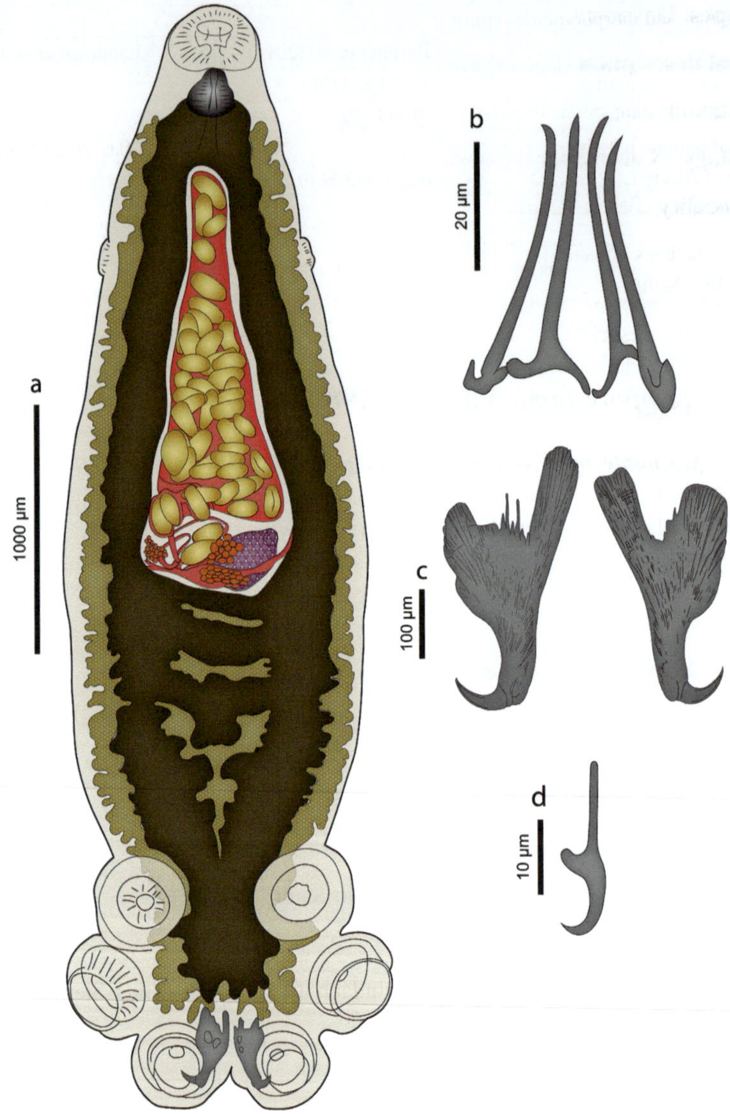

Fig. 4.38 (**a**) *Mesopolystoma samiriense*, (**b**) genital spines, (**c**) hamuli, (**d**) marginal hooklets. Redrawn from Vaucher (1981)

T55803A11364949. https://doi.org/10.2305/IUCN.UK.2010-2.RLTS.T55803A11364949.en. Accessed on 30 January 2022.

Site on host Urinary bladder.

DNA sequence None.

Morphology and morphometrics (μm):		
Source	Vaucher (1981)	Vaucher (1987)
Sample size		1
Body length	4550	3450
Greatest width	1300	1470
Width at vagina		
Haptor length	1120	1120
Haptor width	1430	1510
False oral sucker width	344 × 245	261
Pharynx length	221	251
Pharynx width	205	180
Ovary length		
Ovary width		
Egg length	160–184	179–183
Egg width	86–98	
Egg incubation time		
Number of intra-uterine eggs		
Genital bulb width		
Number of genital spines		
Genital spine length	46	42
Haptoral sucker diameter	319–393	364–458 × 352–401
Hamulus handle length X	368	405
Hamulus guard length Y	409	413
Hamulus hook length		
Marginal hooklet length		
Haptoral L/Body L	0.25	0.32
No of anastomoses	3	

Remarks *Mesopolystoma samiriense* was originally described as *Mesopolystoma samiriensis*. This species has been reported from the same host than *Polystoma napoense* Vaucher, 1987 in Ecuador.

4.4.7 *Metapolystoma* Combes, 1976

4.4.7.1 *Metapolystoma ansuanum* Landman, Verneau, Raharivololoniaina and Du Preez, 2021 (Fig. 4.39)

Collection Parasitic Worm Collection, National Museum, Bloemfontein, South Africa (NMBP)

Holotype NMBP 569.

Paratypes NMBP 570.

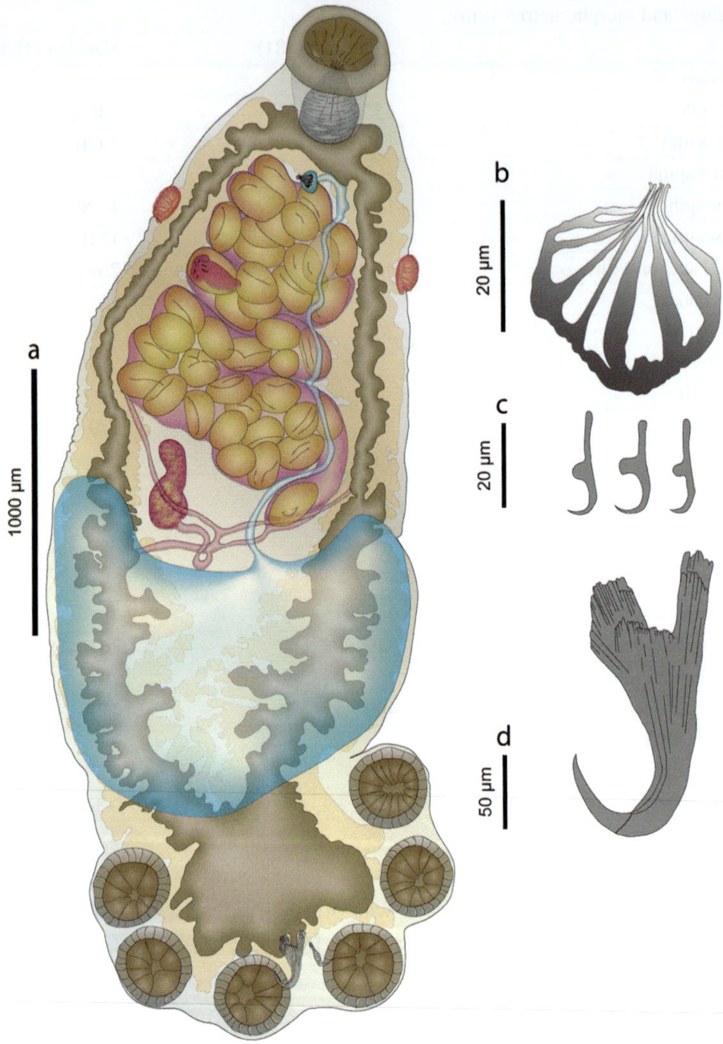

Fig. 4.39 (**a**) *Metapolystoma ansuanum*, (**b**) genital spines, (**c**) marginal hooklets, (**d**) hamuli, (**e**). Redrawn from Landman et al. (2021)

Original description Landman et al. (2021).

Other taxonomic contributions None.

Etymology Named after Mrs Anna-Susan van der Linde, known as Ansu. Klerksdorp, South Africa.

Type locality Cascade des Nymphes, Isalo National Park, Madagascar (22.46977S; 45.260701E).

Other localities None.

Type host *Boophis luteus* (Boulenger, 1882); Ankafana Bright-eyed Frog.

Additional hosts None.

Taxonomy None.

Host geographical distribution *Boophis luteus* is widespread in central and Eastern Madagascar (Frost 2021), where it occurs in both pristine and degraded rainforests.

Host conservation status The Ankafana Bright-eyed Frog is listed as Least Concern by The IUCN Red List of Threatened Species 2016: e.T57410A84163237. https://doi.org/10.2305/IUCN.UK.2016-1.RLTS.T57410A84163237.en. Accessed on 30 January 2022.

Site on host Urinary bladder.

Zoobank registration urn:lsid:zoobank.org:act:4761BF6E-F309-4625-A7D7-CA64F3A3F6F8.

DNA sequence None.

Morphology and morphometrics (μm):	
Source	Landman et al. (2021)
Sample size	2
Body length	2680–3952
Greatest width	1035–1436
Width at vagina	804–1001
Haptor length	669–912
Haptor width	1166–1385
False oral sucker width	307–329
Pharynx length	256–308
Pharynx width	183–222
Testis length	
Testis width	
Ovary length	265–338
Ovary width	98–144
Egg length	196–217 (205)
Egg width	119–137 (127
Egg incubation time	
Number of intra-uterine eggs	41
Genital bulb width	64–73
Number of genital spines	10
Genital spine length	24.5–28.5 (26.19 ± 1.16; 5)
Haptoral sucker diameter	241–305 (273)
Hamulus handle length X	330–420 (380)
Hamulus guard length Y	

(continued)

Morphology and morphometrics (μm):	
Source	Landman et al. (2021)
Hamulus hook length	
Marginal hooklet length	22.3–24.79 (23.5)
Haptoral L/Body L	0.23–0.25
No of anastomoses	0

Remarks None.

4.4.7.2 *Metapolystoma brygoonis* (Euzet and Combes, 1964) (Fig. 4.40)

Collection Claude Combes personal collection, Perpignan, France.

Holotype Unknown.

Paratypes Unknown.

Original description Euzet and Combes (1964).

Other taxonomic contributions Combes (1976).

Etymology Named after Dr. Brygoo, Director, Pasteur Institute, Antananarivo, Madagascar.

Type locality Pasteur Institute, Antananarivo, Madagascar.

Other localities None.

Type host *Ptychadena mascareniensis* (Duméril and Bibron, 1841), Mascarene Ridged Frog. In the original description of the polystome, the host was known as *Rana mascareniensis* Duméril and Bibron, 1841.

Additional hosts None.

Host geographical distribution *Ptychadena mascareniensis* has a wide distribution range occurring in various regions such as Angola, Congo, Egypt, Sudan, South Africa, Zimbabwe, Ethiopia, Madagascar, Liberia, Botswana, Cameroon, Ghana, Kenya, Malawi, Mozambique, Nigeria, Namibia and Tanzania. This species was also introduced to regions such as the Seychelles and Mauritius. It is usually found in marshy regions, grasslands, fields and urban regions (Rödel et al. 2009a).

Host conservation status The Mascarene Ridged Frog has a large population, a wide distribution range as well as a tolerance for a range of habitats. It is listed as Least Concern by The IUCN Red List of Threatened Species 2016: e. T76317565A79825430. https://doi.org/10.2305/IUCN.UK.2016-1.RLTS. T76317565A79825430.en. Accessed on 30 January 2022.

Site on host Urinary bladder.

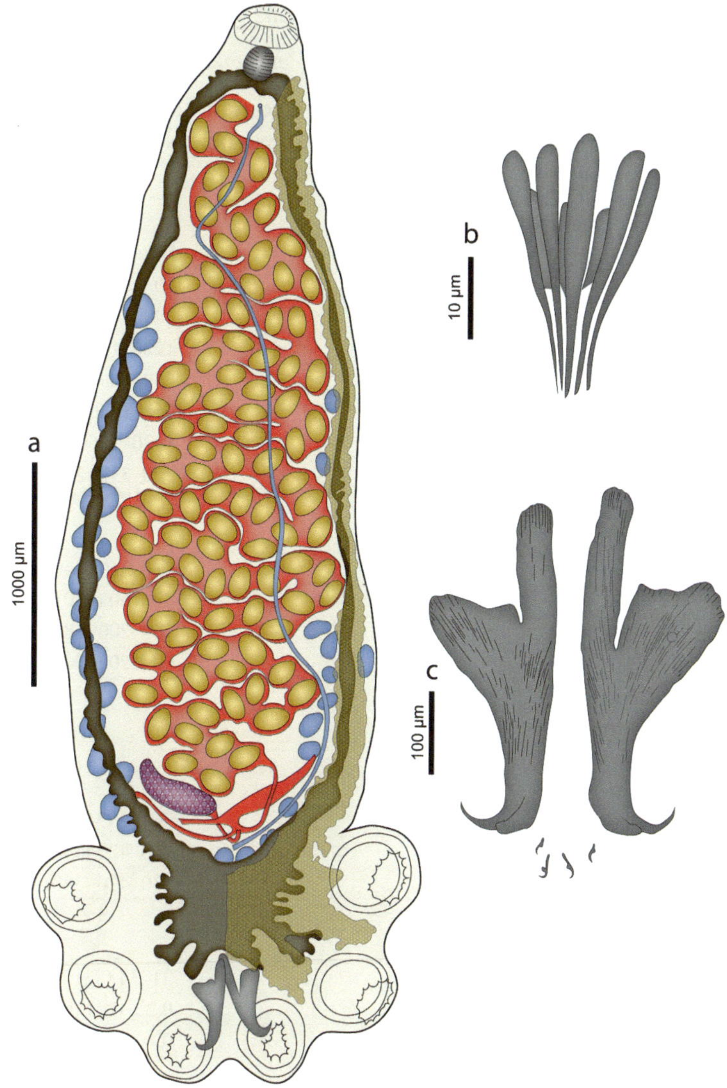

Fig. 4.40 (**a**) *Metapolystoma brygoonis*, (**b**) genital spines, (**c**) 1 and 2. Redrawn from Euzet and Combes (1964)

DNA sequence

- 18S rRNA gene, 614 bp. GenBank Accession: AJ297772
- 18S rRNA gene, 2123 bp. GenBank Accession: FM897287
- 28S rRNA gene, 1452 bp. GenBank Accession: JN800281
- 28S rRNA gene, 1427 bp. GenBank Accession: FM897270
- Cytochrome c Oxidase I, 379 bp. GenBank Accessions: FM897300, FM897297 and JN800284

- Cytochrome c Oxidase I, 396 bp. GenBank Accessions: JN800287 and JN800289
- Cytochrome c Oxidase I, 380 bp. GenBank Accession: JN800288
- Cytochrome c Oxidase I, 390 bp. GenBank Accession: JN800286
- Cytochrome c Oxidase I, 365 bp. GenBank Accession: JN800285
- ITS1, 735 bp. GenBank Accession: AJ310399

Morphology and morphometrics (μm):	
Source	Euzet and Combes (1964)
Sample size	10
Body length	3400–4800 (4200)
Greatest width	1220–1500 (1340)
Width at vagina	650–1030 (800)
Haptor length	760–1070 (900)
Haptor width	1220–1760 (1530)
False oral sucker width	220–340 (270)
Pharynx length	
Pharynx width	
Ovary length	410–460 (440)
Ovary width	90–160 (120)
Egg length	160–200 (180)
Egg width	90–120 (100)
Egg incubation time	
Number of intra-uterine eggs	140
Genital bulb width	
Number of genital spines	8
Genital spine length	30–32 (31)
Haptoral sucker diameter	260–420 (340)
Hamulus handle length X	330–420 (380)
Hamulus guard length Y	
Hamulus hook length	
Marginal hooklet length	31–34 (33)
Haptoral L/Body L	0.22–0.23 (0.22)
No of anastomoses	0

Remarks *Metapolystoma brygoonis* was originally described as *Polystoma brygoonis* Euzet and Combes, 1964. *Ptychadena mascareniensis* in Madagascar is now regarded as a species complex (Zimkus et al. 2017).

4.4.7.3 *Metapolystoma cachani* (Gallien, 1956) (Fig. 4.41)

Collection Specimens stored in author's private collection.

Holotype Unknown.

Paratypes Unknown.

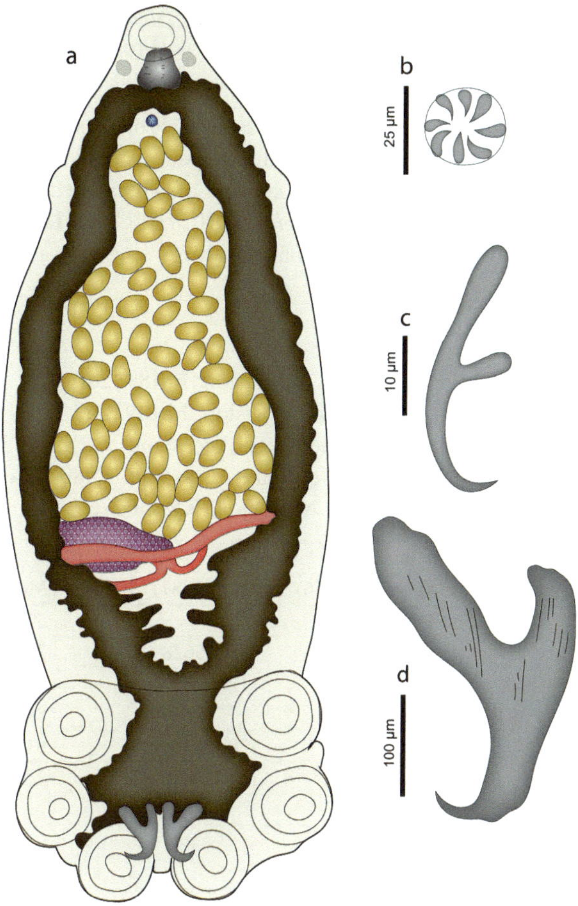

Fig. 4.41 (**a**) *Metapolystoma cachani*, (**b**) genital spines, (**c**) marginal hooklets, (**d**) hamuli, (**e**). Redrawn from Gallien (1956)

Original description Gallien (1956).

Other taxonomic contributions Murith et al. (1977); Murith (1979, 1981); Kulo (1981).

Etymology Named after Mr M. Cachan, Orstom Research Institute, Abidjan, Ivory Coast.

Type locality Adiopodoumé, Ivory Coast.

Other localities Togo.

Type host *Ptychadena longirostris* (Peters, 1870), Snouted Grassland Frog.

Additional hosts *Hoplobatrachus occipitalis* (Günther, 1858).

Host geographical distribution *Ptychadena longirostris* occurs in countries such as Ghana, Liberia, Guinea, Sierra Leone and Nigeria (Frost 2021), where it can be found in rainforest habitats

Host conservation status The Snouted Grassland Frog has a large population, wide distribution and a tolerance for a certain degree of habitat modification. It is listed as Least Concern by The IUCN Red List of Threatened Species 2013: e. T58507A18400287. https://doi.org/10.2305/IUCN.UK.2013-2.RLTS. T58507A18400287.en. Accessed on 30 January 2022.

Site on host Urinary bladder.

DNA sequence
- 18S rRNA gene, 2124 bp. GenBank Accession: FM897280
- 28S rRNA gene, 1454 bp. GenBank Accession: FM897262
- Cytochrome c Oxidase I, 337 bp. GenBank Accession: KR856163
- Cytochrome c Oxidase I, 396 bp. GenBank Accession: JN800294
- 12S rRNA gene, 490 bp. GenBank Accession: KR856076

Morphology and morphometrics (μm):

Source	Gallien (1956)	Kulo (1981)	Murith (1981)
Sample size		3	57
Body length	4000	3880–5628	3680–7250 (5070)
Greatest width	1500	1120–2880	1160–2880 (1980)
Width at vagina			880–1760 (1220)
Haptor length	1	720–1072	910–1500 (1,15)
Haptor width	1500	1248–2800	1300–2350 (1730)
False oral sucker width			
Pharynx length			
Pharynx width			
Ovary length			
Ovary width			
Egg length	190	176–208	0.122–0.159 (0.138)
Egg width	120	96–112	670–780 (720)
Egg incubation time			
Number of intra-uterine eggs		4	8–390 (170)
Genital bulb width			
Number of genital spines			
Genital spine length			
Haptoral sucker diameter	350–400	272–580 × 288–544	300–500 (420)
Hamulus handle length X		280–450	330–520 (430)
Hamulus guard length Y			
Hamulus hook length			
Marginal hooklet length			40
Haptoral L/Body L		1/5	0.15–0.32 (0.23)
No of anastomoses			0

Remarks *Metapolystoma cachani* was originally described as *Polystoma cachani* Gallien, 1956.

4.4.7.4 *Metapolystoma falcatum* Landman, Verneau, Raharivololoniaina and Du Preez, 2021 (Fig. 4.42)

Collection Parasitic Worm Collection, National Museum, Bloemfontein, South Africa (NMBP).

Holotype NMBP 560.

Paratypes NMBP 561–564.

Original description Landman et al. (2021).

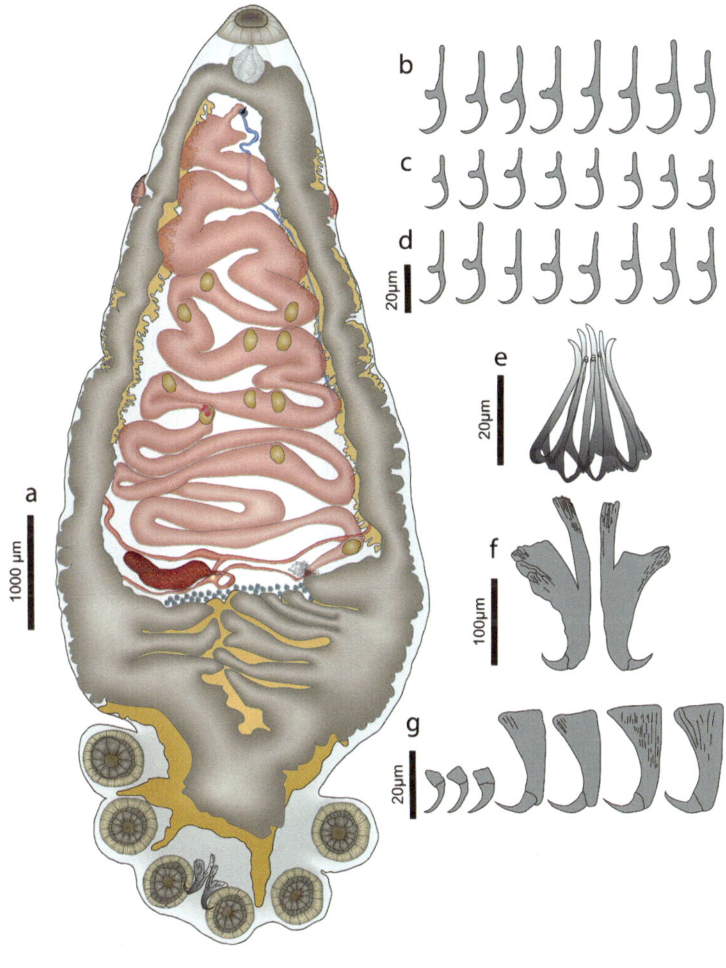

Fig. 4.42 (**a**) *Metapolystoma falcatum*, (**b**) marginal hooklets (1), (**c**) marginal hooklets (2–7), (**d**) marginal hooklets (8), (**e**) genital spines, (**f**) hamuli, (**g**) development of hamuli. Redrawn from Landman et al. (2021)

Other taxonomic contributions None.

Etymology Refers to the sickle shaped tips of the marginal hooklets.

Type locality Ankarafantsika, Madagascar (16.115976S; 47.095631E).

Other localities None.

Type host *Boophis doulioti* (Angel, 1934), no English name.

Additional hosts None.

Host geographical distribution *Boophis doulioti* occurs mostly in western and Southern Madagascar (Frost 2021).

Host conservation status Because of its wide distribution and abundance, this species is listed as Least Concern by The IUCN Red List of Threatened Species 2016: e.T57397A84162240. https://doi.org/10.2305/IUCN.UK.2016-1.RLTS.T57397A84162240.en. Accessed on 30 January 2022.

Site on host Urinary bladder.

Zoobank registration urn:lsid:zoobank.org:act:FB1716E2-5C61–4C53-9A2F-BB364D6782A8.

DNA sequence
- 18S rRNA gene, 2007 bp. GenBank Accession: MW054248
- 18S rRNA gene, 2122 bp. GenBank Accession: FM897286
- 28S rRNA gene, 1452 bp. GenBank Accession: JN800283
- 28S rRNA gene, 1452 bp. GenBank Accession: FM897269
- Cytochrome c Oxidase I, 396 bp. GenBank Accessions: JN800291
- Cytochrome c Oxidase I, 396 bp. GenBank Accessions: MW053458

Morphology and morphometrics (μm):	
Source	Landman et al. (2021)
Sample size	1
Body length	7871
Greatest width	3272
Width at vagina	1783
Haptor length	1492
Haptor width	2636
False oral sucker width	297
Pharynx length	327
Pharynx width	290
Ovary length	772
Ovary width	196
Egg length	161–185 (171)
Egg width	108–123 (117)
Egg incubation time	
Number of intra-uterine eggs	11

(continued)

Morphology and morphometrics (μm):	
Source	Landman et al. (2021)
Genital bulb width	86
Number of genital spines	8
Genital spine length	24–32 (28)
Haptoral sucker diameter	435–498 (468
Hamulus handle length X	393–407
Hamulus guard length Y	
Hamulus hook length	
Marginal hooklet length	33–40 (37)
Haptoral L/Body L	0.19
No of anastomoses	0

Remarks None.

4.4.7.5 *Metapolystoma multuova* Landman, Verneau, Raharivololoniaina and Du Preez, 2021 (Fig. 4.43)

Collection Parasitic Worm Collection, National Museum, Bloemfontein, South Africa (NMBP)

Holotype NMBP 571.

Paratypes NMBP 572.

Original description Landman et al. (2021).

Other taxonomic contributions None.

Etymology Refers to the vast number of in-utero eggs (500+).

Type locality Cascade des Nymphes, Isalo National Park, Madagascar (22.46977S; 45.260701E).

Other localities None.

Type host *Boophis occidentalis* Glaw and Vences, 1994, Western White-lipped Treefrog.

Additional hosts None.

Host geographical distribution *Boophis occidentalis* has widely separated records in western and north-central Madagascar (Frost 2021).

Host conservation status Because of its wide distribution, the Western White-lipped Treefrog listed as Least Concern by The IUCN Red List of Threatened Species 2015: e.T68206465A105178884. https://doi.org/10.2305/IUCN.UK.2015-4.RLTS.T68206465A68200098.en. Accessed on 30 January 2022.

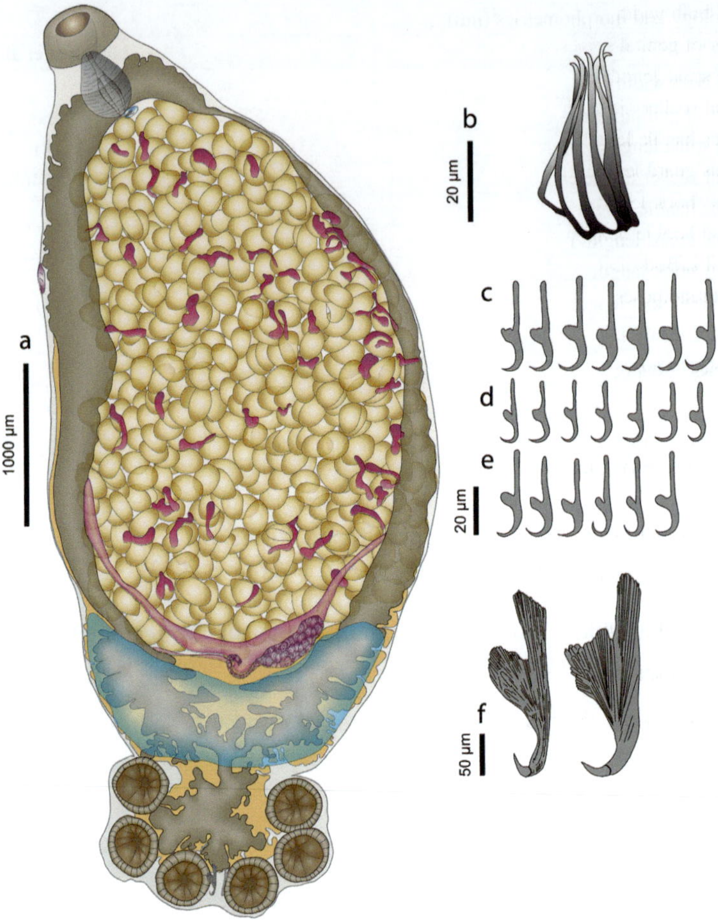

Fig. 4.43 (**a**) *Metapolystoma multiova*, (**b**) genital spines, (**c**) marginal hooklets (1), (**d**) marginal hooklets (2–7), (**e**) marginal hooklets (8), (**f**) hamuli. Redrawn from Landman et al. (2021)

Site on host Urinary bladder.

Zoobank registration urn:lsid: zoobank.org:act:6854087D-44B5–492B-9750-3647188F0F26.

DNA sequence

- 18S rRNA gene, 2122 bp. GenBank Accession: FM897285
- 28S rRNA gene, 1452 bp. GenBank Accession: FM897268
- Cytochrome c Oxidase I, 397 bp. GenBank Accession: FM897301

Morphology and morphometrics (μm):	
Source	Landman et al. (2021)
Sample size	2
Body length	4915–5537
Greatest width	2257–2419
Width at vagina	1469–2086
Haptor length	916
Haptor width	1478
False oral sucker width	248–271
Pharynx length	302–405
Pharynx width	289–320
Ovary length	446
Ovary width	158
Egg length	198–228 (212)
Egg width	144–165 (154)
Egg incubation time	
Number of intra-uterine eggs	499+
Genital bulb width	86
Number of genital spines	6
Genital spine length	40.4–43.7 (42)
Haptoral sucker diameter	329–356 (339
Hamulus handle length X	196–248 (215)
Hamulus guard length Y	121–174 (153)
Hamulus hook length	58.2
Marginal hooklet length	28.6–38.7 (33.8)
Haptoral L/Body L	0.18
No of anastomoses	0

Remarks None.

4.4.7.6 *Metapolystoma ohlerianum* Landman, Verneau, Vences and Du Preez, 2023 (Fig. 4.44)

Collection Parasitic Worm Collection, National Museum, Bloemfontein, South Africa (NMBP).

Holotype NMBP.

Paratypes NMBP.

Original description Landman et al. (2023).

Other taxonomic contributions None.

Etymology Named after Dr. Annemarie Ohler MNHN, Paris, France.

Type locality Andasibe, Madagascar (−18.938753S; 48.414471E).

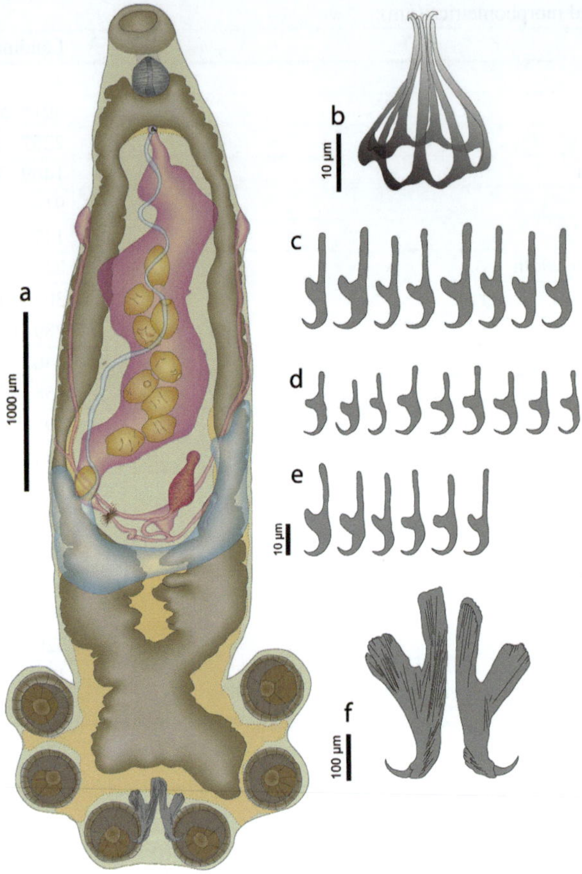

Fig. 4.44 (**a**) *Metapolystoma ohlerianum*, (**b**) genital spines, (**c**) marginal hooklets (1), (**d**) marginal hooklets (2–7), (**e**) marginal hooklets (8), (**f**) hamuli. Redrawn from Landman et al. (2023)

Other localities Mangabe-Ranomena-Sahasarotra, Madagascar.

Type host *Aglyptodactylus madagascariensis* (Duméril, 1853), Madagascar Jumping Frog.

Additional hosts None.

Host geographical distribution Forests of Northern and Eastern Madagascar (Frost 2021).

Host conservation status The Madagascar Jumping Frog is listed as Least Concern by The IUCN Red List of Threatened Species 2016: e.T84500383A84161104. https://doi.org/10.2305/IUCN.UK.2016-1.RLTS.T84500383A84161104.en. Accessed on 26 August 2022.

Site on host Urinary bladder.

DNA sequence

- 18S rRNA gene, 2123 bp. GenBank Accession: FM897283
- 28S rRNA gene, 1452 bp. GenBank Accession: FM897266
- Cytochrome c Oxidase I, 379 bp. GenBank Accession: FM897299

Morphology and morphometrics (μm):	
Source	Landman et al. (2023)
Sample size	13
Body length	3508–5042 (4517)
Greatest width	1209–1758 (1515)
Width at vagina	1041–1340 (1210)
Haptor length	1208–1280 (1239)
Haptor width	1647–2055 (1852)
False oral sucker width	228–284 (250)
Pharynx length	251–297 (282)
Pharynx width	226–284 (257)
Ovary length	349–427 (378)
Ovary width	136–190 (169)
Egg length	187–242 (211)
Egg width	104–182 (127)
Egg incubation time	Intra-uterine development
Number of intra-uterine eggs	61
Genital bulb width	59–72 (66)
Number of genital spines	7
Genital spine length	26.7–32.0 (29.7)
Vagina length	195–270 (229)
Haptoral sucker diameter	296–526 (387)
Hamulus handle length X	340–475 (391)
Hamulus guard length Y	263–392 (315)
Hamulus hook length	57–69 (63)
Marginal hooklet length	C1 29.6–33.6 (31.7); C2–7 17.8–22.2 (20.6); C8 25.9–30.0 (27.5
Haptoral L/Body L	
No of anastomoses	0

Remarks None.

4.4.7.7 *Metapolystoma porosissimae* Du Preez and Kok, 1992 (Fig. 4.45)

Collection Parasitic Worm Collection, National Museum, Bloemfontein, South Africa (NMBP). Parasitic Worm Collection, Natural History Museum, London (NHML).

Holotype NMBP 31.

Paratypes NMBP 32–35.

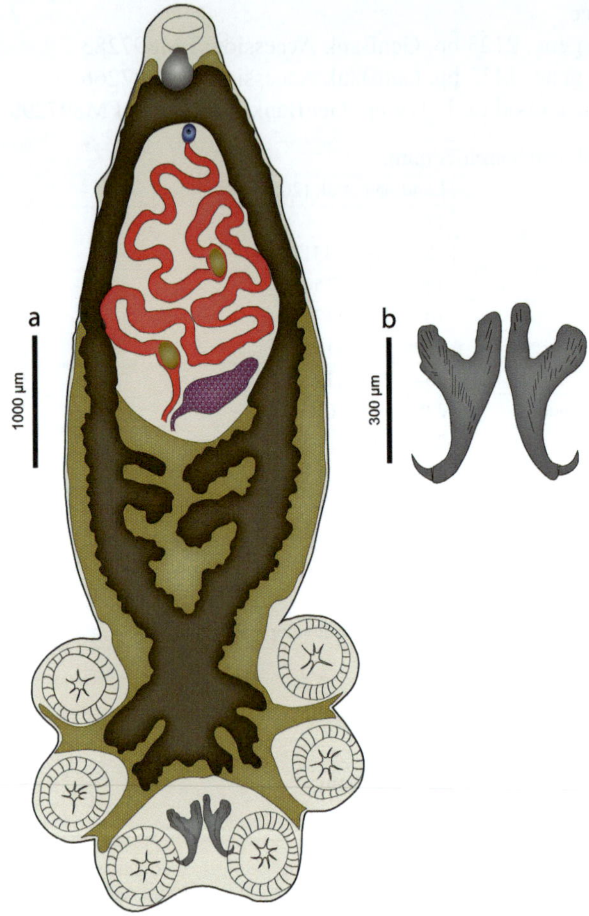

Fig. 4.45 (**a**) *Metapolystoma porosissimae*, (**b**) genital spines. Redrawn from Du Preez and Kok (1992)

Original description Du Preez and Kok (1992).

Other taxonomic contributions None.

Etymology Named after the host.

Type locality KwaZulu Natal, South Africa.

Other localities None.

Type host *Ptychadena porosissima* (Steindachner, 1867), Grassland Ridged Frog.

Additional hosts None.

Host geographical distribution *Ptychadena porosissima* occurs in South Africa, Mozambique, Zimbabwe, Angola, Zaire, East Africa, Ethiopia, Malawi, Tanzania,

Rwanda, Zambia, Congo, Swaziland, Kenya and Uganda (Frost 2021), where it inhabits pans and marshy areas as well as forest, savannah and grassland habitats.

Host conservation status The Grassland Ridged Frog has a large population, a wide distribution and is tolerant for various habitats. It is thus listed as Least Concern by The IUCN Red List of Threatened Species 2013: e.T58519A3070008. https://doi.org/10.2305/IUCN.UK.2013-2.RLTS.T58519A3070008.en. Accessed on 30 January 2022.

Site on host Urinary bladder.

DNA sequence None.

Morphology and morphometrics (μm):	
Source	Du Preez and Kok (1992)
Sample size	10
Body length	4150–6710 (5870)
Greatest width	1570–2360 (2040)
Width at vagina	1120–1700 (1460)
Haptor length	1540–1800 (1670)
Haptor width	2010–2660 (2360)
False oral sucker width	461–566 (521)
Pharynx length	267–348 (310)
Pharynx width	2340–2980 (262)
Ovary length	546–755 (651)
Ovary width	
Egg length	201–229 (214)
Egg width	113–134 (128)
Egg incubation time	
Number of intra-uterine eggs	250
Genital bulb width	
Number of genital spines	9–10
Genital spine length	31
Haptoral sucker diameter	461–566 (521)
Hamulus handle length X	432–472 (447)
Hamulus guard length Y	
Hamulus hook length	
Marginal hooklet length	35
Haptoral L/Body L	0.26–0.33 (0.29)
No of anastomoses	0

Remarks None.

4.4.7.8 *Metapolystoma theroni* Landman, Verneau, Raharivololoniaina and Du Preez, 2021 (Fig. 4.46)

Collection Parasitic Worm Collection, National Museum, Bloemfontein, South Africa.

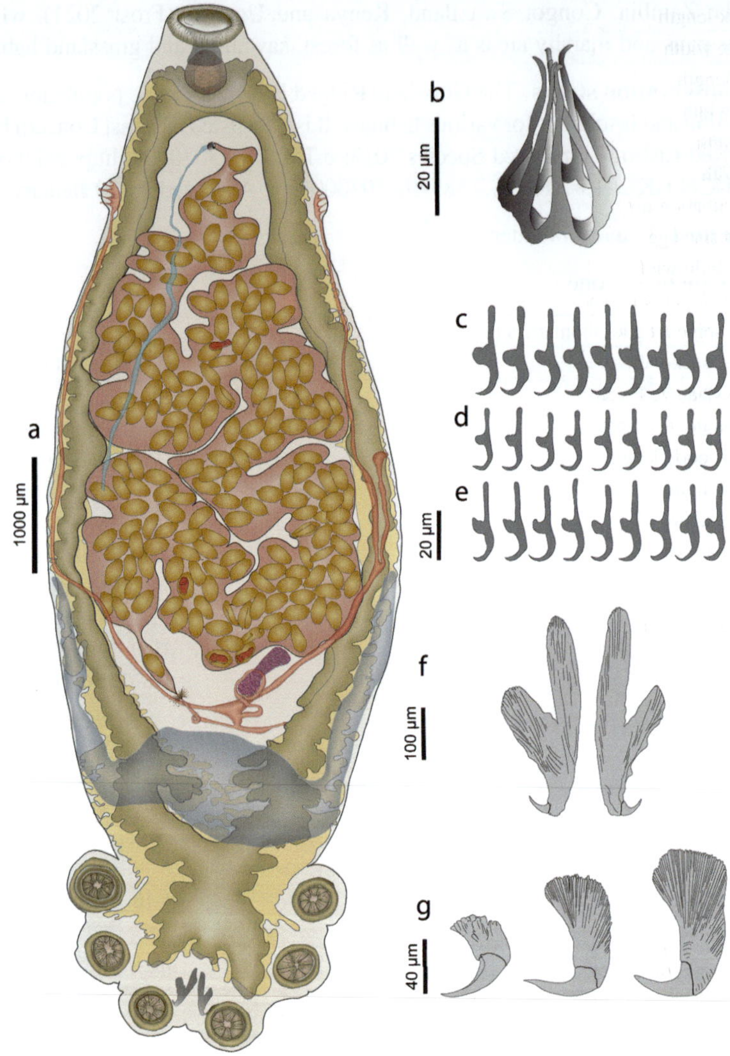

Fig. 4.46 (**a**) *Metapolystoma theroni*, (**b**) genital spines, (**c**) marginal hooklets (1), (**d**) marginal hooklets (2–7), (**e**) marginal hooklets (8), (**f**) hamuli, (**g**) development of hamuli. Redrawn from Landman et al. (2021)

Holotype NMBP 573.

Paratypes NMBP 574–577.

Original description Landman et al. (2021).

Other taxonomic contributions None.

Etymology Named after Prof Pieter Theron, North-West University, South Africa.

Type locality Indri Reserve in Andasibe, Madagascar (18.930856S; 48.413611E).

Other localities None.

Type host *Boophis madagascariensis* (Peters, 1874), Madagascar Bright-eyed Frog.

Additional hosts None.

Host geographical distribution *Boophis madagascariensis* is widespread in Madagascar and can be found in both pristine and disturbed rainforests (Frost 2021).

Host conservation status The Madagascar Bright-eyed Frog is listed as Least Concern by The IUCN Red List of Threatened Species 2016: e. T49545913A67038125. https://doi.org/10.2305/IUCN.UK.2016-1.RLTS. T49545913A67038125.en. Accessed on 30 January 2022.

Site on host Urinary bladder.

Zoobank registration urn:lsid: zoobank.org:act:55203AE0-217D-45D0-B3C8-07DDE53FA9A3.

DNA sequence
- 18S rRNA gene, 2080 bp. GenBank Accession: FM897284
- 18S rRNA gene, 2095 bp. GenBank Accession: MW054249
- 28S rRNA gene, 1452 bp. GenBank Accession: FM897267
- 28S rRNA gene, 1466 bp. GenBank Accession: MW054241
- Cytochrome c Oxidase I, 379 bp. GenBank Accessions: FM897298
- Cytochrome c Oxidase I, 386 bp. GenBank Accessions: JN800293

Morphology and morphometrics (μm):

Source	Landman et al. (2021)
Sample size	1
Body length	9086
Greatest width	3208
Width at vagina	2011
Haptor length	1634
Haptor width	2572
False oral sucker width	486
Pharynx length	401
Pharynx width	385
Ovary length	479
Ovary width	385
Egg length	239–265 (250)
Egg width	136–152 (143)
Egg incubation time	
Number of intra-uterine eggs	176
Genital bulb width	65
Number of genital spines	7

(continued)

Morphology and morphometrics (μm):	
Source	Landman et al. (2021)
Genital spine length	29–35 (33)
Haptoral sucker diameter	404–424 (416)
Hamulus handle length X	392–408
Hamulus guard length Y	252–257
Hamulus hook length	50–55
Marginal hooklet length	34–39 (36)
Haptoral L/Body L	0.18
No of anastomoses	0

Remarks None.

4.4.7.9 *Metapolystoma vencesi* Landman, Verneau, Raharivololoniaina and Du Preez, 2021 (Fig. 4.47)

Collection Parasitic Worm Collection, National Museum, Bloemfontein, South Africa (NMBP).

Holotype NMBP 578.

Paratypes NMBP 579–580 and two immature specimens NMBP 581–582.

Original description Landman et al. (2021).

Other taxonomic contributions None.

Etymology Named after Prof Miguel Vences, Braunschweig, Germany.

Type locality Ambatolahy, Madagascar (21.2438667S; 47.4262167E).

Other localities None.

Type host *Boophis albilabris* (Boulenger, 1888), White-lipped Bright-eyed Frog.

Additional hosts None.

Host geographical distribution *Boophis albilabris* is widely distributed in northeastern, eastern and southeastern Madagascar, 100–1000 m elevation (Frost 2021).

Host conservation status The White-lipped Bright-eyed is listed as Least Concern by The IUCN Red List of Threatened Species 2016: e.T57386A49392305. https://doi.org/10.2305/IUCN.UK.2016-1.RLTS.T57386A49392305.en. Accessed on 30 January.

Site on host Urinary bladder.

Zoobank registration urn:lsid:zoobank.org:act:680E18A9-64D6-40C9-87D9-1AC5F28A1030.

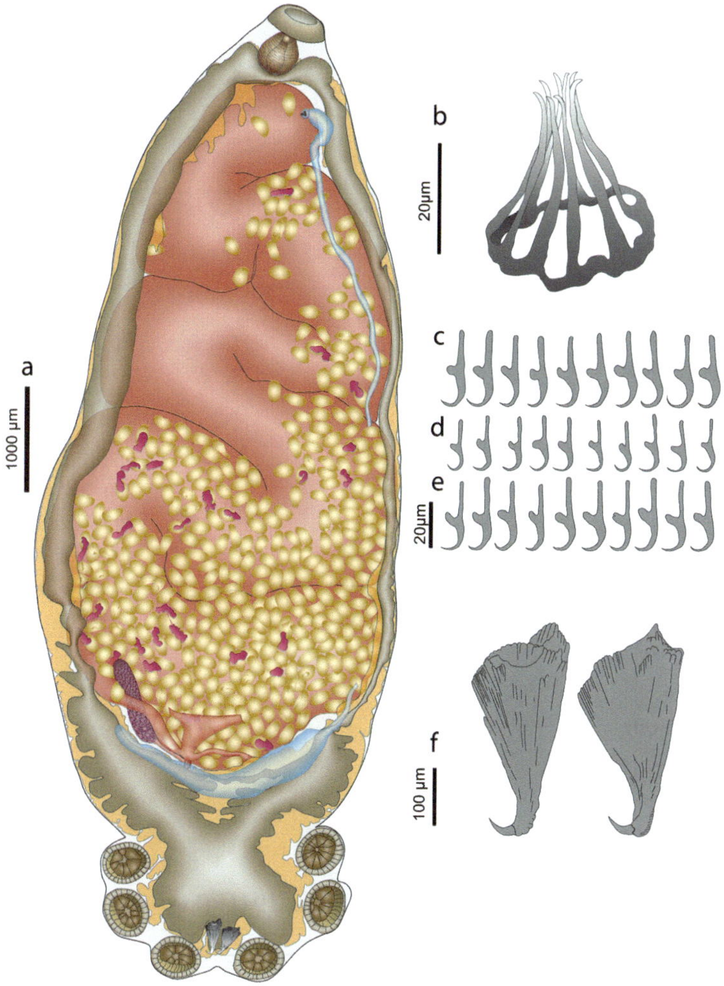

Fig. 4.47 (**a**) *Metapolystoma vencesi,* (**b**) genital spines, (**c**) marginal hooklets (1), (**d**) marginal hooklets (2–7), (**e**) marginal hooklets (8), (**f**) hamuli. Redrawn from Landman et al. (2021)

DNA sequence None.

Morphology and morphometrics:	
Source	Landman et al. (2021)
Sample size	6
Body length	9190
Greatest width	3496
Width at vagina	2315
Haptor length	1437
Haptor width	2344

(continued)

Morphology and morphometrics:	
Source	Landman et al. (2021)
False oral sucker width	391
Pharynx length	302–420
Pharynx width	291–390
Ovary length	861
Ovary width	194
Egg length	210–230 (219)
Egg width	120–147 (130)
Egg incubation time	
Number of intra-uterine eggs	367
Genital bulb width	
Number of genital spines	8
Genital spine length	32–33 (33)
Haptoral sucker diameter	366–452 (424)
Hamulus handle length X	269–332 (299)
Hamulus guard length Y	225–302 (270
Hamulus hook length	40–52 (47)
Marginal hooklet length	30–34 (33)
Haptoral L/Body L	0.16
No of anastomoses	0

Remarks None.

4.4.8 *Neodiplorchis* Yamaguti, 1963

4.4.8.1 *Neodiplorchis scaphiopodis* (Rodgers, 1941) (Fig. 4.48)

Collection United States National Museum (USNM).

Holotype USNM 1337428.

Original description Rodgers (1941).

Other taxonomic contributions Lamothe-Argumedo (1973); Brooks (1976); Tinsley and Earle (1983).

Etymology Named after the host.

Type locality Stillwater, Oklahoma, USA.

Other localities None.

Type host *Spea bombifrons* (Cope, 1863), Plains Spadefoot. In the original description of the polystome, the host was known as *Scaphiopus bombifrons* Cope, 1863.

Additional hosts *Spea multiplicata* (Cope, 1863).

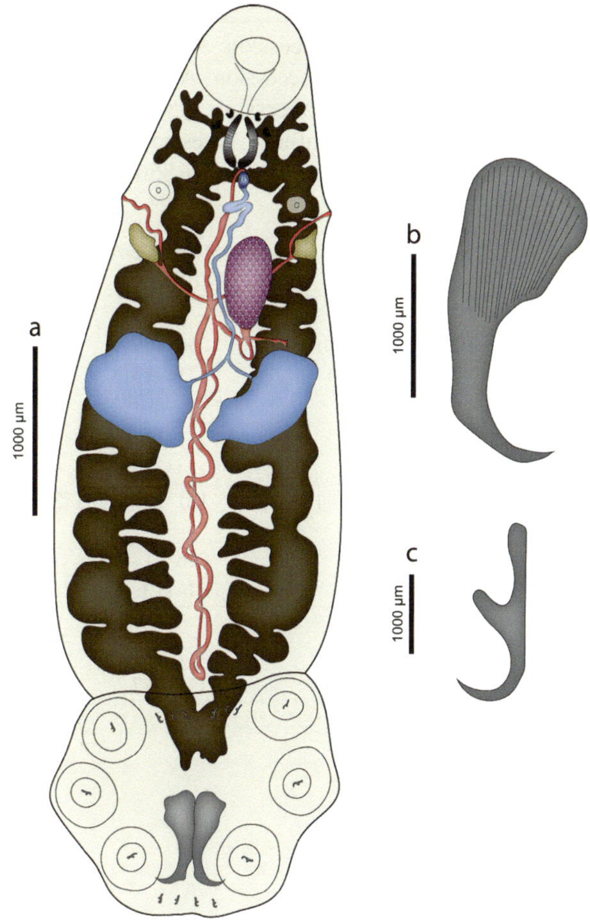

Fig. 4.48 (**a**) *Neodiplorchis scaphiopodis*, (**b**) genital spines, (**c**) marginal hooklets. Redrawn from Rodgers (1941)

Host geographical distribution *Spea bombifrons* is native to Canada, the United States and Mexico, where it is present in various regions including Arizona, Chihuahua, Colorado, Iowa, Missouri, Montana and Texas. This species inhabits terrestrial (Shrublands, semi-desert regions and grasslands) as well as freshwater (water pools caused by rainfalls) ecosystems (Barrera and Hammerson 2004).

Host conservation status The Plains Spadefoot has a wide distribution and a presumably large population that can tolerate a wide range of habitats. It is thus listed as Least Concern by The IUCN Red List of Threatened Species 2015: e. T59044A53972679. https://doi.org/10.2305/IUCN.UK.2015-4.RLTS. T59044A53972679.en. Accessed on 30 January 2022.

Site on host Urinary bladder.

DNA sequence
- 18S rRNA gene, 573 bp. GenBank Accession: AJ297779
- 18S rRNA gene, 1994 bp. GenBank Accession: AM051067
- 28S rRNA gene, 1396 bp. GenBank Accession: AM157201
- Cytochrome c Oxidase I, 395 bp. GenBank Accession: KR856165
- 12S rRNA gene, 498 bp. GenBank Accession: KR856078

Morphology and morphometrics (μm):

Source	Rogers (1941)	Lamothe-Argumedo (1973)	Brooks (1976)
Sample size	17		3
Body length	2150–4350 (3390)	3847–5940	4454–4495
Greatest width	780–1780 (1310)	1336–1513	1465–1554
Width at vagina			
Haptor length	535–959 (763)	917–1127	1000
Haptor width	666–1233 (943)	1244–1529	1336–1406
False oral sucker width	140–219 (173) × 180–280 (228)	161–187 × 241–304	159–188 × 248
Pharynx length	112–213 (163)	206–217	139–218
Pharynx width	58–166 (110)	97–113	128–148
Number of testes	2		
Ovary length	133–466 (209)	177–251	
Ovary width	60–286 (125)	130–177	
Egg length			
Egg width			
Egg incubation time			
Number of intra-uterine eggs			
Genital bulb width	40–52 (46)		
Number of genital spines	8		8
Genital spine length			
Haptoral sucker diameter	159–326 (245)	273–322 × 289–322	267–277
Hamulus handle length X	179–406 (265)		406
Hamulus guard length Y			
Hamulus hook length			40–89
Marginal hooklet length	18–28.5 (23)		
Haptoral L/Body L	0.23	0.21	0.22
No of anastomoses		6–13	

Remarks *Neodiplorchis scaphiopodis* was originally described as *Diplorchis scaphiopi* Rodgers, 1941.

4.4.9 Neoriojatrema Imkongwapang and Tandon, 2010

4.4.9.1 *Neoriojatrema mokokchungense* Imkongwapang and Tandon, 2010 (Fig. 4.49)

Collection Zoological Survey of India, Kolkata. Paratypes deposited in the Helminthological Collection of the Department of Zoology, NEHU, Shillong, India.

Holotype W9247/1.

Paratypes 1.1–1.5.

Original description Imkongwapang and Tandon (2010).

Taxonomic note The original name *Neoriojatrema mokokchungensis* was corrected for gender.

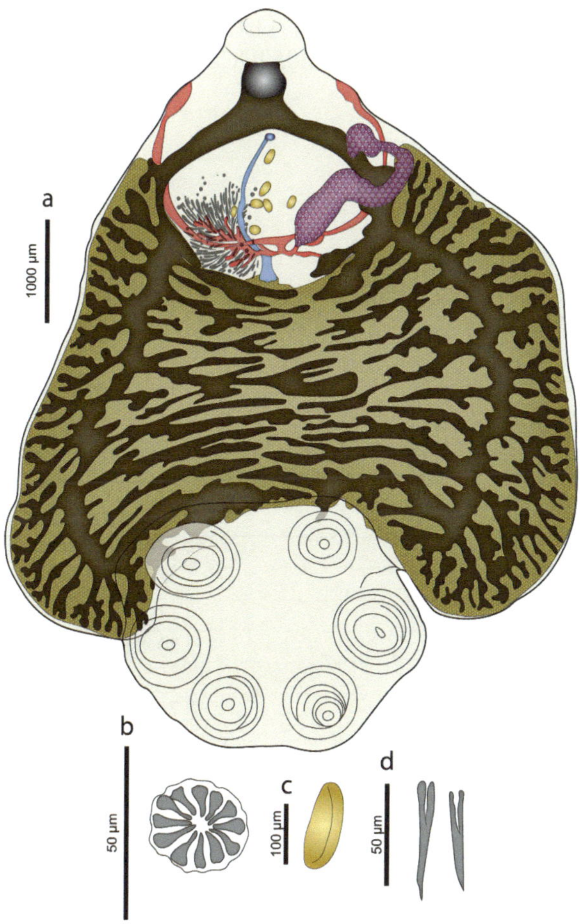

Fig. 4.49 (**a**) *Neoriojatrema mokokchungense*, (**b**) genital spines, (**c**) egg, (**d**) hamuli. Redrawn from Imkongwapang and Tandon (2010)

Other taxonomic contributions None.

Etymology Named after the type locality.

Type locality Mokokchung and Tuensang districts, Nagaland, India.

Other localities None.

Type host *Xenophrys glandulosa* (Fei, Ye and Huang, 1990), Jingdong Spadefoot Toad.

Additional hosts None.

Host geographical distribution *Xenophrys glandulosa* occurs in India and China where it can be found in bushes and shrubs close to streams in montane tropical forest regions (Feng and Ohler 2004).

Host conservation status The Jingdong Spadefoot Toad has a large population and a wide distribution, even though the habitats of this species are declining. It is listed as Least Concern by The IUCN Red List of Threatened Species 2004: e. T57638A11667930. https://doi.org/10.2305/IUCN.UK.2004.RLT

Site on host Urinary bladder.

DNA sequence None.

Morphology and morphometrics (μm):	
Source	Imkongwapang and Tandon (2010)
Sample size	8
Body length	6969–13,340 (9565)
Greatest width	3795–7820 (5911)
Width at vagina	1610–2116 (1932)
Haptor length	2093–4554 (2837)
Haptor width	2093–3358 (2590)
False oral sucker width	667–805 (747) × 299–506 (393)
Pharynx length	345–414 (382)
Pharynx width	299–437 (376)
Ovary length	1495–2484 (2236)
Ovary width	299–460 (391)
Egg length	138–167 (146)
Egg width	80–85 (82)
Egg incubation time	
Intra-uterine eggs (n)	7–49
Genital bulb width	
Number of genital spines	11–12
Genital spine length	1–2 (1)
Haptoral sucker diameter	529–897 × 529–1.012
Hamulus handle length X	40–52 (44)
Hamulus guard length Y	
Hamulus hook length	

(continued)

Morphology and morphometrics (μm):	
Source	Imkongwapang and Tandon (2010)
Marginal hooklet length	
Haptoral L/Body L	0.3
No of anastomoses	0

Remarks *Neoriojatrema mokokchungense* was originally described as *Neoriojatrema mokokchungensis*. Hamuli absent but authors refer to microhooks. Uncertain if these are marginal hooklets or primordial hamuli.

4.4.10 Parapolystoma Ozaki, 1935

4.4.10.1 *Parapolystoma bulliense* (Johnston, 1912) (Fig. 4.50)

Collection Queensland Museum, Australia (QM).

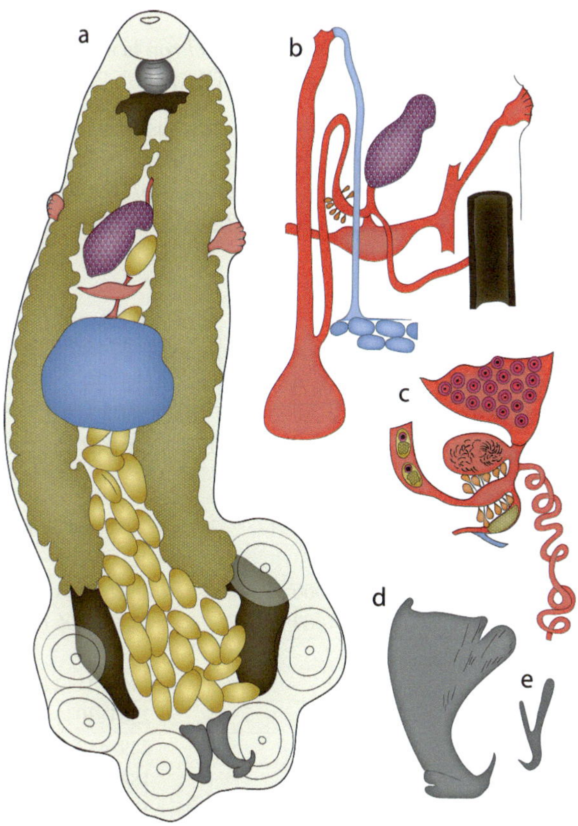

Fig. 4.50 (**a**) *Parapolystoma bulliense*, (**b**) reproductive system, (**c**) reproductive system, (**d**) hamuli, (**e**) marginal hooklets. Redrawn from Johnston (1912)

Holotype QM G206706; AHC S2220.

Paratypes AHC S2217–19.

Original description Johnston (1912).

Other taxonomic contributions Williams (1960a); Pichelin (1995); Barton and Richards (1996); Dutta and Tandon (1998).

Etymology Named after the type locality.

Type locality Tributary of the Loddon River near Bulli, New South Wales, Australia.

Other localities None.

Type host *Ranoidea phyllochroa* (Günther, 1863), Leaf Green Tree Frog. In the original description of the polystome, the host was known as *Litoria phyllochroa* Tyler, 1971.

Additional hosts *Ranoidea lesueurii* (Duméril and Bibron, 1841), *Ranoidea pearsoniana* (Copland 1961) and *Ranoidea genimaculata* (Horst, 1883).

Host geographical distribution *Ranoidea phyllochroa* is native to Australia where it occurs in New South Wales and Queensland (Hero et al. 2004a).

Host conservation status The Leaf Green Treefrog has a large population and wide distribution. It is thus is listed as Least Concern by The IUCN Red List of Threatened Species 2008: e.T41107A10399441. https://doi.org/10.2305/IUCN.UK.2008.RLTS.T41107A10399441.en. Accessed on 30 January 2022.

Site on host Urinary bladder.

DNA sequence
- 18S rRNA gene, 2089 bp. GenBank Accession: AM157186
- 28S rRNA gene, 1450 bp. GenBank Accession: AM157202
- 12S rRNA gene, 489 bp. GenBank Accession: KR856079
- Cytochrome c Oxidase I, 380 bp. GenBank Accession: KR856166

Morphology and morphometrics (μm):		
Source	Johnston (1912)	Dutta and Tandon (1998)
Sample size	16	22
Body length	2930–3000	7344–11,903 (9387)
Greatest width	1000–1007	2493–3396 (2900)
Width at vagina		
Haptor length		1453–2142 (1973)
Haptor width	1400	2264–3044 (2772)
False oral sucker width		229–413 (373) × 382–535 (450)
Pharynx length		184–275 (244)
Pharynx width		153–306 (243)
Ovary length		780–1040 (966)

(continued)

Morphology and morphometrics (μm):		
Source	Johnston (1912)	Dutta and Tandon (1998)
Ovary width		214–291 (269)
Egg length	206	220
Egg width	206 × 95	130–150 (140)
Egg incubation time		
Intra-uterine eggs (n)		
Genital bulb width		
Number of genital spines	8	11
Genital spine length		23–30 (25)
Haptoral sucker diameter	268	
Hamulus handle length X	290	428–703 (550)
Hamulus guard length Y		
Hamulus hook length	50	
Marginal hooklet length		19–34 (25)
Haptoral L/Body L		0.21
No of anastomoses	0	

Remarks *Parapolystoma bulliense* was originally described as *Polystomum bulliense* Johnston, 1912.

4.4.10.2 *Parapolystoma johnstoni* Pichelin, 1995 (Fig. 4.51)

Collection Queensland Museum, Australia (QM).

Holotype QM G211519.

Paratype QM G211520.

Original description Pichelin (1995).

Other taxonomic contributions None.

Etymology Named after Dr SJ Johnston, University of Sydney, Australia.

Type locality Queensland, Australia.

Other localities None.

Type host *Ranoidea nyakalensis* (Liem, 1974), Mountain Mistfrog. In the original description of the polystome, the host was known as *Litoria nyakalensis* Liem, 1974.

Additional hosts None.

Host geographical distribution *Ranoidea nyakalensis* occurs in Australia, where it lives in forests along fast-flowing streams usually on rocks or overhanging vegetation (Hero et al. 2004b).

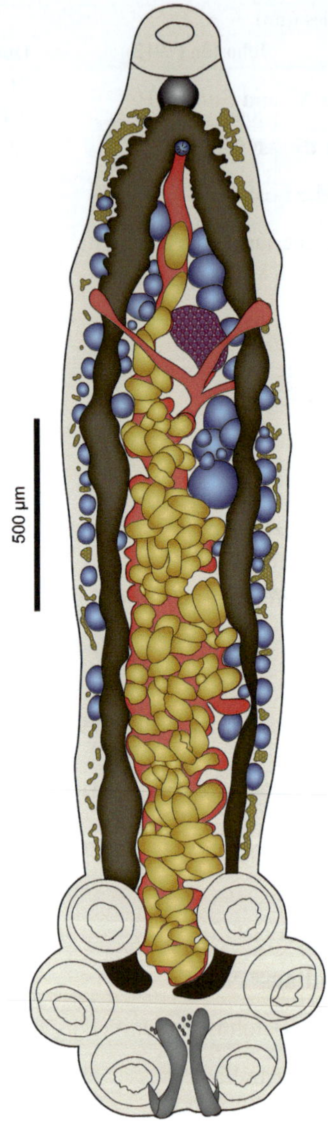

Fig. 4.51 *Parapolystoma johnstoni*. Redrawn from Pichelin (1995)

Host conservation status The Mountain Mistfrog has an extremely small population size. The threats to this species are unknown as the species habitat is protected, however, a potential reason for the population decline may be due to viral infection or chytrid fungus as well as feral pig activity. It is listed as Critically Endangered by The IUCN Red List of Threatened Species 2004: e.T12149A3326835. https://doi.org/10.2305/IUCN.UK.2004.RLTS.T12149A3326835.en. Accessed on 30 January 2022.

Site on host Urinary bladder.

DNA sequence None.

Morphology and morphometrics (μm):	
Source	Pichelin (1995)
Sample size	2
Body length	2860
Greatest width	601
Width at vagina	
Haptor length	620–636
Haptor width	795–822
False oral sucker width	230
Pharynx length	125
Pharynx width	100–110
Ovary length	
Ovary width	
Egg length	115–130 (123)
Egg width	55–60 (58)
Egg incubation time	
Number of intra-uterine eggs	
Genital bulb width	
Number of genital spines	6–7
Genital spine length	
Haptoral sucker diameter	210–212 (211)
Hamulus handle length X	280–285 (282)
Hamulus guard length Y	
Hamulus hook length	
Marginal hooklet length	
Haptoral L/Body L	0.22
No of anastomoses	0

Remarks None.

4.4.11 *Polystoma* Zeder, 1800

4.4.11.1 *Polystoma aeschlimanni* Bourgat and Murith, 1980 (Fig. 4.52)

Collection Claude Combes personal collection, Perpignan, France.

Holotype Unknown.

Paratypes Unknown.

Original description Bourgat and Murith (1980).

Fig. 4.52 (**a**) *Polystoma aeschlimanni*, (**b**) hamuli, (**c**) marginal hooklets. Redrawn from Bourgat and Murith (1980)

Other taxonomic contributions None.

Etymology Named after Prof A. Aeschlimann, University of Fribourg.

Type locality Kanté, Togo.

Other localities None.

Type host *Ptychadena pumilio* (Boulenger, 1920), Medine Grassland Frog.

Additional hosts None.

Host geographical distribution *Ptychadena pumilio* occurs in Cameroon, Congo, Ivory Coast, Senegal, Sierra Leone, Ethiopia, Benin, Nigeria, Central African Republic and Mali (Frost 2021).

Host conservation status The Medine Grassland Frog has a large population, a wide distribution and can tolerate various habitat ranges. It is thus listed as Least Concern by The IUCN Red List of Threatened Species 2013: e.T58521A18401115. https://doi.org/10.2305/IUCN.UK.2013-2.RLTS.T58521A18401115.en. Accessed on 30 January 2022.

Site on host Urinary bladder.

DNA sequence None.

Morphology and morphometrics (μm):	
Source	Bourgat and Murith (1980)
Sample size	3
Body length	5200–6400
Greatest width	2100–2600
Width at vagina	
Haptor length	1100–1500 (1300)
Haptor width	1700–2700 (2200)
False oral sucker width	
Pharynx length	
Pharynx width	
Ovary length	
Ovary width	
Egg length	183–194
Egg width	121–123
Egg incubation time	
Number of intra-uterine eggs	11
Genital bulb width	
Number of genital spines	
Genital spine length	
Haptoral sucker diameter	344–470
Hamulus handle length X	378–495
Hamulus guard length Y	
Hamulus hook length	
Marginal hooklet length	36
Haptoral L/Body L	0.21–0.23
No of anastomoses	1–2

Remarks This species has been reported from the same host as *Polystoma lamottei* Bourgat and Murith, 1980 in Togo.

4.4.11.2 *Polystoma aethiopiense* Meskal, 1970 (Fig. 4.53)

Collection Unknown.

Holotype Unknown.

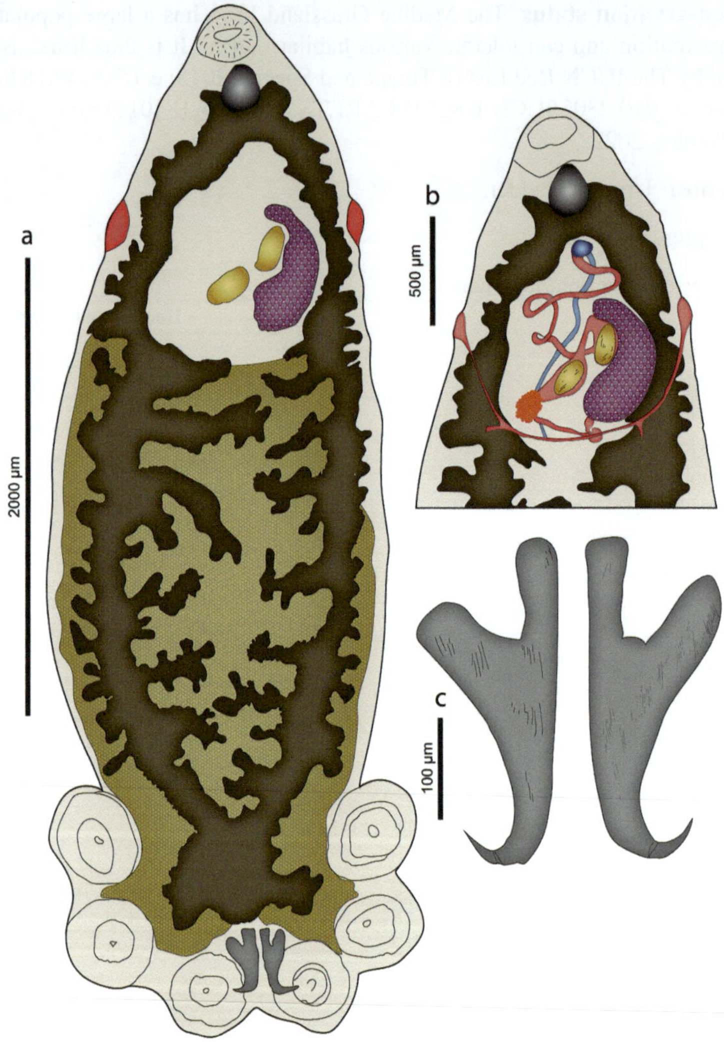

Fig. 4.53 (**a**) *Polystoma aethiopiense*, (**b**) reproductive system, (**c**) hamuli. Redrawn from Meskal (1970)

Paratypes Unknown.

Original description Meskal (1970).

Other taxonomic contributions None.

Etymology Named after type locality country.

Type locality Lake Awassa, Ethiopia.

Other localities Ethiopia: Lake Akaki, Awassa march, Black River and River Huluka.

Type host *Sclerophrys regularis*, Common African Toad. In the original description of the polystome, the host was known as *Bufo regularis*.

Additional hosts *Ptychadena nilotica* (Seetzen, 1855) and *Amietia angolensis* (Bocage 1866).

Host geographical distribution *Sclerophrys regularis* occurs in Egypt, Ethiopia, Sudan, Uganda, Tanzania, Kenya, Democratic Republic of Congo and Angola. The natural terrestrial and freshwater habitats in which this species is present are montane grasslands, rivers and savannahs (dry and moist) (Tandy et al. 2004).

Host conservation status The Common African Toad is widely distributed, has a tolerance for a wide range of habitats and has a presumed large population. It is listed as Least Concern by The IUCN Red List of Threatened Species 2016: e. T54747A107349840. https://doi.org/10.2305/IUCN.UK.2016-3.RLTS. T54747A107349840.en. Accessed on 30 January 2022.

Site on host Urinary bladder.

DNA sequence None.

Morphology and morphometrics (μm):	
Source	Meskal (1970)
Sample size	
Body length	3920–5800 (4708)
Greatest width	1100–1600 (1377)
Width at vagina	
Haptor length	900–1080 (980)
Haptor width	815–1710 (1443)
False oral sucker width	275–300 (282)
Pharynx length	165–220 (195)
Pharynx width	187–250 (208)
Ovary length	450–700 (606)
Ovary width	165–380 (288)
Egg length	180–220 (195)
Egg width	99–132 (103)
Egg incubation time	
Number of intra-uterine eggs	5
Genital bulb width	55–77 (69)
Number of genital spines	8
Genital spine length	33
Haptoral sucker diameter	292–459 (353)
Hamulus handle length X	275–400 (330)
Hamulus guard length Y	
Hamulus hook length	

(continued)

Morphology and morphometrics (µm):	
Source	Meskal (1970)
Marginal hooklet length	
Haptoral L/Body L	0.18–0.23 (0.21)
No of anastomoses	0

Remarks *Polystoma aethiopiense* was originally described as *Polystoma africanum aethiopiense*. This species has been reported from the same host as *E. alluaudi* in Cameroun, Nigeria, Tanzania and Togo, and *P. africanum* Szidat, 1932 in Liberia, Togo and Sierra Leone. Tinsley (1974a) stated that *P. aethiopiense* was conspecific with *Polystoma pricei* Vercammen-Grandjean, 1960 and that the name *P. pricei* should take precedence. Since *P. pricei* was described from *Ptychadena* sp. and *P. aethiopiense* from *S. regularis*, we still consider *P. aethiopiense* as a valid species.

4.4.11.3 *Polystoma africanum* Szidat, 1932 (Fig. 4.54)

Collection Invertebrate Collection of the Professor A.B.M. Egborge Museum, University of Benin City, Nigeria (LPRM), Natural History Museum, London (NHML).

Holotype Unknown.

Paratypes Unknown.

Supplementary specimens LPRM2-0710-8a-d, NHML 2008.11.18.1-6.

Original description Szidat (1932).

Other taxonomic contributions Williams (1969); Tinsley (1974a); Salami-Cadoux (1980); Aisien and Du Preez (2009).

Etymology Named after the continent Africa.

Type locality Libéria.

Other localities Sierra Leone, Ethiopia, Togo, Uganda and Zaire.

Type host *Sclerophrys regularis*, Common African Toad. In the original description of the polystome, the host was known as *Bufo regularis*.

Additional hosts *Ptychadena mascareniensis*.

Site on host Urinary bladder.

Host geographical distribution *Sclerophrys regularis* occurs in Egypt, Ethiopia, Sudan, Uganda, Tanzania, Kenya, Democratic Republic of Congo and Angola. The natural terrestrial and freshwater habitats in which this species is present are montane grasslands, rivers and savannahs (dry and moist) (Tandy et al. 2004).

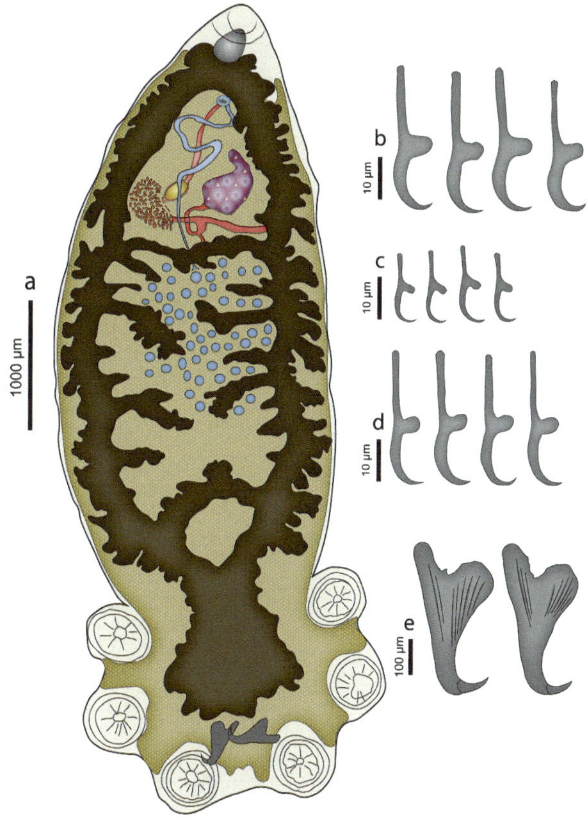

Fig. 4.54 (**a**) *Polystoma africanum*, (**b**) marginal hooklets (1), (**c**) marginal hooklets (2–7), (**d**) marginal hooklets (8), (**e**) hamuli. Redrawn from Szidat (1932)

Host conservation status The Common African Toad is widely distributed, has a tolerance for a wide range of habitats and has a presumed large population. It is listed as Least Concern as it is by The IUCN Red List of Threatened Species 2016: e. T54747A107349840. https://doi.org/10.2305/IUCN.UK.2016-3.RLTS.T54747A107349840.en. Accessed on 30 January 2022.

DNA sequence None.

Morphology and morphometrics (µm):			
Source	Szidat (1932)	Tinsley (1974a)	Aisien and Du Preez (2009)
Sample size		5	20
Body length	4800	4220–5200 (4690)	5162–7481 (6015)
Greatest width	1800	1600–2040 (1790)	1835–2586 (2219)
Width at vagina		900–1240 (1060)	
Haptor length		1220–1430 (1320)	1352–1887 (1572)

(continued)

Morphology and morphometrics (μm):			
Source	Szidat (1932)	Tinsley (1974a)	Aisien and Du Preez (2009)
Haptor width	2200	1790–2310 (1970)	1908–2713 (2245)
False oral sucker width		350–370 (360)	288–577 (431)
Pharynx length	200	160–190 (180)	204–300 (244)
Pharynx width		160–180 (170)	194–325 (232)
Ovary length		431–513 (460)	447–766 (628)
Ovary width		122–203 (154)	154–371 (273)
Egg length		172–205 (189)	174–197 (185)
Egg width		93–115 (104)	132–147 (142)
Egg incubation time			
Number of intra-uterine eggs		2–5	1–3
Genital bulb width		66–78 (73)	
Number of genital spines	9	8–9	7–8
Genital spine length		35–36	28–38 (34)
Haptoral sucker diameter	370	350–430	315–491 (407)
Hamulus handle length X			335–452 (388)
Hamulus guard length Y			
Hamulus hook length			72–93 (85)
Marginal hooklet length			17–38.6
Haptoral L/Body L			0.22–0.28 (0.26)
No of anastomoses	0		0–4

Remarks *Polystoma africanum* was originally described as *Polystomum africanum*. The additional host listed as *P. mascareniensis* is probably *Ptychadena nilotica* in Ethiopia, and yet undescribed cryptic forms in Sierra Leone, Togo, Uganda and Zaire. *Polystoma africanum* most likely consists of a species complex. This species has been reported from the same host as *Eupolystoma alluaudi* in Cameroun, Nigeria, Tanzania and Togo, and *Polystoma aethiopiense* in Ethiopia.

4.4.11.4 *Polystoma andinum* Combes and Laurent, 1978 (Fig. 4.55)

Collection Unknown.

Holotype Unknown.

Paratypes Unknown.

Original description Combes and Laurent (1978).

Other taxonomic contributions Vaira (2004).

Etymology Named after the Andes mountains.

Type locality Calilegua, Argentina.

Other localities None.

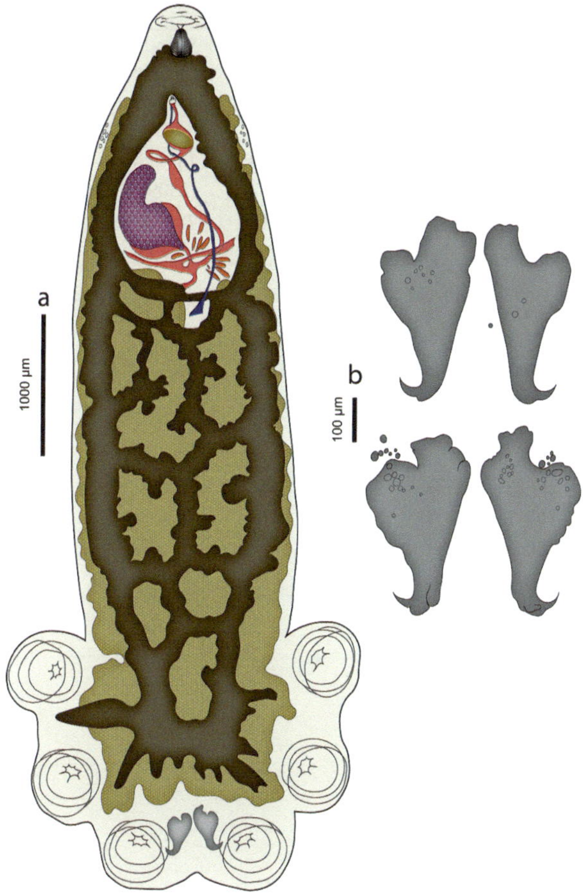

Fig. 4.55 (**a**) *Polystoma andinum*, (**b**) hamuli. Redrawn from Combes and Laurent (1978)

Type host *Melanophryniscus rubriventris* (Vellard, 1947), Yungas Redbelly Toad.

Additional hosts None.

Host geographical distribution *Melanophryniscus rubriventris* occurs in Bolivia, inhabiting dry forest and disturbed habitats, as well as in Argentina, where it inhabits wet montane forests and disturbed habitats (such as roadside ditches) (Cortez et al. 2010).

Host conservation status The Yungas Redbelly Toad has a large population, a wide distribution and is tolerant for habitat modification. It is listed as Least Concern by The IUCN Red List of Threatened Species 2020: e.T54827A101424224. https://doi.org/10.2305/IUCN.UK.2020-3.RLTS.T54827A101424224.en. Accessed on 30 January 2022.

Site on host Urinary bladder.

DNA sequence None.

Morphology and morphometrics (μm):				
Source	Combes and Laurent (1978)	Vaira (2004) Locality: Abra Colorada	Vaira (2004) Locality: Sidras	Vaira (2004) Locality: La Gateada
Sample size	61	10–17	2–14	6–8
Body length	4900–8000 (6100)	1945–4056 (3091)	1414–3203 (2309)	2912–3484 (3212)
Greatest width	1500–2400 (1900)	1144–2038 (1505)	780–1768 (1274)	1300–1612 (1420)
Width at vagina				
Haptor length	1300–2300 (1600)	926–1560 (1249)	946–1404 (1175)	1175–1487 (1281)
Haptor width	1400–3300 (2500)	1175–1716 (1466)	967–1924 (1446)	1425–1737 (1588)
False oral sucker width				
Pharynx length	200–305 (243)			
Pharynx width	195–270 (223)			
Ovary length	570–940 (725)			
Ovary width	340–600 (430)			
Egg length	230–283 (246)			
Egg width	125–135 (133)			
Egg incubation time				
Number of intra-uterine eggs	3			
Genital bulb width				
Number of genital spines	8			
Genital spine length				
Haptoral sucker diameter	380–585 (490)	208–333 (253)	249–260 (255)	218–281 (246)
Hamulus handle length X	370–480 (425)	374–464 (425)	346–445 (399)	390–461 (427)
Hamulus guard length Y		329–390 (361)	243–378 (330)	342–413 (379)
Hamulus hook length		51–70 (61)	51–67 (60)	54–70 (65)
Marginal hooklet length				
Haptoral L/Body L	0.28–0.31	0.39–0.48	0.44–0.67	0.40–0.43
No of anastomoses	>1			

Remarks None.

4.4.11.5 *Polystoma assoulinei* Bourgat, 1975 (Fig. 4.56)

Collection Unknown.

Syntypes 423.0 and 423.1.

Original description Bourgat (1975).

Other taxonomic contributions None.

Etymology Named after Mr MS Assouline, Togo.

Type locality Kandé, North Togo.

Other localities None.

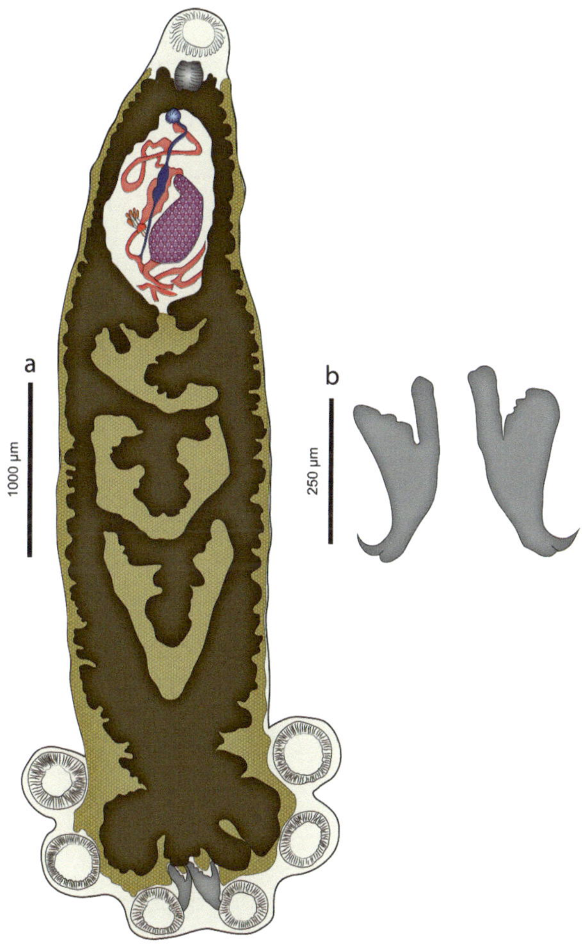

Fig. 4.56 (**a**) *Polystoma assoulinei*, (**b**) hamuli. Redrawn from Bourgat (1975)

Type host *Ptychadena tellinii* (Peracca, 1904), Central Grassland Frog. In the original description of the polystome, the host was known as *Ptychadena huguettae* Inger, 1968.

Additional hosts None.

Host geographical distribution *Ptychadena tellinii* occurs in Togo, Cameroon, Nigeria, Sierra Leone, Ethiopia, Mali, Ghana, Congo and Benin where it inhabits temporary pools, rivers, agricultural land, savannas and forests (Frost 2021).

Host conservation status The Central Grassland Frog has a large population, a wide geographical distribution and is tolerant for various habitats. It is listed as Least Concern by The IUCN Red List of Threatened Species 2013: e.T54357A18402131. https://doi.org/10.2305/IUCN.UK.2013-2.RLTS.T54357A18402131.en. Accessed on 30 January 2022.

Site on host Urinary bladder.

DNA sequence None.

Morphology and morphometrics (µm):	
Source	Bourgat (1975)
Sample size	2
Body length	4700–5300
Greatest width	1450
Width at vagina	950–1100
Haptor length	1900–2150
Haptor width	
False oral sucker width	
Pharynx length	202–216
Pharynx width	187–192
Ovary length	481
Ovary width	255–289
Egg length	
Egg width	
Egg incubation time	
Number of intra-uterine eggs	
Genital bulb width	
Number of genital spines	8–9
Genital spine length	
Haptoral sucker diameter	310–356
Hamulus handle length X	330–353
Hamulus guard length Y	
Hamulus hook length	
Marginal hooklet length	40
Haptoral L/Body L	0.41
No of anastomoses	1–4

Remarks None.

4.4.11.6 *Polystoma australe* Kok and Van Wyk, 1986 (Fig. 4.57)

Collection Parasitic Worm Collection, National Museum, Bloemfontein, South Africa

Holotype NMBP 1.

Paratypes NMBP 2–16.

Original description Kok and Van Wyk (1986).

Taxonomic note The original name *Polystoma australis* was corrected for gender.

Other taxonomic contributions Kok and Du Preez (1989).

Etymology Named after the southern distribution of the species.

Fig. 4.57 (**a**) *Polystoma australe*, (**b**) marginal hooklets, (**c**) hamuli. Redrawn from Kok and van Wyk (1986)

Type locality Farm Greenlands, Harrismith, South Africa (28°17"S; 29°00″E).

Other localities None.

Type host *Semnodactylus wealii (*Boulenger, 1882), Weale's Running Frog.

Additional hosts *Kassina senegalensis* (Duméril and Bibron, 1841).

Host geographical distribution *Semnodactylus wealii* occurs in South Africa, Lesotho and Swaziland where it mainly inhabits grassland habitats (Frost 2021).

Host conservation status The Weale's Running Frog has a large population, a wide distribution and is tolerant for habitat modification. It is listed as Least Concern by The IUCN Red List of Threatened Species 2013: e.T56297A3037487. https://doi.org/10.2305/IUCN.UK.2013-2.RLTS.T56297A3037487.en. Accessed on 30 January.

Site on host Urinary bladder and gills of tadpoles.

DNA sequence

- 18S rRNA gene, 614 bp. GenBank Accession: AJ297771
- 28S rRNA gene, 1448 bp. GenBank Accession: AM913872
- Cytochrome c Oxidase I, 378 bp. GenBank Accession: AM913854
- ITS1, 700 bp. GenBank Accession: AJ310398

Morphology and morphometrics (μm):		
Source	Kok and Van Wyk (1986)	Kok and Du Preez (1989)
Sample size	3	11
Body length	(6730)	3660
Greatest width	2530	1490
Width at vagina	1580	
Haptor length	1740	700
Haptor width	2490	1740
False oral sucker width		
Pharynx length	251	281
Pharynx width	238	311
Ovary length	784	734
Ovary width		206
Egg length	237.4	
Egg width	179.7	
Egg incubation time		
Number of intra-uterine eggs	5–7	
Genital bulb width		
Number of genital spines	7–8 (usually 8)	
Genital spine length	25–30	
Haptoral sucker diameter	376	278–285
Hamulus handle length X	326	
Hamulus guard length Y		

(continued)

Morphology and morphometrics (μm):		
Source	Kok and Van Wyk (1986)	Kok and Du Preez (1989)
Hamulus hook length		
Marginal hooklet length	1: 42; 2–7: 23–26; 8: 38	
Haptoral L/Body L	0.26	0.19
No of anastomoses	0	

Remarks *Polystoma australe* was originally described as *Polystoma australis.*

4.4.11.7 *Polystoma baeri* Maeder, Euzet and Combes, 1970 (Fig. 4.58)

Collection Claude Combes personal collection, Perpignan, France.

Holotype Unknown.

Paratypes Unknown.

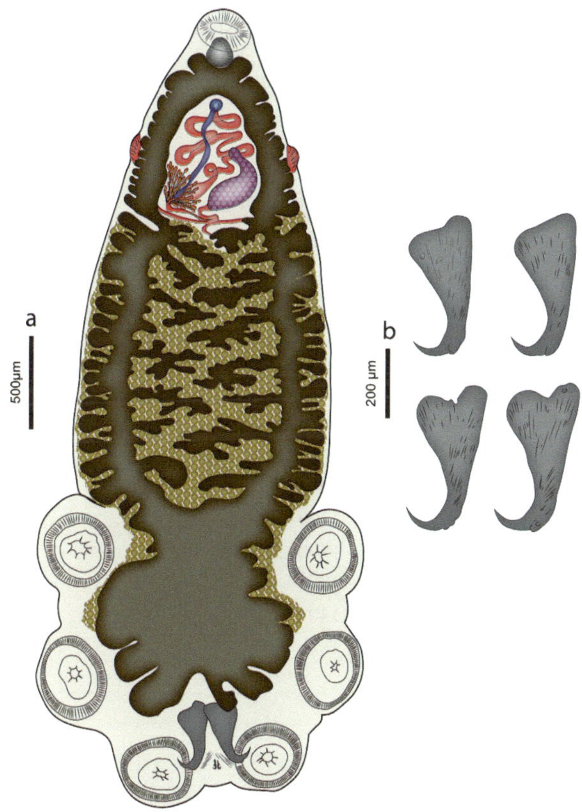

Fig. 4.58 (**a**) *Polystoma baeri*, (**b**) hamuli. Redrawn from Maeder et al. (1970)

Original description Maeder et al. (1970).

Other taxonomic contributions Maeder (1973); Bourgat (1977); Murith (1979, 1981); Murith et al. (1978).

Etymology Named after Prof Jean G Baer, Paris, France.

Type locality Lamto, Ivory Coast.

Other localities Togo and Cameroon.

Type host *Ptychadena bibroni* (Hallowell, 1845), MacCarthy Grassland Frog. In the original description of the polystome, the host was known as *Ptychadena maccarthyensis* Anderson, 1937.

Additional hosts *Hoplobatrachus occipitalis.*

Host geographical distribution *Ptychadena bibroni* occurs in Cameroon, Congo, Chad, Togo, Nigeria, Senegal, Ghana, Gambia and Guinea. This species is usually found in humid, wooded savannahs, degraded former forests and dry savannahs (Frost 2021).

Host conservation status The MacCarthy Grassland Frog has a wide distribution, a large population and is tolerant for a range of habitats. It is listed as Least Concern by The IUCN Red List of Threatened Species 2013: e.T58493A18399180. https://doi.org/10.2305/IUCN.UK.2013-2.RLTS.T58493A18399180.en. Accessed on 30 January 2022.

Site on host Urinary bladder and gills of tadpoles.

DNA sequence

- ITS1, 710 bp. GenBank Accession: AJ310409

Morphology and morphometrics (μm):		
Source	Maeder et al. (1970)	Murith (1981)
Sample size	4	
Body length	3200–4400	4480–9340 (6690)
Greatest width	1000–1500	1000–3100 (2160)
Width at vagina	700–900	970–2170 (1580)
Haptor length	1100–1600	900–2070 (1740)
Haptor width	1500–1700	1380–2830 (2220)
False oral sucker width	300	
Pharynx length	140	
Pharynx width	180	
Ovary length		
Ovary width		
Egg length		200–228 (221)
Egg width		122–167 (137)
Egg incubation time		
Number of intra-uterine eggs		0–4

(continued)

Morphology and morphometrics (μm):		
Source	Maeder et al. (1970)	Murith (1981)
Genital bulb width		
Number of genital spines	8	
Genital spine length	35	
Haptoral sucker diameter	280–400	310–720 (460)
Hamulus handle length X	330–420	430–590 (500)
Hamulus guard length Y		
Hamulus hook length	35–40	
Marginal hooklet length		41
Haptoral L/Body L		0.21–0.39 (0.26)
No of anastomoses	0–3	

Remarks None.

4.4.11.8 *Polystoma batchvarovi* Euzet, Combes and Knoepffler, 1974 (Fig. 4.59)

Collection Claude Combes personal collection, Perpignan, France.

Holotype Unknown.

Paratypes Unknown.

Original description Euzet et al. (1974a).

Other taxonomic contributions Murith (1981); Murith et al. (1978).

Etymology Named after Dr GK Batchvarov, University of Plovdiv, Bulgaria.

Type locality Lamaboké, Central Africa Republic.

Other localities Cameroon.

Type host *Hyperolius tuberculatus* (Mocquard, 1897), Rainforest Reed Frog.

Additional hosts None.

Host geographical distribution *Hyperolius tuberculatus* occurs in Nigeria, Cameroon, Congo, Gabon, Equatorial Guinea and Central African Republic. This species occupies heavily degraded former forests and forest clearings (Frost 2021).

Host conservation status The Rainforest Reed Frog has a presumed large population, a wide distribution range and is tolerant for various habitats. It is listed as Least Concern by The IUCN Red List of Threatened Species 2017: e.T56216A97523116. https://doi.org/10.2305/IUCN.UK.2017-2.RLTS.T56216A97523116.en. Accessed on 30 January 2022.

Site on host Urinary bladder.

DNA sequence None.

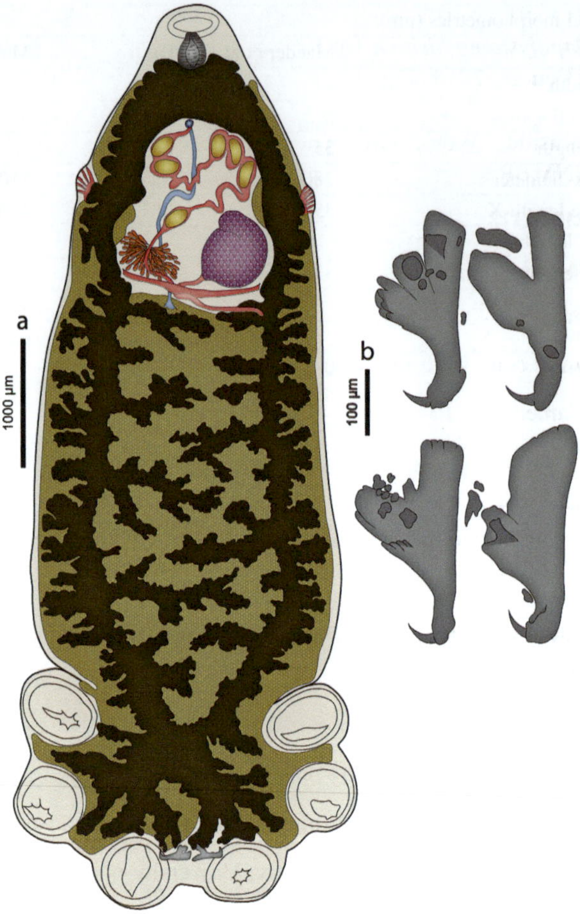

Fig. 4.59 (**a**) *Polystoma batchvarovi*, (**b**) hamuli. Redrawn from Euzet et al. (1974)

Morphology and morphometrics (µm):		
Source	Euzet et al. (1974a)	Murith (1981)
Sample size	4	
Body length	6000–7100	
Greatest width	2200–2500	
Width at vagina	1700–1900	
Haptor length	1400–1500	
Haptor width	2500–2900	
False oral sucker width	400–500	
Pharynx length	350–400	
Pharynx width	300–350	
Ovary length		
Ovary width		

(continued)

Morphology and morphometrics (μm):		
Source	Euzet et al. (1974a)	Murith (1981)
Egg length	215	
Egg width	120	
Egg incubation time		
Number of intra-uterine eggs	1, 2, 5 or 6	
Genital bulb width		
Number of genital spines	8	
Genital spine length	40	
Haptoral sucker diameter	500–520	
Hamulus handle length X	300–350	
Hamulus guard length Y		
Hamulus hook length		
Marginal hooklet length		33
Haptoral L/Body L	0.22–0.25	
No of anastomoses	0	

Remarks None.

4.4.11.9 *Polystoma borellii* Combes and Laurent, 1974 (Fig. 4.60)

Collection Unknown.

Holotype Unknown.

Paratypes Unknown.

Original description Combes and Laurent (1974).

Other taxonomic contributions None.

Etymology Named after the host.

Type locality Tucumán, Argentina.

Other localities None.

Type host *Pleurodema borellii* (Peracca, 1895), Rufous Four-Eyed Frog.

Additional hosts None.

Host geographical distribution *Pleurodema borellii* inhabits montane grasslands in Argentina (Cortez et al. 2004).

Host conservation status The Rufous Four-eyed Frog is widespread, has a large population size and is tolerant for habitat modification. It is listed as Least Concern by The IUCN Red List of Threatened Species 2020: e.T57282A101432254. https://doi.org/10.2305/IUCN.UK.2020-3.RLTS.T57282A101432254.en. Accessed on 30 January 2022.

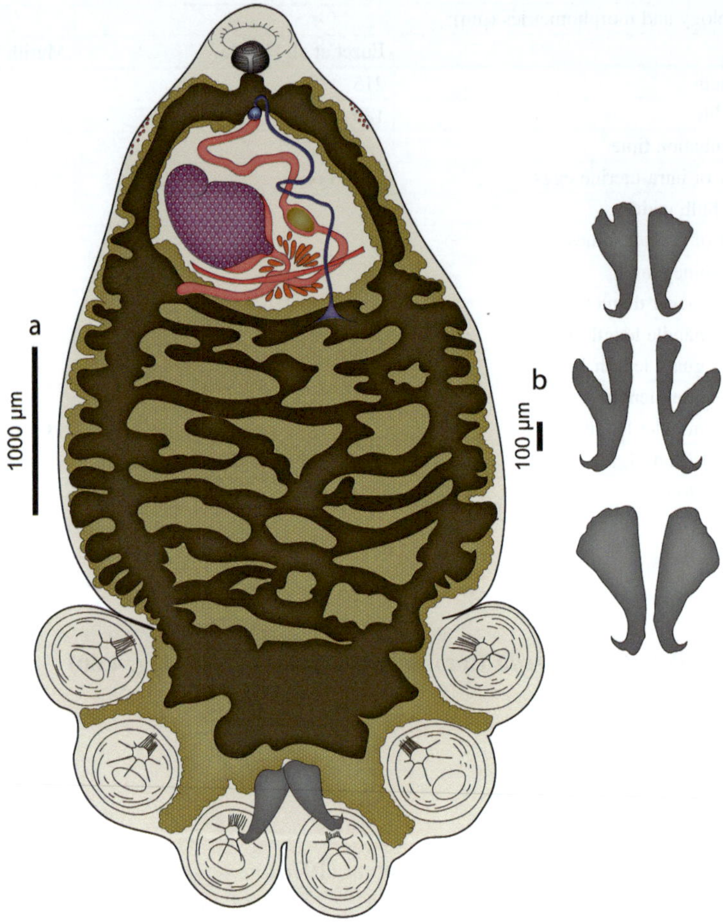

Fig. 4.60 (**a**) *Polystoma borellii*, (**b**) hamuli. Redrawn from Combes and Laurent (1974)

Site on host Urinary bladder.

DNA sequence None.

Morphology and morphometrics (µm):	
Source	Combes and Laurent (1974)
Sample size	11
Body length	4200–5600 (5100)
Greatest width	2000–3200 (2500)
Width at vagina	
Haptor length	1300–1700 (1500)
Haptor width	2300–3200 (2600)
False oral sucker width	

(continued)

Morphology and morphometrics (µm):	
Source	Combes and Laurent (1974)
Pharynx length	229–274 (250)
Pharynx width	200–285 (240)
Ovary length	
Ovary width	
Egg length	
Egg width	120 × 230
Egg incubation time	
Number of intra-uterine eggs	
Genital bulb width	
Number of genital spines	8
Genital spine length	54
Haptoral sucker diameter	510–550
Hamulus handle length X	350–530 (430)
Hamulus guard length Y	
Hamulus hook length	
Marginal hooklet length	
Haptoral L/Body L	0.29
No of anastomoses	Network

Remarks None.

4.4.11.10 *Polystoma channingi* Du Preez, 2013 (Fig. 4.61)

Collection Parasitic Worm Collection, National Museum, Bloemfontein, South Africa (NMBP).

Holotype NMBP 344.

Paratypes NMBP 345–347.

Original description Du Preez (2013).

Other taxonomic contributions None.

Etymology Named after Prof Alan C Channing, Stellenbosch, South Africa.

Type locality Covie and Kenton, South Africa; (33.581298S, 26.665182E).

Other localities None.

Type host *Cacosternum nanum* Boulenger, 1887, Dwarf Dainty Frog and *Cacosternum boettgeri* (Boulenger, 1882), Boettger's Dainty Frog.

Additional hosts None.

Host geographical distribution *Cacosternum nanum* occurs in Swaziland and South Africa, while *Cacosternum boettgeri*, besides these two countries, is also

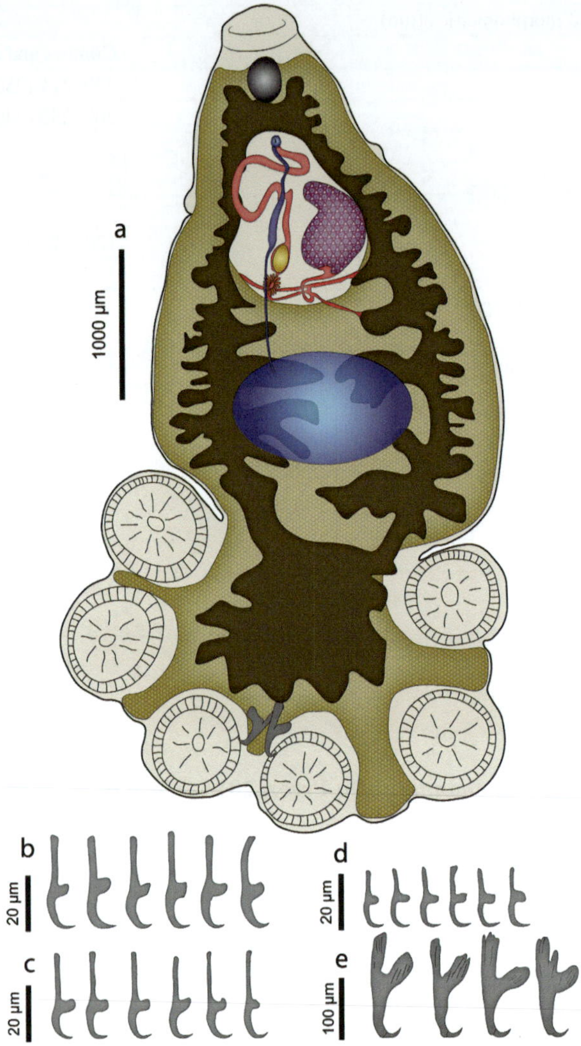

Fig. 4.61 (**a**) *Polystoma channingi*, (**b**) marginal hooklets (1), (**c**) marginal hooklets (8), (**d**) marginal hooklets (2–7), (**e**) hamuli. Redrawn from Du Preez (2013)

found in Botswana, Mozambique, Ethiopia, South Africa, Namibia, Lesotho, Zimbabwe and Zambia. Both frog species are present in grassland and savanna habitats, the former being found in regions with high rainfall such as shrubland, degraded forest and plantations and the latter in open regions with short vegetation such as grassy meadows (Frost 2021).

Host conservation status The Dwarf Dainty Frog and the Boettger's Dainty Frog have a wide distribution, large populations and are tolerant for a range of habitats. They are both listed as Least Concern by The IUCN Red List of Threatened Species

2013: e.T58070A3064515. https://doi.org/10.2305/IUCN.UK.2013-2.RLTS.T58070A3064515.en. Accessed on 30 January 2022.

Site on host Urinary bladder.

DNA sequence None.

Morphology and morphometrics (μm):	
Source	Du Preez (2013)
Sample size	6 mature and 8 immature
Body length	3789–6397 (4960)
Greatest width	884–2427 (1750)
Width at vagina	702–1542 (1299)
Haptor length	1634–2289 (1884)
Haptor width	1284–3114 (2335)
False oral sucker width	277–389 (343)
Pharynx length	200–314 (252)
Pharynx width	170–304 (236)
Ovary length	466–952 (713)
Ovary width	98–441 (274)
Egg length	
Egg width	
Egg incubation time	
Number of intra-uterine eggs	1
Genital bulb width	53–81 (70)
Number of genital spines	7
Genital spine length	29–31 (30)
Haptoral sucker diameter	266–657 (501)
Hamulus handle length X	327–385 (360)
Hamulus guard length Y	228–305 (274)
Hamulus hook length	51–60 (57)
Marginal hooklet length	20–36
Haptoral L/Body L	0.35–0.49 (0.39)
No of anastomoses	0–1

Remarks None.

4.4.11.11 *Polystoma chiromantis* Dupouy and Knoepffler, 1978 (Fig. 4.62)

Collection Unknown.

Holotype Unknown.

Paratypes Unknown.

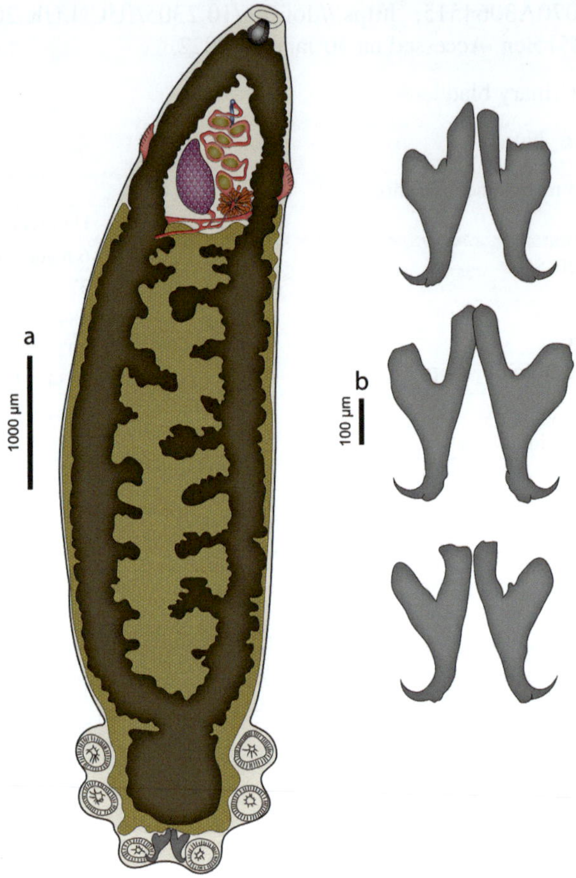

Fig. 4.62 (**a**) *Polystoma chiromantis*, (**b**) hamuli. Redrawn from Dupouy and Knoepffler (1978)

Original description Dupouy and Knoepffler (1978).

Other taxonomic contributions None.

Etymology Named after the host.

Type locality Makokou, Gabon; Lamaboké, Central Africa; Yaoundré, Cameroon.

Other localities None.

Type host *Chiromantis rufescens* (Günther, 1869), African Foam-nest Treefrog.

Additional hosts None.

Host geographical distribution *Chiromantis rufescens* occurs in Central African Republic, Gabon, Cameroon, Uganda, Congo, Equatorial Guinea, Nigeria, Liberia, Ghana, Guinea, Ivory Coast and Sierra Leone where it inhabits forest habitats (Frost 2021).

Host conservation status The African Foam-nest Treefrog has a large population, a wide distribution and is tolerant for habitat modification. It is listed as Least Concern by The IUCN Red List of Threatened Species 2013: e. T58800A18407175. https://doi.org/10.2305/IUCN.UK.2013-2.RLTS. T58800A18407175.en. Accessed on 30 January 2022.

Site on host Urinary bladder.

DNA sequence None.

Morphology and morphometrics (μm):	
Source	Dupouy and Knoepffler (1978)
Sample size	3
Body length	8000–9500
Greatest width	2000–2400
Width at vagina	1400–1700
Haptor length	1320–1750
Haptor width	1880–2250
False oral sucker width	250–380
Pharynx length	210–220
Pharynx width	175–200
Testis length	
Testis width	
Ovary length	675–825
Ovary width	330–430
Egg length	225–245
Egg width	125–145
Egg incubation time	
Number of intra-uterine eggs	1–6
Genital bulb width	
Number of genital spines	8
Genital spine length	
Haptoral sucker diameter	310–380
Hamulus handle length X	400–440
Hamulus guard length Y	
Hamulus hook length	
Marginal hooklet length	
Haptoral L/Body L	0.17–0.18
No of anastomoses	0

Remarks None.

4.4.11.12 *Polystoma cinereum* Du Preez, Verneau and Gross, 2007 nom. nov. (Fig. 4.63)

Collection United States National Museum (USNM); Parasitic Worm Collection, National Museum, Bloemfontein, South Africa.

Holotype USNM 1395407.

Paratypes USNM 1395408–13; NMBP 303–308.

Original description Du Preez et al. (2007).

Taxonomic note When corrected for gender *Polystoma floridana* Du Preez, Verneau and Gross, 2007 becomes a junior homonym of *Polystoma floridanum*

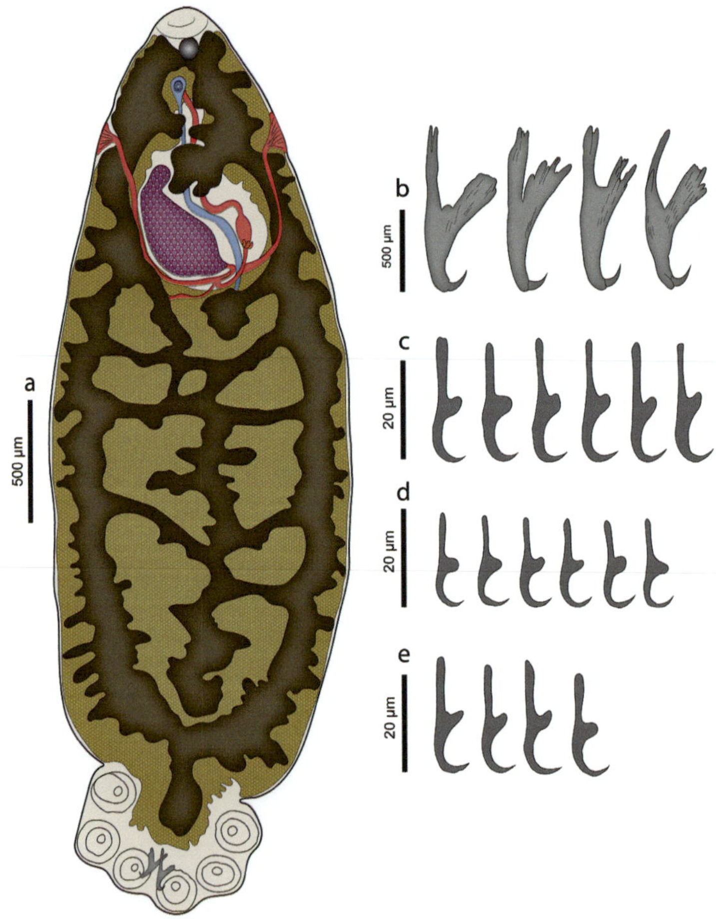

Fig. 4.63 (**a**) *Polystoma cinereum*, (**b**) hamuli, (**c**) marginal hooklets (1), (**d**) marginal hooklets (2), (**e**) marginal hooklets (8). Redrawn from Du Preez et al. (2007)

Stunkard, 1924, which was considered as a junior synonym of *Neopolystoma orbiculare* Price, 1939. According to the International code of Zoological Nomenclature (ICZN) articles 53.3, 60.1 and 60.3, the junior homonym is hereby rejected and replaced with *Polystoma cinereum* nom. nov.

Other taxonomic contributions None.

Etymology Named after the host.

Type locality Gainesville (Florida), USA.

Other localities None.

Type host *Dryophytes cinereus*, North American Green Treefrog. In the original description of the polystome, the host was known as *Hyla cinerea*.

Additional hosts None.

Host geographical distribution *Dryophytes cinereus* occurs in the United States and Puerto Rico where it has been introduced. It commonly inhabits areas where there is an abundance of floating vegetation such as streams, ponds, swamps, lakes and marshes (Hammerson and Hedges 2004).

Host conservation status The North American Green Treefrog has a large population and a wide distribution. It is listed as Least Concern by The IUCN Red List of Threatened Species 2021: e.T55449A118978218. https://doi.org/10.2305/IUCN.UK.2021-3.RLTS.T55449A118978218.en. Accessed on 30 January 2022.

Site on host Urinary bladder.

DNA sequence

- 18S rRNA gene, 2095 bp. GenBank Accession: AM157188
- 28S rRNA gene, 1458 bp. GenBank Accessions: AM157211 and AM913878
- 28S rRNA gene, 999 bp. GenBank Accession: AM913879
- Cytochrome c Oxidase I, 333 bp. GenBank Accessions: AM913869 to AM913871
- 12S rRNA gene, 494 bp. GenBank Accession: KR856083

Morphology and morphometrics (μm):	
Source	Du Preez et al. (2007)
Sample size	24
Body length	4680–9080 (6881)
Greatest width	1480–2940 (2415)
Width at vagina	
Haptor length	640–1300 (917)
Haptor width	640–1880 (1569)
False oral sucker width	155–303 (239)
Pharynx length	163–311 (247)
Pharynx width	156–319 (209)
Ovary length	606–1151 (906)

(continued)

Morphology and morphometrics (μm):	
Source	Du Preez et al. (2007)
Ovary width	326–591 (460)
Egg length	185–210 (195)
Egg width	161–169 (163)
Egg incubation time	
Number of intra-uterine eggs	1
Genital bulb width	
Number of genital spines	8–11
Genital spine length	21–24 (22)
Haptoral sucker diameter	241–420 (340)
Hamulus handle length X	202–350 (281)
Hamulus guard length Y	
Hamulus hook length	33–51 (40)
Marginal hooklet length	23.3–25.7 (24.4)
Haptoral L/Body L	0.9–0.22 (0.14)
No of anastomoses	0–6

Remarks *Polystoma cinereum* was originally described as *Polystoma floridana*.

4.4.11.13 *Polystoma claudecombesi* Du Preez and Kok, 1995 (Fig. 4.64)

Collection Parasitic Worm Collection, National Museum, Bloemfontein, South Africa (NMBP).

Holotype NMBP 46.

Paratypes NMBP 47–49.

Original description Du Preez and Kok (1995).

Other taxonomic contributions None.

Etymology Named after Prof Claude Combes, Perpignan, France.

Type locality Witsieshoek Mountain Resort, Qwa Qwa, South Africa.

Other localities None.

Type host *Amietia delalandi* (Duméril and Bibron, 1841), Common River Frog. In the original description of the polystome, the host was known as *Rana angolensis* Barboza du Bocage, 1866.

Additional hosts None.

Host geographical distribution *Amietia delalandi* occurs in Lesotho and in South Africa where it inhabits permanent streams in forests and grassland habitats (Frost 2021).

Host conservation status The Common River Frog is adaptable with a large population. It is listed as Least Concern by The IUCN Red List of Threatened

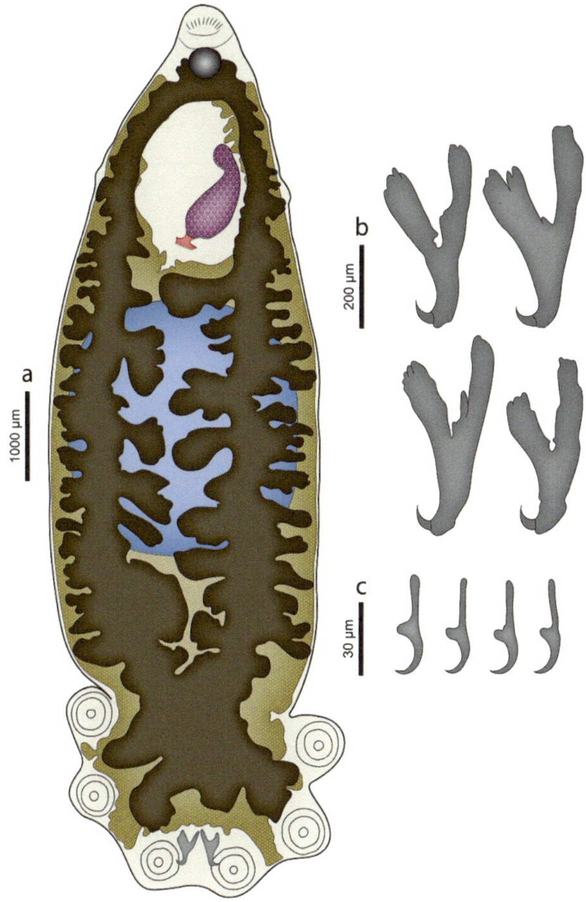

Fig. 4.64 (**a**) *Polystoma claudecombesi*, (**b**) hamuli, (**c**) marginal hooklets. Redrawn from Du Preez and Kok (1995)

Species 2017: e.T113263541A113263686. https://doi.org/10.2305/IUCN.UK.2017-2.RLTS.T113263541A113263686.en. Accessed on 30 January 2022.

Site on host Urinary bladder.

DNA sequence

- 18S rRNA gene, 2103 bp. GenBank Accession: FM897281
- 28S rRNA gene, 1433 bp. GenBank Accession: FM897263

Morphology and morphometrics (µm):	
Source	Du Preez and Kok (1995)
Sample size	6
Body length	8300–10,510 (9390)
Greatest width	2940–3350 (3090)

(continued)

Morphology and morphometrics (μm):	
Source	Du Preez and Kok (1995)
Width at vagina	1920–2130 (2060)
Haptor length	1720–2130 (1950)
Haptor width	3030–3410 (3140)
False oral sucker width	
Pharynx length	344–417 (381)
Pharynx width	349–423 (387)
Ovary length	865–1357 (1127)
Ovary width	
Egg length	241–275 (263)
Egg width	182–206 (196)
Egg incubation time	12–14 days
Number of intra-uterine eggs	0–1
Genital bulb width	
Number of genital spines	7–8
Genital spine length	30–33
Haptoral sucker diameter	442–541 (487)
Hamulus handle length X	413–501 (458)
Hamulus guard length Y	
Hamulus hook length	
Hamulus x/y ratio	1.21
Marginal hooklet length	28–41
Haptoral L/Body L	0.19–0.24 (0.21)
No of anastomoses	0

Remarks None.

4.4.11.14 *Polystoma combesi* Batchvarov, 1982 (Fig. 4.65)

Collection Unknown.

Holotype Unknown.

Paratypes Unknown.

Original description Batchvarov (1982).

Other taxonomic contributions None.

Etymology Named after Prof Claude Combes, Perpignan, France.

Type locality Towns of Trun, Godech and Kyustendil, Bulgaria.

Other localities None.

Type host *Rana graeca* Boulenger, 1891, Greek Frog.

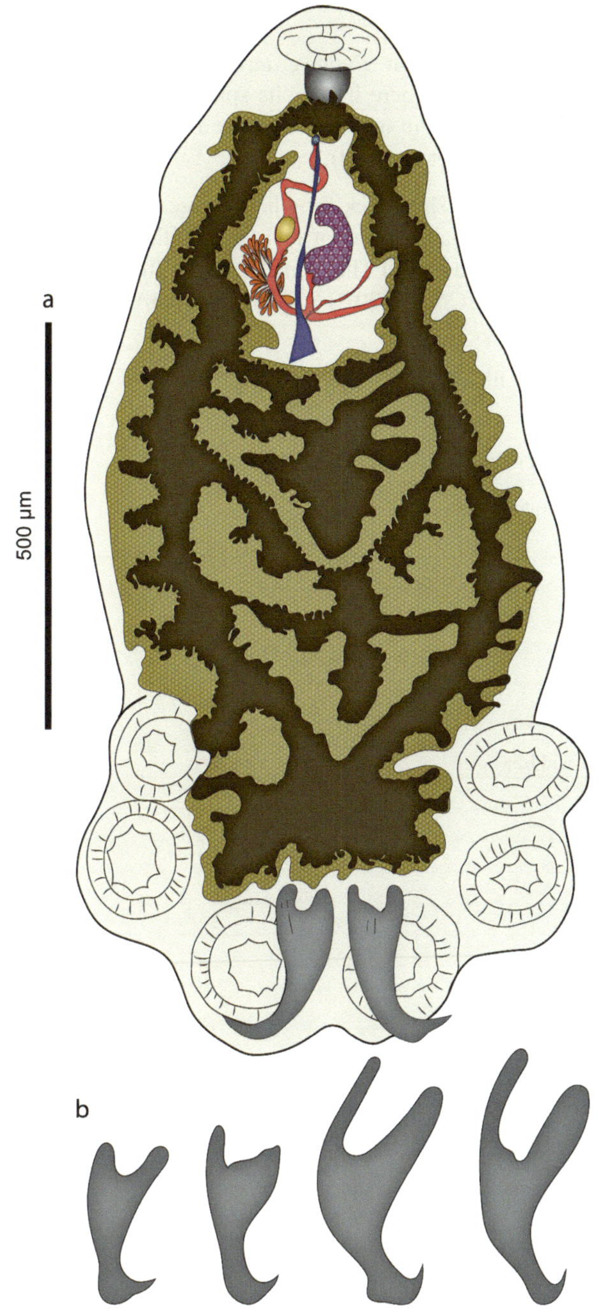

Fig. 4.65 (**a**) *Polystoma combesi*, (**b**) hamuli. Redrawn from Batchvarov (1982)

Additional hosts None.

Host geographical distribution *Rana graeca* occurs in Macedonia, Albania, Bulgaria, Greece and Serbia where it inhabits springs, streams and small rivers in forest habitat. It is also found in meadows, moors as well as in and around glacial pools in lakes (Lymberakis et al. 2009).

Host conservation status The Greek Frog has a large population, a wide distribution and can tolerate a certain degree of habitat modification. It is listed as Least Concern by The IUCN Red List of Threatened Species 2020: e.T58605A89705363. https://doi.org/10.2305/IUCN.UK.2020-3.RLTS.T58605A89705363.en. Accessed on 30 January 2022.

Site on host Urinary bladder.

DNA sequence

- ITS1, 756 bp. GenBank Accession: AJ310403

Morphology and morphometrics (μm):	
Source	Batchvarov (1982)
Sample size	8
Body length	1300–3300 (2300)
Greatest width	599–1500 (1000)
Width at vagina	
Haptor length	1000–1300
Haptor width	753–2000
False oral sucker width	143–500 (243)
Pharynx length	135–337 (217)
Pharynx width	143–362 (231)
Ovary length	330–880 (550)
Ovary width	130–440
Egg length	
Egg width	
Egg incubation time	
Number of intra-uterine eggs	
Genital bulb width	
Number of genital spines	
Genital spine length	
Haptoral sucker diameter	
Hamulus handle length X	173–240
Hamulus guard length Y	157–175
Hamulus hook length	
Marginal hooklet length	
Haptoral L/Body L	
No of anastomoses	3

Remarks None.

4.4.11.15 *Polystoma cuvieri* Vaucher, 1990 (Fig. 4.66)

Collection Natural History Museum, Geneva, Switzerland (MHNG).

Holotype MHNG 987.429 (0030754).

Paratypes MHNG 987.430 (0030755).

Original description Vaucher (1990).

Other taxonomic contributions Santos and Amato (2012).

Etymology Named after the host.

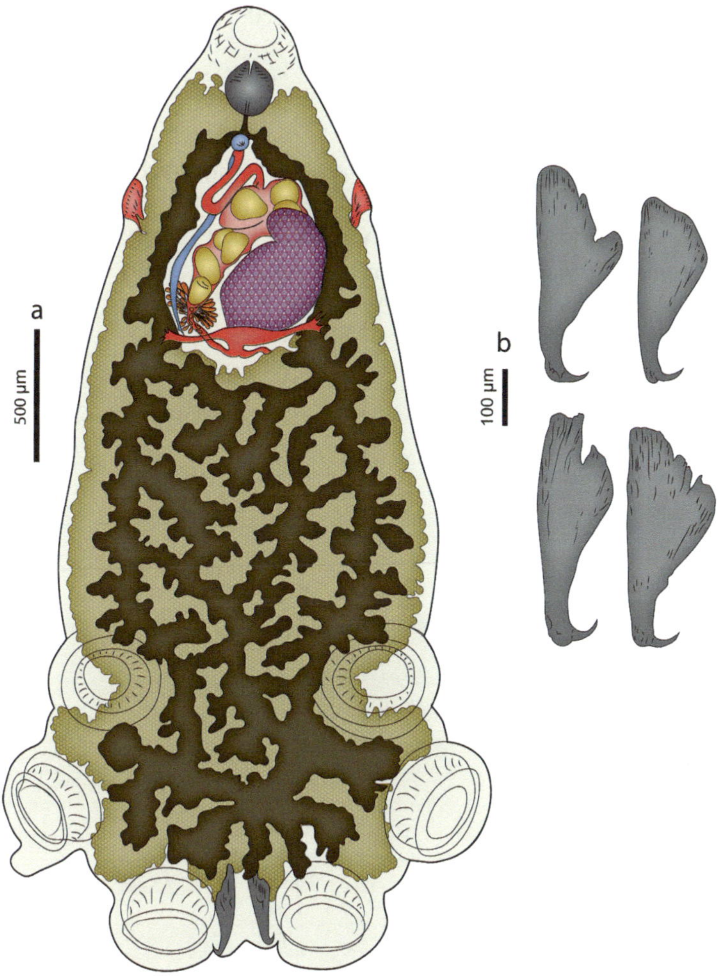

Fig. 4.66 (**a**) *Polystoma cuvieri*, (**b**) hamuli. Redrawn from Vaucher (1990)

Type locality San Carlos, Paraguay.

Other localities Brazil.

Type host *Physalaemus cuvieri* Fitzinger, 1826, Cuvier's Foam Froglet.

Additional hosts None.

Host geographical distribution *Physalaemus cuvieri* occurs in Paraguay, Argentina and Brazil. It inhabits grassland habitats near water bodies (puddles), pastureland and flooded savannah habitats (Mijares et al. 2010; Kwet et al. 2010).

Host conservation status The Cuvier's Foam Froglet has a large population, a wide distribution and is tolerant for various habitats. It is listed as Least Concern by The IUCN Red List of Threatened Species 2010: e.T57250A11609155. https://doi.org/10.2305/IUCN.UK.2010-2.RLTS.T57250A11609155.en. Accessed on 30 January 2022.

Site on host Urinary bladder.

DNA sequence
- 18S rRNA gene, 2092 bp. GenBank Accession: AM051068
- 18S rRNA gene, 626 bp. GenBank Accession: AJ297774
- 28S rRNA gene, 1453 bp. GenBank Accession: AM157203
- Cytochrome c Oxidase I, 384 bp. GenBank Accession: AM913862
- 12S rRNA gene, 491 bp. GenBank Accession: KR856080
- ITS1, 729 bp. GenBank Accession: AJ301691

Morphology and morphometrics (μm):		
Source	Vaucher (1990)	Santos and Amato (2012)
Sample size	17	17
Body length	2400–4230 (3600)	2000–4800 (3100)
Greatest width	900–1700 (1400)	630–2210 (1240)
Width at vagina		
Haptor length	900–1400 (1300)	766–1718 (1179)
Haptor width	1200–2100 (1800)	780–2271
False oral sucker width	205–327 (270)	40–210 (182)
Pharynx length	164–245 (214)	150–270 (188)
Pharynx width	131–205 (183)	145–280 (191)
Ovary length		400–960 (637)
Ovary width		100–514 (292)
Egg length	165	160–280 (220)
Egg width	90–106	95–100 (98)
Egg incubation time		
Number of intra-uterine eggs	1–7	1

(continued)

Morphology and morphometrics (μm):		
Source	Vaucher (1990)	Santos and Amato (2012)
Genital bulb width		
Number of genital spines	8	8
Genital spine length		13–28 (18)
Haptoral sucker diameter	Anterior (393 × 375), middle (377), Posterior (370)	Anterior (320 × 334), middle (294 × 319), posterior (293 × 314)
Hamulus handle length X	278–413 (360)	210–345 (285)
Hamulus guard length Y		210–325 (260)
Hamulus X/Y ratio		0.9–1.2 (1.1)
Hamulus hook length		48–68 (59)
Marginal hooklet length		
Haptoral L/Body L	0.36	0.38–0.36 (0.38)
No of anastomoses	Network	Network

Remarks Santos and Amato (2012) reported an average haptor width of 62. This clearly is a mistake. We therefore excluded this value from the table.

4.4.11.16 *Polystoma dawiekoki* Du Preez, Vaucher and Mariaux, 2002 (Fig. 4.67)

Collection Parasitic Worm Collection, National Museum, Bloemfontein, South Africa (NMBP); Parasitic Worm Collection, National History Museum, London (NHML).

Holotype NMBP 256.

Paratypes NMBP 257–263; NHML 2001.7.26.1–2.

Original description Du Preez et al. (2002).

Other taxonomic contributions None.

Etymology Named after Prof Dawid J Kok, Bloemfontein, South Africa.

Type locality Kumahlala Pan, Mkuze Game reserve, KwaZulu-Natal, South Africa (27°35′50”S, 32°13′12″E).

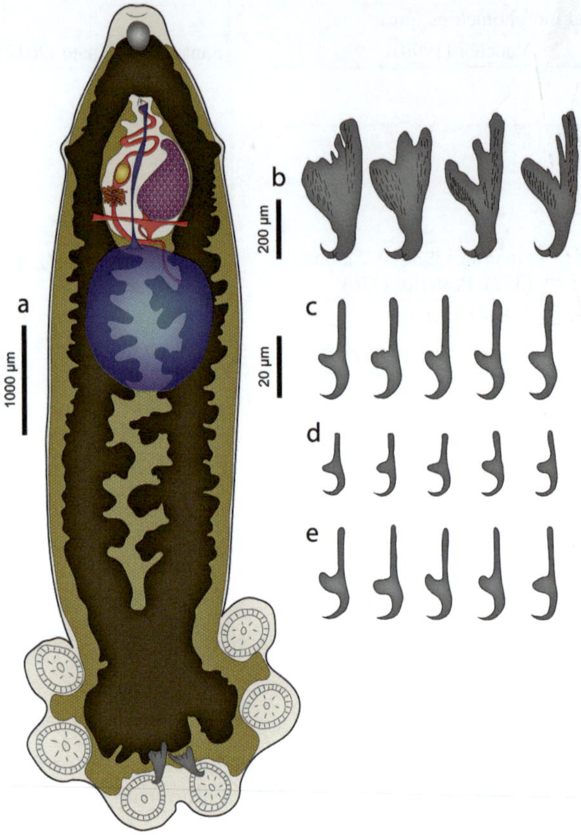

Fig. 4.67 (**a**) *Polystoma dawiekoki*, (**b**) hamuli, (**c**) marginal hooklets (1), (**d**) marginal hooklets (2–7), (**e**) marginal hooklets(8). Redrawn from Du Preez et al. (2002)

Other localities Bulwa, Tanga, Tanzania (5°3'S, 38°38′E).

Type host *Ptychadena anchietae* (Bocage, 1868), Anchieta's Ridged Frog.

Additional hosts None.

Host geographical distribution *Ptychadena anchietae* occurs in South Africa, Tanzania, Congo, Angola, Swaziland, Botswana, Zimbabwe, Kenya, Uganda, Congo, Somalia, Zambia, Malawi, Ethiopia, South Sudan, Namibia and Mozambique. This species inhabits forest clearings, agricultural areas, woodland, grassland, open country and savannah habitats usually close to permanent water sources (Frost 2021).

Host conservation status The Anchieta's Ridged Frog has a large population and a wide distribution across various habitats. It is thus listed as Least Concern by The IUCN Red List of Threatened Species 2016: e.T58490A87737574. https://doi.org/10.2305/IUCN.UK.2016-1.RLTS.T58490A87737574.en. Accessed on 30 January 2022.

Site on host Urinary bladder.

DNA sequence

- 18S rRNA gene, 2124 bp. GenBank Accession: AM051069
- 28S rRNA gene, 1450 bp. GenBank Accession: AM913875
- 28S rRNA gene, 1439 bp. GenBank Accession: AM157204
- Cytochrome c Oxidase I, 369 bp. GenBank Accessions: AM913856 and AM913857
- 12S rRNA gene, 482 bp. GenBank Accession: KR856081
- ITS1, 695 bp. GenBank Accessions: AJ310405 and AJ310406

Morphology and morphometrics (μm):		
Source	Du Preez et al. (2002) South African specimens fixed under coverslip pressure	Du Preez et al. (2002) Tanzanian specimens not flat fixed
Sample size	26	6
Body length	4050–8200 (7096)	2980–4140
Greatest width	1250–2425 (2031)	1630–1690
Width at vagina		
Haptor length	1000–1825 (1582)	1330–1370
Haptor width	1325–2750 (2432)	1570–1730
False oral sucker width	160–329 (286)	270–335
Pharynx length	170–314 (270)	188–209
Pharynx width	177–280 (245)	
Ovary length	534–911 (760)	
Ovary width	281–475 (410)	
Egg length	180–219 (201)	172–205
Egg width	116–140 (127)	110–131
Egg incubation time		
Number of intra-uterine eggs	1–9	1
Genital bulb width		
Number of genital spines	7–8	
Genital spine length	29–34 (32)	
Haptoral sucker diameter	252–543 (455)	344–507
Hamulus handle length X	281–485 (440)	384–507
Hamulus guard length Y		
	62–71 (66)	

(continued)

Morphology and morphometrics (μm):		
Source	Du Preez et al. (2002) South African specimens fixed under coverslip pressure	Du Preez et al. (2002) Tanzanian specimens not flat fixed
Hamulus hook length		
Hamulus x/y ratio	1.36	
Marginal hooklet length	19–33 (26)	C1: 31–33 (32); C2–7: 19–20 (20); C8: 28–29 (28)
Haptoral L/Body L	0.34	
No of anastomoses	0–2	

Remarks The C1–8 reported for the marginal hooklet length refer to the protocol of numbering used by Murith (1981) where C1 is the posteriormost hooklet and C8 the anteriormost.

4.4.11.17 *Polystoma diptychi* Vaucher, 1986 (Fig. 4.68)

Collection Geneva Natural History Museum (MHNG).

Holotype MHNG 0015585.

Paratypes Unknown.

Original description Vaucher (1986).

Other taxonomic contributions None.

Etymology Named after the host.

Type locality Rio Alto Parana, Puerto Presidente Stroessner, Paraguay.

Other localities None.

Type host *Rhinella diptycha* (Cope, 1862), Cope's Toad. In the original description of the polystome, the host was known as *Bufo diptychus* Cope, 1862.

Additional hosts None.

Host geographical distribution *Rhinella diptycha* has been recorded in Paraguay and can be found in dry savanna habitats (Scott et al. 2004).

Host conservation status As there are uncertainties regarding its taxonomic status, the Cope's toad is listed as Data Deficient by The IUCN Red List of Threatened

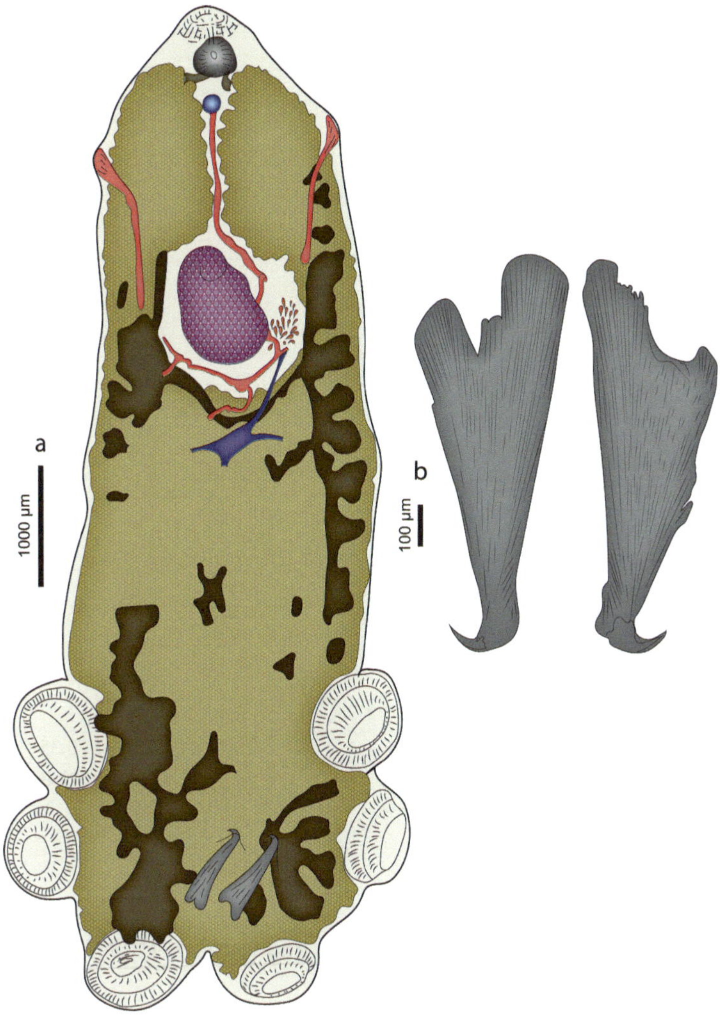

Fig. 4.68 (**a**) *Polystoma diptychi*, (**b**) hamuli. Redrawn from Vaucher (1986)

Species 2004: e.T54628A11177973. https://doi.org/10.2305/IUCN.UK.2004.RLTS.T54628A11177973.en. Accessed on 30 January 2022.

Site on host Urinary bladder.

DNA sequence None.

Morphology and morphometrics (µm):	
Source	Vaucher (1986)
Sample size	1
Body length	8300

(continued)

Morphology and morphometrics (μm):	
Source	Vaucher (1986)
Greatest width	2600
Width at vagina	
Haptor length	2900
Haptor width	3400
False oral sucker width	
Pharynx length	
Pharynx width	330
Ovary length	
Ovary width	
Egg length	
Egg width	
Egg incubation time	
Number of intra-uterine eggs	
Genital bulb width	
Number of genital spines	
Genital spine length	
Haptoral sucker diameter	755–836
Hamulus handle length X	970–980
Hamulus guard length Y	
Hamulus hook length	
Marginal hooklet length	
Haptoral L/Body L	0.35
No of anastomoses	Network

Remarks None.

4.4.11.18 *Polystoma dorsale* Maeder, Euzet and Combes, 1970 (Fig. 4.69)

Collection Claude Combes personal collection, Perpignan, France.

Holotype Unknown.

Paratypes Unknown.

Original description Maeder et al. (1970).

Taxonomic note The original name *Polystoma dorsalis* was corrected for gender.

Other taxonomic contributions Maeder (1973); Murith (1979, 1981).

Etymology Named after the host.

Type locality Lamco, Liberia.

Other localities None.

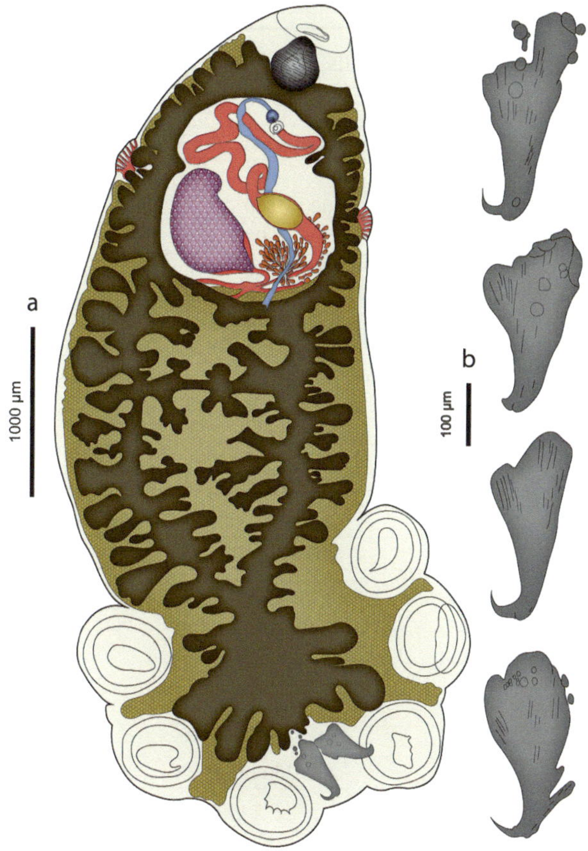

Fig. 4.69 (**a**) *Polystoma dorsale*, (**b**) hamuli. Redrawn from Maeder et al. (1970)

Type host *Afrixalus dorsalis* (Peters, 1875), Striped Spiny Reed Frog.

Additional hosts *Hoplobatrachus occipitalis*.

Host geographical distribution *Afrixalus dorsalis* occurs in Angola, Gabon, Liberia, Guinea, Nigeria, Cameroon, Ghana and Congo where it inhabits grassy vegetation, bush land and cultivated land (Frost 2021).

Host conservation status The Striped Spiny Reed Frog has a wide distribution, large population and tolerance for a range of habitats. It is thus listed as Least Concern by The IUCN Red List of Threatened Species 2013: e. T56060A18369144. https://doi.org/10.2305/IUCN.UK.2013-2.RLTS.T56060A18369144.en. Accessed on 30 January 2022.

DNA sequence None.

Site on host Urinary bladder and gills of tadpoles.

Morphology and morphometrics (μm):		
Source	Maeder et al. (1970)	Murith (1981)
Sample size	6	6
Body length	4000–5500	1350–1890 (1700)
Greatest width	1300–1900	530–610 (570)
Width at vagina	1000–1500	
Haptor length	1000	340–400 (370)
Haptor width	2000	560–590 (580)
False oral sucker width	300	
Pharynx length		
Pharynx width	220	
Ovary length		
Ovary width		
Egg length	240	119–168 (144)
Egg width	140	68–147 (108)
Egg incubation time		
Number of intra-uterine eggs	1	0–1
Genital bulb width		
Number of genital spines	8–9	
Genital spine length	30	
Haptoral sucker diameter	350–450	
Hamulus handle length X	250–350 (310)	
Hamulus guard length Y		
Hamulus hook length		
Marginal hooklet length		42
Haptoral L/Body L	0.18–0.25	0.22
No of anastomoses	0–2	1

Remarks *Polystoma dorsale* was originally described as *Polystoma dorsalis*.

4.4.11.19 *Polystoma ebriense* Maeder, 1973 (Fig. 4.70)

Collection Deposited in the Helminthological collection in the Royal Museum of Central Africa, Tervuren (MRAC).

Holotype MRAC 36.523.

Paratypes MRAC 36.524,36.525.

Original description Maeder (1973).

Taxonomic note The original name *Polystoma ebriensis* was corrected for gender.

Other taxonomic contributions Dupouy (1978); Murith (1979, 1981).

Fig. 4.70 *Polystoma ebriense*. Redrawn from Maeder 1973

Etymology Unknown.

Type locality Ivory Coast.

Other localities None.

Type host *Ptychadena aequiplicata* (Werner, 1898), Victoria Grassland Frog.

Additional hosts *Hoplobatrachus occipitalis*, *Ptychadena bibroni*.

Host geographical distribution *Ptychadena aequiplicata* occurs in Benin, Togo, Congo, Guinea, Nigeria, Liberia, Cameroon, Gabon and Ghana where it is most commonly found in lowland, moist forests (Rödel et al. 2009b).

Host conservation status The Victoria Grassland Frog has a wide distribution, large population size and tolerance for habitat modification. It is listed as Least Concern by The IUCN Red List of Threatened Species 2019: e.T58489A18398710. https://doi.org/10.2305/IUCN.UK.2019-2.RLTS.T58489A18398710.en. Accessed on 30 January 2022.

Site on host Urinary bladder and gills of tadpoles.

DNA sequence None.

Morphology and morphometrics (μm):			
Source	Maeder (1973)	Dupouy (1978)	Murith (1981)
Sample size	4	5	10
Body length	2450–2920	1840–2920	2520–4080 (3450)
Greatest width	620–850	360–850	710–1410 (1110)
Width at vagina	550–670	340–670	
Haptor length	600–900	460–900	410–680 (540)
Haptor width	670–1130	590–1130	440–1090 (710)
False oral sucker width	96–100	77–100	
Pharynx length	177–235		
Pharynx width	158–187	163–187 × 134–235	
Ovary length	254	134–254	
Ovary width	100	48–100	
Egg length	216–240 (228)	216–240 (228)	170–187 (181)
Egg width	100–156 (128)	100–156 (128)	119–170 (136)
Egg incubation time			
Number of intra-uterine eggs	1	1	0–1
Genital bulb width			
Number of genital spines	8	8	
Genital spine length	24	24	
Haptoral sucker diameter	144–283	141–283 × 115–271	
Hamulus handle length X	370–404	270–404	
Hamulus guard length Y			
Hamulus hook length			
Marginal hooklet length	43		43
Haptoral L/Body L	0.24–0.31	0.24–0.31	0.16
No of anastomoses	0–1	0–1	

Remarks *Polystoma ebriense* was originally described as *Polystoma ebriensis*.

4.4.11.20 *Polystoma ezoense* Uchida, Machida, Uchida and Itagaki, 1988 (Fig. 4.71)

Collection Personal Collection of K Uchida.

Holotype Unknown.

Fig. 4.71 (**a**) *Polystoma ezoense*, (**b**) hamuli. Redrawn from Uchida et al. (1988)

Paratypes Unknown.

Original description Uchida et al. (1988).

Taxonomic note The original name *Polystoma ezoensis* was corrected for gender.

Other taxonomic contributions None.

Etymology Unknown.

Type locality Sagamihara, Kanagawa Prefecture, Japan.

Other localities Sapporo and Teshio, Hokkaido, Japan.

Type host *Rana japonica* Boulenger, 1879, Japanese Brown Frog.

Additional hosts *Rana chensinensis* David, 1875.

Host geographical distribution *Rana japonica* occurs in Japan where it is found in grasslands while *R. chensinensis* occurs in Mongolia and China where it is present in forests close to water (rivers, ponds and streams), in marshes and cultivated fields (Kuzmin et al. 2004b; Kaneko and Matsui 2004c).

Host conservation status Both species have a wide distribution and large populations. They are listed as Least Concern by The IUCN Red List of Threatened Species 2021: e.T58625A179266101. https://doi.org/10.2305/IUCN.UK.2021-1.RLTS.T58625A179266101.en. Accessed on 30 January 2022.

Site on host Urinary bladder.

DNA sequence None.

Morphology and morphometrics (μm):	
Source	Uchida et al. (1988)
Sample size	
Body length	2500–5200
Greatest width	1800–2300
Width at vagina	
Haptor length	1000–1600
Haptor width	600–900
False oral sucker width	
Pharynx length	260–290
Pharynx width	220–260
Ovary length	320–630
Ovary width	280–320
Egg length	
Egg width	
Egg incubation time	
Number of intra-uterine eggs	
Genital bulb width	
Number of genital spines	8
Genital spine length	26–41
Haptoral sucker diameter	250–320
Hamulus handle length X	
Hamulus guard length Y	345–480
Hamulus hook length	
Marginal hooklet length	
Haptoral L/Body L	0.4–0.31
No of anastomoses	Network

Remarks *Polystoma ezoense* was originally described as *Polystoma ezoensis*.

4.4.11.21 *Polystoma fuscum* Biserkov and Hadjinikolova, 1993 (Fig. 4.72)

Collection Collections of the Institute of Parasitology, Bulgarian Academy of Sciences (IPBAS); Natural History Museum, London (NHML).

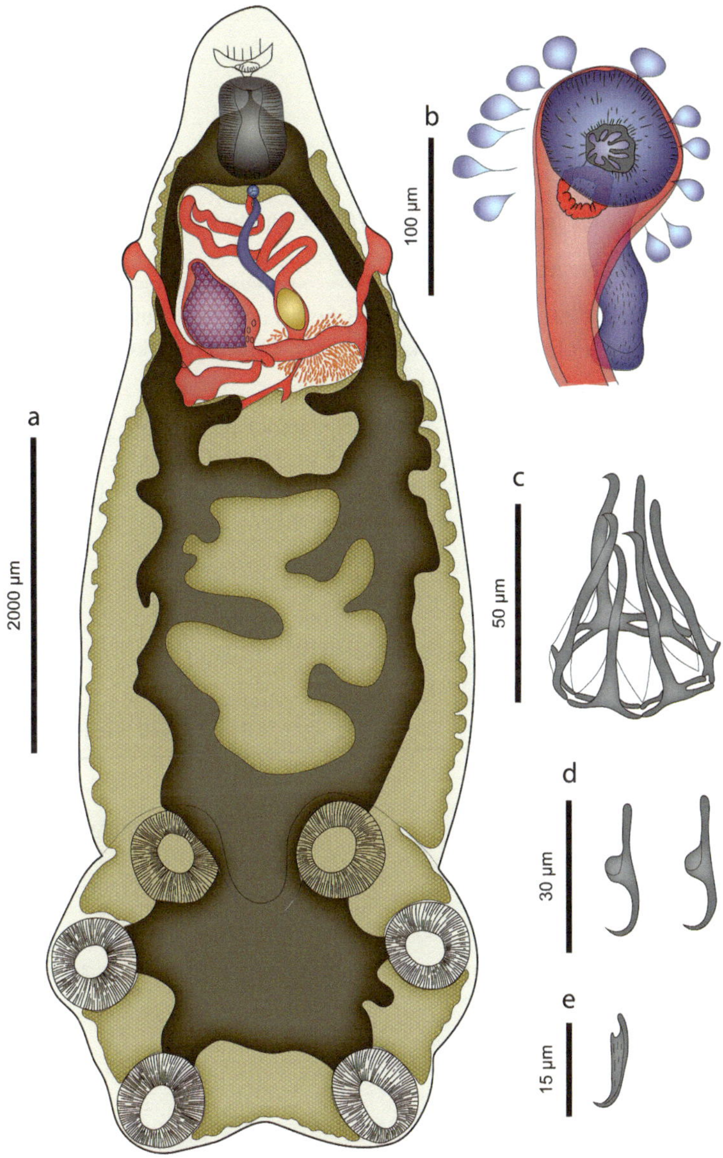

Fig. 4.72 (**a**) *Polystoma fuscum*, (**b**) genital bulb, (**c**) genital spines, (**d**) marginal hooklets, (**e**) hamuli. Redrawn from Biserkov and Hadjinikolova (1993)

Holotype IPBAS 14303–1.

Paratypes IPBAS 12115, 14002, 14006–14009, 14106–14109, 14116, 14303, 14305, 14325, 14327; NHML 1991.5.2.1–2.

Original description Biserkov and Hadjinikolova (1993).

Taxonomic note The original name *Polystoma fuscus* was corrected for gender.

Other taxonomic contributions None.

Etymology Named after the host.

Type locality Bulgaria.

Other localities None.

Type host *Pelobates fuscus* (Laurenti, 1768), Common Spadefoot.

Additional hosts None.

Host geographical distribution *Pelobates fuscus* occurs in Bulgaria, Italy, France, Belgium, Sweden, Poland, Denmark, Romania, Germany, Ukraine and Hungary (Frost 2021).

Host conservation status The Common Spadefoot has a large population and a wide distribution. It is listed as Least Concern by The IUCN Red List of Threatened Species 2009: e.T16498A5951455. https://doi.org/10.2305/IUCN.UK.2009.RLTS.T16498A5951455.en. Accessed on 30 January 2022.

Site on host Urinary bladder.

DNA sequence
- ITS1, 720 bp. GenBank Accession: AJ310401

Morphology and morphometrics (μm):	
Source	Biserkov and Hadjinikolova (1993)
Sample size	39
Body length	5405–8378 (6574)
Greatest width	1703–2756 (2285)
Width at vagina	1270–2162 (1753)
Haptor length	1432–2432 (1920)
Haptor width	1432–3405 (2667)
False oral sucker width	
Pharynx length	440–812 (455)
Pharynx width	338–609 (451)
Ovary length	
Ovary width	
Egg length	205–276 (237)
Egg width	147–169 (156)
Egg incubation time	

(continued)

Morphology and morphometrics (μm):	
Source	Biserkov and Hadjinikolova (1993)
Number of intra-uterine eggs	1–2
Genital bulb width	41–85 (65)
Number of genital spines	6–9 (8)
Genital spine length	42–45
Haptoral sucker diameter	380–736
Hamulus handle length X	15
Hamulus guard length Y	
Hamulus hook length	
Marginal hooklet length	
Haptoral L/Body L	0.29
No of anastomoses	Infrequent

Remarks *Polystoma fuscum* was originally described as *Polystoma fuscus*.

4.4.11.22 *Polystoma gabonense* Euzet, Combes and Knoepffler, 1966 (Fig. 4.73)

Collection Claude Combes personal collection, Perpignan, France.

Holotype Unknown.

Paratypes Unknown.

Original description Euzet et al. (1966).

Taxonomic note The original name *Polystoma gabonensis* was corrected for gender.

Other taxonomic contributions Euzet et al. (1974a); Murith et al. (1978).

Etymology Named after the type locality country.

Type locality Makokou, Gabon.

Other localities Cameroon and Central African Republic.

Type host *Amnirana albolabris* (Hallowell, 1856), White-lipped Frog. In the original description of the polystome, the host was known as *Hylarana albolabris albolabris* Laurent, 1954.

Additional hosts *Amnirana amnicola* (Perret, 1977).

Host geographical distribution *Amnirana albolabris* occurs in Angola, Kenya, Uganda, Cameroon, Congo, Ghana, Gabon, Nigeria, Liberia and Tanzania (Frost 2021).

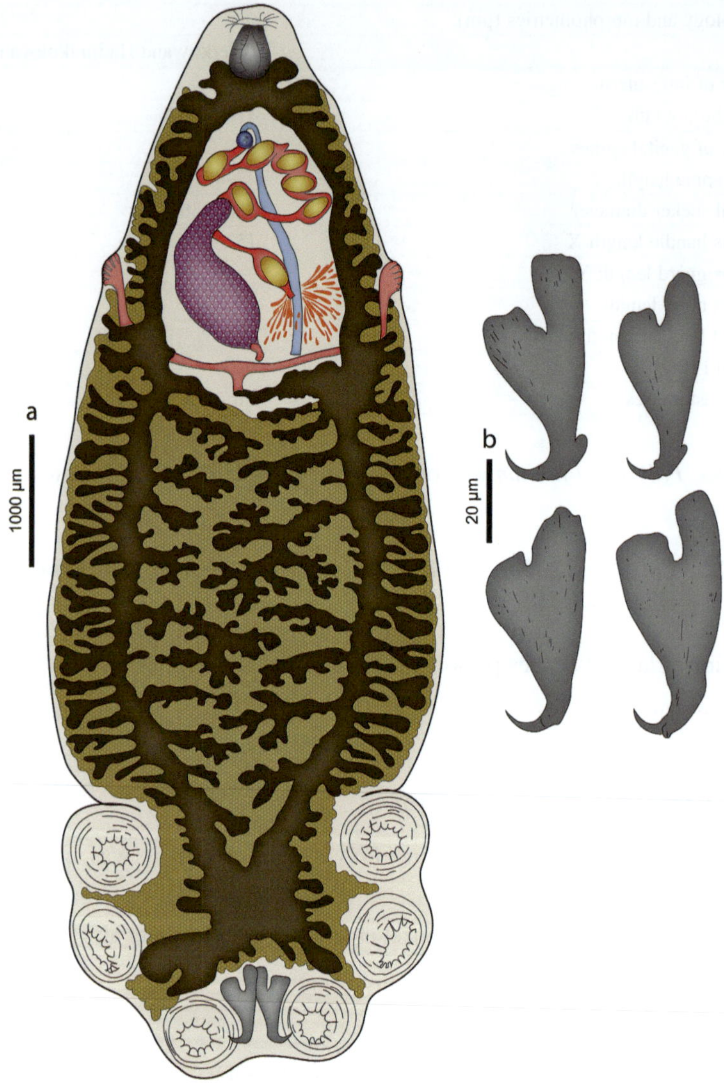

Fig. 4.73 (**a**) *Polystoma gabonense*, (**b**) hamuli. Redrawn from Euzet et al. (1966)

Host conservation status The White-lipped Frog has a large population size, a wide distribution range and is tolerant for habitat modification. It is listed as Least Concern by The IUCN Red List of Threatened Species 2013: e.T58192A89360738. https://doi.org/10.2305/IUCN.UK.2013-2.RLTS.T58192A18405823.en. Accessed on 30 January 2022.

DNA sequence None.

Site on host Urinary bladder and gills of tadpoles.

Morphology and morphometrics (μm):		
Source	Euzet et al. (1966)	Euzet et al. (1974a)
Sample size	8	8
Body length	5000–8000	6600–9000
Greatest width	2200–2800	
Width at vagina	2000–2200	
Haptor length	1300–2000	
Haptor width	2000–2500	
False oral sucker width		
Pharynx length		
Pharynx width		
Ovary length		
Ovary width		
Egg length	211–268 (230)	
Egg width	103–131 (120)	
Egg incubation time		
Intra-uterine eggs (n)	1–7	
Genital bulb width		
Number of genital spines	8	
Genital spine length	40–42	
Haptoral sucker diameter	500	
Hamulus handle length X	500–550	430–460
Hamulus guard length Y		
Hamulus hook length		
Marginal hooklet length		
Haptoral L/Body L	0.25–0.26	
No of anastomoses	0–7	0–7

Remarks *Polystoma gabonense* was originally described as *Polystoma africanum gabonensis*. This species has been reported from the same host species as *Polystoma perreti* Maeder, 1973 in Ivory Coast and Togo.

4.4.11.23 *Polystoma galamense* Euzet, Bourgat and Salami-Cadoux, 1974 (Fig. 4.74)

Collection Unknown.

Holotype Unknown.

Paratypes Unknown.

Original description Euzet et al. (1974b).

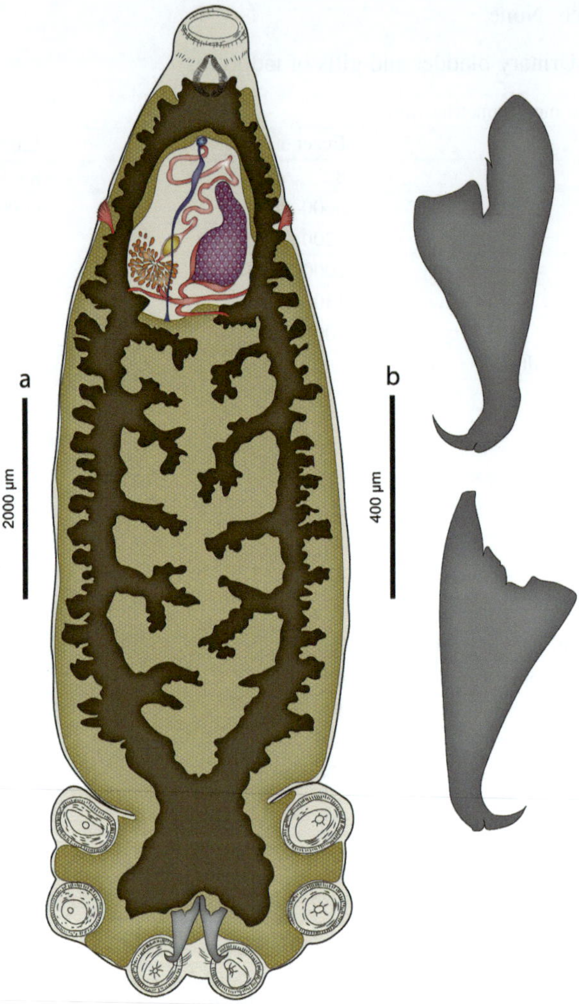

Fig. 4.74 (**a**) *Polystoma galamense*, (**b**) hamuli. Redrawn from Euzet et al. (1974)

Taxonomic note The original name *Polystoma galamensis* was corrected for gender.

Other taxonomic contributions Bourgat (1977); Aisien et al. (2004).

Etymology Named after the host.

Type locality Klouto, Togo.

Other localities Nigeria (see Aisien et al. 2004).

Type host *Amnirana galamensis* (Duméril and Bibron, 1841), Galam White-lipped Frog. In the original description of the polystome, the host was known as *Rana galamensis* Duméril and Bibron, 1841.

Additional hosts None.

Host geographical distribution *Amnirana galamensis* occurs in Uganda, Tanzania, Senegal, Nigeria, Cameroon, Ethiopia, Congo, Gambia, Kenya, Mali, Malawi, Sierra Leone, Mozambique, Ghana, Somalia and Zambia (Frost 2021).

Host conservation status The Galam White-lipped Frog has a wide distribution range, a large population and is tolerant for various habitats. It is listed as Least Concern by The IUCN Red List of Threatened Species 2014: e.T58196A89360210. https://doi.org/10.2305/IUCN.UK.2014-3.RLTS.T58196A18406295.en. Accessed on 30 January 2022.

Site on host Urinary bladder.

DNA sequence None.

Morphology and morphometrics (μm):

Source	Euzet et al. (1974b)
Sample size	20
Body length	8000–10,000
Greatest width	2600–3200
Width at vagina	1900–2200
Haptor length	1700–1900
Haptor width	
False oral sucker width	550–600
Pharynx length	320–400
Pharynx width	300–400
Ovary length	1000
Ovary width	500
Egg length	190–220
Egg width	140–150
Egg incubation time	
Number of intra-uterine eggs	1–7
Genital bulb width	
Number of genital spines	
Genital spine length	35
Haptoral sucker diameter	550–650
Hamulus handle length X	650–700
Hamulus guard length Y	
Hamulus hook length	
Marginal hooklet length	35
Haptoral L/Body L	0.19–0.21
No of anastomoses	0–1

Remarks *Polystoma galamense* was originally described as *Polystoma galamensis.*

4.4.11.24 *Polystoma gallieni* Price, 1939 (Fig. 4.75)

Collection Unknown.

Holotype Unknown.

Paratypes Unknown.

Original description Price (1939).

Other taxonomic contributions Gallien (1947); Combes (1967, 1968); Euzet and Combes (1975); Badets et al. (2009).

Etymology Named after Prof Louis Gallien, Toulouse, France.

Type locality Toulouse, France.

Other localities Morocco.

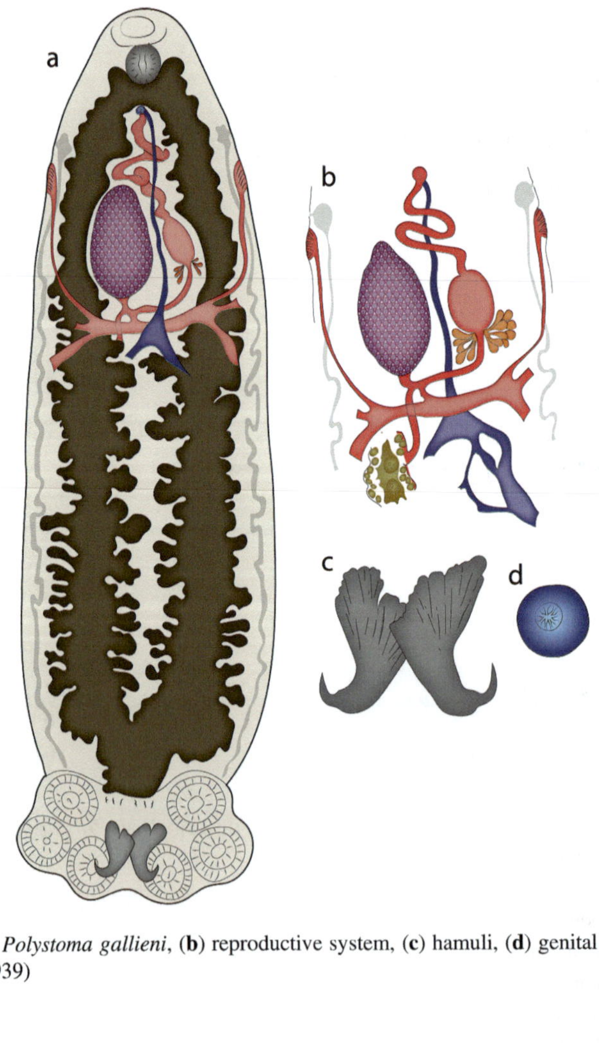

Fig. 4.75 (**a**) *Polystoma gallieni*, (**b**) reproductive system, (**c**) hamuli, (**d**) genital bulb. Redrawn from Price (1939)

Type host *Hyla meridionalis* Boettger, 1874, Stripeless Tree Frog.

Additional hosts None.

Host geographical distribution *Hyla meridionalis* occurs in Algeria, Spain, Portugal, Morocco, Tunisia, France, Italy, Monaco and Gibraltar. It inhabits trees, orchards, grasses, vineyards and shrubs as well as ponds, springs, lagoons and meadows (Barroso et al. 2009).

Host conservation status The Stripeless Tree Frog has a certain tolerance for a wide range of habitats, and a presumed large population. It is listed as Least Concern by The IUCN Red List of Threatened Species 2009: e.T55557A11317442. Accessed on 30 January 2022.

Site on host Urinary bladder and gills of tadpoles.

DNA sequence
- 18S rRNA gene, 2123 bp. GenBank Accession: AM051070
- 18S rRNA gene, 627 bp. GenBank Accession: AJ287989
- 28S rRNA gene, 1449 bp. GenBank Accession: AM157205
- 28S rRNA gene, 1223 bp. GenBank Accession: AF382064
- Cytochrome c Oxidase I, 380 bp. GenBank Accession: JF699305
- 12S rRNA gene, 482 bp. GenBank Accession: KR856084
- ITS1, 688 bp. GenBank Accession: AJ301687
- Abd-A/Lox4 gene, 1761 bp. GenBank Accession: FN597719
- Abd-A/Lox4 gene, allele A, 75 bp. GenBank Accession: FN298170
- Abd-A/Lox4 gene, allele B, 75 bp. GenBank Accession: FN298169

Morphology and morphometrics (μm):

Source	Gallien (1947)	Combes (1967)	Badets et al. (2009)
Sample size			
Body length	3200–4200	6200	
Greatest width	1000–1,400	1400	
Width at vagina			
Haptor length		800	
Haptor width		1560	
False oral sucker width			
Pharynx length			
Pharynx width		285	
Testis length		570	
Testis width		250	
Ovary length		850	
Ovary width		230	
Egg length		180	
Egg width		100	
Egg incubation time			12–15 days
Number of intra-uterine eggs			

(continued)

Morphology and morphometrics (μm):			
Source	Gallien (1947)	Combes (1967)	Badets et al. (2009)
Genital bulb width		50	
Number of genital spines			
Genital spine length		12	
Haptoral sucker diameter		325	
Hamulus handle length X		290	
Hamulus guard length Y			
Hamulus hook length			
Marginal hooklet length			
Haptoral L/Body L		0.13	
Number of anastomoses		0	

Remarks *Polystoma gallieni* was originally described as *Polystomum gallieni.* Note that Gallien (1938) reported that polystome as *Polystomum* sp.

4.4.11.25 *Polystoma goeldii* Sales, Du Preez, Verneau and Domingues, 2023 (Fig. 4.76)

Collection Museu Paraense Emilio Goeldi, Belém, Pará State, Brazil (MPEG).

Holotype MPEG.

Paratypes MPEG.

Original description Sales et al. (2023).

Other taxonomic contributions None.

Etymology Named after Emílio Augusto Goeldi, Belem, Brazil.

Type locality Ramal do Maneta, Itabocal Village, municipality of Irituia, Pará, Brazil (−186382S, −47,41,287 W).

Other localities Municipality of Igarapé-Açu, Pará, Brazil (−1.1315S, −47.6825 W).

Type host *Physalaemus ephippifer* (Steindachner, 1864), Steindachner's Dwarf Frog.

Additional hosts None.

Host geographical distribution Central and eastern Brazilian Amazonia and the Guianas (Frost 2021).

Host conservation status The Steindachner's Dwarf Frog is listed as Least Concern by the IUCN Red List of Threatened Species 2004: e.T57252A11609456. https://doi.org/10.2305/IUCN.UK.2004.RLTS.T57252A11609456.en. Accessed on 26 August 2022.

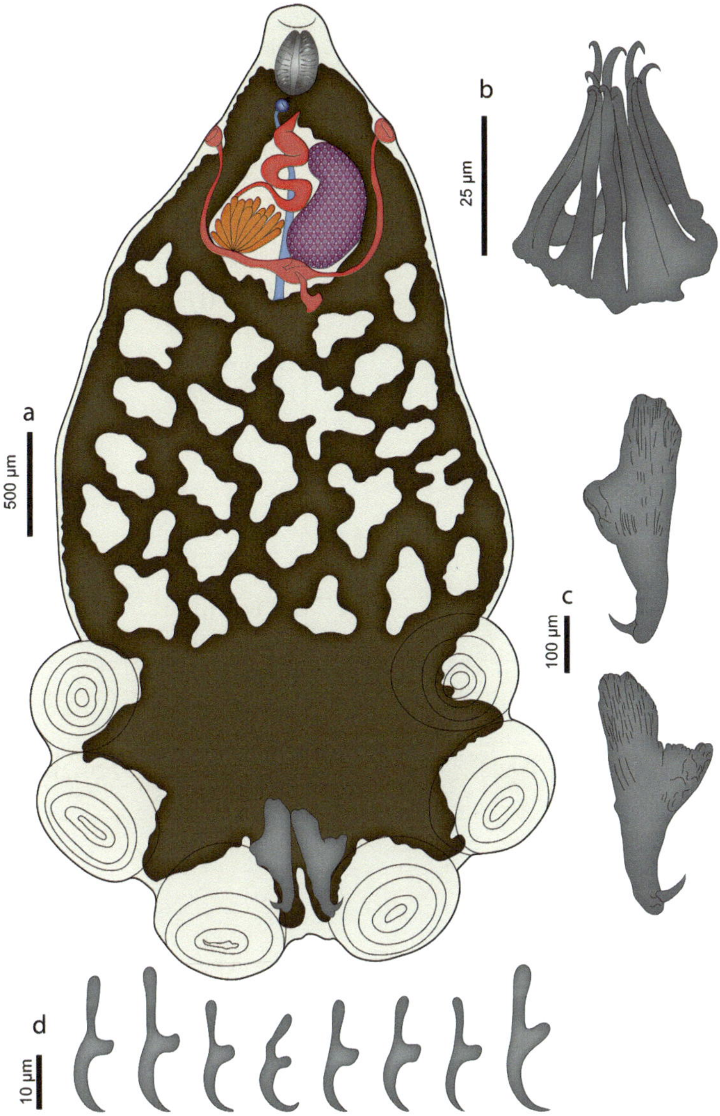

Fig. 4.76 (**a**) *Polystoma goeldii* (**b**) genital bulb, (**c**) hamuli, (**d**) marginal hooklets (1–8). Redrawn from Sales et al. (2023)

Site on host Urinary bladder

DNA sequence

- 18S rRNA gene, 2082 bp. GenBank Accession: OP538667
- 28S rRNA gene, 1445 bp. GenBank Accession: OP538666
- Cytochrome c Oxidase I, 393 bp. GenBank Accession: OP537251 and OP537252

Morphology and morphometrics (μm):	
Source	Sales et al. (2023)
Sample size	7
Body length	2371–4893 (3327)
Greatest width	786–2403 (1356)
Width at vagina	515–940 (661)
Haptor length	747–1706 (1027)
Haptor width	1237–2852 (1712)
False oral sucker width	199–277 (242)
Pharynx length	150–189 (173)
Pharynx width	115–171 (147)
Ovary length	296–564 (429)
Ovary width	155–278 (242)
Egg length	
Egg width	
Egg incubation time	
Number of intra-uterine eggs	
Genital bulb width	
Number of genital spines	9
Genital spine length	31–35 (33)
Haptoral sucker diameter	247–361 (306)
Hamulus handle length X	339–424 (331)
Hamulus guard length Y	253–312 (284)
Hamulus hook length	50–66 (58)
Marginal hooklet length	C1 27–29 (28); C2–7 20–21 (20); C8 25
Haptoral L/Body L	0.23–0.39 (0.31)
No of anastomoses	Network

Remarks None.

4.4.11.26 *Polystoma grassei* Euzet, Combes and Knoepffler, 1966 (Fig. 4.77)

Collection Claude Combes personal collection, Perpignan, France.

Holotype Unknown.

Paratypes Unknown.

Original description Euzet et al. (1966).

Other taxonomic contributions Maeder (1973); Dupouy and Combes (1977); Murith et al. (1978).

Etymology Unknown.

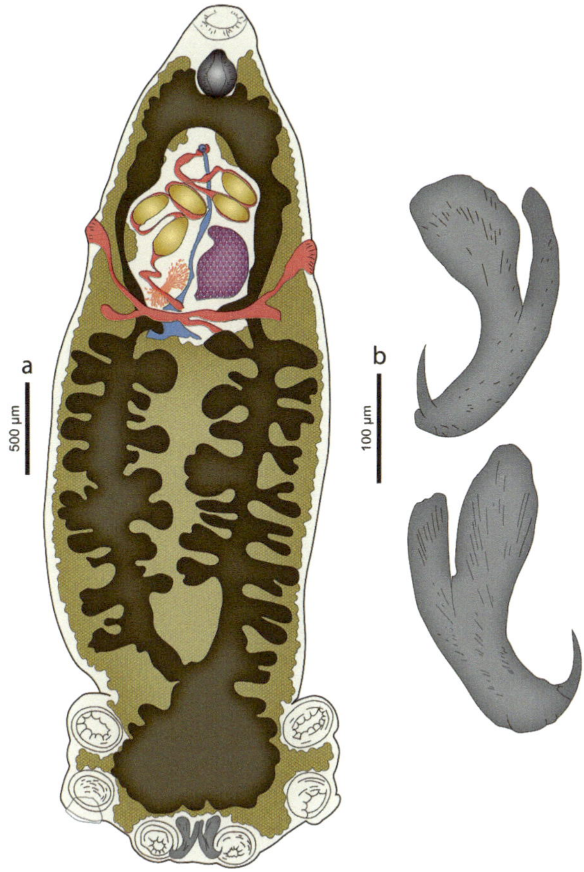

Fig. 4.77 (**a**) *Polystoma grassei*, (**b**) hamuli. Redrawn from Euzet et al. (1966)

Type locality Gabon.

Other localities Ivory Coast, Cameroon.

Type host *Leptopelis calcaratus* (Boulenger, 1906), Efulen Forest Treefrog.

Additional hosts *Leptopelis viridis* (Günther, 1869) and *Leptopelis ocellatus* (Mocquard, 1902).

Host geographical distribution *Leptopelis calcaratus* occurs in Nigeria, Gabon, Cameroon and Congo where it inhabits montane rainforests and lowland regions (Frost 2021).

Host conservation status The Efulen Forest Treefrog has a large population size and a wide distribution range. It is listed as Least Concern by The IUCN Red List of Threatened Species 2013: e.T56250A18387924. https://doi.org/10.2305/IUCN.UK.2013-2.RLTS.T56250A18387924.en. Accessed on 30 January 2022.

Site on host Urinary bladder and gills of tadpoles.

DNA sequence None.

Morphology and morphometrics (μm):		
Source	Euzet et al. (1966)	Maeder (1973)
Sample size	6	
Body length	2500	3900–7000
Greatest width	1100	
Width at vagina		900–1700
Haptor length	1000	800–1400
Haptor width	1600	1200–2000
False oral sucker width		
Pharynx length		195–250
Pharynx width		155–240
Ovary length	400	
Ovary width	285	
Egg length	240–280	230–285
Egg width	120–135	110–140
Egg incubation time		
Number of intra-uterine eggs	5–10	0–4
Genital bulb width		
Number of genital spines	8	
Genital spine length	30	
Haptoral sucker diameter	320	250–360
Hamulus handle length X		250–320
Hamulus guard length Y		
Hamulus hook length		
Marginal hooklet length	15–25	17–25
Haptoral L/Body L	0.4	0.20
No of anastomoses		0–2

Remarks None.

4.4.11.27 *Polystoma guevarai* Combes and Laurent, 1979 (Fig. 4.78)

Collection Unknown.

Holotype Unknown.

Paratypes Unknown.

Original description Combes and Laurent (1979).

Other taxonomic contributions None.

Etymology Unknown.

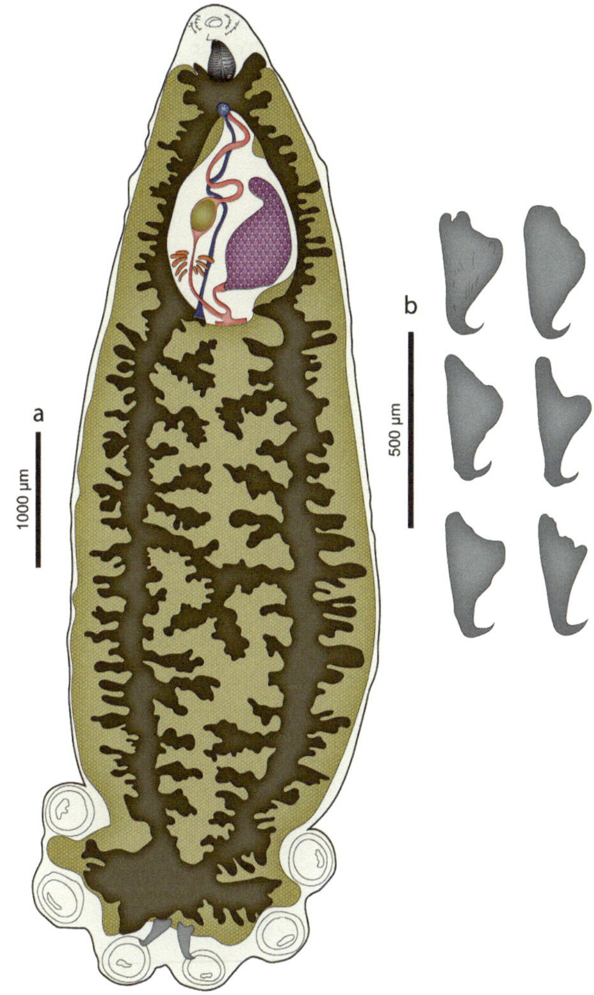

Fig. 4.78 (**a**) *Polystoma guevarai*, (**b**) hamuli. Redrawn from Combes and Laurent (1979)

Type locality Argentine.

Other localities None.

Type host *Boana pulchella* (Duméril and Bibron, 1841), Montevideo Treefrog. In the original description of the polystome, the host was known as *Hyla pulchella* Duméril and Bibron, 1841.

Additional hosts None.

Host geographical distribution *Boana pulchellua* occurs in Uruguay, Brazil, Argentina and Paraguay where it can be found in forests, flooded savannas and grasslands (Kwet et al. 2004).

Host conservation status The Montevideo Treefrog has a large population and a wide distribution. It is listed as Least Concern by The IUCN Red List of Threatened Species 2004: e.T55619A11340514. https://doi.org/10.2305/IUCN.UK.2004.RLTS.T55619A11340514.en. Accessed on 30 January 2022.

Site on host Urinary bladder.

DNA sequence None.

Morphology and morphometrics (μm):	
Source	Combes and Laurent (1979)
Sample size	5
Body length	6790–7880
Greatest width	2050–2390
Width at vagina	
Haptor length	960–1330
Haptor width	1940–2180
False oral sucker width	
Pharynx length	286–342
Pharynx width	230–274
Testis length	
Testis width	
Ovary length	
Ovary width	
Egg length	
Egg width	
Egg incubation time	
Number of intra-uterine eggs	
Genital bulb width	
Number of genital spines	
Genital spine length	
Haptoral sucker diameter	316–401
Hamulus handle length X	298–348
Hamulus guard length Y	
Hamulus hook length	
Marginal hooklet length	
Haptoral L/Body L	0.1
No of anastomoses	0–2

Remarks None.

4.4.11.28 *Polystoma integerrimum* (Fröhlich, 1791) (Fig. 4.79)

Collection Unknown.

Holotype Unknown.

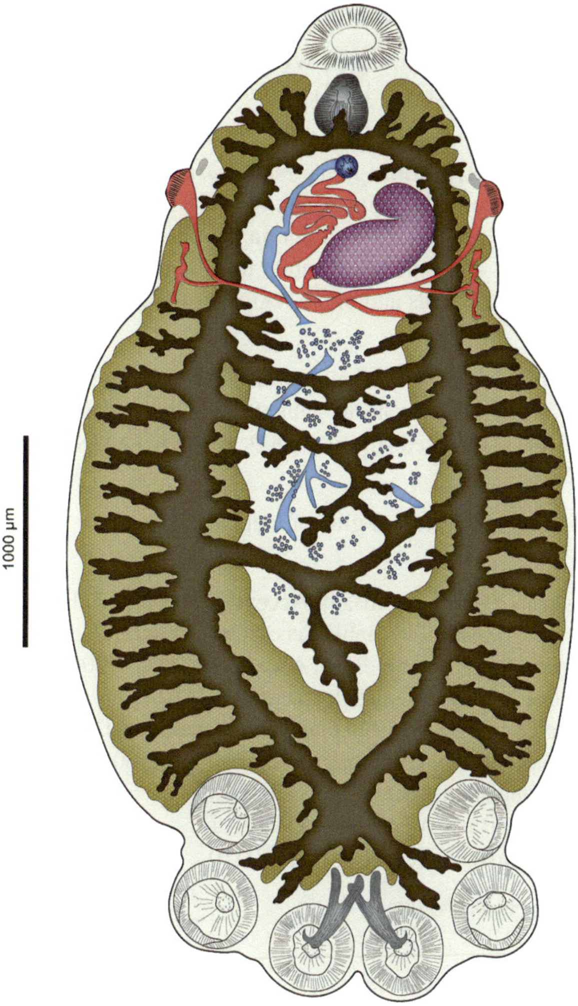

Fig. 4.79 *Polystoma integerrimum*. Redrawn from Fröhlich (1791)

Paratypes Unknown.

Original description Fröhlich von (1791).

Other taxonomic contributions Macé E (1880); Gallien (1932, 1935, 1947); Ozaki (1935); Park (1938); Williams (1960a, b); Combes (1968); Frandsen (1974); Al-Sorkhy and Amr (2003).

Etymology Unknown.

Type locality Europe.

Other localities Europe (Denmark, England, France, Switzerland), Japan, Jordan and Korea.

Type host *Rana temporaria* Linnaeus, 1758, Common Frog.

Additional hosts *Rana ornativentris,* Werner 1903, *Rana arvalis* Nilsson, 1842, *Pelophylax lessonae* (Camerano, 1882), *Bufotes viridis* (Laurenti, 1768) and *Hyla savignyi* Audouin, 1827.

Host geographical distribution *Rana temporaria* occurs in United Kingdom, Greece, Sweden, Belgium, Albania, Spain, Netherlands, Germany, France, Romania, Denmark, Norway, Switzerland, Bulgaria, Hungary, Finland, Italy, Poland and Ireland among others. This species inhabits forests, meadows, grasslands, glades, gardens, marshes, parks, fields and urban regions as well as rivers, lakes and ponds (Frost 2021).

Host conservation status The Common Frog has a large population, a wide distribution and is tolerant for various habitats. It is listed as Least Concern by The IUCN Red List of Threatened Species 2009: e.T58734A11833856. Accessed on 30 January 2022.

Site on host Urinary bladder and gills of tadpoles.

DNA sequence
- 18S rRNA gene, 2094 bp. GenBank Accession: AM051071
- 18S rRNA gene, 627 bp. GenBank Accession: AJ287990
- 28S rRNA gene, 1438 bp. GenBank Accession: AM157206
- 28S rRNA gene, 316 bp. GenBank Accession: AF131719
- Cytochrome c Oxidase I, 375 bp. GenBank Accession: JF699306
- 12S rRNA gene, 491 bp. GenBank Accession: KR856086
- ITS1, 728 bp. GenBank Accession: AJ301688
- ITS1, 756 bp. GenBank Accession: AJ310411
- ITS1, 738 bp. GenBank Accession: AJ310410

Morphology and morphometrics (μm):

Source	Mace (1880)	Gallien (1947)	Combes (1968)	Combes and Euzet (1966)
Sample size				
Body length	5000	8000–10,000	10,100	
Greatest width	1000	3000–4000	3400	
Width at vagina				
Haptor length				
Haptor width				
False oral sucker width				
Pharynx length				
Pharynx width				
Testis length				
Testis width				
Ovary length				
Ovary width				

(continued)

Morphology and morphometrics (μm):				
Source	Mace (1880)	Gallien (1947)	Combes (1968)	Combes and Euzet (1966)
Egg length				
Egg width				
Egg incubation time				
Number of intra-uterine eggs				
Genital bulb width				
Number of genital spines			8	
Genital spine length			30	
Haptoral sucker diameter				
Hamulus handle length X			524	400–632 (524)
Hamulus guard length Y				
Hamulus hook length				
Marginal hooklet length			C1,28–29; C2–8, 25–26; C8, 22–23	
Haptoral L/Body L				
No of anastomoses		>1		

Remarks *Polystoma integerrimum* was originally described as *Linguatula integerrima* Fröhlich, 1791.

4.4.11.29 *Polystoma ivindoi* Euzet, Combes and Knoepffler, 1966 (Fig. 4.80)

Collection Claude Combes personal collection, Perpignan, France.

Holotype Unknown.

Paratypes Unknown.

Original description Euzet et al. (1966).

Taxonomic note It was originally described as *Polystoma africanum ivindoi.*

Other taxonomic contributions Murith et al. (1978).

Etymology Probably named after the Ivindo National Park in Gabon.

Type locality Makokou, Gabon.

Other localities Cameroon.

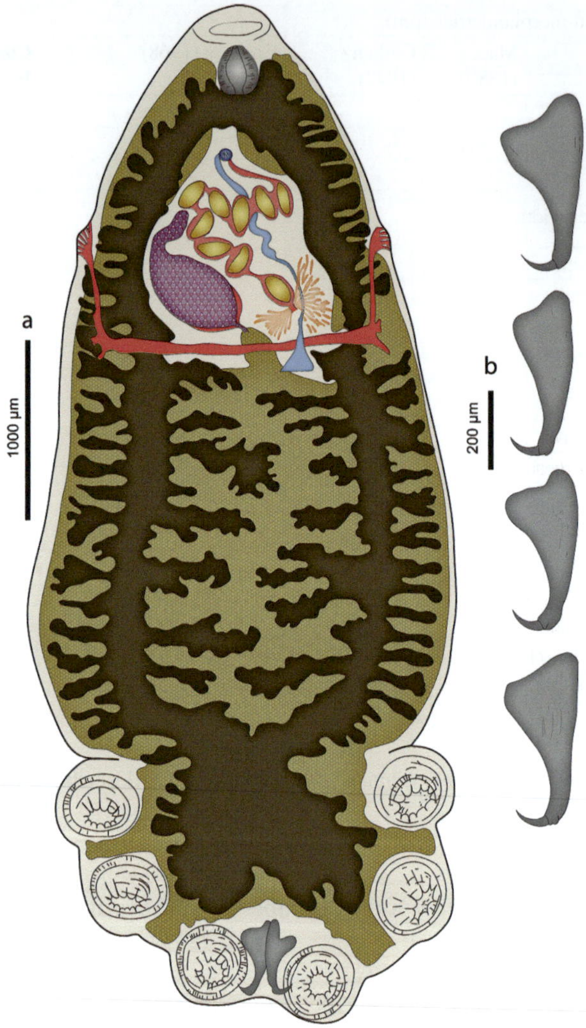

Fig. 4.80 (**a**) *Polystoma ivindoi*, (**b**) hamuli. Redrawn from Euzet et al. (1966)

Type host *Amnirana lepus* (Andersson, 1903), Andersson's Cameroon Frog. In the original description of the polystome, the host was known as *Hylarana lepus lepus* Perret, 1959.

Additional hosts None.

Host geographical distribution *Amnirana lepus* occurs in Gabon, Cameroon and the Equatorial Guinea (Frost 2021). This species is common in various forests including lowland forests, gallery forests and degraded former forests.

Host conservation status The Andersson's Cameroon Frog is adaptable, has a large population size, a wide distribution range and is tolerant for various habitats. It is thus listed as Least Concern by The IUCN Red List of Threatened Species 2013: e. T58198A89361810. https://doi.org/10.2305/IUCN.UK.2013-2.RLTS.T58198A18406665.en. Accessed on 30 January 2022.

Site on host Urinary bladder.

DNA sequence None.

Morphology and morphometrics (μm):	
Source	Euzet et al. (1966)
Sample size	46
Body length	3000–6000
Greatest width	1500–2500
Width at vagina	1500–2000
Haptor length	1200–1500
Haptor width	1800–2200
False oral sucker width	400
Pharynx length	150
Pharynx width	80
Ovary length	740
Ovary width	330
Egg length	190–230 (205)
Egg width	115–130 (120)
Egg incubation time	
Number of intra-uterine eggs	1–10
Genital bulb width	
Number of genital spines	8
Genital spine length	40–42
Haptoral sucker diameter	400
Hamulus handle length X	200–260
Hamulus guard length Y	
Hamulus hook length	
Marginal hooklet length	
Haptoral L/Body L	0.25–0.40
No of anastomoses	0–2

Remarks *Polystoma ivindoi* was originally described as *Polystoma africanum ivindo*.

4.4.11.30 *Polystoma knoffi* Du Preez and Domingues, 2019 (Fig. 4.81)

Collection Helminthological Collection, Instituto Oswaldo Cruz, Rio de Janeiro, Brazil (CHIOC); Parasitic Worm Collection, National Museum, Bloemfontein, South Africa (NMBP)

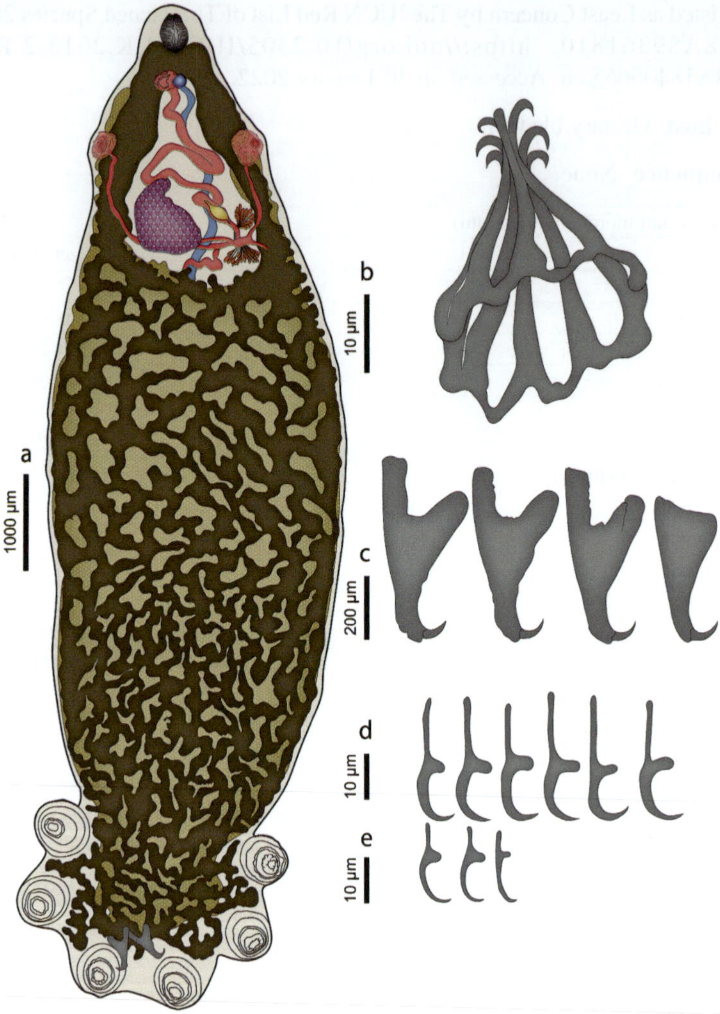

Fig. 4.81 (**a**) *Polystoma knoffi*, (**b**) genital spines, (**c**) hamuli, (**d**) marginal hooklets (1), (**e**) marginal hooklets (2–7). Redrawn from Du Preez and Domingues (2019)

Holotype CHIOC 25702a.

Paratypes CHIOC 25701a-c; 25702b; NMBP 506–508.

Original description Du Preez and Domingues (2019).

Other taxonomic contributions Kohn et al. (1978).

Etymology Named after Dr Marcelo Knoff, Instituto Oswaldo Cruz, Rio de Janeiro, Brazil.

Type locality Jacarepagua, Rio de Janeiro, Brazil.

Other localities None.

Type host *Trachycephalus nigromaculatus* Tschudi, 1838, Black-spotted Casque-headed Treefrog.

Additional hosts None.

Taxonomy Original material was collected from *Trachycephalus geographicus* but this species is regarded as a junior synonym of *T. nigromaculatus*.

Host geographical distribution *Trachycephalus nigromaculatus* is a low altitude species and it is distributed from southern Brazil from Espírito Santo to São Paulo; interior in Bahia, Minas Gerais, and Goiás to southern Piauí (Frost 2021).

Host conservation status The Black-spotted Casque-headed Treefrog is widespread and tolerates a certain degree of habitat modification. It is listed as Least Concern by The IUCN Red List of Threatened Species 2010: e.T56050A11417783. https://doi.org/10.2305/IUCN.UK.2010-2.RLTS.T56050A11417783.en. Accessed on 30 January 2022.

Site on host Urinary bladder.

Zoobank registration urn:lsid:zoobank.org:act: 5F0454AD-0BDB-4D9B-AC17-A7E6DB142C54.

DNA sequence None.

Morphology and morphometrics (μm):	
Source	Du Preez and Domingues (2019)
Sample size	8
Body length	5198–10,625 (7386)
Greatest width	1590–3409 (2494)
Width at vagina	852–1818 (1339)
Haptor length	1191–1818 (1473)
Haptor width	1654–2840 (2207)
False oral sucker width	207–345 (264)
Pharynx length	240–430 (318)
Pharynx width	210–325 (266)
Ovary length	235–1095 (577)
Ovary width	132–670 (378).
Egg length	
Egg width	
Egg incubation time	
Number of intra-uterine eggs	1
Genital bulb width	75–130 (107)
Number of genital spines	8
Genital spine length	

(continued)

Morphology and morphometrics (μm):	
Source	Du Preez and Domingues (2019)
Haptoral sucker diameter	320–470 (398)
Hamulus handle length X	420–571 (509)
Hamulus guard length Y	386–571 (451)
Hamulus hook length	84–100 (94)
Marginal hooklet length	31–34 (31)
Haptoral L/Body L	0.17–0.25 (0.21)
No of anastomoses	Network

Remarks None.

4.4.11.31 *Polystoma lamottei* Bourgat and Murith, 1980 (Fig. 4.82)

Collection Unknown.

Holotype Unknown.

Paratypes Unknown.

Original description Bourgat and Murith (1980).

Other taxonomic contributions None.

Etymology Named after Prof M. Lamotte, Paris, France.

Type locality Kanté, Togo.

Other localities None.

Type host *Ptychadena pumilio*, Medine Grassland Frog.

Additional hosts None.

Taxonomy None.

Host geographical distribution *Ptychadena pumilio* occurs in Cameroon, Congo, Ivory Coast, Senegal, Sierra Leone, Ethiopia, Benin, Nigeria, Central African Republic and Mali (Frost 2021). This species inhabits habitats close to water, such as the banks of pools and rivers, as well as grassland and forest habitats.

Host conservation status The Medine Grassland Frog has a large population, a wide distribution and is tolerant of various habitat ranges. It is listed as Least Concern by The IUCN Red List of Threatened Species 2013: e. T58521A18401115. https://doi.org/10.2305/IUCN.UK.2013-2.RLTS. T58521A18401115.en. Accessed on 30 January 2022.

Site on host Urinary bladder.

DNA sequence None.

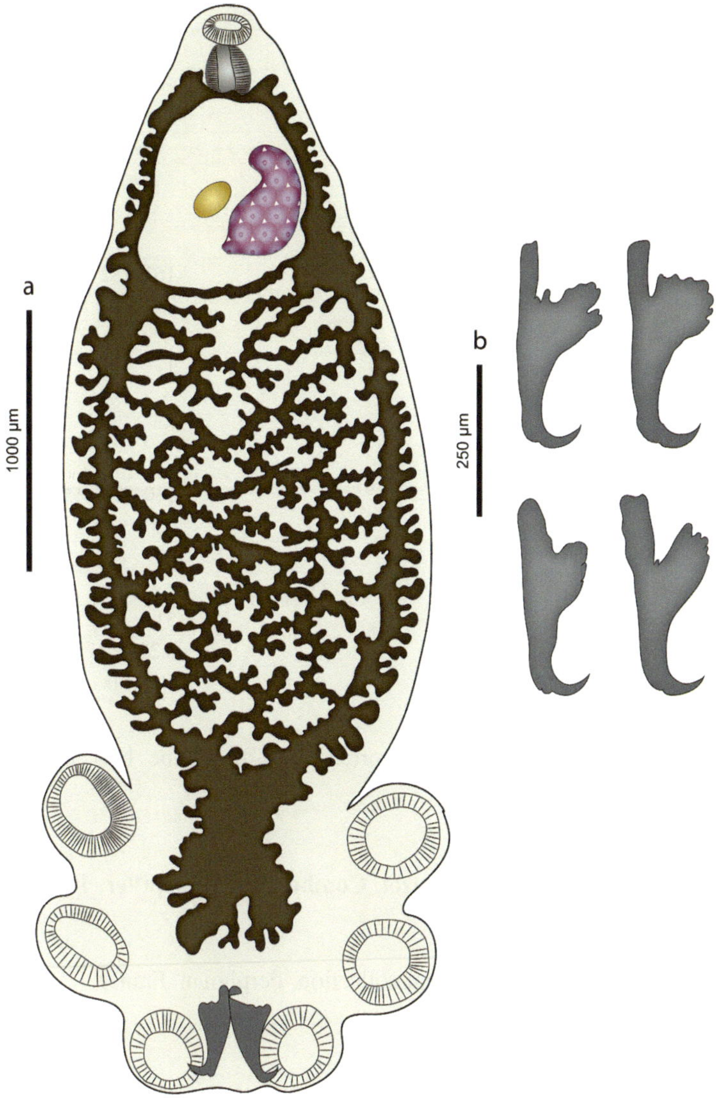

Fig. 4.82 (**a**) *Polystoma lamottei*, (**b**) hamuli. Redrawn from Bourgat and Murith (1980)

Morphology and morphometrics (µm):	
Source	Bourgat and Murith (1980)
Sample size	3
Body length	4000–4500
Greatest width	1300–1900
Width at vagina	
Haptor length	900–1100

(continued)

Morphology and morphometrics (µm):	
Source	Bourgat and Murith (1980)
Haptor width	1600–2200
False oral sucker width	
Pharynx length	
Pharynx width	
Ovary length	
Ovary width	
Egg length	167–182
Egg width	85–104
Egg incubation time	
Number of intra-uterine eggs	1
Genital bulb width	
Number of genital spines	
Genital spine length	
Haptoral sucker diameter	310–378
Hamulus handle length X	336–395
Hamulus guard length Y	
Hamulus hook length	
Marginal hooklet length	31
Haptoral L/Body L	0.22–0.24
No of anastomoses	Network

Remarks This species has been reported from the same host as *Polystoma aeschlimanni* in Togo.

4.4.11.32 *Polystoma llewellyni* Euzet, Combes and Knoepffler, 1974 (Fig. 4.83)

Collection Claude Combes personal collection, Perpignan, France.

Holotype Unknown.

Paratypes Unknown.

Original description Euzet et al. (1974a).

Other taxonomic contributions Murith et al. (1978).

Etymology Named after Prof Jack Llewellyn, University of Birmingham, UK.

Type locality Lamaboké, Republic Central Africa.

Other localities Cameroon.

Type host *Afrixalus fulvovittatus* (Cope, 1861), Banded Banana Frog.

Additional hosts None.

Taxonomy None.

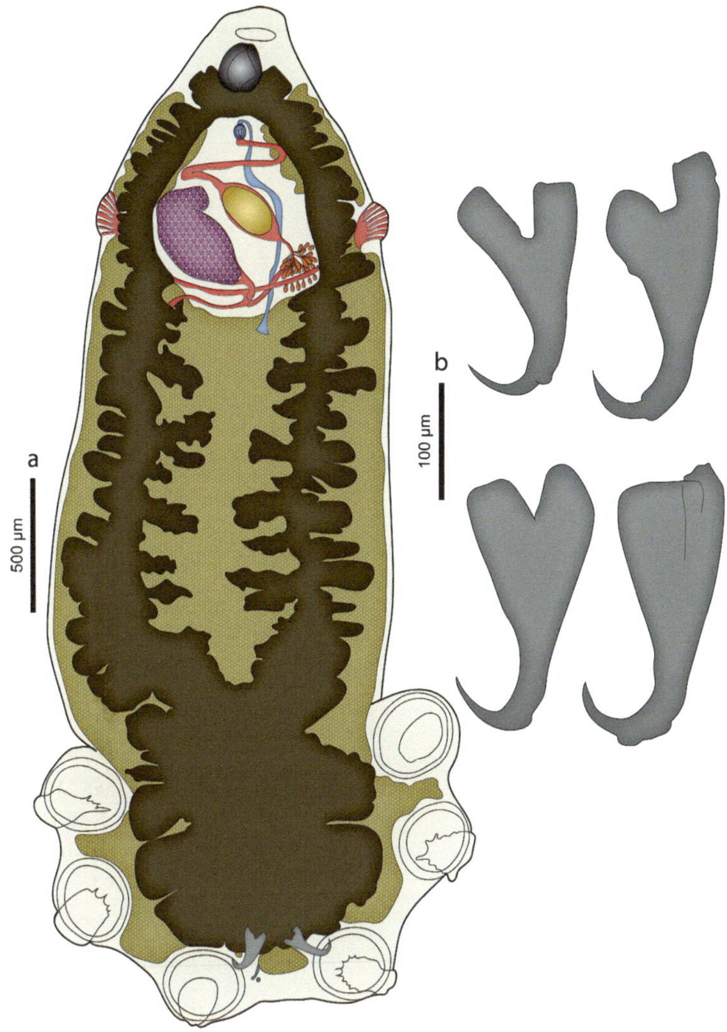

Fig. 4.83 (**a**) *Polystoma llewellyni*, (**b**) hamuli. Redrawn from Euzet et al. (1974)

Host geographical distribution *Afrixalus fulvovittatus* occurs in Sierra Leone, Guinea, Ghana, Liberia, Ivory Coast and Guinea where it inhabits degraded former forests (in the forest zone) (Frost 2021).

Host conservation status The Banded Banana Frog has a large population, a wide distribution range and is tolerant to habitat modification. It is listed as Least Concern by The IUCN Red List of Threatened Species 2013: e.T56064A18369490. https://doi.org/10.2305/IUCN.UK.2013-2.RLTS.T56064A18369490.en. Accessed on 30 January 2022.

Site on host Urinary bladder.

DNA sequence None.

Morphology and morphometrics (μm):	
Source	Euzet et al. (1974a)
Sample size	3
Body length	3500–3800
Greatest width	1200–1500
Width at vagina	1100–1200
Haptor length	800–1000
Haptor width	
False oral sucker width	250
Pharynx length	160–180
Pharynx width	150 × 170
Ovary length	200
Ovary width	140
Egg length	
Egg width	
Egg incubation time	
Number of intra-uterine eggs	1
Genital bulb width	
Number of genital spines	6–8
Genital spine length	35
Haptoral sucker diameter	300–310
Hamulus handle length X	200–230
Hamulus guard length Y	
Hamulus hook length	200–230
Marginal hooklet length	22–40
Haptoral L/Body L	0.23–0.26
No of anastomoses	0–1

Remarks None.

4.4.11.33 *Polystoma lopezromani* Combes and Laurent, 1979 (Fig. 4.84)

Collection Unknown.

Holotype Unknown.

Paratypes Unknown.

Original description Combes and Laurent (1979).

Other taxonomic contributions Vaucher (1983).

Etymology Unknown.

Type locality Province de Salta, Argentine.

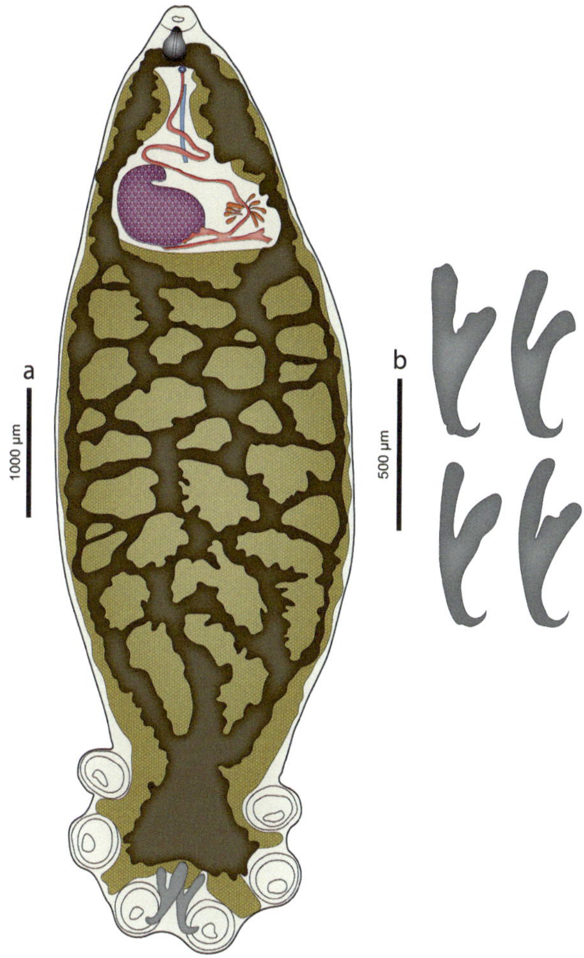

Fig. 4.84 (**a**) *Polystoma lopezromani*, (**b**) hamuli. Redrawn from Combes and Laurent (1979)

Other localities Paraguay.

Type host *Trachycephalus typhonius* (Linnaeus, 1758), Pepper Treefrog. In the original description of the polystome, the host was known as *Phrynohyas venulosa* (Laurenti, 1768).

Additional hosts None.

Host geographical distribution *Trachycephalus typhonius* occurs in Argentina, Brazil, Mexico, Peru, French Guiana, Bolivia, Colombia, Ecuador, Costa Rica, Guyana, El Salvador, Panama, Belize, Venezuela, Paraguay, Nicaragua, Honduras and Guatemala. This species usually inhabits forests, savannas, disturbed habitats (human dwellings and plantations) and other open habitats (La Marca et al. 2010b).

Host conservation status The Pepper Treefrog has a large population, a wide distribution and is tolerant for various habitats. It is listed as Least Concern by The IUCN Red List of Threatened Species 2010: e.T55824A95508779. https://doi.org/10.2305/IUCN.UK.2010-2.RLTS.T55824A11373788.en. Accessed on 30 January 2022.

Site on host Urinary bladder.

DNA sequence

- 18S rRNA gene, 2086 bp. GenBank Accession: AM051072
- 28S rRNA gene, 1441 bp. GenBank Accession: AM157207
- Cytochrome c Oxidase I, 372 bp. GenBank Accession: AM913863
- 12S rRNA gene, 491 bp. GenBank Accession: KR856087
- ITS1, 684 bp. GenBank Accession: AJ301690

Morphology and morphometrics (μm):	
Source	Combes and Laurent (1979)
Sample size	5
Body length	6990–8160
Greatest width	2220–2730
Width at vagina	
Haptor length	1160–1430
Haptor width	1600–2110
False oral sucker width	
Pharynx length	292–330
Pharynx width	201–241
Ovary length	
Ovary width	
Egg length	
Egg width	
Egg incubation time	
Number of intra-uterine eggs	
Genital bulb width	
Number of genital spines	
Genital spine length	
Haptoral sucker diameter	316–401
Hamulus handle length X	544–606
Hamulus guard length Y	
Hamulus hook length	
Marginal hooklet length	
Haptoral L/Body L	0.18
No of anastomoses	Network

Remarks None.

4.4.11.34 *Polystoma luohetong* Fan, Xu, Jia, Netherlands and Du Preez, 2020 (Fig. 4.85)

Collection Parasitic Worm Collection, National Museum, Bloemfontein, South Africa (NMBP); School of Life Science of Yunnan Normal University, Kunming, China (YNU).

Holotype NMBP 554.

Paratypes NMBP 555–559; YNU 200505C015, 200505C018–2, 200505C023 and 200505C070.

Original description Fan et al. (2020).

Fig. 4.85 (**a**) *Polystoma luohetong*, (**b**) hamuli, (**c**) marginal hooklets (1), (**d**) marginal hooklets (2–8). Redrawn from Fan et al. (2020)

Other taxonomic contributions Héritier et al. (2015).

Etymology Named after Professors Y.D. Luo, H.M. He and J.R. Tong, China.

Type locality Lufeng County, Yunnan Province in southwest of China (25.160930 N, 101.895574E).

Other localities None.

Type host *Rana chaochiaoensis* Liu, 1946, Chaochiao Frog.

Additional hosts None.

Host geographical distribution *Rana chaochiaoensis* is known from the mountains of Yunnan Province, southern Sichuan Province and western Guizhou Province in China (Datong and Shunqing 2004b).

Host conservation status Because of its wide distribution, the Chaochiao Frog is listed as Least Concern by The IUCN Red List of Threatened Species 2020: e. T58569A63877176. https://doi.org/10.2305/IUCN.UK.2020-1.RLTS.T58569A63877176.en. Accessed on 30 January 2022.

Site on host Urinary bladder.

DNA sequence
- 18S rRNA gene, 1522 bp. GenBank Accession: EU734834
- 18S rRNA gene, 2062 bp. GenBank Accession: KR856125

Morphology and morphometrics (μm):	
Source	Fan et al. (2020)
Sample size	22
Body length	1876–7610 (4215)
Greatest width	788–2980 (1960)
Width at vagina	509–2034 (1033)
Haptor length	465–2560 (1126)
Haptor width	800–2610 (1734)
False oral sucker width	
Pharynx length	119–279 (164)
Pharynx width	103–188 (148)
Ovary length	99–778 (495)
Ovary width	54–549 (338)
Egg length	118–261 (220)
Egg width	90–151 (115)
Egg incubation time	
Number of intra-uterine eggs	13
Genital bulb width	
Number of genital spines	8–9
Genital spine length	15.0–49.8 (33.3)
Haptoral sucker diameter	206–474 (340)

(continued)

Morphology and morphometrics (μm):	
Source	Fan et al. (2020)
Hamulus handle length X	221–513 (344)
Hamulus guard length Y	188–396 (282)
Hamulus hook length	38–64 (51)
Marginal hooklet length	C1: 26.9–32.2 (29.2); C2–8 18.8–24.5 (22.1)
Haptoral L/Body L	
No of anastomoses	Network

Remarks *Polystoma dianxiensis* referred to in Héritier et al. (2015) was never formally described under this name but briefly described as *Polystoma chaochiaoensis* in a symposium abstract. It is not a valid taxon since the description was incomplete and no types assigned. *Polystoma chaochiaoensis* was regarded as *nomen nudum* and described as *P. luohetong*.

4.4.11.35 *Polystoma macrocnemis* Biserkov, Yildirimhan, Buchvarov and Ugurtas, 2001 (Fig. 4.86)

Collections Natural History Museum, London (NHML); Collection of V. Biserkov (VB); Collection of H.S. Yildirimhan (HSY).

Holotype NHML 2000.6.26.1.

Paratypes NHML 2000.6.26.2; VB 199912121–4; HSY 199910121–4.

Original description Biserkov et al. (2001).

Other taxonomic contributions Yildirimhan et al. (1997).

Etymology Named after the host.

Type locality Turkey.

Other localities None.

Type host *Rana macrocnemis* Boulenger, 1885, Iranian Long-Legged Frog.

Additional hosts None.

Host geographical distribution *Rana macrocnemis* occurs in Turkey, Armenia, Georgia, Azerbaijan and Iran. It inhabits swamps, forests, meadows and steppes as well as permanent waterbodies (surrounded by vegetation), such as springs, brooks, lakes and rivers in dry regions (Kuzmin et al. 2009a).

Host conservation status The Iranian Long-Legged Frog has a large population, a wide distribution and various habitats. However, the population in the Caucasus is decreasing due to deforestation and predation by raccoons. It is listed as Least Concern by The IUCN Red List of Threatened Species 2009: e. T58651A11820145. https://doi.org/10.2305/IUCN.UK.2009.RLTS.T58651A11820145.en. Accessed on 30 January 2022.

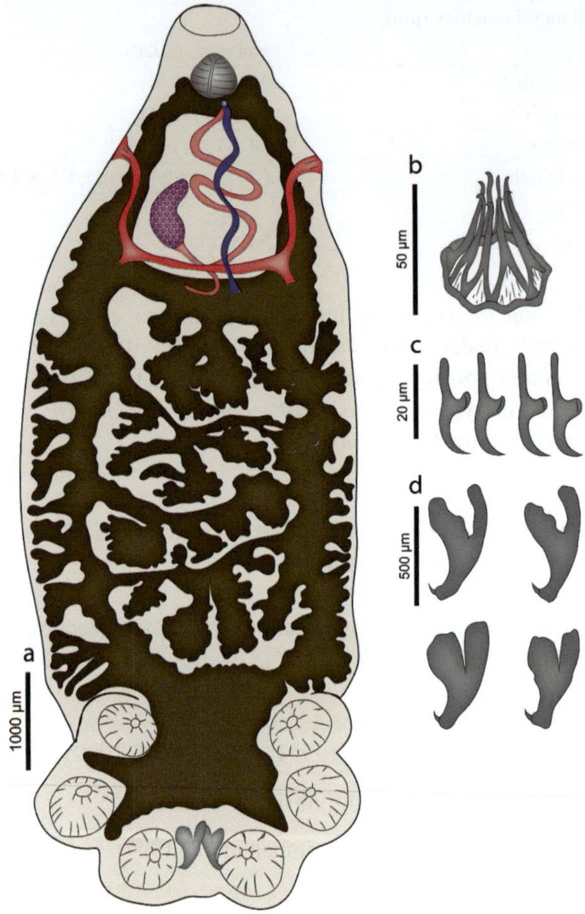

Fig. 4.86 (**a**) *Polystoma macrocnemis*, (**b**) genital spines, (**c**) marginal hooklets, (**d**) hamuli. Redrawn from Biserkov et al. (2001)

Site on host Urinary bladder.

DNA sequence None.

Morphology and morphometrics (µm):	
Source	Biserkov et al. (2001)
Sample size	11
Body length	3550–12,400 (9480)
Greatest width	1600–4510 (3060)
Width at vagina	1110–2880 (1980)
Haptor length	742–3360 (2063)
Haptor width	1500–3840 (2813)

(continued)

Morphology and morphometrics (μm):	
Source	Biserkov et al. (2001)
False oral sucker width	
Pharynx length	263–1048 (434)
Pharynx width	257–936 (418)
Ovary length	
Ovary width	
Egg length	
Egg width	
Egg incubation time	
Number of intra-uterine eggs	
Genital bulb width	56–88 (74)
Number of genital spines	6–8
Genital spine length	30–58 (46)
Haptoral sucker diameter	319–726 × 344–726
Hamulus handle length X	
Hamulus guard length Y	319–315 (405)
Hamulus hook length	
Marginal hooklet length	26.0–28.5 (27.5)
Haptoral L/Body L	0.22
No of anastomoses	2–4

Remarks None.

4.4.11.36 *Polystoma makereri* Tinsley, 1973 (Fig. 4.87)

Collection Natural History Museum of London (NHML), National Museum of Natural History, France (MNHN).

Cotype nos NHML 1973.2.5.1–2.

Original description Tinsley (1973).

Other taxonomic contributions None.

Etymology Named after Makerere University, Uganda.

Type locality Uganda.

Other localities None.

Type host *Phrynobatrachus* sp.

Additional hosts None.

Host geographical distribution Unknown.

Host conservation status Unknown.

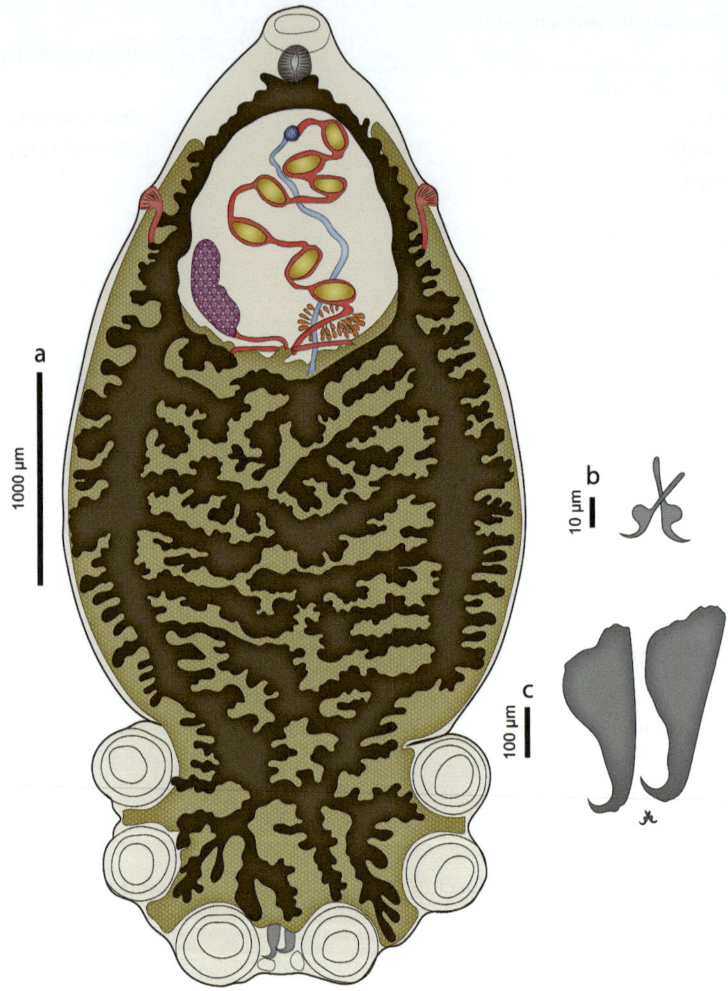

Fig. 4.87 (**a**) *Polystoma makereri*, (**b**) marginal hooklets, (**c**) hamuli. Redrawn from Tinsley (1973)

Site on host Urinary bladder.

DNA sequence None.

Morphology and morphometrics (µm):	
Source	Tinsley (1973)
Sample size	2
Body length	4500–5300
Greatest width	1500–2200
Width at vagina	1000–1300
Haptor length	1400

(continued)

Morphology and morphometrics (μm):	
Source	Tinsley (1973)
Haptor width	2100
False oral sucker width	310
Pharynx length	
Pharynx width	250
Ovary length	400–490
Ovary width	140–210
Egg length	175–210
Egg width	92–107
Egg incubation time	
Number of intra-uterine eggs	7–48
Genital bulb width	68
Number of genital spines	8
Genital spine length	23
Haptoral sucker diameter	410–490 (450)
Hamulus handle length X	390–425 (405)
Hamulus guard length Y	
Hamulus hook length	
Marginal hooklet length	21–34
Haptoral L/Body L	0.26–0.31
No of anastomoses	Network

Remarks None.

4.4.11.37 *Polystoma mangenoti* Gallien, 1956 (Fig. 4.88)

Collection Specimens are stored in the author's private collection.

Holotype Unknown.

Paratypes Unknown.

Original description Gallien (1956).

Other taxonomic contributions Maeder et al. (1970); Maeder (1973); Murith (1979, 1981).

Etymology Named after Prof Mangenot, Orstom Research Institute, Abidjan, Ivory Coast.

Type locality Adiopodoumé, Ivory Coast.

Other localities Liberia.

Type host *Ptychadena superciliaris* (Günther, 1858), Savana Ridged Frog.

Additional hosts *Hoplobatrachus occipitalis*.

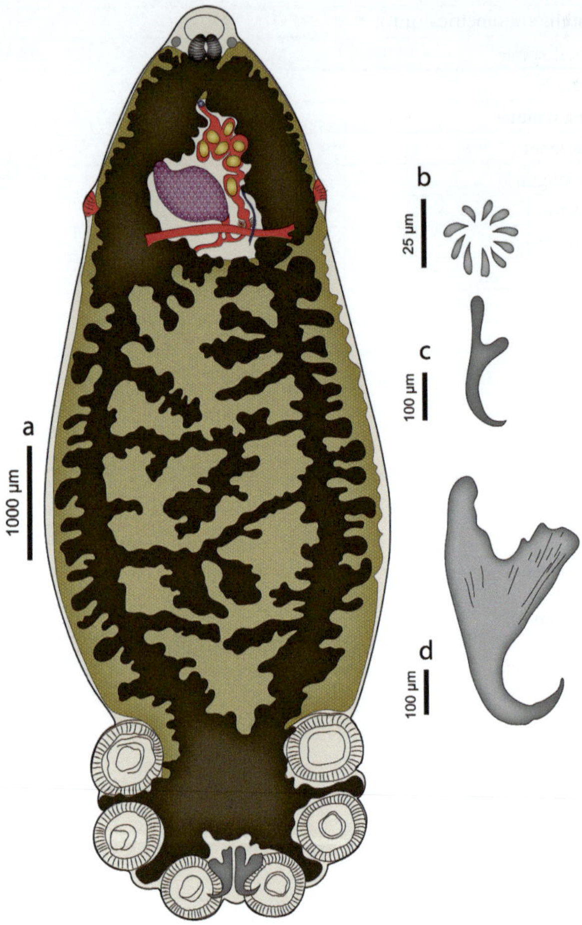

Fig. 4.88 (**a**) *Polystoma mangenoti*, (**b**) genital spines, (**c**) marginal hooklets, (**d**) hamuli. Redrawn from Gallien (1956)

Host geographical distribution *Ptychadena superciliaris* occurs in Ghana, Liberia, Guinea and Sierra Leone, especially in rainforest habitats (Rödel and Schiøtz 2004).

Host conservation status The Savana Ridged Frog is undergoing a decrease of its habitat range and quality because of logging, agricultural development and expanding human settlements. It is listed as Least Concern by The IUCN Red List of Threatened Species 2019: e.T58528A16929881. https://doi.org/10.2305/IUCN.UK.2019-1.RLTS.T58528A16929881.en. Accessed on 30 January 2022.

Site on host Urinary bladder and gills of tadpoles.

DNA sequence

- ITS1, 693 bp. GenBank Accession: AJ310408

Morphology and morphometrics (μm):				
Source	Gallien (1956)	Maeder et al. (1970)	Maeder (1973)	Murith (1981)
Sample size		5	2	
Body length	8000	4000–7000	4000– 5000	3820–9700 (5990)
Greatest width	3000	1100–2300	1370–1840	1480–3240 (2360)
Width at vagina		800–1600	840–1320	1170–2310 (1780)
Haptor length	1750	1100–1400	1070–1380	1240–2170 (1640)
Haptor width	2500	1500– 2300	1500–1800	1590–3070 (2240)
False oral sucker width		350–400	170–290	
Pharynx length			185	
Pharynx width		230	150–185	
Ovary length				
Ovary width				
Egg length	220	180–235 (210)	190	194–222 (204)
Egg width	120	100–120 (110)	128	128–172 (141)
Egg incubation time				
Number of intra-uterine eggs	7	1–5		0–7 (2, 5)
Genital bulb width				
Number of genital spines	10	7–9 (8)	8–9	
Genital spine length		35		
Haptoral sucker diameter	500	350–460	300–450 × 300–450	
Hamulus handle length X	550	470–530 (494)	400–500	310–550 (410)
Hamulus guard length Y				
Hamulus hook length				
Marginal hooklet length	35	32–36	35–38	
Haptoral L/Body L			1.07–1.38	0.22–0.33 (0.28)
No of anastomoses			1–4	

Remarks This species has been reported from the same host as that of *Polystoma vaucheri* Maeder 1973 in Ivory Coast (Maeder 1973). Murith (1981) regarded *P. vaucheri* as a junior synonym of *Polystoma mangenoti*.

4.4.11.38 *Polystoma marmorati* Van Niekerk, Kok and Seaman, 1993 (Fig. 4.89)

Collections Parasitic Worm Collection, National Museum, Bloemfontein, South Africa (NMBP). Parasitic Worm Collection, Natural History Museum, London (NHML).

Holotype NMBP 41.

Paratypes NMBP 42–45; NHML 1992.6.5.1–2.

Original description Van Niekerk et al. (1993).

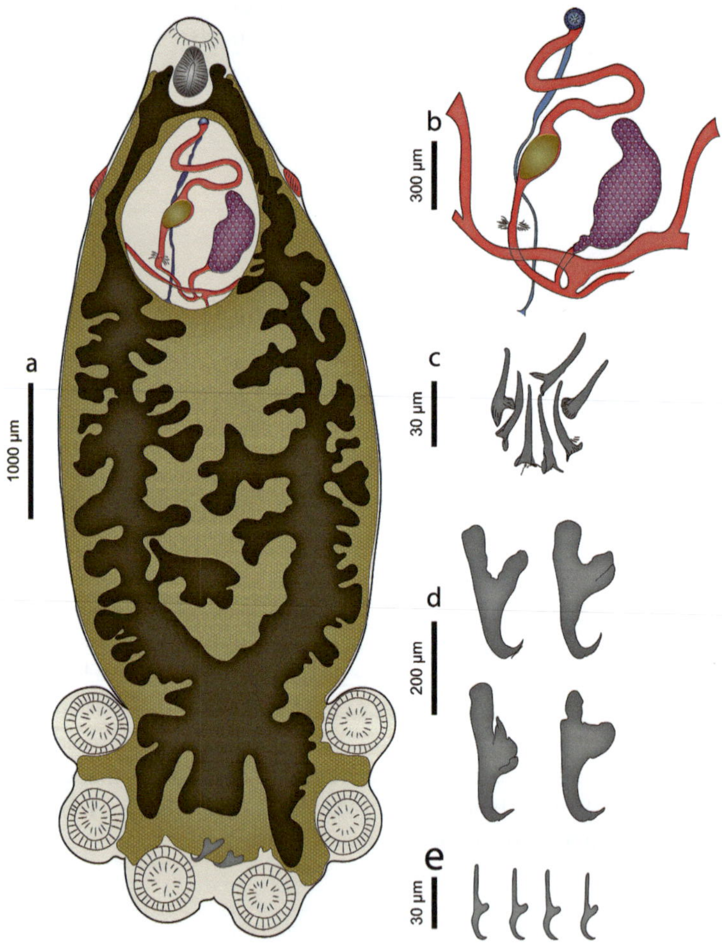

Fig. 4.89 (**a**) *Polystoma marmorati*, (**b**) reproductive system, (**c**) genital spines, (**d**) hamuli, (**e**) marginal hooklets. Redrawn from Van Niekerk et al. (1993)

Other taxonomic contributions Kok and Du Preez (1998).

Etymology Named after the host.

Type locality Vernon Crookes Nature Reserve, KwaZulu-Natal, South Africa.

Other localities None.

Type host *Hyperolius marmoratus* Rapp, 1842, Painted Reed Frog.

Additional hosts None.

Host geographical distribution *Hyperolius marmoratus* occurs in South Africa, Malawi, Zimbabwe, Swaziland and Mozambique (Frost 2021), where it lives in savannah, rivers and lakes, bush land, swamps and grassland habitats. This species also inhabits human-modified habitats such as gardens and cultivated land.

Host conservation status The Painted Reed Frog has a large population, a wide distribution and is tolerant for various habitats. It is listed as Least Concern by The IUCN Red List of Threatened Species 2013: e.T56163A3035416. https://doi.org/10.2305/IUCN.UK.2013-2.RLTS.T56163A3035416.en. Accessed on 30 January 2022.

Site on host Urinary bladder and gills of tadpoles.

DNA sequence
- 18S rRNA gene, 2095 bp. GenBank Accession: AM051073
- 28S rRNA gene, 1448 bp. GenBank Accession: AM157208
- Cytochrome c Oxidase I, 375 bp. GenBank Accession: AM913858
- Cytochrome c Oxidase I, 378 bp. GenBank Accession: AM913859
- 12S rRNA gene, 490 bp. GenBank Accession: KR856088
- ITS1, 686 bp. GenBank Accession: AJ310396

Morphology and morphometrics (μm):

Source	Van Niekerk et al. (1993)
Sample size	18
Prevalence	14.5–47.4%
Mean intensity	1.0–1.1
Body length	4400–7800 (6250)
Greatest width	1700–3000 (2260)
Width at vagina	1500–1200 (1510)
Haptor length	1000–1900 (1460)
Haptor width	1800–3100 (2320)
False oral sucker width	
Pharynx length	240–400 (332)
Pharynx width	240–400 (303)
Ovary length	264–1128 (723)
Ovary width	
Egg length	211–277 (224)
Egg width	155–188 (171)

(continued)

Morphology and morphometrics (μm):	
Source	Van Niekerk et al. (1993)
Egg incubation time	
Number of intra-uterine eggs	9
Genital bulb width	
Number of genital spines	7–8
Genital spine length	(33)
Haptoral sucker diameter	363–576 (477)
Hamulus handle length X	228–390 (323)
Hamulus guard length Y	
Hamulus hook length	
Marginal hooklet length	37–46 (41)
Haptoral L/Body L	0.16–0.27 (0.23)
No of anastomoses	One in 18 has an anastomose

Remarks None.

4.4.11.39 *Polystoma mashoni* Beverley-Burton, 1962 (Fig. 4.90)

Collection Natural History Museum of London (NHML).

Syntypes NHML 1962.8.3.35–37.

Original description Beverley-Burton (1962).

Other taxonomic contributions None.

Etymology Named after Mashonaland region, Zimbabwe.

Type locality Harare, Zimbabwe.

Other localities South Africa.

Type host *Sclerophrys gutturalis*, Guttural Toad. In the original description of the polystome, the host was known as *Bufo regularis.*

Additional hosts None.

Host geographical distribution *Sclerophrys gutturalis* is known from Kenya and Tanzania, south through Mozambique to South Africa, Lesotho, Swaziland, Botswana, Zimbabwe, northern Namibia, Angola as well as Reunion and Mauritius Islands (introduced) (Frost 2021).

Host conservation status The Guttural Toad is widely distributed, has a tolerance for a wide range of habitats and has a presumed large population. It is listed as Least Concern by The IUCN Red List of Threatened Species 2016: e. T54659A107346817. https://doi.org/10.2305/IUCN.UK.2016-3.RLTS. T54659A107346817.en. Accessed on 10 May 2022.

Site on host Urinary bladder.

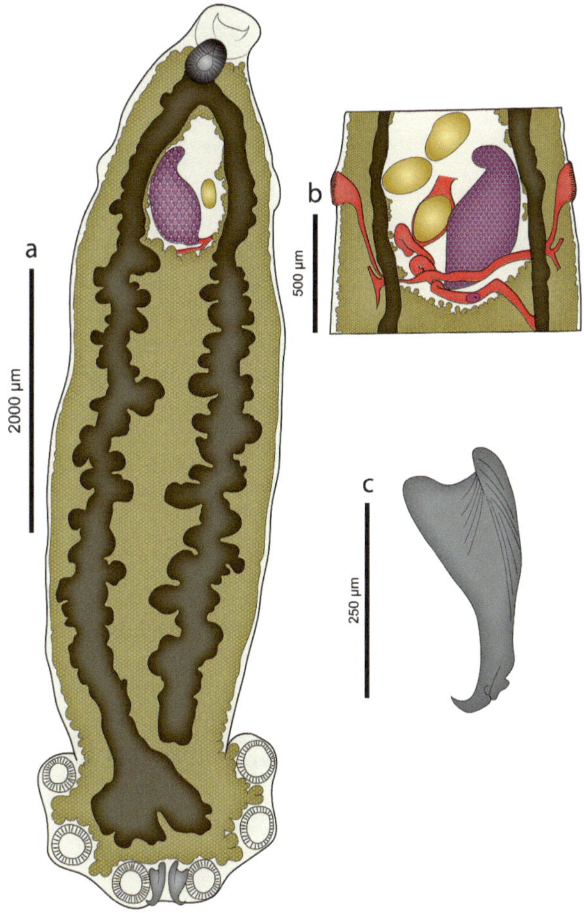

Fig. 4.90 (**a**) *Polystoma mashoni*, (**b**) reproductive system, (**c**) hamuli. Redrawn from Beverley-Burton (1962)

DNA sequence None.

Morphology and morphometrics (µm):	
Source	Beverley-Burton (1962)
Sample size	52
Body length	3500–6920
Greatest width	1190–2000
Width at vagina	
Haptor length	690–1380
Haptor width	1040–1920
False oral sucker width	400–460
Pharynx length	230–380
Pharynx width	190–250

(continued)

Morphology and morphometrics (μm):	
Source	Beverley-Burton (1962)
Ovary length	570–950
Ovary width	320–440
Egg length	230–250
Egg width	140–150
Egg incubation time	18–24 hr
Number of intra-uterine eggs	27–37
Genital bulb width	90
Number of genital spines	7–9 (8)
Genital spine length	
Haptoral sucker diameter	210–350
Hamulus handle length X	380
Hamulus guard length Y	
Hamulus hook length	
Marginal hooklet length	32
Haptoral L/Body L	0.20
No of anastomoses	0

Remarks This species has been reported from the same host as that of *Eupolystoma alluaudi* in Zimbabwe.

4.4.11.40 *Polystoma mazurmovici* Buchvarov, 1980 (Fig. 4.91)

Collection Unknown.

Holotype Unknown.

Paratypes Unknown.

Original description Buchvarov (1980).

Other taxonomic contributions Rubtsova (2018).

Etymology Unknown.

Type locality Plovdiv, Bulgaria.

Other localities None.

Type host *Rana dalmatina* Fitzinger, 1838, Agile Frog.

Additional hosts None.

Host geographical distribution *Rana dalmatina* occurs in Northern France and extreme southern Sweden to northeastern Spain, Sicily, Greece, Carpathian Ukraine, the Balkans south to Greece, Turkey, Jersey and Guernsey (Frost 2021). This species is commonly found in glades and deciduous (oak and beech) woodland habitats

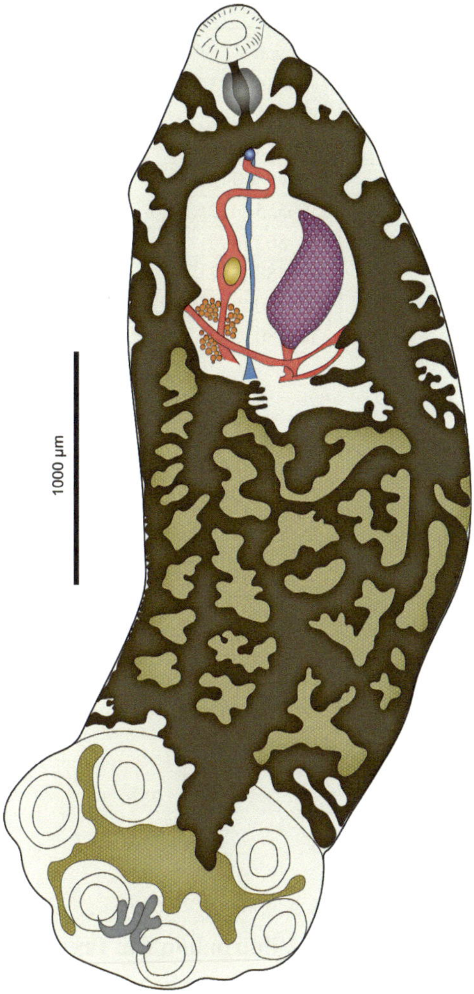

Fig. 4.91 *Polystoma mazurmovici*. Redrawn from Buchvarov (1980)

Host conservation status The Agile Frog has a large population. It is listed as Least Concern by The IUCN Red List of Threatened Species 2009: e.T58584A11790916. Accessed on 30 January 2022.

Site on host Urinary bladder.

DNA sequence None.

Morphology and morphometrics (μm):		
Source	Buchvarov (1980)	Rubtsova (2018)
Sample size	42	1
Body length	5400–11,900 (8400)	3800

(continued)

Morphology and morphometrics (μm):		
Source	Buchvarov (1980)	Rubtsova (2018)
Greatest width	2300–4600 (3500)	1550
Width at vagina		
Haptor length	1300–2300 (1800)	1150
Haptor width	2700–4400 (3400)	1800
False oral sucker width	236–555 (412) × 338–911 (647)	250
Pharynx length	422–562 (496)	320
Pharynx width	343–488 (312)	190
Ovary length	590–1500 (1100)	590–1500 (1100)
Ovary width	330–880 (560)	330–880 (560)
Egg length		
Egg width		
Egg incubation time		
Number of intra-uterine eggs		
Genital bulb width		700
Number of genital spines		11
Genital spine length		40
Haptoral sucker diameter		500
Hamulus handle length X		
Hamulus guard length Y	465	430
Hamulus hook length		
Marginal hooklet length		
Haptoral L/Body L	0.24–0.139 (0.214)	
No of anastomoses		4

Remarks None.

4.4.11.41 *Polystoma nacialtuneli* Yildirimhan, Du Preez and Verneau, 2012 (Fig. 4.92)

Collection Parasitic Worms Collection, Natural History Museum, London (NHML); Parasitic Worm Collection, National Museum, Bloemfontein, South Africa (NMBP); Helminth collection of Uludag University, Museum of Zoology, Bursa, Turkey.

Holotype NHML 2011.2.28.1.

Paratypes NHML 2011.2.28.2; NMBP 326–327.

Original description Yildirimhan et al. (2012).

Other taxonomic contributions None.

Etymology Named after Prof Naci Altunel, University of Uludag, Turkey.

Type locality Seydisehir, Turkey (37°20'N, 32°06′E).

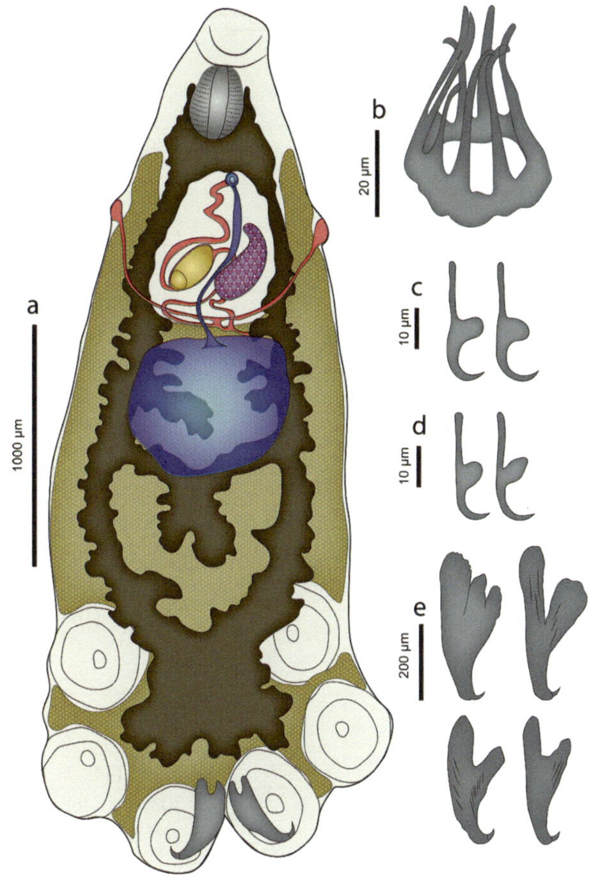

Fig. 4.92 (**a**) *Polystoma nacialtuneli*, (**b**) genital spines, (**c**) marginal hooklets (1), (**d**) marginal hooklets (2–8), (**e**) hamuli. Redrawn from Yildirimhan et al. (2012)

Other localities None.

Type host *Pelobates syriacus* Boettger, 1889, Eastern Spadefoot.

Additional hosts None.

Host geographical distribution *Pelobates syriacus* occurs in Turkey, Bulgaria, Lebanon, Georgia, Romania, Iran, Greece, Israel, Russia, Serbia, Armenia and Azerbaijan (Frost 2021). It is now extinct in Jordan. This species inhabits a variety of soil types (as it is a fossorial species) in open woodlands, rocky areas, semi-desert regions as well as steppe and steppe-like regions

Host conservation status The eastern spadefoot has a wide distribution and a large population. It is listed as Least Concern by The IUCN Red List of Threatened Species 2021: e.T172313717A89702569. https://doi.org/10.2305/IUCN.UK.2021-3.RLTS.T172313717A89702569.en. Accessed on 30 January 2022.

Site on host Urinary bladder.

DNA sequence None.

Morphology and morphometrics (μm):	
Source	Yildirimhan et al. (2012)
Sample size	7
Body length	3412–7013 (5067)
Greatest width	1228–2771 (1815)
Width at vagina	887–1631 (1230)
Haptor length	993–2061 (1472)
Haptor width	1476–3286 (2190)
False oral sucker width	470–768 (565)
Pharynx length	319–521 (393)
Pharynx width	275–507 (371)
Testis length	
Testis width	
Ovary length	339–666 (522)
Ovary width	155–331 (268)
Egg length	234–236 (235)
Egg width	156–180 (168)
Egg incubation time	
Number of intra-uterine eggs	1
Genital bulb width	
Number of genital spines	7
Genital spine length	42–45 (44)
Haptoral sucker diameter	303–658 (468)
Hamulus handle length X	275–545 (410)
Hamulus guard length Y	
Hamulus hook length	40–51 (45)
Marginal hooklet length	C1: 30–32 (31); C2–8: 21–26 (24)
Haptoral L/Body L	0.29
No of anastomoses	0–2

Remarks None.

4.4.11.42 *Polystoma naevius* Caballero and Cerecero, 1941 (Fig. 4.93)

Collection Helminthological Collection, Institute of Biology, Mexico (UNAM); United States National Museum (USNM).

Holotype Unknown.

Cotype no USNM 36803.

Original description Caballero and Cerecero (1941).

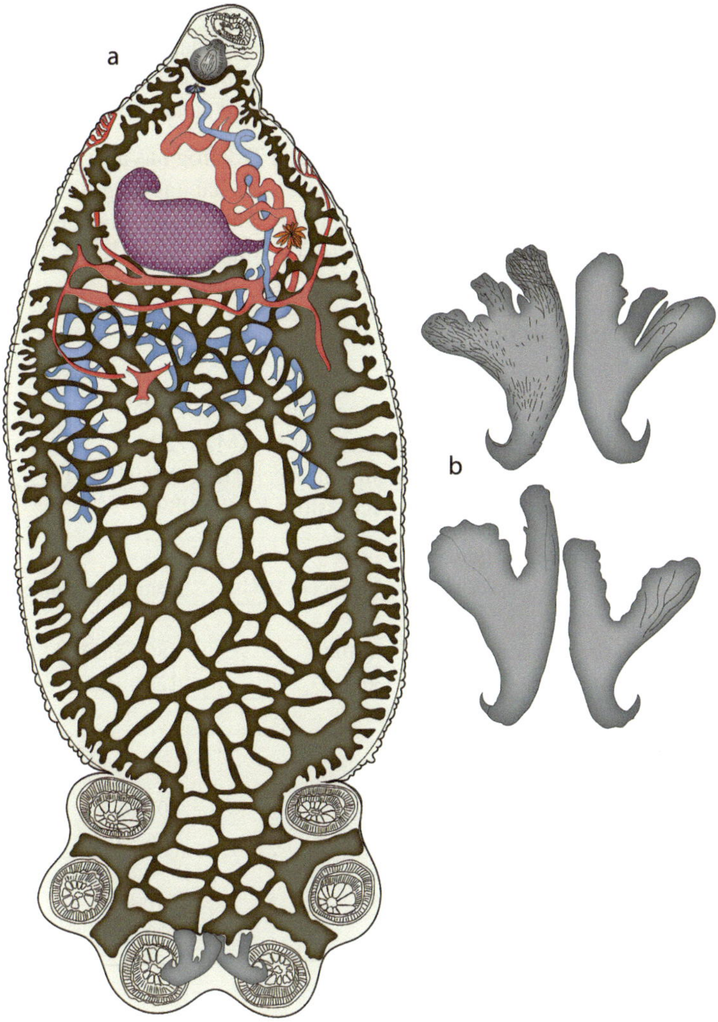

Fig. 4.93 (**a**) *Polystoma naevius*, (**b**) hamuli. Redrawn from Caballero and Cerecero (1941)

Other taxonomic contributions Lamothe-Argumedo (1976).

Etymology Unknown.

Type locality Potrero Viejo, Veracruz, Mexico.

Other localities None.

Type host *Smilisca baudinii* (Duméril and Bibron, 1841), Common Mexican Treefrog.

Additional hosts None.

Host geographical distribution *Smilisca baudinii* occurs in the United States, Mexico, Costa Rico, El Salvador, Belize and Guatemala. It is usually found in lowlands and foothills as well as in savannahs, humid evergreen forests (under tree bark, in tree holes or in leaf axils), ponds, pools and canals (Barrera et al. 2010).

Host conservation status The Common Mexican Treefrog has a large distribution, a large population size and tolerance for a variety of habitat ranges. It is listed as Least Concern by The IUCN Red List of Threatened Species 2021: e. T143847252A53962488. https://doi.org/10.2305/IUCN.UK.2021-3.RLTS. T143847252A53962488.en. Accessed on 30 January 2022.

Site on host Urinary bladder.

DNA sequence
- 18S rRNA gene, 2099 bp. GenBank Accession: AM157187
- 28S rRNA gene, 1454 bp. GenBank Accession: AM157209
- Cytochrome c Oxidase I, 396 bp. GenBank Accession: AM913854
- 12S rRNA gene, 493 bp. GenBank Accession: KR856089

Morphology and morphometrics (μm):		
Source	Caballero and Cerecero 1941	Lamothe-Argumedo (1976)
Sample size		
Body length	7344–10,368 (8100)	3864–5876
Greatest width	3024–4158 (3078)	1225–1625
Width at vagina	1890–2916 (2160)	
Vagina length	300	
Vagina to anterior tip	10,040–1620	
Haptor length		805–982
Haptor width	1458–1890 (3132)	1062–1658
False oral sucker width	320–600	
Pharynx length	420–580	161–402
Pharynx width	400–480	128–300
Ovary length	1100–1640	
Ovary width	400–820	
Egg length	180–200	
Egg width	90–100	
Egg incubation time		
Intra-uterine eggs (n)		
Genital bulb width		
Number of genital spines	8	
Genital spine length		
Anterior to genital bulb	760–1060	
Haptoral sucker diameter	480–560	273–305 × 305–370
Hamulus handle length X	458–600	
Hamulus guard length Y		
Hamulus hook length		

(continued)

Morphology and morphometrics (µm):		
Source	Caballero and Cerecero 1941	Lamothe-Argumedo (1976)
Marginal hooklet length		
Haptoral L/Body L		0.18
No of anastomoses	4	3

Remarks Scale bars in type drawing are incorrect.

4.4.11.43 *Polystoma napoense* Vaucher, 1987 (Fig. 4.94)

Collection Natural History Museum, Geneva, Switzerland (MHNG).

Holotype MHNG 985.650. (0027690).

Paratypes Four paratypes MHNG 986.651 (0027691).

Original description Vaucher (1987).

Taxonomic note The original name *Polystoma napoensis* was corrected for gender.

Other taxonomic contributions None.

Etymology Named after the type locality province.

Type locality San Pablo de Kantesiya, Napo Province, Ecuador.

Other localities None.

Type host *Osteocephalus taurinus*, Manaus Slender-Legged Treefrog.

Additional hosts *Osteocephalus leprieurii* (Duméril and Bibron, 1841), Cayenne Slender-legged Treefrog.

Host geographical distribution *Osteocephalus taurinus* and *O. leprieurii* are both present in Colombia, Peru, Venezuela, Bolivia, Guyana, French Guiana, Brazil and Suriname. The two species can be found on tree trunks or on the banks of ponds and along rivers in forest habitats. Additionally *O. taurinus* can also be found in Ecuador (La Marca et al. 2010a).

Host conservation status Both frog species have large populations, wide distributions and are tolerant for habitat modification. They are listed as Least Concern by The IUCN Red List of Threatened Species 2010: e.T55803A11364949. https://doi.org/10.2305/IUCN.UK.2010-2.RLTS.T55803A11364949.en. Accessed on 30 January 2022.

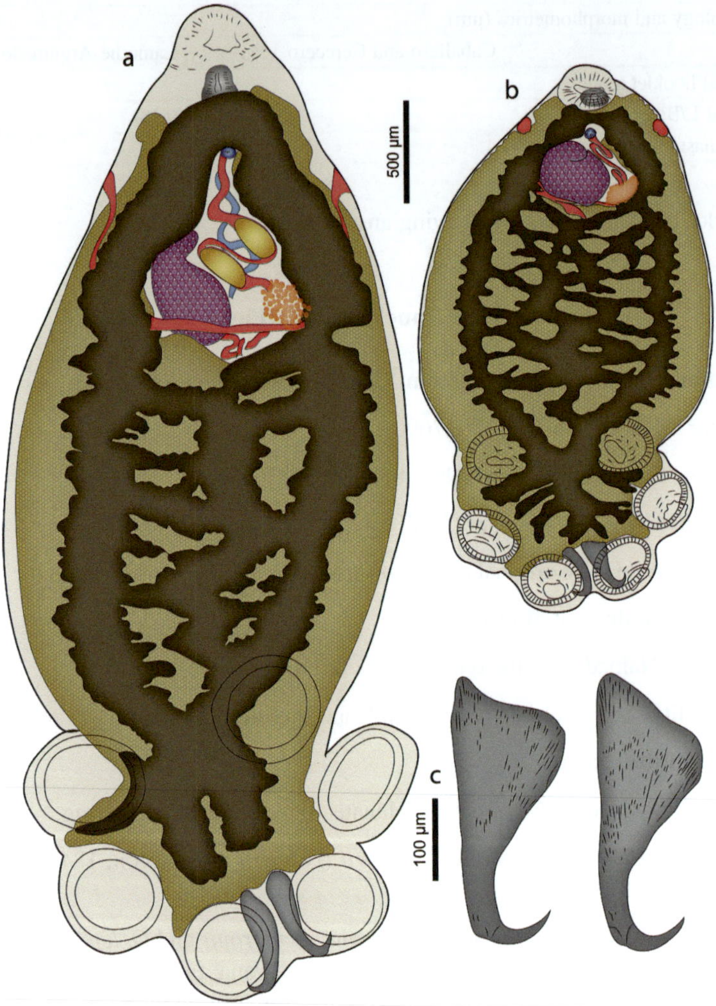

Fig. 4.94 (**a**) *Polystoma napoense* (Type), (**b**) Paratype, (**c**) hamuli. Redrawn from Vaucher (1987)

Site on host Urinary bladder.

DNA sequence None.

Morphology and morphometrics (µm):	
Source	Vaucher (1987)
Sample size	
Body length	3120–3470
Greatest width	1290–1490
Width at vagina	
Haptor length	1000–1220

(continued)

Morphology and morphometrics (μm):	
Source	Vaucher (1987)
Haptor width	1200–1410
False oral sucker width	303–360 × 205–262
Pharynx length	186–209
Pharynx width	139–153
Ovary length	
Ovary width	
Egg length	82–102
Egg width	180–210
Egg incubation time	
Number of intra-uterine eggs	
Genital bulb width	
Number of genital spines	
Genital spine length	
Haptoral sucker diameter	286–403
Hamulus handle length X	286–368
Hamulus guard length Y	
Hamulus hook length	
Marginal hooklet length	
Haptoral L/Body L	0.34
No of anastomoses	Network

Remarks *Polystoma napoense* was originally described as *Polystoma napoensis*. This species has been reported from the same host as that of *Mesopolystoma samiriense* in Peru.

4.4.11.44 *Polystoma natalense* Combes and Channing, 1978 (Fig. 4.95)

Collection Claude Combes personal collection, Perpignan, France.

Holotype Unknown.

Paratypes Unknown.

Original description Combes and Channing (1978).

Taxonomic note The original name *Polystoma natalensis* was corrected for gender.

Other taxonomic contributions Kok and Seaman (1987).

Etymology Named after the type locality provence.

Type locality Cathedral Peak; KwaZulu-Natal, South Africa (28°45'S; 29°01E).

Other localities None.

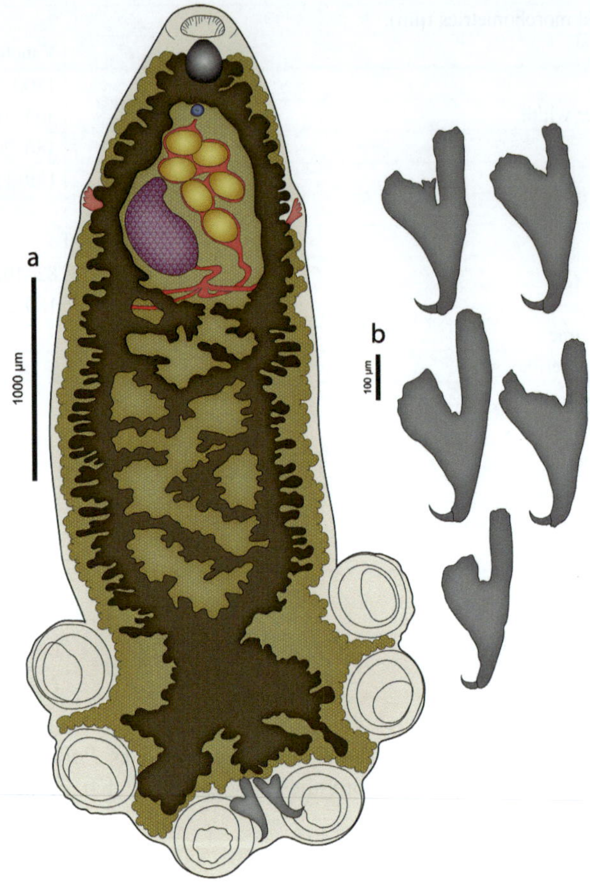

Fig. 4.95 (**a**) *Polystoma natalense*, (**b**) hamuli. Redrawn from Combes and Channing (1978)

Type host *Strongylopus grayii* (Smith, 1849), Gray's Stream Frog.

Additional hosts None.

Host geographical distribution *Strongylopus grayii* occurs in Swaziland, Lesotho and South Africa where it is found in all biomes, including disturbed habitats but excluding arid regions (Frost 2021).

Host conservation status The Gray's Stream Frog has a large population, a wide distribution and is tolerant for various habitats. It is listed as Least Concern by The IUCN Red List of Threatened Species 2013: e.T58767A3073333. https://doi.org/10.2305/IUCN.UK.2013-2.RLTS.T58767A3073333.en. Accessed on 30 January 2022.

Site on host Urinary bladder.

DNA sequence None.

Morphology and morphometrics (μm):		
Source	Combes and Channing (1978)	Kok and Seaman (1987)
Sample size	6	
Body length	1720–5570	4050
Greatest width	790–1820	1420
Width at vagina		
Haptor length	680–1580	1250
Haptor width	1100–2030	1750
False oral sucker width	180 × 450	
Pharynx length	120–230	
Pharynx width	135–260	
Ovary length	203–719	481
Ovary width	130–359	
Egg length	207	207
Egg width	123	123
Egg incubation time		
Number of intra-uterine eggs	10	
Genital bulb width	70–90	
Number of genital spines	8	
Genital spine length		
Haptoral sucker diameter	210–390	329
Hamulus handle length X	218–437	345
Hamulus guard length Y		
Hamulus hook length		
Marginal hooklet length		
Haptoral L/Body L	0.28–0.40 (0.34)	0.33
No of anastomoses	0–4	3–4

Remarks *Polystoma natalense* was originally described as *Polystoma natalensis*.

4.4.11.45 *Polystoma nearcticum* Paul, 1935 (Fig. 4.96)

Collection American Museum of Natural History (AMNH).

Holotype AMNH 281.

Paratypes AMNH 282.

Original description Paul (1935).

Other taxonomic contributions Paul (1938); Campbell (1967, 1969); Olsen (1974); Meyer and Olsen (1972).

Etymology Named after the Nearctic Realm.

Type locality Minnesota, United States of America.

Other localities None.

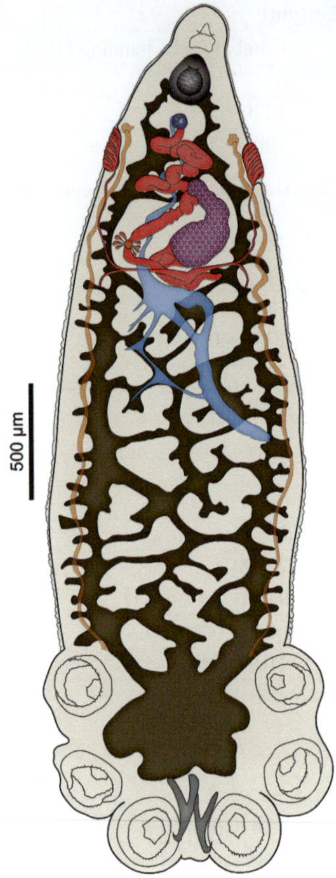

Fig. 4.96 *Polystoma nearcticum*. Redrawn from Paul (1935)

Type host *Dryophytes versicolor* (LeConte, 1825), Eastern Gray Treefrog. In the original description of the polystome, the host was known as *Hyla versicolor* Le Conte, 1825.

Additional hosts *Dryophytes cinereus* (Schneider 1799).

Taxonomy None.

Host geographical distribution *Dryophytes versicolor* occurs in the United States and Canada. It inhabits tree holes, rotten logs and can be found under bark, leaves and roots in wooded and forested habitats (Hammerson 2004c).

Host conservation status The Eastern Gray Treefrog is widely distributed, presumably has a large population and is highly tolerant for habitat modification. It is listed as Least Concern by The IUCN Red List of Threatened Species 2017: e. T55687A112715618. https://doi.org/10.2305/IUCN.UK.2017-1.RLTS. T55687A112715618.en. Accessed on 30 January 2022.

Site on host Urinary bladder and gills of tadpoles.

DNA sequence

- 18S rRNA gene, 2118 bp. GenBank Accession: AM051074
- 18S rRNA gene, 626 bp. GenBank Accession: AJ297775
- 28S rRNA gene, 1439 bp. GenBank Accession: AM157210
- Cytochrome c Oxidase I, 369 bp. GenBank Accession: AM913865
- 12S rRNA gene, 493 bp. GenBank Accession: KR856090
- ITS1, 692 bp. GenBank Accession: AJ301692

Morphology and morphometrics (μm):			
Source	Paul (1935)	Meyer and Olsen (1972)	Olsen (1974)
Sample size	12		
Level of infection	.		
Body length	2500–4500 (3600)	4500	2500–4500
Greatest width	900–1500 (1200)	1500	900–1500
Width at vagina			
Haptor length	500–800		
Haptor width	800–1700		
False oral sucker width			
Pharynx length			
Pharynx width	150–200		
Ovary length	250–750		
Ovary width			
Egg length	300	300	300
Egg width	150	150	150
Number of intra-uterine eggs			
Genital bulb width			
Number of genital spines	8–9		
Genital spine length			
Haptoral sucker diameter	150–300 (260)		
Hamulus handle length X			
Hamulus guard length Y			
Hamulus hook length			
Marginal hooklet length			
Haptoral L/Body L	0.18		
No of anastomoses		Network	

Remarks *Polystoma nearcticum* was originally described as *Polystoma integerrimum nearcticum*.

4.4.11.46 *Polystoma occipitalis* Maeder, 1973 (Fig. 4.97)

Collection Unknown.

Holotype Unknown.

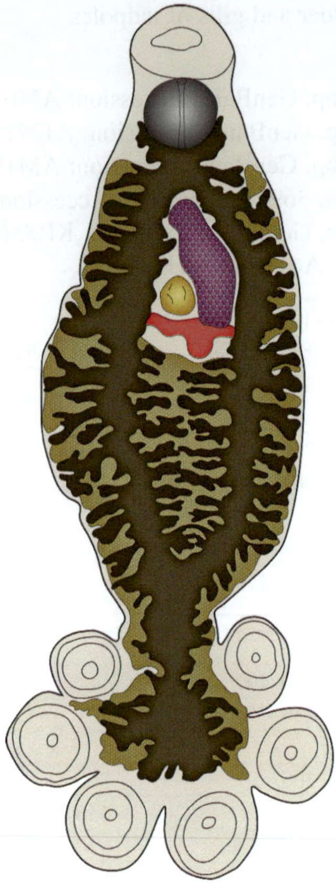

Fig. 4.97 *Polystoma occipitalis*. Redrawn from Maeder (1973)

Paratypes Unknown.

Original description Maeder (1973).

Other taxonomic contributions Murith (1979).

Etymology Named after the host.

Type locality Ivory Coast.

Other localities None.

Type host *Hoplobatrachus occipitalis*, Crowned Bullfrog.

Additional hosts None.

Host geographical distribution *Hoplobatrachus occipitalis* occurs in Algeria, Uganda, Zambia, Sahara, Congo, Nigeria, Morocco, Gabon, Liberia, Kenya, Ethiopia, Mali and Mauritius among others. It inhabits dry savannahs as well as lowland and disturbed forest regions (Rödel et al. 2004).

Host conservation status The Crowned Bullfrog is undergoing a decrease of its habitat range and quality because of logging, agricultural development and expanding human settlements. It is still listed as Least Concern by The IUCN Red List of Threatened Species 2019: e.T58528A16929881. https://doi.org/10.2305/IUCN.UK.2019-1.RLTS.T58528A16929881.en. Accessed on 30 January 2022.

DNA sequence The sequences below were reported from *P. occipitalis* infecting *Hemisus marmoratus* (Peters 1854) in Ivory Coast.

- 18S rRNA gene, 2095 bp. GenBank Accession: AM051075
- 18S rRNA gene, 613 bp. GenBank Accession: AJ297776
- 28S rRNA gene, 1450 bp. GenBank Accession: FM897264
- ITS1, 711 bp. GenBank Accession: AJ301686

Morphology and morphometrics (μm):	
Source	Maeder (1973)
Sample size	9
Body length	1500–3500
Greatest width	410–670
Width at vagina	
Haptor length	370–750
Haptor width	500–1000
False oral sucker width	180–330
Pharynx length	
Pharynx width	145–235
Ovary length	440–860
Ovary width	
Egg length	130–180
Egg width	100–135
Egg incubation time	
Number of intra-uterine eggs	
Genital bulb width	
Number of genital spines	7–8
Genital spine length	25–27
Haptoral sucker diameter	120–280 × 110–250
Hamulus handle length X	
Hamulus guard length Y	
Hamulus hook length	45
Marginal hooklet length	17–43
Haptoral L/Body L	0.22
No of anastomoses	>1

Remarks None.

4.4.11.47 *Polystoma okomuense* Aisien, Du Preez and Imasuen, 2011 (Fig. 4.98)

Collection Parasitic Worm Collection of Professor A.B.M. Egborge Museum, University of Benin, Benin City, Nigeria (LPRM). Parasitic Worms Collection of the Natural History Museum, London (NHML).

Fig. 4.98 (**a**) *Polystoma okumuense*, (**b**) hamuli, (**c**) marginal hooklets (1), (**d**) marginal hooklets (2–7), (**e**) marginal hooklets (8). Redrawn from Aisien et al. (2011)

Holotype LPRM15–0707–01.

Paratypes LPRM15–0905– 03, LPRM15–0905–05; NHML 2010.1.28.1–2.

Original description Aisien et al. (2011).

Taxonomic note The original name *Polystoma okomuensis* was corrected for gender.

Other taxonomic contributions *Polystoma okomuensis*.

Etymology Named after the type locality.

Type locality Okumu Reserve, Nigeria.

Other localities None.

Type host *Hylambates boulengeri* (Perret, 1986), Boulenger's Striped Frog.

Additional hosts None.

Host geographical distribution *Hylambates boulengeri* occurs in Nigeria, Cameroon, Ghana, Ivory Coast, Liberia, Equatorial Guinea and Guinea. It is usually found in forests and evenly in heavily degraded former forests (Frost 2021).

Host conservation status The Boulenger's Striped Frog has a wide distribution and is tolerant for a range of habitats. It is listed as Least Concern by The IUCN Red List of Threatened Species 2013: e.T56293A18364476. https://doi.org/10.2305/IUCN.UK.2013-2.RLTS.T56293A18364476.en. Accessed on 30 January 2022.

Site on host Urinary bladder.

DNA sequence None.

Morphology and morphometrics (μm):	
Source	Aisien et al. (2011)
Sample size	9
Body length	5570–7597 (6513)
Greatest width	1882–3207 (2334)
Width at vagina	1188–1853 (1565)
Haptor length	1141–2153 (1857)
Haptor width	1871–2765 (2459)
False oral sucker width	324–592 (472)
Pharynx length	218–277 (242)
Pharynx width	204–253 (225)
Ovary length	675–1040 (775)
Ovary width	169–402 (267)
Egg length	256–297 (249)
Egg width	145–192 (167)
Egg incubation time	
Number of intra-uterine eggs	1–7

(continued)

Morphology and morphometrics (μm):	
Source	Aisien et al. (2011)
Genital bulb width	62–97 (79)
Number of genital spines	7–8
Genital spine length	28.3–32.0 (29.7)
Haptoral sucker diameter	324–592 (403)
Hamulus handle length X	325–456 (369)
Hamulus guard length Y	
Hamulus X/Y ratio	1.43
Hamulus hook length	51–67 (56)
Marginal hooklet length	C1: 40–45 (43); C2–7: 21–33 (28); C8 37–0 (39)
Haptoral L/Body L	0.20–0.31 (0.28)
No of anastomoses	0–2

Remarks *Polystoma okomuense* was originally described as *Polystoma okomuensis*.

4.4.11.48 *Polystoma ozakii* Price, 1939 (Fig. 4.99)

Collection Unknown.

Holotype Unknown.

Paratypes Unknown.

Original description Price (1939).

Other taxonomic contributions Ozaki (1935).

Etymology Named after Prof Y Ozaki, Japan.

Type locality Japan.

Other localities None.

Type host *Rana ornativentris* Werner 1903, Montane Brown Frog. In the original description of the polystome, the host was known as *Rana temporaria ornativentris* Stejneger, 1924.

Additional hosts None.

Host geographical distribution *Rana ornativentris* is native to Japan, where it is found in Kyushu, Honshu, Shikoku and Sado. It inhabits forests, ponds and paddy fields (Kaneko and Matsui 2004d).

Host conservation status The Montane Brown Frog has a certain tolerance for a wide range of habitats, has a presumed large population and is widely distributed. It is listed as Least Concern by The IUCN Red List of Threatened Species 2021: e. T58688A179361616. https://doi.org/10.2305/IUCN.UK.2021-1.RLTS.T58688A179361616.en. Accessed on 31 January 2022.

Fig. 4.99 *Polystoma ozakii*. Redrawn from Price (1939)

Site on host Urinary bladder.

DNA sequence None.

Morphology and morphometrics (μm):	
Characteristics	Ozaki (1935)
Sample size	
Body length	8000–10,000
Greatest width	2800–4500

(continued)

Morphology and morphometrics (μm):	
Characteristics	Ozaki (1935)
Width at vagina	
Haptor length	
Haptor width	
False oral sucker width	
Pharynx length	
Pharynx width	
Testis length	
Testis width	
Ovary length	
Ovary width	
Egg length	
Egg width	
Egg incubation time	
Number of intra-uterine eggs	
Genital bulb width	
Number of genital spines	
Genital spine length	
Haptoral sucker diameter	
Hamulus handle length X	
Hamulus guard length Y	
Hamulus X/Y ratio	
Hamulus hook length	
Marginal hooklet length	
Haptoral L/Body L	
No of anastomoses	Network

Remarks None.

4.4.11.49 *Polystoma palancai* López-Neyra, 1952 (Fig. 4.100)

Collection Unknown.

Holotype Unknown.

Paratypes Unknown.

Original description López-Neyra (1952).

Other taxonomic contributions None.

Etymology Unknown.

Type locality Grenade, Spain.

Other localities None.

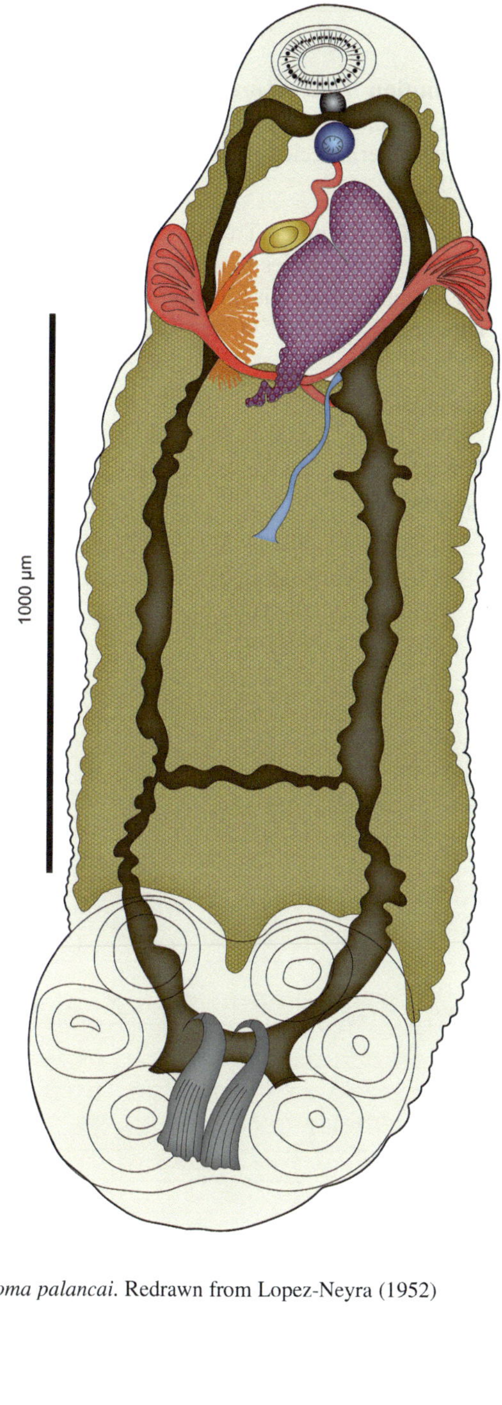

Fig. 4.100 *Polystoma palancai*. Redrawn from Lopez-Neyra (1952)

Type host *Hyla arborea* (Linnaeus, 1758), Common Tree Frog.

Additional hosts None.

Host geographical distribution *Hyla arborea* occurs in Belgium, France, Georgia, Italy, Poland, Romania, Greece, Germany, Portugal and Spain among others. This terrestrial and aquatic (freshwater) species inhabits forests, meadows, gardens, parks and lake shores (Frost 2021).

Host conservation status The Common Tree Frog has a large distribution range and population size. However, this species is sensitive to any changes in its habitat, such as loss of forests or fragmentation. It is however listed as Least Concern by The IUCN Red List of Threatened Species 2009: e.T10351A3197050. Accessed on 31 January 2022.

Site on host Urinary bladder.

DNA sequence None.

Morphology and morphometrics (µm):	
Source	López-Neyra (1952)
Sample size	
Body length	2200
Greatest width	730
Width at vagina	
Haptor length	
Haptor width	
False oral sucker width	425–689
Pharynx length	
Pharynx width	
Ovary length	
Ovary width	
Egg length	
Egg width	
Egg incubation time	
Number of intra-uterine eggs	
Genital bulb width	
Number of genital spines	
Genital spine length	
Haptoral sucker diameter	189–165 × 196–151
Hamulus handle length X	166–260
Hamulus guard length Y	
Hamulus hook length	
Marginal hooklet length	
Haptoral L/Body L	
Anastomosis	

Remarks None.

4.4.11.50 *Polystoma pelobatis* Euzet and Combes, 1966 Fig. 4.101)

Collection National Museum of Natural History, Paris (MNHN); Natural History Museum of London (NHML).

Holotype Unknown.

Paratypes NHML 1970.10.13.2.

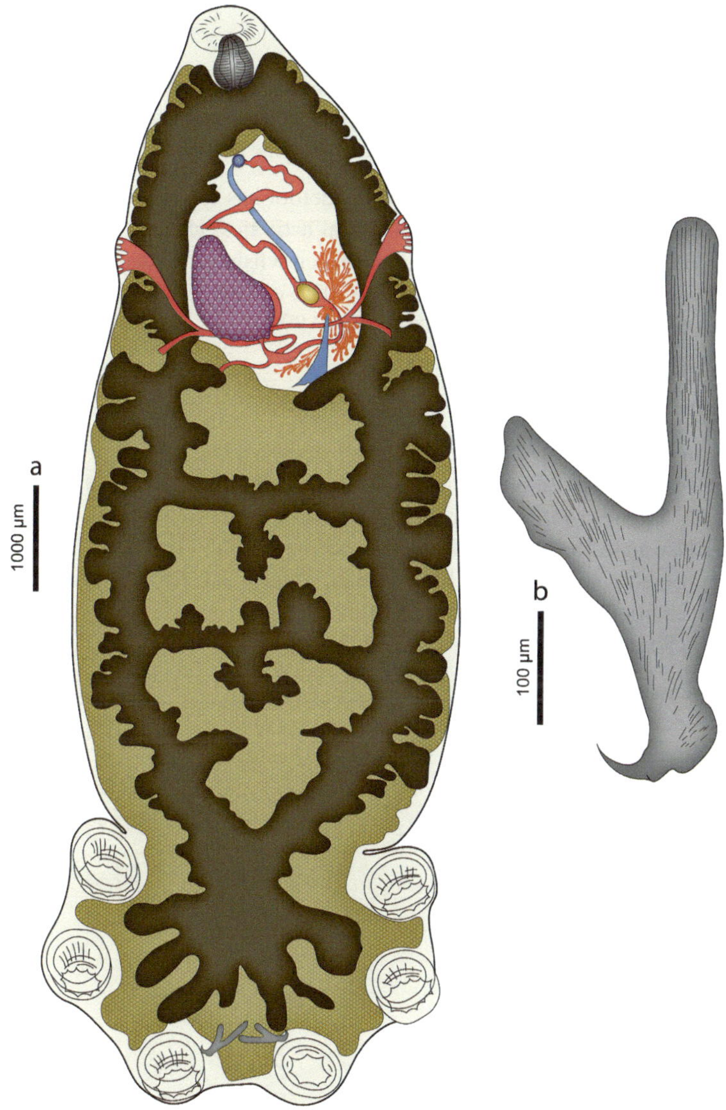

Fig. 4.101 (**a**) *Polystoma pelobatis*, (**b**) hamuli. Redrawn from Euzet and Combes (1966)

Original description Euzet and Combes (1966).

Other taxonomic contributions Combes (1968).

Etymology Named after the host.

Type locality Perpignan, France.

Other localities None.

Type host *Pelobates cultripes* (Cuvier, 1829), Western Spadefoot.

Additional hosts None.

Host geographical distribution *Pelobates cultripes* occurs in Spain, France and Portugal where it is commonly present in coastal regions, dunes, scrub, oak forest and cultivated land. In Gibraltar the species has gone extinct (Beja et al. 2009).

Host conservation status The Western Spadefoot has undergone a severe loss of its habitat range due to pollution, agricultural activities, tourism development and invasive predators in the form of the Louisiana crawfish. It is listed as Vulnerable by The IUCN Red List of Threatened Species 2020: e.T58052A89708267. https://doi.org/10.2305/IUCN.UK.2020-3.RLTS.T58052A89708267.en. Accessed on 31 January 2022.

Site on host Urinary bladder and gills of tadpoles.

DNA sequence
- 18S rRNA gene, 2122 bp. GenBank Accession: AM051076
- 28S rRNA gene, 1418 bp. GenBank Accession: KR856144
- 12S rRNA gene, 460 bp. GenBank Accession: KR856091
- Cytochrome c Oxidase I, 354 bp. GenBank Accession: KR856168
- ITS1, 715 bp. GenBank Accession: FR821519
- ITS1, 717 bp. GenBank Accession: AJ310404

Morphology and morphometrics (μm):	
Source	Euzet and Combes (1966)
Sample size	
Body length	8200–11,500 (10,200)
Greatest width	3300–4200 (3800)
Width at vagina	
Haptor length	
Haptor width	
False oral sucker width	
Pharynx length	
Pharynx width	
Ovary length	
Ovary width	
Egg length	
Egg width	

(continued)

Morphology and morphometrics (μm):	
Source	Euzet and Combes (1966)
Egg incubation time	
Number of intra-uterine eggs	
Genital bulb width	
Number of genital spines	
Genital spine length	
Haptoral sucker diameter	
Hamulus handle length X	460–586 (514)
Hamulus guard length Y	
Hamulus hook length	
Marginal hooklet length	
Haptoral L/Body L	
No of anastomoses	1–4

Remarks *Polystoma pelobatis* was originally described as *Polystoma integerrimum pelobatis*.

4.4.11.51 *Polystoma perreti* Maeder, 1973 (Fig. 4.102)

Collection Natural History Museum, Geneva, Switzerland.

Syntype nos 0038142; 0039799; 0039799.

Original description Maeder (1973).

Other taxonomic contributions Bourgat (1977); Murith (1981).

Etymology Named after Dr JL Perret, Natural History Museum, Geneva, Switzerland.

Type locality Anguédédou, Ivory Coast.

Other localities Togo.

Type host *Amnirana albolabris*, White-lipped Frog. In the original description of the polystome, the host was known as *Hylarana albolabris*.

Additional hosts None.

Host geographical distribution *Amnirana albolabris* occurs in Angola, Kenya, Uganda, Cameroon, Congo, Ghana, Gabon, Nigeria, Liberia and Tanzania. It is commonly present in forests (Frost 2021).

Host conservation status The White-lipped Frog has a large population size, a wide distribution range and is tolerant for habitat modification. It is listed as Least Concern by The IUCN Red List of Threatened Species 2013: e.T58192A89360738. https://doi.org/10.2305/IUCN.UK.2013-2.RLTS.T58192A18405823.en. Accessed on 31 January 2022.

Fig. 4.102 (**a**) *Polystoma perreti*, (**b**) hamuli. Redrawn from Maeder (1973)

Site on host Urinary bladder and gills of tadpoles.

DNA sequence None.

Morphology and morphometrics (μm):		
Source	Maeder (1973)	Murith (1981)
Sample size	5	5
Body length	4500–5800	4680–11,450 (6890)
Greatest width	1380–2300	1600–3240 (2440)
Width at vagina	1150–1600	1320–2550 (1820)
Haptor length	800–1180	960–2310 (1540)
Haptor width	1320–2240	1720–2890 (2210)
False oral sucker width	345 × 495	
Pharynx length	380–480	
Pharynx width	253–300	
Testis length		
Testis width		
Ovary length		
Ovary width		
Egg length	195–240 (219)	216–244 (233)
Egg width	103–150 (120)	161–183 (172)
Egg incubation time		
Number of intra-uterine eggs	1–3	0, 1, 2, 6 (0,2)
Genital bulb width		
Number of genital spines	7–8	
Genital spine length	34–37	
Haptoral sucker diameter	253–391 × 230–414	360–500 (420)
Hamulus handle length X	347–402	340–410 (380)
Hamulus guard length Y		
Hamulus hook length		
Marginal hooklet length	16–35	34
Haptoral L/Body L		0.17–0.28 (0.23)
No of anastomoses	0–1	0–1

Remarks This species has been reported from the same host as that of *Polystoma gabonense* in Gabon and Central Africa Republic.

4.4.11.52 *Polystoma persicum* Dollfus, Euzet and Combes, 1965 (Fig. 4.103)

Collection Unknown.

Holotype Unknown.

Paratypes Unknown.

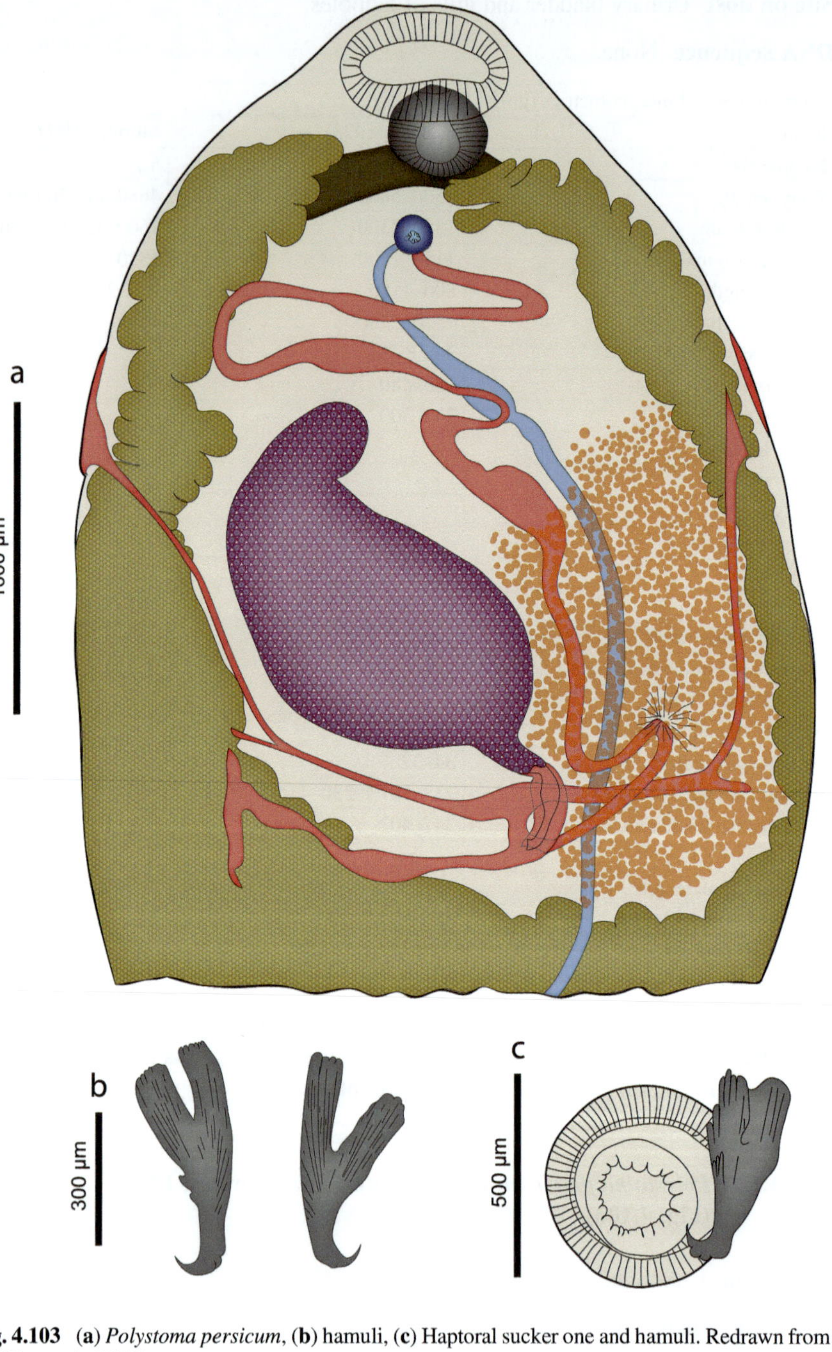

Fig. 4.103 (**a**) *Polystoma persicum*, (**b**) hamuli, (**c**) Haptoral sucker one and hamuli. Redrawn from Dollfus et al. (1965)

Original description Dollfus et al. (1965).

Other taxonomic contributions None.

Etymology Named after the type locality province.

Type locality Ouest Razi and Akinlou, Iran.

Other localities None.

Type host *Bufotes surdus* (Boulenger, 1891), Iranian Toad. In the original description of the polystome, the host was known as *Bufo persicus* Steindachner, 1867.

Additional hosts None.

Host geographical distribution *Bufotes surdus* occurs in Iran, Pakistan and Iraq where it is usually found in temporary waterbodies (Khan et al. 2009).

Host conservation status The Iranian Toad has a large population size, wide distribution and is tolerant for various habitats. It is thus listed as Least Concern by The IUCN Red List of Threatened Species 2015: e.T54772A74505632. https://doi.org/10.2305/IUCN.UK.2015-1.RLTS.T54772A74505632.en. Accessed on 30 January 2022.

Site on host Urinary bladder.

DNA sequence None.

Morphology and morphometrics (μm):	
Source	Dollfus et al. (1965)
Sample size	
Body length	8000–9000
Greatest width	3000–4000
Width at vagina	
Haptor length	
Haptor width	
False oral sucker width	346 × 583
Pharynx length	363
Pharynx width	330
Ovary length	8100
Ovary width	680
Egg length	240
Egg width	116
Egg incubation time	
Number of intra-uterine eggs	
Genital bulb width	
Number of genital spines	
Genital spine length	52
Haptoral sucker diameter	500
Hamulus handle length X	485–520

(continued)

Morphology and morphometrics (μm):	
Source	Dollfus et al. (1965)
Hamulus guard length Y	
Hamulus hook length	
Marginal hooklet length	
Haptoral L/Body L	
No of anastomoses	

Remarks *Polystoma persicum* was originally described as *Polystoma integerrimum persicum.*

4.4.11.53 *Polystoma praecox* Combes and Laurent, 1978 (Fig. 4.104)

Collection Unknown.

Holotype Unknown.

Paratypes Unknown.

Original description Combes and Laurent (1978).

Other taxonomic contributions None.

Etymology Unknown.

Type locality Alto Calilegua, Jujuy Province, Argentina.

Other localities None.

Type host *Telmatobius oxycephalus* Vellard, 1946, Red-headed Water Frog.

Additional hosts None.

Host geographical distribution *Telmatobius oxycephalus* occurs in Argentina where it can be found in montane forests and streams. This species is absent in disturbed or modified habitats (Lavilla 2004).

Host conservation status The Red-headed Water Frog is extremely rare and only found in protected regions. It is threatened by mining operations, alteration of watersheds, selective logging and the introduction of predatory fishes. It is listed as Endangered by The IUCN Red List of Threatened Species 2019: e. T57353A101434020. https://doi.org/10.2305/IUCN.UK.2019-1.RLTS. T57353A101434020.en. Accessed on 31 January 2022.

Site on host Urinary bladder.

DNA sequence None.

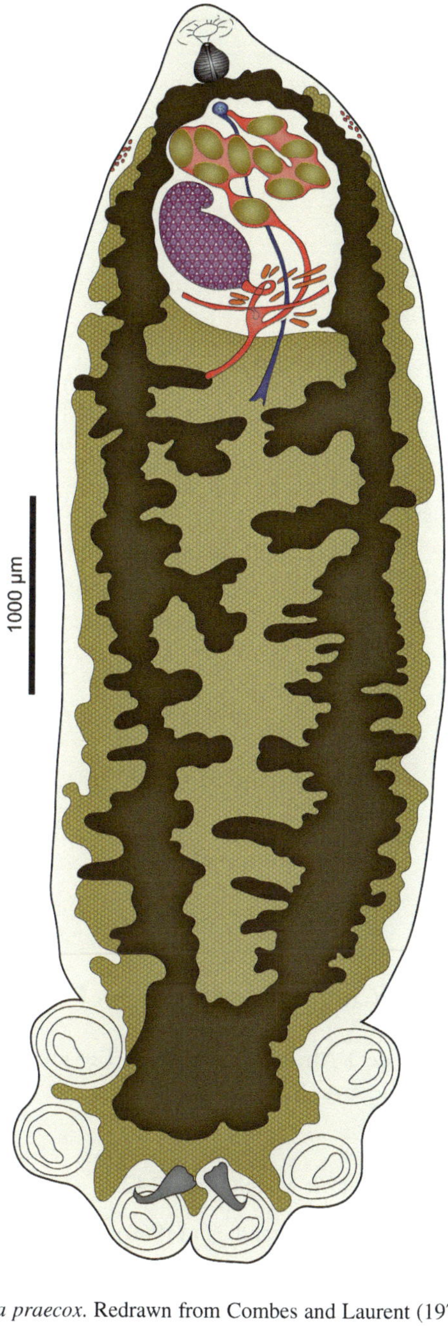

Fig. 4.104 *Polystoma praecox*. Redrawn from Combes and Laurent (1978)

Morphology and morphometrics (μm):

Source	Combes and Laurent (1978)
Sample size	17
Body length	3000–6400 (4600)
Greatest width	700–1900 (1200)
Width at vagina	
Haptor length	900–1200 (1000)
Haptor width	1000–1900 (1300)
False oral sucker width	
Pharynx length	110–220 (200)
Pharynx width	140–210 (170)
Ovary length	320–620 (425)
Ovary width	150–350 (205)
Egg length	170–195 (180)
Egg width	91–114 (110)
Egg incubation time	
Number of intra-uterine eggs	5
Genital bulb width	
Number of genital spines	
Genital spine length	
Haptoral sucker diameter	260–410 (330)
Hamulus handle length X	350–377
Hamulus guard length Y	
Hamulus hook length	
Marginal hooklet length	
Haptoral L/Body L	0.22–0.27
No of anastomoses	0

Remarks None.

4.4.11.54 *Polystoma pricei* Vercammen-Grandjean, 1960 (Fig. 4.105)

Collection Musée Royal de l'Afrique Centrale at Tervuren Belgium (MRAC); Natural History Museum, Geneva, Switzerland (MHNG).

Holotype MRAC 32127.

Paratypes MRAC 32128, 32,129; MHNG 0038142.

Original description Vercammen-Grandjean (1960).

Other taxonomic contributions Tinsley (1974b).

Etymology Named after Prof Emmet Price, USA.

Type locality Lac Kivu, Congo.

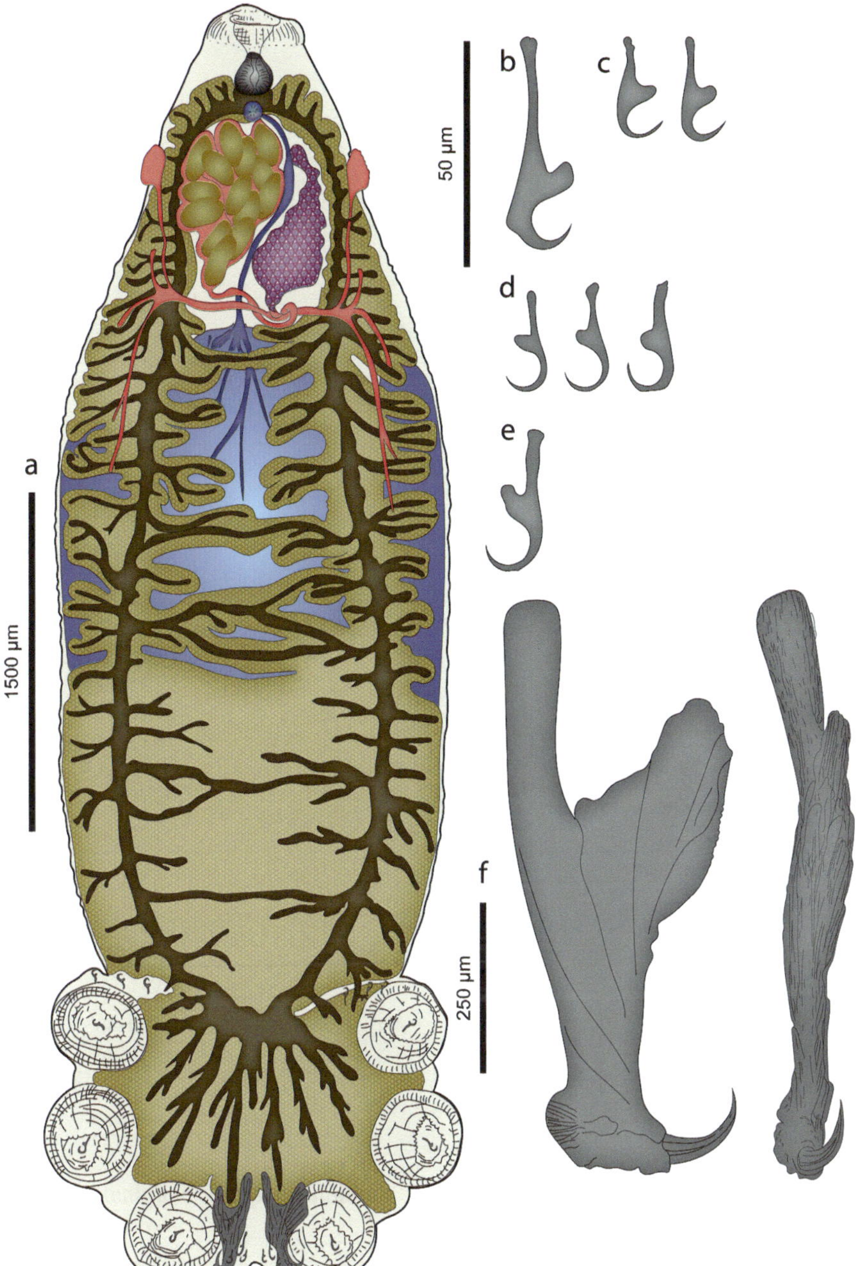

Fig. 4.105 (**a**) *Polystoma pricei*, (**b**) marginal hooklets (1), (**c**) marginal hooklets (3–5), (**d**) marginal hooklets (6–7), (**e**) marginal hooklets (8), (**f**) hamuli. Redrawn from Vercammen-Grandjean (1960)

Other localities None.

Type host *Ptychadena* sp.

Site on host Urinary bladder.

Host geographical distribution Unknown.

Host conservation status Unknown.

DNA sequence None.

Morphology and morphometrics (μm):		
Characteristics	Vercammen-Grandjean (1960)	Tinsley (1974b)
Sample size	2	1
Body length	4250	4780 (5480)
Greatest width	1780	1740 (1780)
Width at vagina		1010
Haptor length	1230	1260 (1230)
Haptor width	2120	1730 (2120)
False oral sucker width	400	380 (400)
Pharynx length		180
Pharynx width	160	170 (160)
Ovary length	680	557 (680)
Ovary width	280	314 (280)
Egg length	224	181–202 (224)
Egg width	100	122–109 (100)
Egg incubation time		
Number of intra-uterine eggs		11
Genital bulb width	82	73 (82)
Number of genital spines		8
Genital spine length		
Haptoral sucker diameter	400	350–380 (370)
Hamulus handle length X	425	400–425
Hamulus guard length Y		
Hamulus hook length		39–40
Marginal hooklet length	40	19–40
Haptoral L/Body L		0.26 (0.22)
No of anastomoses	>1	0

Remarks The type drawings indicate an additional pair of marginal hooklets between sucker pair 1. This needs to be verified.

4.4.11.55 *Polystoma prudhoei* Saoud, 1967 (Fig. 4.106)

Collection Natural History Museum of London (NHML).

Holotype NHML 1998.5.5.2.

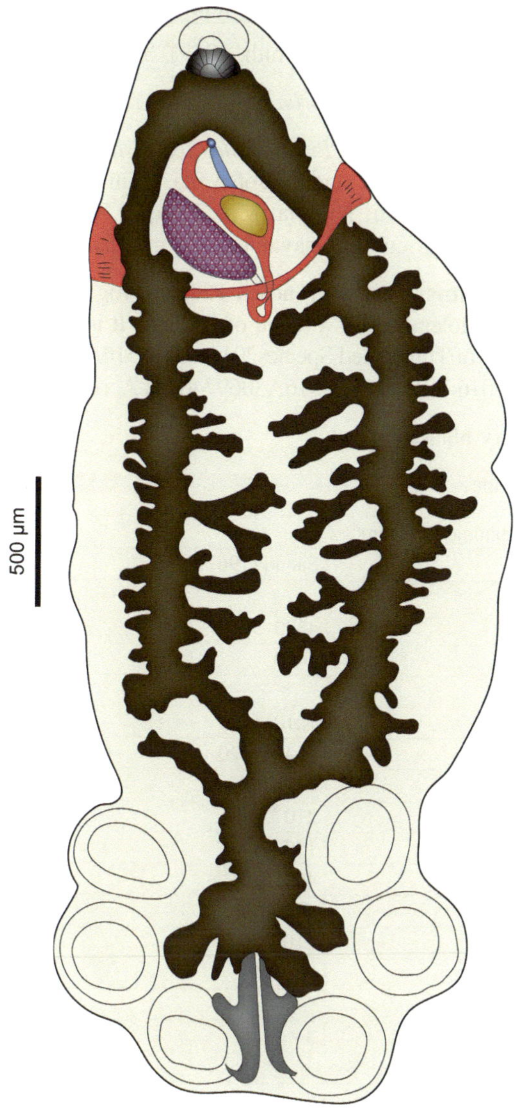

Fig. 4.106 *Polystoma prudhoei*. Redrawn from Saoud (1967)

Original description Saoud (1967).

Other taxonomic contributions Euzet et al. (1974a); Bourgat (1977); Murith (1978); Aisien et al. (2004).

Etymology Named after Dr Stephen Prudhoe, British Museum, London, UK.

Type locality Kumba, Cameroon.

Other localities Central African Republic, Cameroon, Nigeria and Togo.

Type host *Ptychadena oxyrhynchus* (Smith, 1849), Sharp-nosed Ridged Frog.

Additional hosts *Sclerophrys regularis.*

Host geographical distribution *Ptychadena oxyrhynchus* occurs in Angola, Uganda, Zambia, Zimbabwe, Congo, Botswana, Cameroon, Gambia, Ghana, Kenya, Namibia, Malawi, Mali, Nigeria, Mozambique, Tanzania, Swaziland and South Africa. It inhabits agricultural, savannah and marshy regions (Frost 2021).

Host conservation status The Sharp-nosed Ridged Frog has a wide distribution, large population and a tolerance for a range of habitats. It is listed as Least Concern by The IUCN Red List of Threatened Species2016: e.T58516A3069746. https://doi.org/10.2305/IUCN.UK.2016-1.RLTS.T58516A3069746.en. Accessed on 31 January 2022.

Site on host Urinary bladder.

DNA sequence None.

Morphology and morphometrics (µm):		
Source	Saoud (1967)	Euzet et al. (1974a)
Sample size		4
Body length	4320–5460	4700–5600
Greatest width	1400	
Width at vagina		
Haptor length	1210–1350	
Haptor width	1640–1720	
False oral sucker width	420	
Pharynx length	210	
Pharynx width	150	
Ovary length	490–530	
Ovary width	350–390	
Egg length	210–230	
Egg width	140–150	
Egg incubation time		
Number of intra-uterine eggs	3	
Genital bulb width		
Number of genital spines	8	
Genital spine length		
Haptoral sucker diameter	400–460 × 330–390	
Hamulus handle length X	550–620	420–480
Hamulus guard length Y		
Hamulus hook length		
Marginal hooklet length		43 (Murith 1981)
Haptoral L/Body L	0.25–0.28	
No of anastomoses	0	

Remarks None.

4.4.11.56 *Polystoma ragnari* Maeder, Euzet and Combes, 1970 (Fig. 4.107)

Collection Claude Combes personal collection, Perpignan, France.

Holotype Unknown.

Paratypes Unknown.

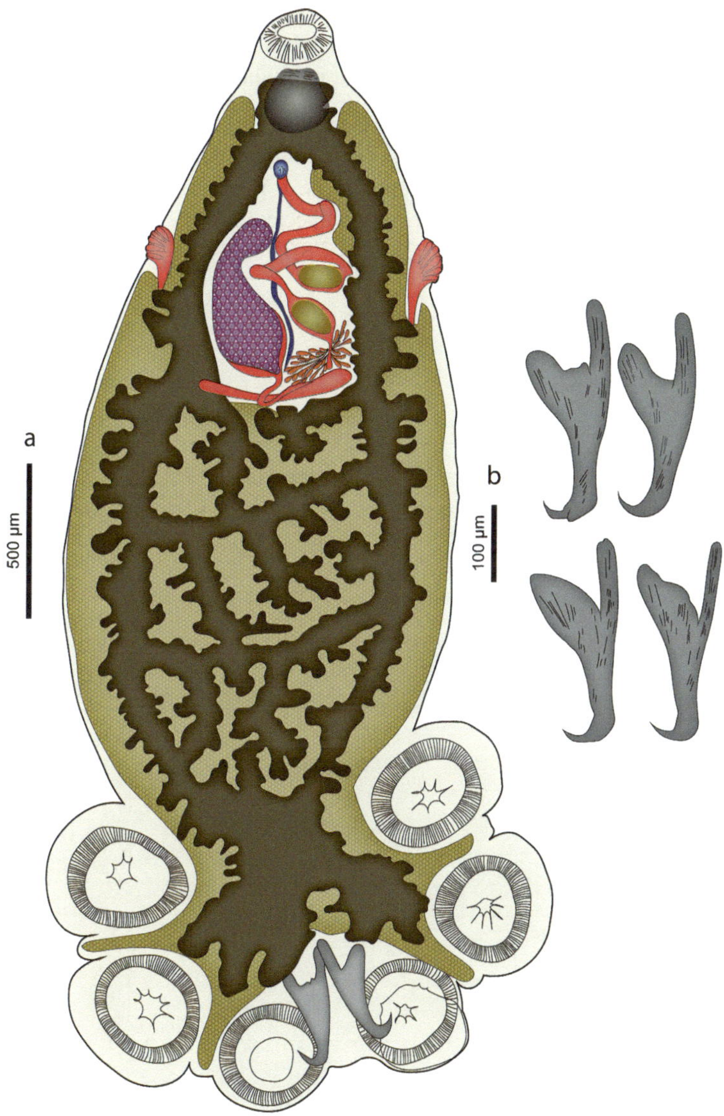

Fig. 4.107 (**a**) *Polystoma ragnari*, (**b**) hamuli. Redrawn from Maeder et al. (1970)

Original description Maeder et al. (1970).

Other taxonomic contributions Maeder (1973); Bourgat (1977).

Etymology Named after Ragnar, son of Prof L.Hh. Knoepffler, France.

Type locality Yapo and Bolo, Ivory Coast.

Type host *Phrynobatrachus alleni* Parker, 1936, Allen's River Frog.

Additional hosts *Phrynobatrachus latifrons* Ahl, 1924.

Host geographical distribution *Phrynobatrachus alleni* occurs in Sierra Leone, Ghana, Nigeria and Guinea where it is found in lowland rainforests (Frost 2021).

Host conservation status The Allen's River Frog is dependent on undisturbed forest habitats whose quality and extent are declining due to logging, agricultural development and expanding human settlements. It is however listed as Least Concern by The IUCN Red List of Threatened Species 2020: e.T58090A176939299. https://doi.org/10.2305/IUCN.UK.2020-3.RLTS.T58090A176939299.en. Accessed on 31 January 2022.

Site on host Urinary bladder.

DNA sequence None.

Morphology and morphometrics (μm):	
Source	Maeder et al. (1970)
Sample size	4
Body length	3500–4000
Greatest width	1200–1500
Width at vagina	800–900
Haptor length	900–1300
Haptor width	1400–1700
False oral sucker width	250–300
Pharynx length	
Pharynx width	150–200
Ovary length	
Ovary width	
Egg length	165–190
Egg width	75–100
Egg incubation time	
Number of intra-uterine eggs	3
Genital bulb width	
Number of genital spines	6–8
Genital spine length	30

(continued)

Morphology and morphometrics (μm):	
Source	Maeder et al. (1970)
Haptoral sucker diameter	330–420
Hamulus handle length X	360–400
Hamulus guard length Y	
Hamulus hook length	
Marginal hooklet length	15–30
Haptoral L/Body L	0.26–0.33
No of anastomoses	Network

Remarks None.

4.4.11.57 *Polystoma skrjabini* Buchvarov, 1984 (Fig. 4.108)

Collection Unknown.

Holotype Unknown.

Paratypes Unknown.

Original description Buchvarov (1984a).

Other taxonomic contributions None.

Etymology Unknown.

Type locality Petrich, Bulgaria.

Other localities None.

Type host *Hyla arborea*, Common Tree Frog.

Additional hosts None.

Taxonomy None.

Host geographical distribution *Hyla arborea* occurs in Belgium, France, Georgia, Italy, Poland, Romania, Greece, Germany, Portugal and Spain among others. This terrestrial and aquatic (freshwater) species inhabits forests, meadows, gardens, parks and lake shores (Frost 2021).

Host conservation status The Common Tree Frog has a large distribution range and population size. However, this species is sensitive to any changes in its habitat, such as loss of forests or fragmentation. It is listed as Least Concern by The IUCN Red List of Threatened Species 2009: e.T10351A3197050. Accessed on 31 January 2022.

Fig. 4.108 (**a**) *Polystoma skrjabini*, (**b**) hamuli. Redrawn from Buchvarov (1984)

Site on host Urinary bladder.

DNA sequence None.

Morphology and morphometrics (μm):	
Source	Buchvarov (1984a)
Sample size	4
Body length	2964–4275 (3426)
Greatest width	1235–1651 (1401)
Width at vagina	
Haptor length	760–950
Haptor width	1235–1273
False oral sucker width	198–247 × 300–418
Pharynx length	72–50
Pharynx width	102–126
Ovary length	510–665
Ovary width	285–323
Egg length	
Egg width	
Egg incubation time	
Number of intra-uterine eggs	
Genital bulb width	
Number of genital spines	
Genital spine length	
Haptoral sucker diameter	
Hamulus handle length X	
Hamulus guard length Y	217
Hamulus hook length	
Marginal hooklet length	
Haptoral L/Body L	0.22–0.26
No of anastomoses	5

Remarks *Polystoma skrjabini* was described from the same host as *Polystoma palancai* without parasite comparison. This therefore requires further investigation for species validation.

4.4.11.58 *Polystoma skuratovitchi* Buchvarov, 1984 (Fig. 4.109)

Collection Unknown.

Holotype Unknown.

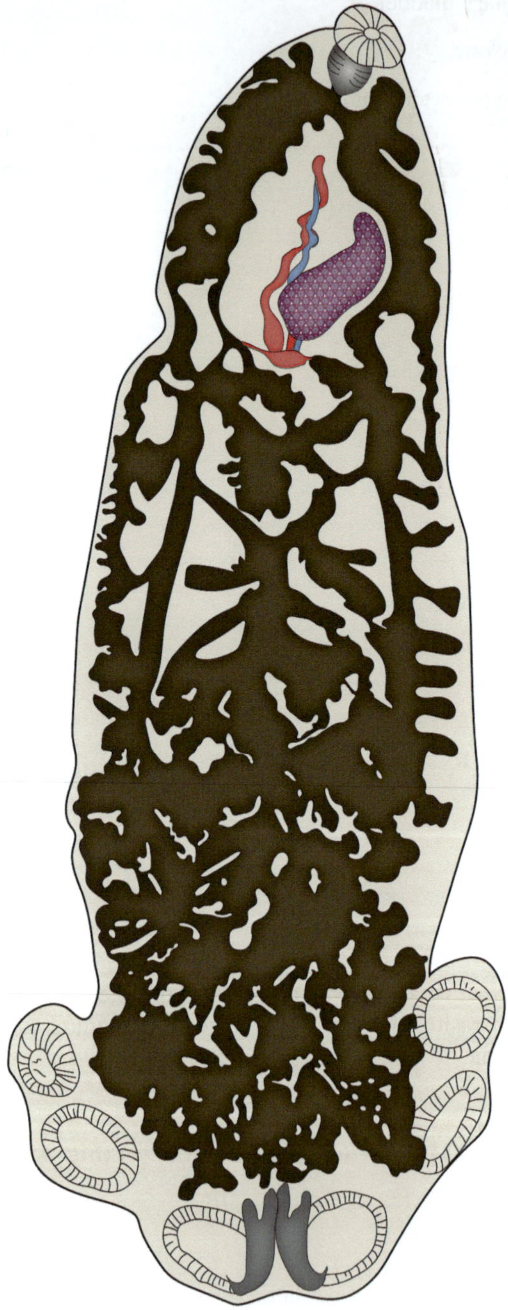

Fig. 4.109 *Polystoma skuratovitchi*. Redrawn from Buchvarov (1984b)

Paratypes Unknown.

Original description Buchvarov (1984b).

Other taxonomic contributions Rubtsova (2018).

Etymology Unknown.

Type locality Poland.

Other localities None.

Type host *Rana arvalis*, Moor Frog.

Additional hosts None.

Host geographical distribution *Rana arvalis* occurs in China, Austria, Romania, Croatia, Norway, Finland, Germany, France, Belgium, Belarus, Ukraine, Latvia, Slovakia, Netherlands, Poland, Slovenia and Hungary. It has gone extinct in Switzerland (Kuzmin et al. 2009b). It inhabits a variety of habitats such as forests, tundra, swamps, gardens, meadows, semi-deserts and fields.

Host conservation status The Moor Frog has a large population, a wide distribution and is tolerant for various habitat ranges. However, its habitat is threatened by tourism, urbanization, industrial activities, drought and predation of spawn by waterfowl. It is listed as Least Concern by The IUCN Red List of Threatened Species 2009: e.T58548A86232114. https://doi.org/10.2305/IUCN.UK.2009.RLTS.T58548A11800564.en. Accessed on 31 January 2022.

Site on host Urinary bladder.

DNA sequence None.

Morphology and morphometrics (μm):		
Source	Buchvarov (1984b)	Rubstova (2018)
Sample size		5
Body length	2400	2300–4300 (2900)
Greatest width	1100	600–1700 (990)
Width at vagina		
Haptor length	294–1560 (887)	550–1500 (870)
Haptor width	480–1700 (1261)	1000–1350 (1180)
False oral sucker width	90–402 (197) × 114–462 (294)	100–270 (170) × 200–450 (280)
Pharynx length	30–306 (189)	230–720 (360)
Pharynx width	66–234 (170)	220–600 (350)
Ovary length	60–1380 (476)	200–700 (390)
Ovary width	30–560 (203)	100–300 (170)
Egg length		
Egg width		
Egg incubation time		
Number of intra-uterine eggs		

(continued)

Morphology and morphometrics (μm):		
Source	Buchvarov (1984b)	Rubstova (2018)
Genital bulb width		700–900 (800)
Number of genital spines		9
Genital spine length		35–53 (40)
Haptoral sucker diameter	280 × 253	200–600 (320)
Hamulus handle length X		
Hamulus guard length Y	287	260–320 (286)
Hamulus hook length		
Marginal hooklet length		
Haptoral L/Body L	0.37	0.239–0.348 (0.3)
No of anastomoses	Network	Network

Remarks None.

4.4.11.59 *Polystoma sodwanense* Du Preez and Kok, 1992 (Fig. 4.110)

Collection Parasitic Worm Collection, National Museum, Bloemfontein, South Africa (NMBP); Parasitic Worm Collection, Natural History Museum, London (NHML).

Holotype NMBP 26.

Paratypes NMBP 27–30; NHML 1991.1.2.4–5.

Original description Du Preez and Kok (1992).

Taxonomic note The original name *Polystoma sodwanensis* was corrected for gender.

Other taxonomic contributions None.

Etymology Named after the type locality.

Type locality Sodwana Bay, KwaZulu-Natal, South Africa.

Other localities None.

Type host *Ptychadena porosissima* (Steindachner, 1867), Grassland Ridged Frog.

Additional hosts None.

Host geographical distribution *Ptychadena porosissima* occurs in South Africa, Mozambique, Zimbabwe, Angola, Zaire, East Africa, Ethiopia, Malawi, Tanzania, Rwanda, Zambia, Congo, Swaziland, Kenya and Uganda. It inhabits pans and marshy areas as well as forest, savannah and grassland habitats (Frost 2021).

Host conservation status The Grassland Ridged Frog has a large population, a wide distribution and is tolerant for various habitats. It is listed as Least Concern by

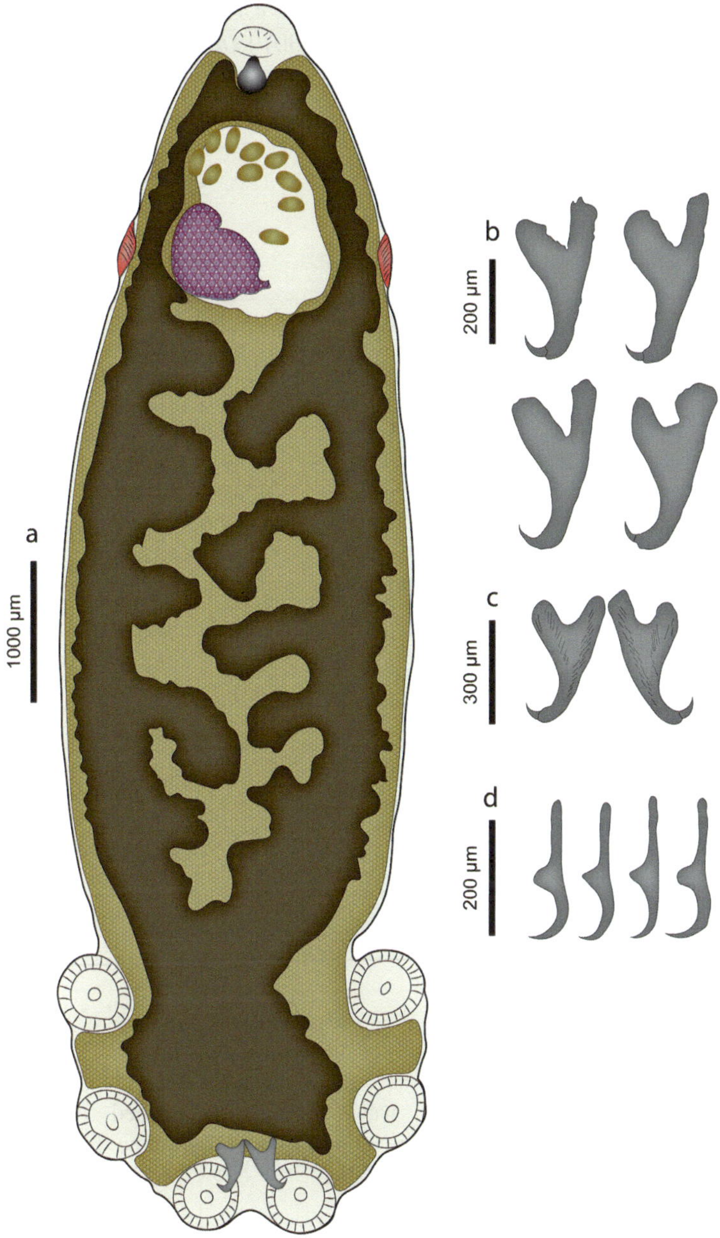

Fig. 4.110 (**a**) *Polystoma sodwanense*, (**b**) hamuli, (**c**) hamuli of holotype, (**d**) marginal hooklets. Redrawn from Du Preez and Kok (1992)

The IUCN Red List of Threatened Species 2013: e.T58519A3070008. https://doi.org/10.2305/IUCN.UK.2013-2.RLTS.T58519A3070008.en. Accessed on 31 January 2022.

Site on host Urinary bladder.

DNA sequence None.

Morphology and morphometrics (μm):	
Source	Du Preez and Kok (1992)
Sample size	6
Body length	6420–7720 (6960)
Greatest width	1800–2430 (2190)
Width at vagina	1420–1920 (1710)
Haptor length	1540–1740 (1600)
Haptor width	2200–2430 (2370)
False oral sucker width	
Pharynx length	283–330 (305)
Pharynx width	248–279 (270)
Ovary length	546–745 (675)
Ovary width	
Egg length	192–228 (212)
Egg width	156–177 (167)
Egg incubation time	10–12 days
Number of intra-uterine eggs	10
Genital bulb width	
Number of genital spines	9–10
Genital spine length	031
Haptoral sucker diameter	422–526 (482)
Hamulus handle length X	407–442 (429)
Hamulus guard length Y	
Hamulus hook length	
Marginal hooklet length	38
Haptoral L/Body L	0.20–0.24 (0.23)
No of anastomoses	0

Remarks *Polystoma sodwanense* was originally described as *Polystoma sodwanensis*. This species has been reported from the same host as that of *Metapolystoma porrosisimae* in South Africa.

4.4.11.60 *Polystoma stellai* Pérez-Vigueras, 1955 (Fig. 4.111)

Collection Unknown.

Holotype Unknown.

Paratypes Unknown.

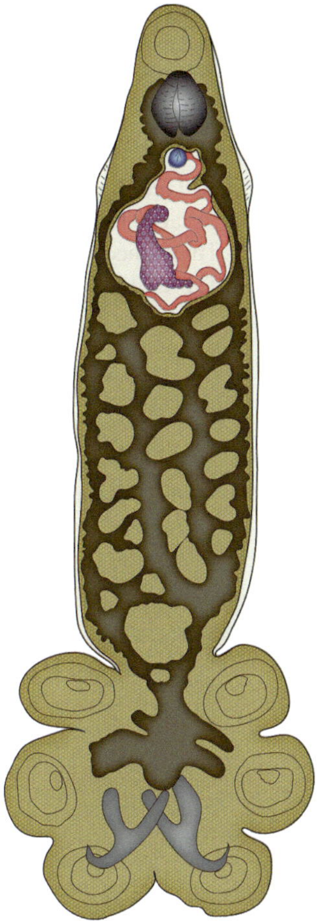

Fig. 4.111 *Polystoma stellai*. Redrawn from Pérez-Vigueras (1955)

Original description Pérez-Vigueras (1955).

Other taxonomic contributions Stunkard (1959).

Etymology Unknown.

Type locality Havana, Cuba.

Other localities None.

Type host *Osteopilus septentrionalis* (Duméril and Bibron, 1841), Cuban Treefrog. In the original description of the polystome, the host was known as *Hyla septentrionalis* Duméril and Bibron, 1841.

Additional hosts None.

Host geographical distribution *Osteopilus septentrionalis* is native to Cuba, the Cayman Islands and the Bahamas and has also been introduced into a wide range of regions including the United States, Costa Rica, Turks and Anguilla where it is found in disturbed habitats (houses and towns), mangroves, brackish water, forests and tree trunks (Hedges et al. 2010).

Host conservation status The Cuban Treefrog has a wide distribution, a presumed large population and is tolerant for a large range of habitats. It is listed as Least Concern by The IUCN Red List of Threatened Species 2021: e.T55811A3032751. https://doi.org/10.2305/IUCN.UK.2021-1.RLTS.T55811A3032751.en. Accessed on 31 January 2022.

Site on host Urinary bladder.

DNA sequence None.

Morphology and morphometrics (μm):		
Source	Pérez-Vigueras (1955)	Stunkard (1959)
Sample size		1
Body length	7100	4300
Greatest width	2100–2600	950
Width at vagina		
Haptor length	1400	
Haptor width	2100	
False oral sucker width		
Pharynx length		
Pharynx width		
Ovary length	880	
Ovary width	530	
Egg length		
Egg width		
Egg incubation time		
Number of intra-uterine eggs		
Genital bulb width	170	
Number of genital spines		
Genital spine length	21	
Haptoral sucker diameter	350–380	290
Hamulus handle length X	480	
Hamulus guard length Y		
Hamulus hook length		
Marginal hooklet length		
Haptoral L/Body L	0.20	
No of anastomoses	Network	

Remarks None.

4.4.11.61 *Polystoma testimagnum* Du Preez and Kok, 1993 (Fig. 4.112)

Collection Parasitic Worm Collection, National Museum, Bloemfontein, South Africa (NMBP); Parasitic Worm Collection, Natural History Museum, London (NHML).

Holotype NMBP 36.

Paratypes NMBP 37–40; NHML 2010.8.27.3.

Original description Du Preez and Kok (1993).

Taxonomic note The original name *Polystoma testimagna* was corrected for gender.

Other taxonomic contributions None.

Etymology Refers to the large testis observed in this species.

Type locality KwaZulu-Natal, South Africa.

Other localities None.

Type host *Strongylopus fasciatus* (Smith, 1849), Striped Stream Frog.

Additional hosts None.

Host geographical distribution *Strongylopus fasciatus* occurs in Zimbabwe, South Africa, Mozambique, Zambia and Swaziland. It inhabits wet open grassy areas near streams and ponds and marshy seeps. It is also found in forest, savannah, thickets, fynbos, heathland and well-vegetated man-made ponds, dams and irrigation canals (Frost 2021).

Host conservation status The Striped Stream Frog has a large population, a wide distribution and is tolerant for various habitats. It is listed as Least Concern by The IUCN Red List of Threatened Species 2013: e.T58765A3073186. https://doi.org/10.2305/IUCN.UK.2013-2.RLTS.T58765A3073186.en. Accessed on 31 January 2022.

Site on host Urinary bladder.

DNA sequence
- 18S rRNA gene, 2099 bp. GenBank Accession: AM157194
- 18S rRNA gene, 627 bp. GenBank Accession: AJ297770
- 28S rRNA gene, 1448 bp. GenBank Accession: AM157217
- 12S rRNA gene, 490 bp. GenBank Accession: KR856092
- Cytochrome c Oxidase I, 375 bp. GenBank Accession: AM913860
- ITS1, 697 bp. GenBank Accession: AJ310397

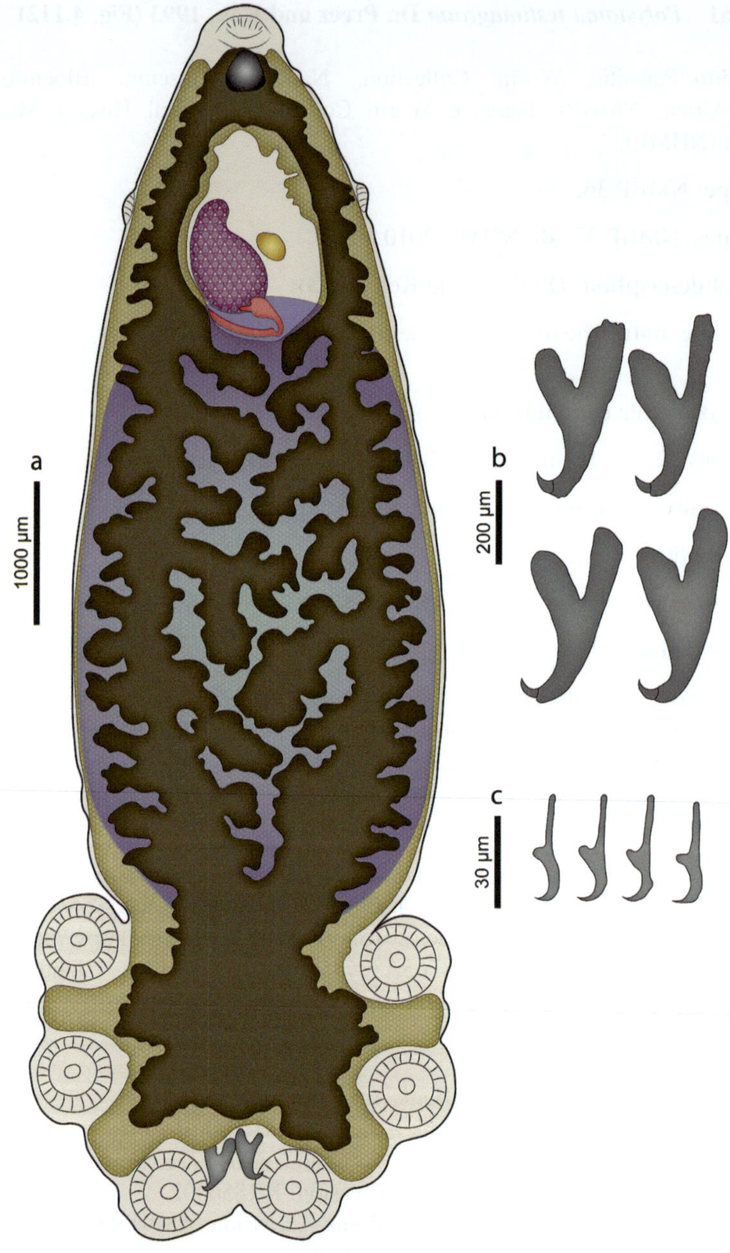

Fig. 4.112 (**a**) *Polystoma testimagnum*, (**b**) hamuli, (**c**) marginal hooklets. Redrawn from Du Preez and Kok (1993)

Morphology and morphometrics (μm):	
Source	Du Preez and Kok (1993)
Sample size	10
Body length	6100–8800 (7850)
Greatest width	2200–2800 (2580)
Width at vagina	1400–1800 (1680)
Haptor length	1500–2400 (2050)
Haptor width	2500–3300 (2860)
False oral sucker width	
Pharynx length	275–319 (293)
Pharynx width	255–324 (294)
Testis length	
Testis width	
Ovary length	619–992 (836)
Ovary width	
Egg length	201–246 (228)
Egg width	150–186 (169)
Egg incubation time	10–12 days
Number of intra-uterine eggs	
Genital bulb width	
Number of genital spines	9–10
Genital spine length	33
Haptoral sucker diameter	462–648 (553)
Hamulus handle length X	334–442 (411)
Hamulus guard length Y	
Hamulus X/Y ratio	(1.24)
Hamulus hook length	
Marginal hooklet length	C1: 38; C2–7 24; C8: 35
Marginal hooklet total length to handle length ratio	1.79
Haptoral L/Body L	0.245–0.272 (0.26)
No of anastomoses	0–2

Remarks *Polystoma testimagnum* was originally described as *Polystoma testimagna.*

4.4.11.62 *Polystoma togoense* Bourgat, 1977 (Fig. 4.113)

Collection Unknown.

Holotype Unknown.

Paratypes Unknown.

Original description Bourgat (1977).

Taxonomic note The original name *Polystoma togoensis* was corrected for gender.

Fig. 4.113 (**a**) *Polystoma togoense*, (**b**) hamuli. Redrawn from Bourgat (1977)

Other taxonomic contributions Murith (1981); Murith et al. (1978).

Etymology Named after type locality country.

Type locality Lomé and Kovié, Togo.

Other localities Cameroon and Ivory Coast.

Type host *Ptychadena hylaea* Schmidt and Inger, 1959, no English name noted.

Additional hosts *Ptychadena bibroni.*

Site on host Urinary bladder and gills of tadpoles.

Host geographical distribution *Ptychadena hylaea* occurs in Gambia and from southern Mauritania to northeastern Democratic Republic of Congo, presumably also in South Sudan (Frost 2021).

Host conservation status The species has a wide distribution, is tolerant for a broad range of habitats and is abundant. Not listed in the IUCN red list.

DNA sequence None.

Morphology and morphometrics (μm):		
Source	Bourgat (1977)	Murith (1981)
Sample size	36	12
Body length	3630–6840 (5390)	3020–6790 (4640)
Greatest width		1140–2550 (1810)
Width at vagina		930–1620 (1320)
Haptor length	640–1410 (1030)	930–1720 (1360)
Haptor width		1390–2520 (1940)
False oral sucker width		
Pharynx length		
Pharynx width		
Ovary length		
Ovary width		
Egg length	200	172–250 (217)
Egg width		117–194 (138)
Egg incubation time		
Number of intra-uterine eggs	1–19	1–10 (1,5)
Genital bulb width		
Number of genital spines	8–9	
Genital spine length		
Haptoral sucker diameter	258–439 (352)	340–480 (400)
Hamulus handle length X	258–534 (402)	400–540 (470)
Hamulus guard length Y		
Hamulus hook length		
Marginal hooklet length		40
Haptoral L/Body L	0.16–0.23 (0.19)	0.250–0.380 (0.300)
No of anastomoses		0–1

Remarks *Polystoma togoense* was originally described as *Polystoma togoensis.*

4.4.11.63 *Polystoma touzeti* Vaucher, 1987 (Fig. 4.114)

Collection Natural History Museum, Geneva, Switzerland (MHNG).

Holotype MHNG 985.560. 0015594.

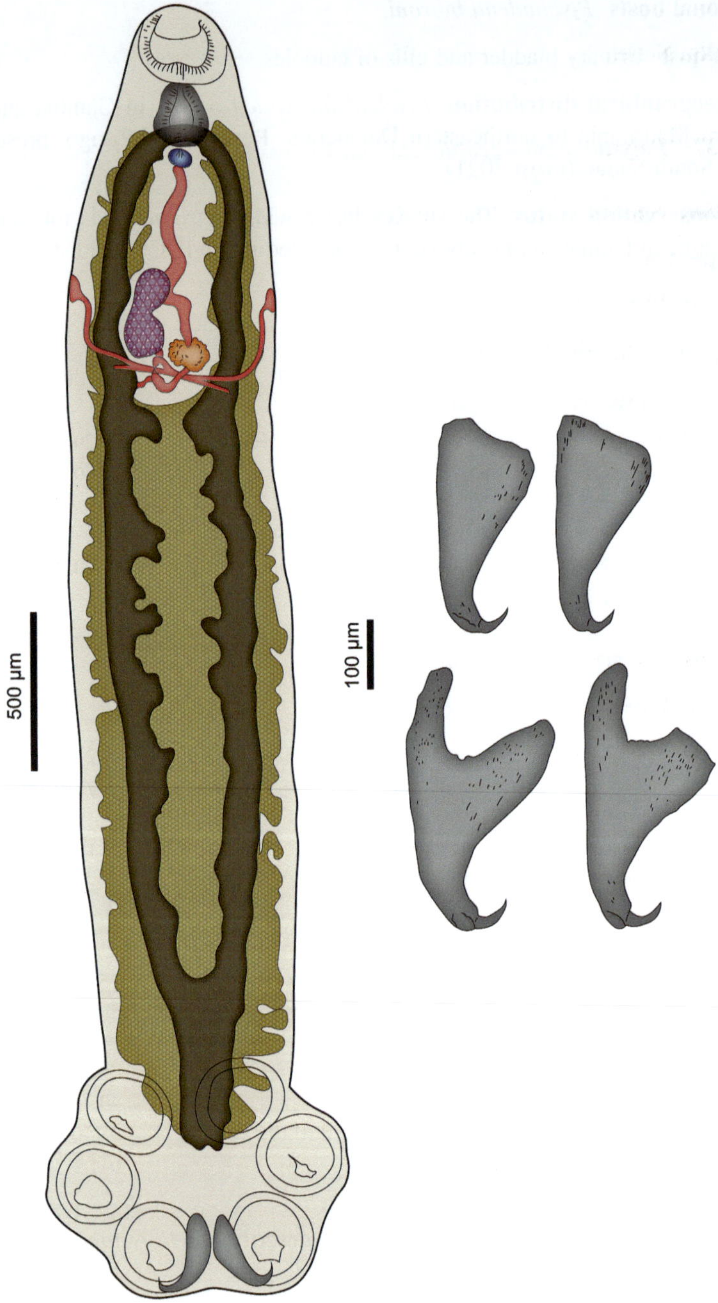

Fig. 4.114 (**a**) *Polystoma touzeti*, (**b**) hamuli. Redrawn from Vaucher (1987)

Paratypes MHNG 985.561.

Original description Vaucher (1987).

Other taxonomic contributions None.

Etymology Named after Jean-Marc Touzet.

Type locality San Pablo de Kantesiya, Napo Province, Ecuador.

Other localities 24 km E Pto Francisco de Orellana, Napo Province, Ecuador.

Type host *Gastrotheca riobambae* (Fowler, 1913), Riobamba Marsupial Frog.

Additional hosts None.

Host geographical distribution *Gastrotheca riobambae* occurs in Ecuador where it is found on vegetation in montane meadows, montane forests and corn fields as well as near water sources (streams, irrigation ditches and drainage) (Coloma et al. 2004a).

Host conservation status The Riobamba Marsupial Frog's population is declining due to the loss of its habitat, which may be the result of pollution and climate change. It is listed as Endangered by The IUCN Red List of Threatened Species 2004: e. T55357A11298345. https://doi.org/10.2305/IUCN.UK.2004.RLTS.T55357A11298345.en. Accessed on 31 January 2022.

Site on host Urinary bladder.

DNA sequence None.

Morphology and morphometrics (µm):	
Source	Vaucher (1987)
Sample size	
Body length	4180
Greatest width	755
Width at vagina	
Haptor length	815
Haptor width	1020
False oral sucker width	245 × 286
Pharynx length	
Pharynx width	176 × 213
Ovary length	
Ovary width	
Egg length	
Egg width	
Egg incubation time	
Number of intra-uterine eggs	
Genital bulb width	
Number of genital spines	

(continued)

Morphology and morphometrics (μm):	
Source	Vaucher (1987)
Genital spine length	
Haptoral sucker diameter	270–310 × 254–311
Hamulus handle length X	315–319
Hamulus guard length Y	
Hamulus hook length	
Marginal hooklet length	
Haptoral L/Body L	0.20
No of anastomoses	0

Remarks None.

4.4.11.64 *Polystoma travassosi* Du Preez and Domingues, 2019 (Fig. 4.115)

Collection Helminthological Collection, Instituto Oswaldo Cruz, Rio de Janeiro, Brazil (CHIOC).

Holotype CHIOC 10151.

Paratypes CHIOC 31420, 31,421, 31422a, 31422b, 31422c.

Original description Du Preez and Domingues (2019).

Other taxonomic contributions Kohn et al. (1978).

Etymology Named after Dr Lauro Travassos, Rio de Janeiro, Brazil.

Type locality Angra dos Reis, Rio de Janeiro, Brazil.

Other localities None.

Type host *Trachycephalus mesophaeus* (Hensel, 1867), Porto Alegre Golden-eyed Treefrog.

Additional hosts None.

Host geographical distribution *Trachycephalus mesophaeus* occurs in Atlantic forests of eastern Brazil, from Alagoas to Rio Grande do Sul, inland to include eastern and central Minas Gerais (Frost 2021).

Host conservation status The Porto Alegre Golden-eyed Treefrog has a wide distribution and a tolerance for habitat modification. It is presumed having a large population. It is thus listed as Least Concern by The IUCN Red List of Threatened Species 2004: e.T55822A11372732. https://doi.org/10.2305/IUCN.UK.2004.RLTS.T55822A11372732.en. Accessed on 31 January 2022.

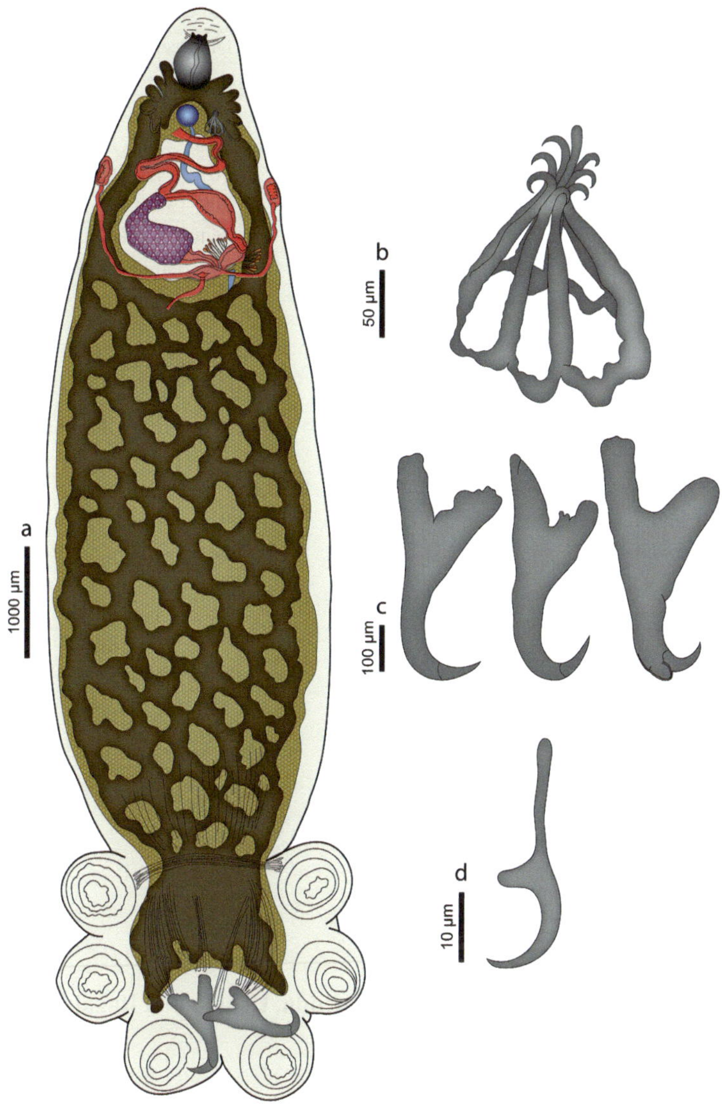

Fig. 4.115 (**a**) *Polystoma travassosi*, (**b**) genital spines, (**c**) hamuli, (**d**) marginal hooklets. Redrawn from Du Preez and Domingues (2019)

Site on host Urinary bladder.

Zoobank registration urn:lsid:zoobank.org:act: D08CB059-8E32-4B44-A033-55E508A9C145.

DNA sequence None.

Morphology and morphometrics (μm):	
Source	Du Preez and Domingues (2019)
Sample size	6
Body length	4980–7820 (5869)
Greatest width	1500–2360 (1852)
Width at vagina	1012–1420 (1268)
Haptor length	800–1373 (968)
Haptor width	1640–2000 (1741)
False oral sucker width	190–350 (259)
Pharynx length	280–325 (303)
Pharynx width	212–270 (237)
Ovary length	710–1360 (915)
Ovary width	310–580 (396)
Egg length	
Egg width	
Egg incubation time	
Number of intra-uterine eggs	0
Genital bulb width	90–210 (128)
Number of genital spines	8
Genital spine length	41–45 (44)
Haptoral sucker diameter	315–360 (338)
Hamulus handle length X	390–505 (436)
Hamulus guard length Y	360–457 (387)
Hamulus hook length	75–98 (84)
Marginal hooklet length	23–27 (25)
Haptoral L/Body L	0.12–0.17 (0.15)
No of anastomoses	Network

Remarks None.

4.4.11.65 *Polystoma uchidai* Uchida, Machida, Uchida and Itagaki, 1988 nom. nov. (Fig. 4.116)

Collection Unknown.

Holotype Unknown.

Paratypes Unknown.

Original description Uchida et al. (1988).

Taxonomic note *Polystoma ozakii* Uchida, Machida, Uchida and Itagaki, 1988 is regarded as a primary junior homonym of *Polystoma ozakii* Price, 1939. According to the International code of Zoological Nomenclature (ICZN) article 53.3, two taxa erected within a genus with the same epithet constitute primary homonymy.

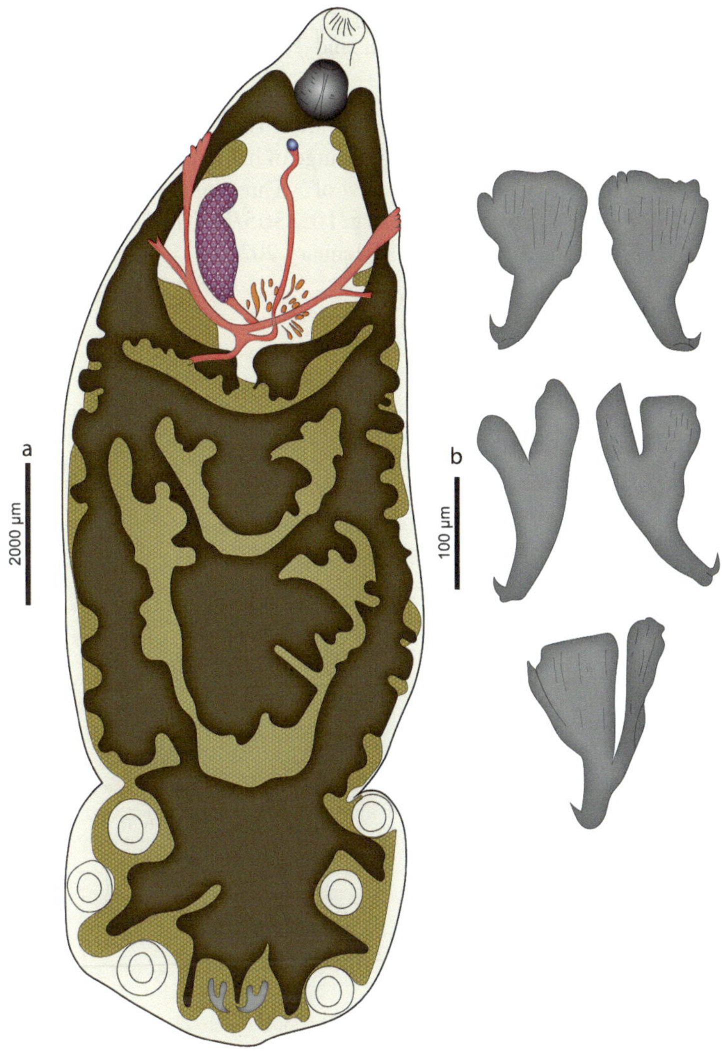

Fig. 4.116 (**a**) *Polystoma uchidai*, (**b**) hamuli. Redrawn from Uchida et al. (1988)

According to ICZN articles 60.1 and 60.3, the junior homonym is hereby rejected and replaced with *Polystoma uchidai* nom. nov.

Other taxonomic contributions None.

Etymology Named after Prof Uchida, Japan.

Type locality Japan.

Type host *Bufo japonicus* Temminck and Schlegel 1838, Japanese Common Toad.

Additional hosts None.

Host geographical distribution *Bufo japonicus* is endemic to Japan where it is distributed throughout Kyushu, Shikoku and Honshu and their adjacent islands (Frost 2021).

Host conservation status The Japanese Common Toad has a wide distribution, has a large population and is tolerant for a wide range of habitats. It is thus listed as Least Concern by The IUCN Red List of Threatened Species 2021: e. T54673A177177294. https://doi.org/10.2305/IUCN.UK.2021-1.RLTS.T54673A177177294.en. Accessed on 31 January 2022.

Site on host Urinary bladder.

Other localities None.

Additional hosts None.

Site on host Urinary bladder.

DNA sequence None.

Morphology and morphometrics:	
Source	Uchida et al. (1988)
Sample size	
Body length	
Greatest width	3500–6000
Width at vagina	
Haptor length	3800–4500
Haptor width	7500–9200
False oral sucker width	
Pharynx length	420–620
Pharynx width	600–720
Ovary length	1230–1480
Ovary width	140–620
Egg length	
Egg width	
Egg incubation time	
Number of intra-uterine eggs	
Genital bulb width	
Number of genital spines	8
Genital spine length	42
Haptoral sucker diameter	420–580
Hamulus handle length X	
Hamulus guard length Y	325–367 but in text 367–383
Hamulus hook length	
Marginal hooklet length	
Haptoral L/Body L	
No of anastomoses	Network

Remarks *Polystoma uchidai* was originally described as *Polystoma ozakii.*

4.4.11.66 *Polystoma umthakathi* Kok and Seaman, 1987 (Fig. 4.117)

Collection Parasitic Worm Collection, National Museum, Bloemfontein, South Africa (NMBP).

Holotype NMBP 17.

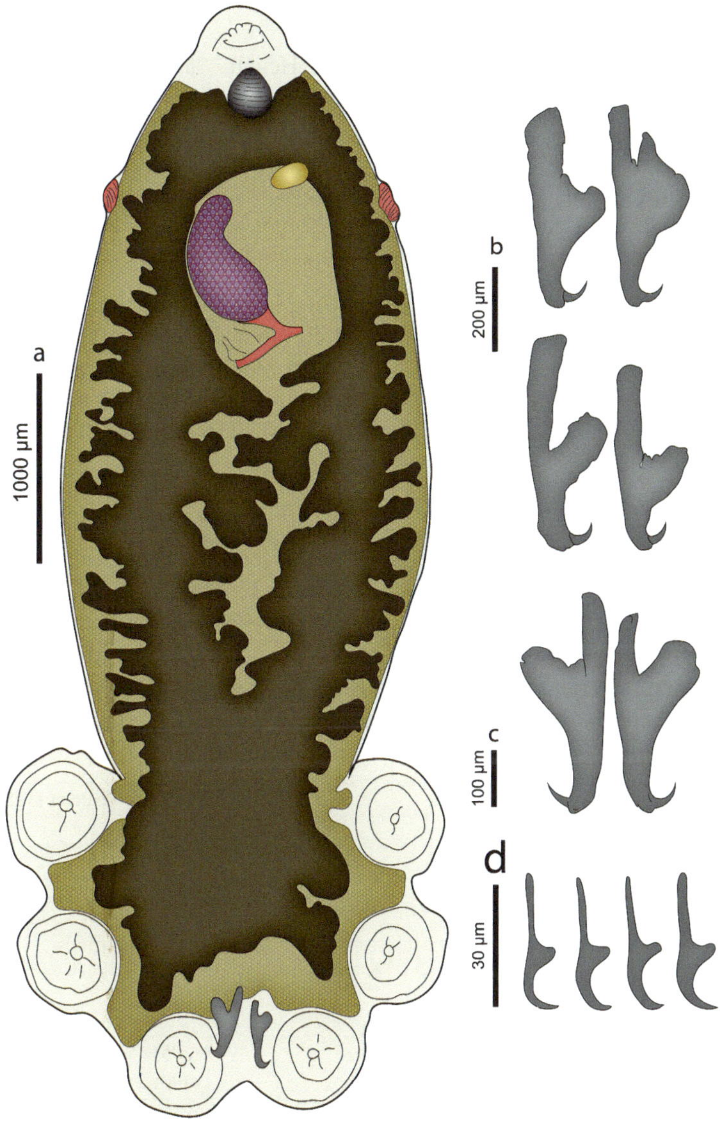

Fig. 4.117 (**a**) *Polystoma umthakathi*, (**b**) hamuli, (**c**) hamuli of holotype, (**d**) marginal hooklets. Redrawn from Kok and Seaman (1987)

Paratypes NMBP 18 to P23.

Original description Kok and Seaman (1987).

Other taxonomic contributions None.

Etymology Named after the umThakathi forest at the type locality.

Type locality Vernon Crookes Reserve, KwaZulu-Natal, South Africa.

Other localities None.

Type host *Natalobatrachus bonebergi* Hewitt and Methuen, 1912, Boneberg's Frog.

Additional hosts None.

Host geographical distribution *Natalobatrachus bonebergi* occurs below 900 m elevation in forested canyons from Dwesa Nature Reserve in the Eastern Cape Province of South Africa east to southern and central KwaZulu-Natal (Frost 2021).

Host conservation status The Boneberg's Frog is under threat as its forest habitat range has been lost or severely fragmented due to agricultural activities, such as sugarcane cultivation, woodcutting, urbanization, pollution and afforestation. It is listed as Endangered by The IUCN Red List of Threatened Species 2016: e. T58076A77159820. https://doi.org/10.2305/IUCN.UK.2016-3.RLTS. T58076A77159820.en. Accessed on 31 January 2022.

Site on host Urinary bladder.

DNA sequence
- 18S rRNA gene, 2121 bp. GenBank Accession: FM897282
- 18S rRNA gene, 614 bp. GenBank Accession: AJ297769
- 28S rRNA gene, 1448 bp. GenBank Accessions: AM913874 and FM897265
- Cytochrome c Oxidase I, 369 bp. GenBank Accession: AM913861
- ITS1, 699 bp. GenBank Accession: AJ301685

Morphology and morphometrics (μm):	
Source	Kok and Seaman (1987)
Sample size	7
Body length	(7290)
Greatest width	(2450)
Width at vagina	(1770)
Haptor length	(2060)
Haptor width	(2760)
False oral sucker width	
Pharynx length	(295)
Pharynx width	(300)
Ovary length	(862)
Ovary width	

(continued)

Morphology and morphometrics (μm):	
Source	Kok and Seaman (1987)
Egg length	(219)
Egg width	(155)
Egg incubation time	
Number of intra-uterine eggs	1–3
Genital bulb width	
Number of genital spines	7–9
Genital spine length	33–40
Haptoral sucker diameter	(474)
Hamulus handle length X	(408)
Hamulus guard length Y	
Hamulus hook length	
Marginal hooklet length	C1: 33; C2–7: 22–24; C8: 31
Haptoral L/Body L	0.28
No of anastomoses	0–1

Remarks None.

4.4.11.67 *Polystoma unicinulatum* Macé, 1880 (Fig. 4.118)

Collection Unknown.

Holotype Unknown.

Paratypes NMBP Unknown.

Original description Macé (1880).

Other taxonomic contributions None.

Etymology Unknown.

Type locality Unknown

Other localities None.

Type host *Pelophylax lessonae* (Camerano, 1882), Pool Frog. In the original description of the polystome, the host was known as *Rana viridis* Garsault, 1764.

Host geographical distribution *Pelophylax lessonae* is distributed over much of Europe, ranging as far east as the southern Ural Mountains of Russia (Frost 2021).

Host conservation status The Pool Frog has a wide distribution, is tolerant to a broad range of habitats. It is listed as Least Concern by The IUCN Red List of Threatened Species 2009: e.T58643A86643256. https://doi.org/10.2305/IUCN.UK.2009.RLTS.T58643A11818386.en. Accessed on 31 January 2022.

Site on host Urinary bladder.

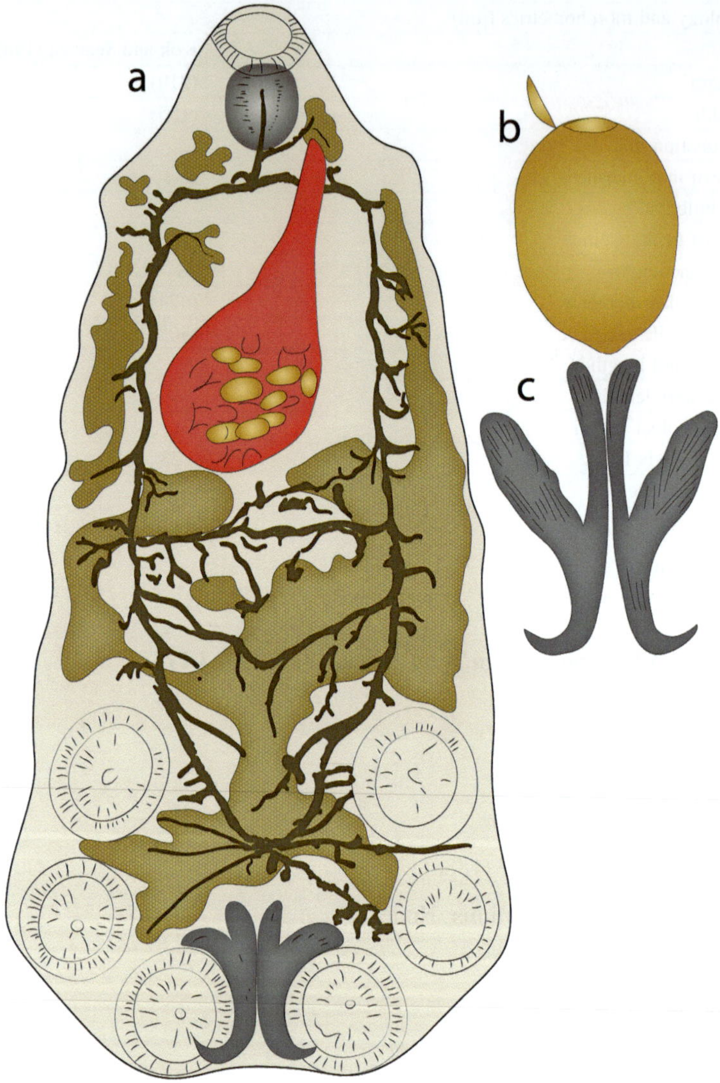

Fig. 4.118 (**a**) *Polystoma unicinulatum*, (**b**) egg, (**c**) hamuli. Redrawn from Macé (1880)

DNA sequence None

Morphology and morphometrics (μm):		
Source		Macé, 1880
Sample size		6
Body length		4200
Greatest width		1600

(continued)

Morphology and morphometrics (μm):		
Source		Macé, 1880
Width at vagina		
Haptor length		
Haptor width		
False oral sucker width		
Pharynx length		
Pharynx width		
Testis length		
Testis width		
Ovary length		
Ovary width		
Egg length		
Egg width		
Egg incubation time		
Number of intra-uterine eggs		
Genital bulb width		
Number of genital spines		
Genital spine length		
Haptoral sucker diameter		
Hamulus handle length X		
Hamulus guard length Y		
Hamulus hook length		
Marginal hooklet length		
Haptoral L/Body L		
No of anastomoses		

Remarks *Polystoma unicinulatum* was originally described as *Polystomum unicinulatum.*

4.4.11.68 *Polystoma vernoni* Du Preez, 2011 (Fig. 4.119)

Collection Parasitic Worm Collection National Museum, Bloemfontein, South Africa (NMBP); Parasitic Worms Collection, Natural History Museum, London (NHML).

Holotype NMBP 312.

Paratypes NMBP 313–317; NHML 2010.8.27.1–2.

Original description Du Preez (2011).

Other taxonomic contributions None.

Etymology Named after the type locality.

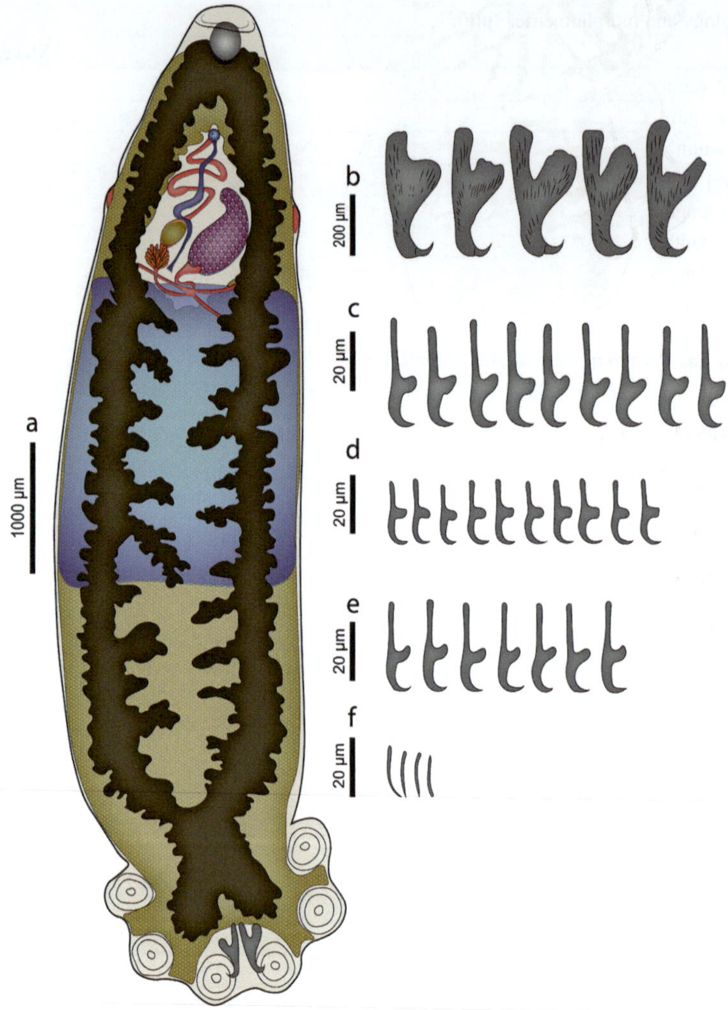

Fig. 4.119 (**a**) *Polystoma vernoni*, (**b**) hamuli, (**c**) marginal hooklets (1), (**d**) marginal hooklets (2–7), (**e**) marginal hooklets (8), (**f**) genital spines. Redrawn from Du Preez (2011)

Type locality Vernon Crookes Reserve, Kwa-Zulu Natal, South Africa.

Other localities None.

Type host *Ptychadena oxyrhynchus*, Sharp-nosed Ridged Frog.

Additional hosts None.

Host geographical distribution *Ptychadena oxyrhynchus* occurs in Angola, Uganda, Zambia, Zimbabwe, Congo, Botswana, Cameroon, Gambia, Ghana, Kenya, Namibia, Malawi, Mali, Nigeria, Mozambique, Tanzania, Swaziland and

South Africa. It inhabits agricultural regions, savannah and marshy regions (Frost 2021).

Host conservation status The Sharp-nosed Ridged Frog has a wide distribution, has a large population and inhabits a broad range of habitats. It is listed as Least Concern by The IUCN Red List of Threatened Species 2016: e.T58516A3069746. https://doi.org/10.2305/IUCN.UK.2016-1.RLTS.T58516A3069746.en. Accessed on 31 January 2022.

Site on host Urinary bladder.

DNA sequence None.

Morphology and morphometrics (μm):	
Source	Du Preez (2011)
Sample size	13
Body length	5067–9628 (7244)
Greatest width	956–2578 (1816)
Width at vagina	
Haptor length	1008–2291 (1407)
Haptor width	1119–2404 (1796)
False oral sucker width	303–531 (422)
Pharynx length	207–312 (279)
Pharynx width	182–249 (223)
Ovary length	487–959 (719)
Ovary width	146–360 (254)
Egg length	180–219 (201)
Egg width	116–140 (127)
Egg incubation time	
Number of intra-uterine eggs	1–10
Genital bulb width	71–90 (84)
Number of genital spines	8–9
Genital spine length	28–29 (29)
Haptoral sucker diameter	255–444 (356)
Hamulus handle length X	332–459 (395)
Hamulus guard length Y	291–381 (338)
Hamulus X/Y ratio	1.16
Hamulus hook length	70–77 (73)
Marginal hooklet length	C1: 34–38 (36); C2–7: 20–23 (21); C8: 30–31 (31)
Haptoral L/Body L	0.19
No of anastomoses	0

Remarks None.

4.4.11.69 *Polystoma viridis* Euzet, Combes and Buchvarov, 1974 (Fig. 4.120)

Collection Unknown.

Holotype Unknown.

Paratypes Unknown.

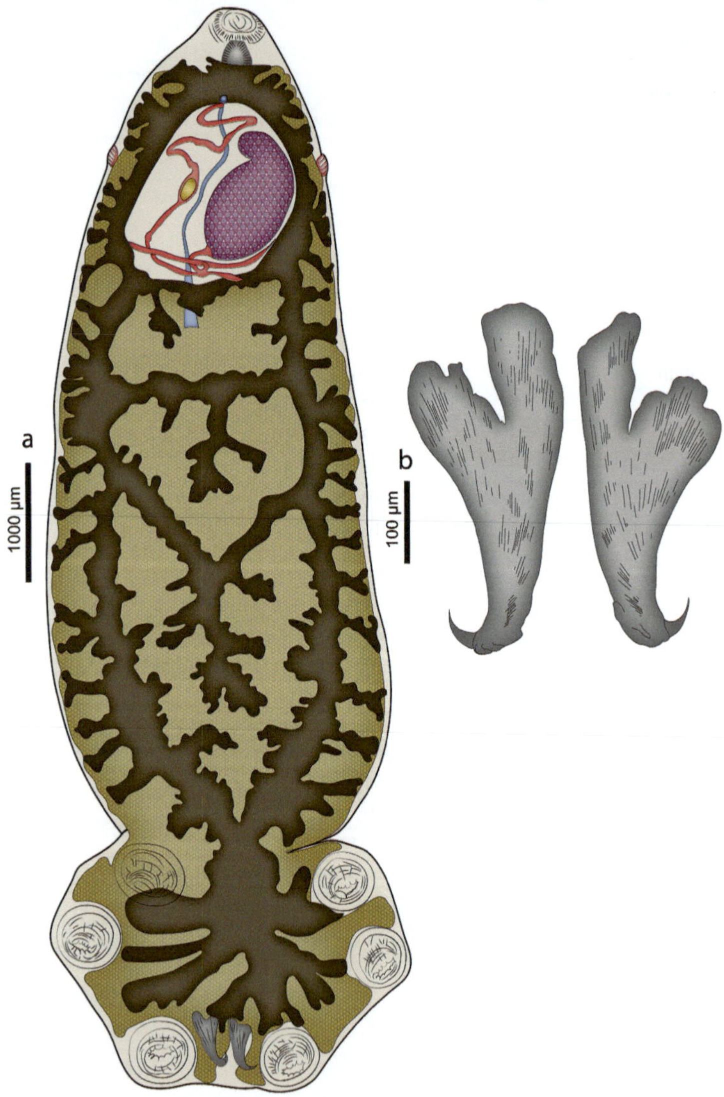

Fig. 4.120 (**a**) *Polystoma viridis*, (**b**) hamuli. Redrawn from Euzet et al. (1974c)

Original description Euzet et al. (1974c).

Other taxonomic contributions None.

Etymology Named after the host.

Type locality Corsica and Bulgaria.

Other localities None.

Type host *Bufotes viridis*, Green Toad. In the original description of the polystome, the host was known as *Bufo viridis* Laurenti, 1768.

Additional hosts None.

Host geographical distribution *Bufotes viridis* occurs in Bulgaria, Poland, Austria, Belarus, Serbia, Ukraine, Croatia, Germany, France, Italy, Greece, Romania, Hungary, Slovakia, Albania, Montenegro and Slovenia. It is considered extinct in Switzerland and has been introduced in Spain. This species inhabits forests, grasslands, scrublands and disturbed habitats, such as gardens (Frost 2021).

Host conservation status The Green Toad has a large population size and is widely distributed. It is thus listed as Least Concern by The IUCN Red List of Threatened Species 2015: e.T155333A74516784. Accessed on 31 January 2022.

Site on host Urinary bladder.

DNA sequence None.

Morphology and morphometrics (μm):	
Source	Euzet et al. (1974c)
Sample size	7
Body length	8000–9600 (9100)
Greatest width	2900–3100 (3000)
Width at vagina	
Haptor length	2000–2300 (2100)
Haptor width	3000–3200 (3100)
False oral sucker width	
Pharynx length	310–390 (356)
Pharynx width	300–355 (343)
Ovary length	1200–1480 (1260)
Ovary width	600–700 (660)
Egg length	230
Egg width	110
Egg incubation time	
Number of intra-uterine eggs	
Genital bulb width	
Number of genital spines	8–9
Genital spine length	40
Haptoral sucker diameter	520–570 (550)

(continued)

Morphology and morphometrics (μm):	
Source	Euzet et al. (1974c)
Hamulus handle length X	360–510 (470)
Hamulus guard length Y	
Hamulus hook length	
Marginal hooklet length	
Haptoral L/Body L	0.23
No of anastomoses	0–5

Remarks None.

4.4.11.70 *Polystoma xinpingense* Gao, Chen and Fan, 2012 (Fig. 4.121)

Collection School of Life Sciences of Yunnan Normal University.

Holotype H9–2.

Paratypes H9–2–1, H4–2.

Original description Gao et al. (2012).

Taxonomic note The original *Polystoma xinpingensis* was corrected for gender.

Other taxonomic contributions None.

Etymology Named after the type locality.

Type locality Xinping County, Yunnan Province, China.

Other localities None.

Type host *Hyla annectans* (Jerdon, 1870), Assam Treefrog. In the original description of the polystome, the host was known as *Hyla annectans annectans* Yang, Su and Li, 1983.

Additional hosts None.

Host geographical distribution *Hyla annectans* occurs in northern Myanmar, Vietnam, northeast India, southwestern and central China and northern Thailand. It may also occur in Southeast Asia inhabiting grasslands and forests (van Dijk et al. 2004d).

Host conservation status The Assam Treefrog has a large distribution range, a big population size and occurs in a wide range of habitats. However, it may be threatened by habitat degradation such as water pollution. It is listed as Least Concern by The IUCN Red List of Threatened Species 2004: e.T55391A86174841. https://doi.org/10.2305/IUCN.UK.2004.RLTS.T55391A11291178.en. Accessed on 31 January 2022.

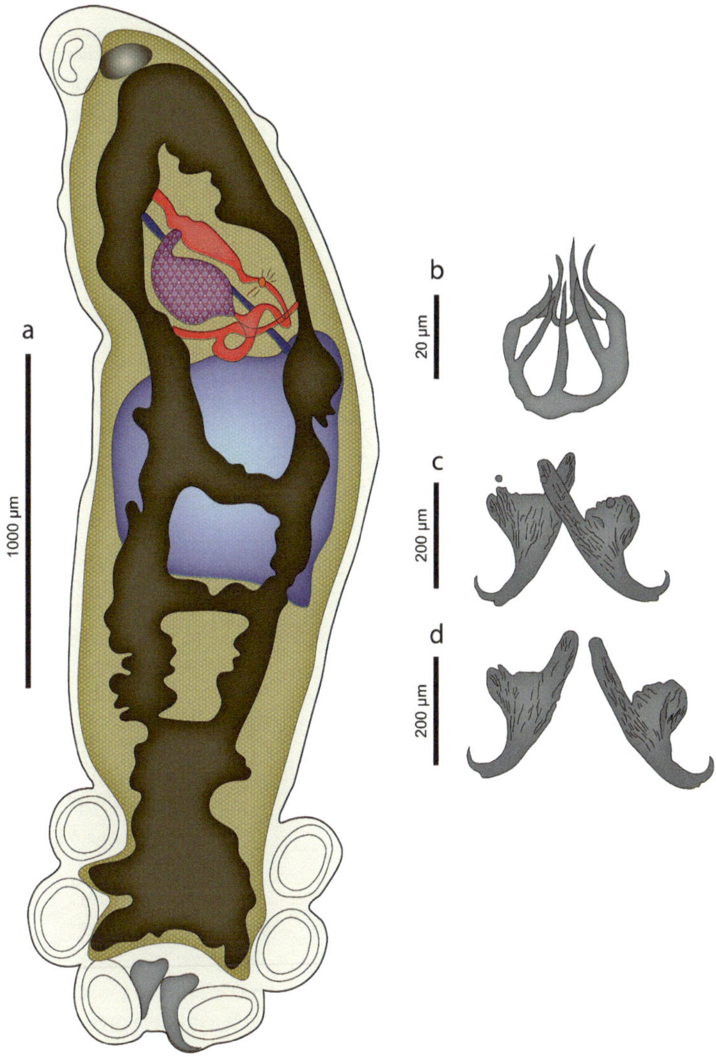

Fig. 4.121 (**a**) *Polystoma xinpingense*, (**b**) genital spines, (**c**) hamuli, (**d**) hamuli. Redrawn from Gao et al. (2012)

Site on host Urinary bladder.

DNA sequence None.

Morphology and morphometrics (µm):	
Source	Gao et al. (2012)
Sample size	11
Body length	2082–3940 (2967)

(continued)

Morphology and morphometrics (μm):	
Source	Gao et al. (2012)
Greatest width	1187
Width at vagina	904
Haptor length	532–1501 (801)
Haptor width	867–1451 (1211)
False oral sucker width	
Pharynx length	199.5
Pharynx width	153.7
Ovary length	414.2
Ovary width	
Egg length	250.6
Egg width	90.7
Egg incubation time	
Number of intra-uterine eggs	1
Genital bulb width	
Number of genital spines	7–9
Genital spine length	24.5–37.4 (29.8)
Haptoral sucker diameter	229–323 (275)
Hamulus handle length X	
Hamulus guard length Y	
Hamulus X/Y ratio	1.27
Hamulus hook length	
Marginal hooklet length	
Haptoral L/Body L	0.27
No of anastomoses	>1

Remarks None.

4.4.12 Protopolystoma Bychowsky, 1957

4.4.12.1 *Protopolystoma fissile* Tinsley and Jackson, 1998 (Fig. 4.122)

Collection Parasitic Worms Collection, Natural History Museum of London (NHML).

Holotype NHML 1997.7.14.6.

Paratypes NHML 1997.7.14.7–10.

Original description Tinsley and Jackson (1998).

Taxonomic note The original name *Protopolystoma fissilis* was corrected for gender.

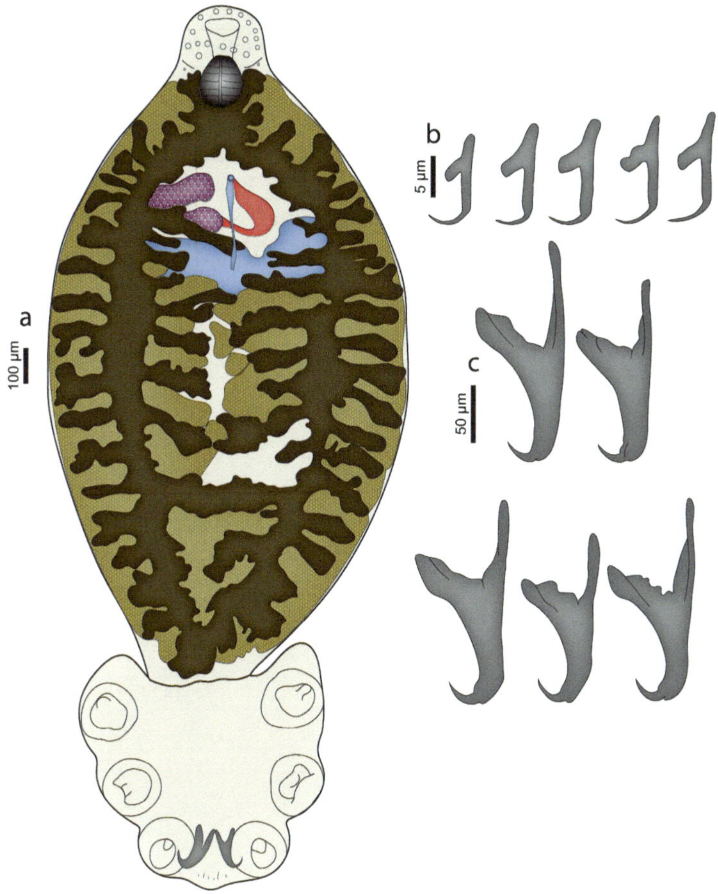

Fig. 4.122 (**a**) *Protopolystoma fissile*, (**b**) marginal hooklets, (**c**) hamuli. Redrawn from Tinsley and Jackson (1998)

Other taxonomic contributions None.

Etymology Unknown.

Type locality Ebisha, Democratic Republic of the Congo.

Other localities Cameroon.

Type host *Xenopus fraseri* Boulenger, 1905, Fraser's Clawed Frog.

Additional hosts *Xenopus wittei* Tinsley, Kobel and Fischberg 1979 in Rwanda and *Xenopus pygmaeus* Loumont 1986 in the Democratic Republic of the Congo.

Site on host Urinary bladder.

Host geographical distribution *Xenopus fraseri* occurs in Congo, Gabon, Angola, Equatorial Guinea and Cameroon. This species, which is water dependent, is usually found in pools in lowland rainforests (Tinsley et al. 2004; Tinsley and Measey 2004).

Host conservation status The Fraser's Clawed Frog has a large population, a wide distribution and is tolerant for habitat modification. It is listed as Least Concern by The IUCN Red List of Threatened Species2017: e.T89257302A18397804. https://doi.org/10.2305/IUCN.UK.2017-2.RLTS.T89257302A18397804.en. Accessed on 01 February 2022.

DNA sequence None.

Morphology and morphometrics (μm):	
Source	Tinsley and Jackson (1998)
Sample size	
Body length	2650
Greatest width	1110
Width at vagina	
Haptor length	610
Haptor width	760
False oral sucker width	
Pharynx length	
Pharynx width	142
Ovary length	
Ovary width	
Egg length	
Egg width	
Egg incubation time	
Number of intra-uterine eggs	
Genital bulb width	38
Number of genital spines	
Genital spine length inner circle	19–24
Genital spine length outer circle	13–16
Haptoral sucker diameter	167
Hamulus handle length X	126
Hamulus guard length Y	
Hamulus hook length	26–33 (29)
Marginal hooklet length	13
Haptoral L/Body L	0.23
No of anastomoses	0–2

Remarks *Protopolystoma fissile* was originally described as *Protopolystoma fissilis*.

4.4.12.2 *Protopolystoma occidentale* Tinsley and Jackson, 1998 (Fig. 4.123)

Collection Parasitic Worms Collection, Natural History Museum, London (NHML).

Holotype NHML 1997.7.16.5.

Paratypes NHML 1997.7.16.6–9.

Original description Tinsley and Jackson (1998).

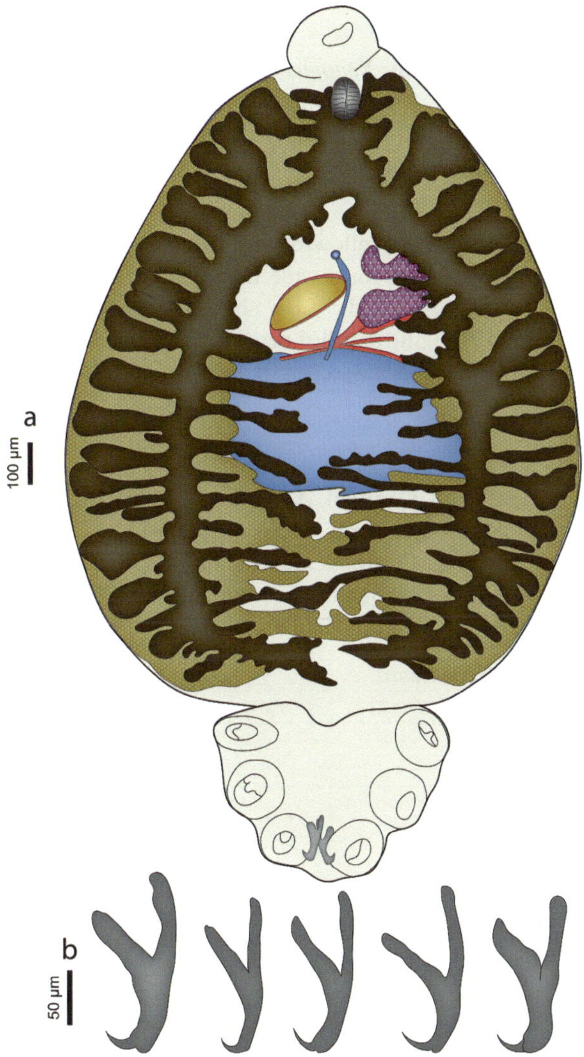

Fig. 4.123 (**a**) *Protopolystoma occidentale*, (**b**) hamuli. Redrawn from Tinsley and Jackson (1998)

Taxonomic note The original name *Protopolystoma occidentalis* was corrected for gender.

Other taxonomic contributions Avery (1971); Bourgat and Combes (1976).

Etymology Unknown.

Type locality Bolgatanga, Ghana.

Other localities Togo, Nigeria and Cameroon.

Type host *Xenopus muelleri* (Peters 1844), Yellow-bellied Platanna (western form).

Additional hosts None.

Host geographical distribution *Xenopus muelleri* occurs in South Africa, Ghana, Zambia, Cameroon, Angola, Congo, Nigeria, South Sudan, Zimbabwe, Tanzania, Namibia, Botswana, Benin, Mozambique, Kenya, Swaziland, Malawi, Uganda and Togo. This primarily aquatic species inhabits ponds, rivers and streams in altered habitats (Frost 2021).

Host conservation status The Yellow-bellied Platanna has a large population, a wide distribution and is tolerant for a wide range of habitats. It is listed as Least Concern by The IUCN Red List of Threatened Species2017: e. T107580991A107581762. https://doi.org/10.2305/IUCN.UK.2017-2.RLTS.T107580991A107581762.en. Accessed on 01 February 2022.

Site on host Urinary bladder.

DNA sequence
- 18S rRNA gene, 2024 bp. GenBank Accession: AM051077
- 18S rRNA gene, 588 bp. GenBank Accession: AJ297780
- 28S rRNA gene, 1643 bp. GenBank Accession: KR856160
- 12S rRNA gene, 441 bp. GenBank Accession: KR856121
- Cytochrome c Oxidase I, 368 bp. GenBank Accession: KR856179

Morphology and morphometrics (μm):	
Source	Tinsley and Jackson (1998)
Sample size	
Body length	2270
Greatest width	1370
Width at vagina	
Haptor length	420
Haptor width	640
False oral sucker width	
Pharynx length	
Pharynx width	130
Ovary length	
Ovary width	

(continued)

Morphology and morphometrics (μm):	
Source	Tinsley and Jackson (1998)
Egg length	
Egg width	
Egg incubation time	
Number of intra-uterine eggs	
Genital bulb width	35
Number of genital spines	
Genital spine length inner circle	16–20
Genital spine length outer circle	12–14
Haptoral sucker diameter	137
Hamulus handle length X	137
Hamulus guard length Y	
Hamulus hook length	
Marginal hooklet length	13–14
Haptoral L/Body L	0.185
No of anastomoses	0–3

Remarks *Protopolystoma occidentale* was originally described as *Protopolystoma occidentalis*.

4.4.12.3 *Protopolystoma orientale* Tinsley and Jackson, 1998 (Fig. 4.124)

Collection Parasitic Worms Collection, Natural History Museum of London (NHML).

Holotype NHML 1997.7.16.10.

Paratypes NHML 1997.7.16.11–14.

Original description Tinsley and Jackson (1998).

Taxonomic note The original name *Protopolystoma orientalis* was corrected for gender.

Other taxonomic contributions None.

Etymology Unknown.

Type locality Ndumu, South Africa.

Other localities Tanzania and Zimbabwe.

Type host *Xenopus muelleri*, Yellow-bellied Platanna (eastern form).

Additional hosts None.

Host geographical distribution *Xenopus muelleri* occurs in South Africa, Ghana, Zambia, Cameroon, Angola, Congo, Nigeria, South Sudan, Zimbabwe, Tanzania,

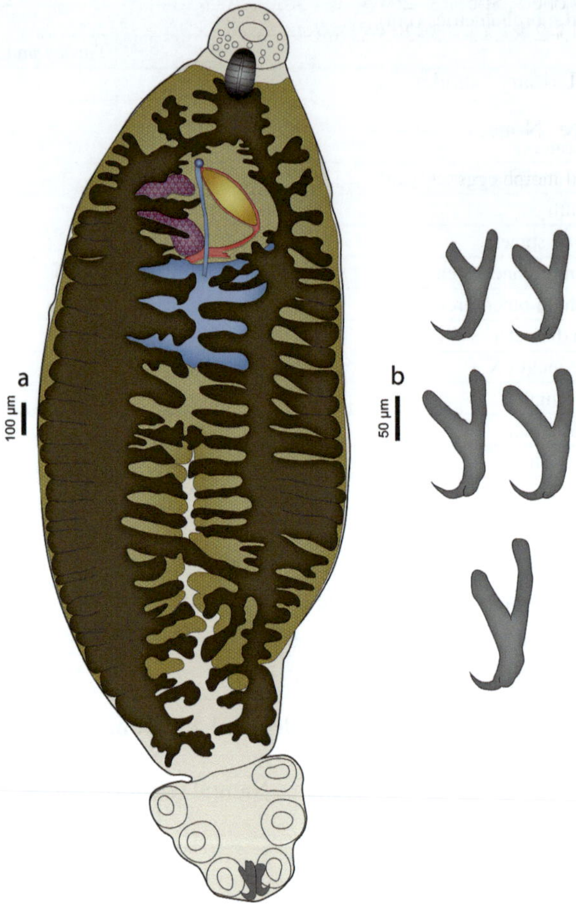

Fig. 4.124 (**a**) *Protopolystoma orientale*, (**b**) hamuli. Redrawn from Tinsley and Jackson (1998)

Namibia, Botswana, Benin, Mozambique, Kenya, Swaziland, Malawi, Uganda and Togo. This primarily aquatic species inhabits ponds, rivers and streams in altered habitats (Frost 2021).

Host conservation status The Yellow-bellied Platanna has a large population, a wide distribution and is tolerant for a wide range of habitats. It is listed as Least Concern by The IUCN Red List of Threatened Species2017: e. T107580991A107581762. https://doi.org/10.2305/IUCN.UK.2017-2.RLTS.T107580991A107581762.en. Accessed on 01 February 2022.

Site on host Urinary bladder.

DNA sequence

- Cytochrome c Oxidase I, 306 bp. GenBank Accession: EF380003

Morphology and morphometrics (μm):	
Source	Tinsley and Jackson (1998)
Sample size	
Body length	3340
Greatest width	1460
Width at vagina	
Haptor length	580
Haptor width	550
False oral sucker width	
Pharynx length	
Pharynx width	148
Ovary length	
Ovary width	
Egg length	
Egg width	
Egg incubation time	
Number of intra-uterine eggs	
Genital bulb width	37
Number of genital spines	
Genital spine length inner circle	16–19
Genital spine length inner circle	12–13
Haptoral sucker diameter	172
Hamulus handle length X	126
Hamulus guard length Y	
Hamulus hook length	32–49 (41)
Marginal hooklet length	12
Haptoral L/Body L	0.173
No of anastomoses	0–2

Remarks *Protopolystoma orientale* was originally described as *Protopolystoma orientalis*.

4.4.12.4 *Protopolystoma ramulosum* Tinsley and Jackson, 1998 (Fig. 4.125)

Collection Parasitic Worms Collection, Natural History Museum, London (NHML).

Holotype NHML 1997.7.14.11.

Paratypes NHML 1997.7.14.12–15.

Original description Tinsley and Jackson (1998).

Taxonomic note The original *Protopolystoma ramulosus* was corrected for gender.

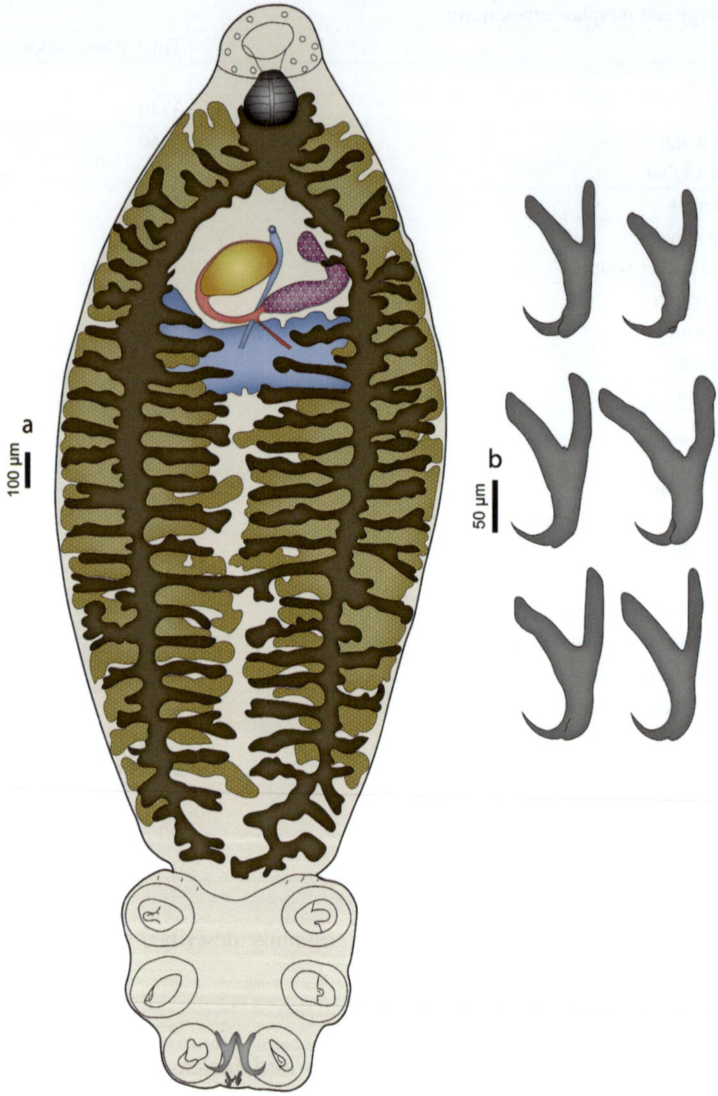

Fig. 4.125 (**a**) *Protopolystoma ramulosum*, (**b**) hamuli. Redrawn from Tinsley and Jackson (1998)

Other taxonomic contributions None.

Etymology Unknown.

Type locality Ebisha, Democratic Republic of the Congo.

Other localities None.

Type host *Xenopus fraseri*, Fraser's Clawed Frog.

Additional hosts None.

Host geographical distribution *Xenopus fraseri* occurs in Congo, Gabon, Angola, Equatorial Guinea and Cameroon. This primarily aquatic species can be found in pools in lowland rainforests (Tinsley et al. 2004).

Host conservation status The Fraser's Clawed Frog has a large population, a wide distribution and is tolerant for habitat modification. It is listed as Least Concern by The IUCN Red List of Threatened Species2017: e.T89257302A18397804. https://doi.org/10.2305/IUCN.UK.2017-2.RLTS.T89257302A18397804.en. Accessed on 01 February 2022.

Site on host Urinary bladder.

DNA sequence None.

Morphology and morphometrics (μm):	
Source	Tinsley and Jackson (1998)
Sample size	12
Body length	580–3390 (3390)
Greatest width	280–1670 (1370)
Width at vagina	
Haptor length	490–790 (620)
Haptor width	690–940 (730)
False oral sucker width	
Pharynx length	
Pharynx width	66–186 (168)
Ovary length	
Ovary width	
Egg length	
Egg width	
Egg incubation time	
Number of intra-uterine eggs	
Genital bulb width	24
Number of genital spines	
Genital spine length inner circle	17–21
Genital spine length outer circle	12–14
Haptoral sucker diameter	67–240 (197)
Hamulus handle length X	87–218 (144)
Hamulus guard length Y	
Hamulus hook length	46–48 (47)
Marginal hooklet length	12–13 (13)
Haptoral L/Body L	0.23–0.844 (0.182)
No of anastomoses	1–2

Remarks *Protopolystoma ramulosum* was originally described as *Protopolystoma ramulosus*.

4.4.12.5 *Protopolystoma simplicis* Tinsley and Jackson, 1998 (Fig. 4.126)

Collection Parasitic Worms Collection, Natural History Museum, London (NHML).

Holotype NHML 1997.7.16.1.

Paratypes NHML 1997.7.16.2–4.

Original description Tinsley and Jackson (1998).

Other taxonomic contributions Tinsley (1973).

Etymology Unknown.

Fig. 4.126 (**a**) *Protopolystoma simplicis*, (**b**) hamuli. Redrawn from Tinsley and Jackson (1998)

Type locality Cyamudongo forest, Rwanda.

Other localities Burundi, Democratic Republic of the Congo and Uganda.

Type host *Xenopus wittei*, Witte's Clawed Frog.

Additional hosts *Xenopus victorianus* Ahl, 1924 in Democratic Republic of the Congo, Kenya, Rwanda and Uganda, *Xenopus vestitus* Laurent, 1972 in Uganda, *Xenopus poweri,* Hewitt 1927 in Democratic Republic of the Congo and *Xenopus* sp. in Kenya.

Host geographical distribution *Xenopus wittei*, which is water dependent, occurs in the Democratic Republic of the Congo, Uganda and Rwanda (Frost 2021).

Host conservation status The Witte's Clawed Frog has a large population, a wide distribution and is tolerant for habitat modification. It is listed as Least Concern by The IUCN Red List of Threatened Species 2013: e.T58182A18398471. https://doi.org/10.2305/IUCN.UK.2013-2.RLTS.T58182A18398471.en. Accessed on 01 February 2022.

Site on host Urinary bladder.

DNA sequence
- Cytochrome c Oxidase I, 306 bp. GenBank Accessions: EF380000 to EF380002

Morphology and morphometrics (μm):	
Source	Tinsley and Jackson (1998)
Sample size	
Body length	2600
Greatest width	1570
Width at vagina	
Haptor length	580
Haptor width	810
False oral sucker width	
Pharynx length	
Pharynx width	174
Ovary length	
Ovary width	
Egg length	
Egg width	
Egg incubation time	
Number of intra-uterine eggs	
Genital bulb width	31
Number of genital spines	
Genital spine length inner circle	15–22
Genital spine length outer circle	12–15
Haptoral sucker diameter	161
Hamulus handle length X	151

(continued)

Morphology and morphometrics (μm):	
Source	Tinsley and Jackson (1998)
Hamulus guard length Y	
Hamulus hook length	32–45 (41)
Marginal hooklet length	12
Haptoral L/Body L	0.22
No of anastomoses	0–3

Remarks None.

4.4.12.6 *Protopolystoma xenopodis* (Price, 1943) (Fig. 4.127)

Collection United States National Museum (USNM); Natural History Museum of London (NHML).

Holotype USNM 1344766.

Paratypes NHML 2007.1.31.1.

Original description Price (1943).

Other taxonomic contributions Williams (1960a); Beverley-Burton (1962); Pritchard (1964) Maeder et al. (1970); Macnae et al. (1973); Tinsley and Owen (1975); Tinsley and Jackson (1998).

Etymology Named after the host.

Type locality Cape Town, South Africa.

Other localities Zimbabwe.

Type host *Xenopus laevis* (Daudin, 1802), Common Platanna.

Additional hosts *Xenopus victorianus* in Kenya and Rwanda and *Xenopus poweri* in Cameroon and the Democratic Republic of the Congo.

Host geographical distribution *Xenopus laevis* occurs in South Africa, Congo, Botswana, Angola, Cameroon, Namibia, Zimbabwe, Malawi, Lesotho and Mozambique, where it is found in ponds and streams. Among regions where it has been introduced, this species occurs in Chile, France and Italy (Tinsley et al. 2009).

Host conservation status The Common Platanna has a large population, a wide distribution and is tolerant for a wide range of habitats. It is listed as Least Concern by The IUCN Red List of Threatened Species 2020: e.T110466172A3066881. https://doi.org/10.2305/IUCN.UK.2020-3.RLTS.T110466172A3066881.en. Accessed on 01 February 2022.

Site on host Urinary bladder.

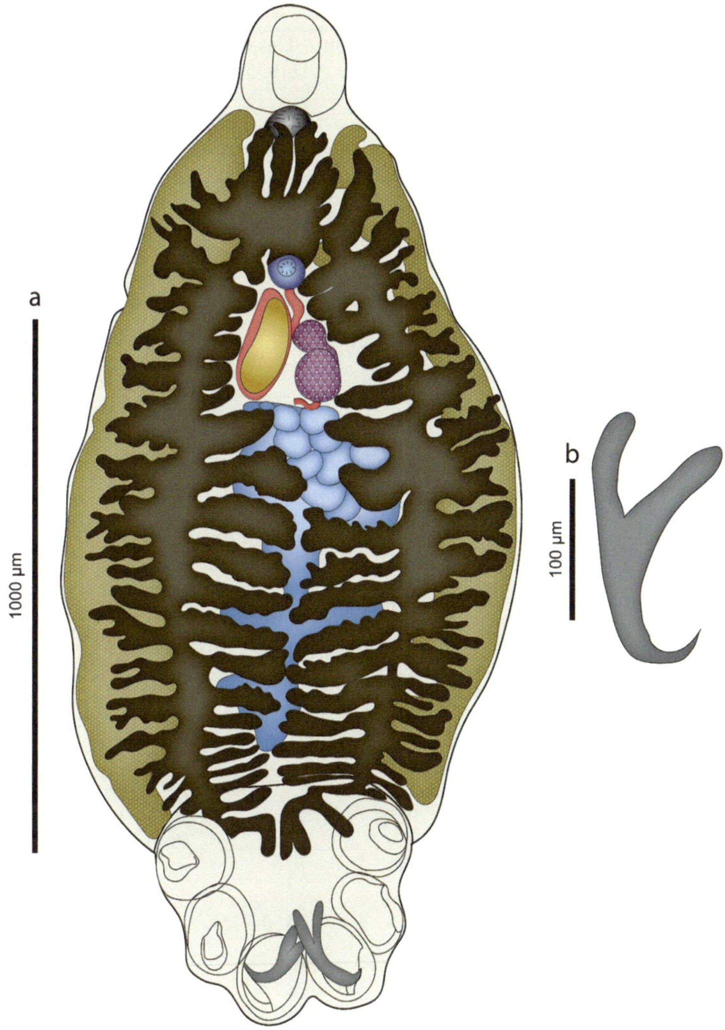

Fig. 4.127 (**a**) *Protopolystoma xenopodis*, (**b**) hamuli. Redrawn from Price (1943)

DNA sequence

- 18S rRNA gene, 1991 bp. GenBank Accession: AM051078
- 28S rRNA gene, 1701 bp. GenBank Accession: AM157218
- 12S rRNA gene, 489 bp. GenBank Accession: KR856096
- Cytochrome c Oxidase I, 367 bp. GenBank Accession: KR856172
- Cytochrome c Oxidase I, 306 bp. GenBank Accessions: EF380004 to EF380008
- Dfd/Hox4 gene, 75 bp. GenBank Accession: FN298160
- Lab/Hox1 gene, 75 bp. GenBank Accession: FN298148

Morphology and morphometrics (μm):				
Source	Price (1943)	Beverley-Burton (1962)	Tinsley and Owen (1975)	Tinsley and Jackson (1998)
Sample size		5		
Body length	2120	1500–1980		790–2460
Greatest width	870	390–560		520–1670
Width at vagina				
Haptor length	420	370–510		270–470
Haptor width	530			340–800
False oral sucker width		110–170		
Pharynx length		90–120		
Pharynx width	105	90		76–145
Ovary length		90		
Ovary width		50		
Egg length	235			230–280
Egg width	95			95–135
Egg incubation time			20–24 days	
Intra-uterine eggs (n)	1		1	1
Genital bulb width	72	300–380		
Genital bulb to anterior	500			
Number of genital spines	14	16		16
Genital spine length inner circle		14		16–30
Genital spine length outer circle				
Haptoral sucker diameter	150	120–180		94–154
Hamulus handle length X	162	161	150–175	113–166
Hamulus guard length Y				
Hamulus hook length				
Marginal hooklet length			13–14	11–13
Haptoral L/Body L		0.246–0.257		
No of anastomoses		3		0–4

Remarks *Protopolystoma xenopodis* was originally described as *Polystoma xenopi* Price, 1943.

4.4.13 *Pseudodiplorchis* Yamaguti, 1963

4.4.13.1 *Pseudodiplorchis americanus* (Rodgers and Kuntz, 1940) (Fig. 4.128)

Collection United States National Museum (USNM).

Holotype and paratypes USNM 1337427.

Original description Rodgers and Kuntz (1940).

Other taxonomic contributions Yamaguti (1963); Lamothe–Argumedo (1985); Tinsley and Earle (1983).

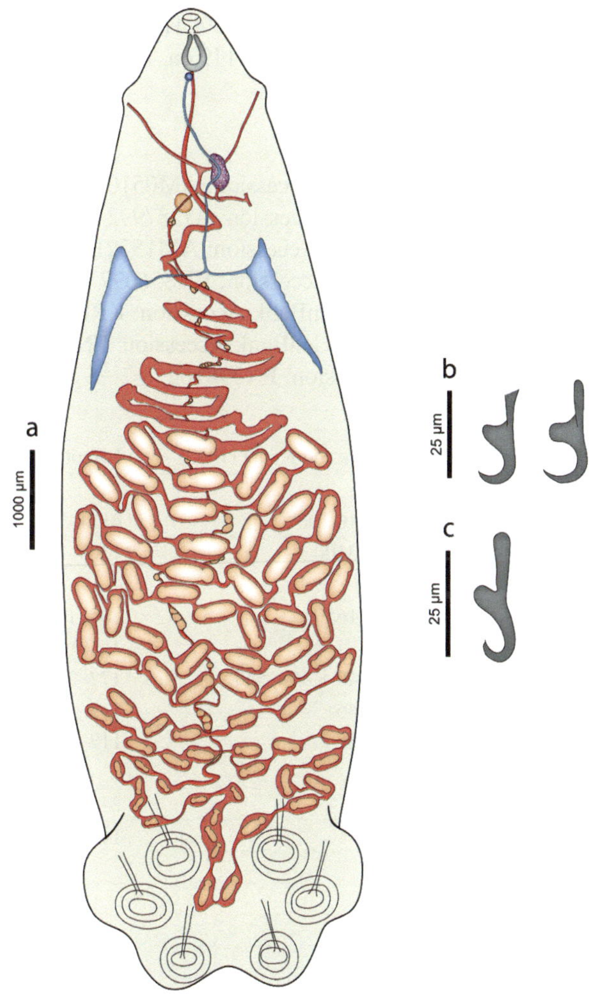

Fig. 4.128 (**a**) *Pseudodiplorchis americanus*, (**b**) marginal hooklets (adult), (**c**) marginal hooklets (embryo). Redrawn from Rodgers and Kuntz (1940)

Etymology Named after the type locality country.

Type locality Lawton, Oklahoma, USA.

Other localities Mexico.

Type host *Scaphiopus couchii* Baird, 1854, Couch's Spadefoot.

Additional hosts None.

Host geographical distribution *Scaphiopus couchii* occurs in the United States as well as in Mexico where it usually lives in forests, savannah, cultivated areas, short grass plains, creosote bush desert and shrubland habitats (Frost 2021).

Host conservation status The Couch's Spadefoot has a large population, a wide distribution and is tolerant for various habitats. It is listed as Least Concern by The IUCN Red List of Threatened Species 2004: e.T59041A11874119. https://doi.org/10.2305/IUCN.UK.2004.RLTS.T59041A11874119.en. Accessed on 10 May 2022.

Site on host Urinary bladder.

DNA sequence
- 18S rRNA gene, 2020 bp. GenBank Accession: AM051079
- 18S rRNA gene, 573 bp. GenBank Accession: AJ287992
- 28S rRNA gene, 1397 bp. GenBank Accession: AM157219
- 12S rRNA gene, 463 bp. GenBank Accession: KR856097
- Cytochrome c Oxidase I, 368 bp. GenBank Accession: KR856173
- Abd-A/Lox4 gene, allele A, 75 bp. GenBank Accession: FN298168
- Hox5 gene, 75 bp. GenBank Accession: FN298162

Morphology and morphometrics (μm):		
Source	Rodgers and Kuntz (1940)	Lamothe-Argumedo (1985)
Sample size	2	
Body length	8000–10,100	4121–7518
Greatest width	2900–3100	1304–2720
Width at vagina		
Haptor length	1260–1650	805–1493
Haptor width		1.320–2.447
False oral sucker width	265–340 × 400–546	177–257 × 305–434
Pharynx length	233–246	117–305
Pharynx width	133–166	112–161
Number of testes	2	
Ovary length	226	241–386
Ovary width	133	144–209
Egg length		
Egg width		
Largest embryo length	500–600	
Largest embryo width	215–200	
Egg incubation time		

(continued)

Morphology and morphometrics (μm):		
Source	Rodgers and Kuntz (1940)	Lamothe-Argumedo (1985)
Number of intra-uterine eggs		
Genital bulb width	133	10–12
Number of genital spines	6	
Genital spine length		18–20
Haptoral sucker diameter	(500)	322–418 × 322–434
Hamulus handle length X		
Hamulus guard length Y		
Hamulus hook length		
Marginal hooklet length	30	30–33
Haptoral L/Body L		0.20
No of anastomoses		0–9

Remarks *Pseudodiplorchis americanus* was originally described as *Diplorchis americanus* Rodgers and Kuntz, 1940.

4.4.14 *Riojatrema* Lamothe-Argumedo, 1963

4.4.14.1 *Riojatrema bravoae* Lamothe-Argumedo, 1963 (Fig. 4.129)

Collection Helminthological collection in the Biology Institute, in the National University of Mexico.

Holotype and paratypes 219–3.

Original description Lamothe-Argumedo (1963).

Other taxonomic contributions None.

Etymology Unknown.

Type locality Morelos, Mexico.

Other localities None.

Type host *Eleutherodactylus nitidus* (Peters,1870), Shiny Peeping Frog and *Rhinella spinulosa* (Wiegmann, 1834), Warty Toad. In the original description of the polystome, the hosts were known as *Tomodactylus amulae* Günther, 1900 and *Bufo simus* Schmidt, 1857, respectively.

Host geographical distribution *Eleutherodactylus nitidus* is found in Southern Tlaxcala, southern Puebla, Morelos and southwestern México into northern Guerrero, southern Puebla and the highlands of Oaxaca, Mexico (Frost 2021). *Rhinella spinulosa* occurs in Argentina, Chile, Bolivia and Peru where it is most commonly found in grasslands and scrublands (Frost 2021).

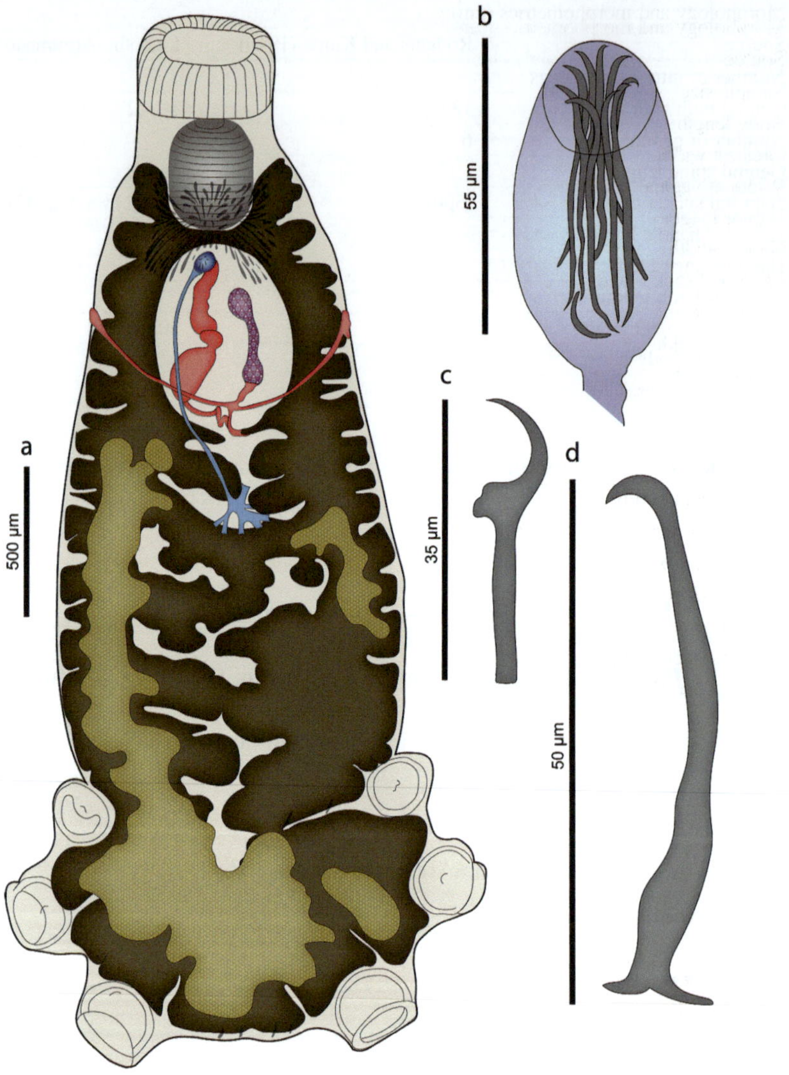

Fig. 4.129 (**a**) *Riojatrema bravoae*, (**b**) genital bulb, (**c**) marginal hooklets, (**d**) genital spines. Redrawn from Lamothe-Argumedo (1963)

Host conservation status The Shiny Peeping Frog has a small extent of occurrence (EOO) of less than 5000 km^2, is found in only four threat-defined locations and is in decline. It is listed as Endangered by The IUCN Red List of Threatened Species 2020: e.T55307A53951846. https://doi.org/10.2305/IUCN.UK.2020-3.RLTS.T55307A53951846.en. Accessed on 01 February 2022. The Warty Toad is listed as Least Concern by The IUCN Red List of Threatened Species 2020: e.T88992858A101436432. https://doi.org/10.2305/IUCN.UK.2020-1.RLTS.T88992858A101436432.en. Accessed on 10 May 2022.

Site on host Urinary bladder.

DNA sequence None.

Morphology and morphometrics (μm):	
Source	Lamothe-Argumedo (1963)
Sample size	3
Body length	1616–8790
Greatest width	1017–3703
Width at vagina	
Haptor length	
Haptor width	
False oral sucker width	145–209 × 265–450
Pharynx length	330–740
Pharynx width	288–515
Ovary length	262–1094
Ovary width	82–515
Egg length	
Egg width	
Egg incubation time	
Number of intra-uterine eggs	
Genital bulb width	
Number of genital spines	
Genital spine length	
Haptoral sucker diameter	242–483
Hamulus handle length X	
Hamulus guard length Y	
Hamulus hook length	
Marginal hooklet length	
Haptoral L/Body L	
No of anastomoses	0–4

Remarks None.

4.4.14.2 *Riojatrema cerrocoloradense* (Nasir and Fuentes-Zambrano, 1983) n. comb. (Fig. 4.130)

Collection Specimens deposited in the National Parasite collection in the USA.

Holotype 75600.

Paratypes Unknown.

Original description Nasir and Fuentes-Zambrano (1983).

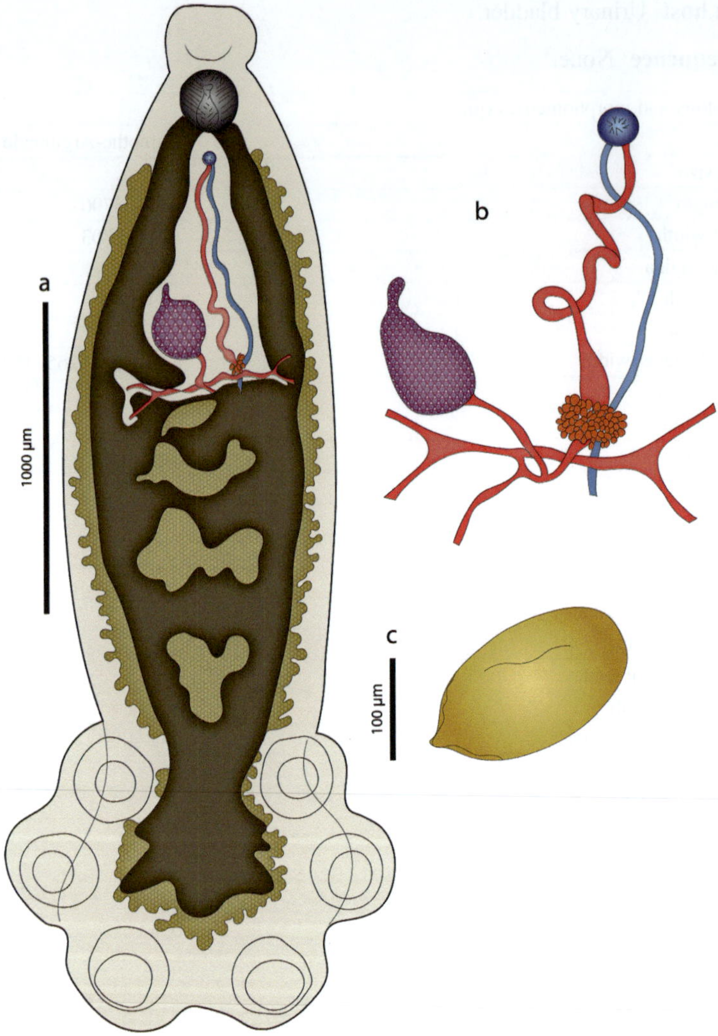

Fig. 4.130 (**a**) *Riojatrema cerrocoloradense*, (**b**) reproductive system, (**c**) egg. Redrawn from Nasir and Fuentes Zambrano (1983)

Taxonomic note This species was originally described as *Parapseudopolystoma cerrocoloradensis* Nasir and Fuentes Zambrano, 1983. We regard it as *Riojatrema cerrocoloradense* (Nasir and Fuentes Zambrano, 1983) n. comb. as explained earlier. Note that the original name *Parapseudopolystoma cerrocoloradensis* was also corrected for gender.

Other taxonomic contributions None.

Etymology Named after the type locality.

Type locality Cerro, Colorado and Cumana, Venezuela.

Other localities None.

Type host *Rhinella marina* (Linnaeus, 1758), Marine Toad. In the original description of the polystome, the host was known as *Bufo marinus*.

Additional hosts None.

Host geographical distribution *Rhinella marina* can be found in various countries including Venezuela, Haiti, Costa Rica, Brazil, El Salvador, Bolivia, Peru, Ecuador, Panama, Mexico and Colombia. This frog has also been introduced into a number of other regions such as the United States, Jamaica, Saint Lucia, Japan, Puerto Rico, Virgin Islands, Guam, Taiwan and Province of China. It inhabits well-watered gardens, cane fields, forest and savanna regions. This species is generally found in disturbed and/or open habitats (Solís et al. 2009).

Host conservation status The Marine Toad has a large population, a wide distribution and can tolerate various range of habitats. It is thus listed as Least Concern by The IUCN Red List of Threatened Species 2009: e.T41065A10382424. https://doi.org/10.2305/IUCN.UK.2009-2.RLTS.T41065A10382424.en. Accessed on 30 January 2022.

Site on host Urinary bladder.

DNA sequence None.

Morphology and morphometrics (μm):	
Source	Nasir and Fuentes-Zambrano (1983)
Sample size	3
Body length	2835–4520
Greatest width	770–1575
Width at vagina	
Haptor length	980–1575
Haptor width	1190–2176
False oral sucker width	
Pharynx length	164–297
Pharynx width	164–245
Ovary length	
Ovary width	
Egg length	
Egg width	172 × 90
Egg incubation time	
Number of intra-uterine eggs	
Genital bulb width	
Number of genital spines	8–9
Genital spine length	12
Haptoral sucker diameter	251–406
Hamulus handle length X	

(continued)

Morphology and morphometrics (μm):	
Source	Nasir and Fuentes-Zambrano (1983)
Hamulus guard length Y	
Hamulus hook length	
Marginal hooklet length	
Haptoral L/Body L	0.35
No of anastomoses	5

Remarks *Riojatrema cerrocoloradense* was originally described as *Parapseudopolystoma cerrocoloradensis* Nasir and Fuentes-Zambrano, 1983.

4.4.14.3 *Riojatrema ecuadorense* Dyer, 1985 (Fig. 4.131)

Collection United States National Museum.

Holotype USNM 1373838.

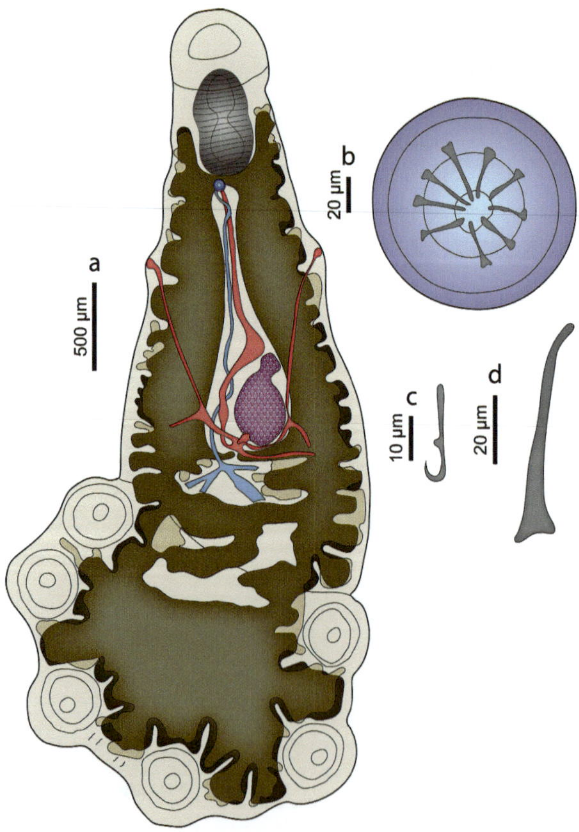

Fig. 4.131 (**a**) *Riojatrema ecuadorense*, (**b**) genital bulb, (**c**) marginal hooklets, (**d**) genital spines. Redrawn from Dyer (1985)

Paratypes USNM 1373839.

Original description Dyer (1985).

Taxonomic note The original name *Riojatrema ecuadorensis* was corrected for gender.

Other taxonomic contributions Vaucher (1987).

Etymology Named after the type locality country.

Type locality Ecuador.

Other localities None.

Type host *Rhinella margaritifera* (Laurenti, 1768), South American Common Toad. In the original description of the polystome, the host was known as *Bufo typhonius* Schneider, 1799.

Additional hosts None.

Host geographical distribution *Rhinella margaritifera* occurs in Colombia, Panama, Bolivia, Venezuela, Peru, Ecuador, Brazil, Guyana, French Guiana and Suriname. This species inhabits open regions and forest habitats (Solís et al. 2010).

Host conservation status The South American Common Toad has a large population, a wide distribution and is tolerant for various habitat ranges. It is listed as Least Concern by The IUCN Red List of Threatened Species 2010: e. T54701A11185916. https://doi.org/10.2305/IUCN.UK.2010-2.RLTS.T54701A11185916.en. Accessed on 01 February 2022.

Site on host Urinary bladder.

DNA sequence None.

Morphology and morphometrics (μm):		
Source	Dyer (1985)	Vaucher (1987)
Sample size	3	2
Body length	3450–4720 (4250)	1980
Greatest width	1100–1530 (1370)	1110
Width at vagina		
Haptor length	900–1910 (1400)	1150
Haptor width	1580–2300 (1790)	990–1390
False oral sucker width	400–650 (537)	393 × 205
Pharynx length	470–600 (520)	303–442
Pharynx width	300–340 (307)	270–295
Ovary length	490–690 (590)	
Ovary width	280–390 (350)	
Egg length		
Egg width		
Egg incubation time		

(continued)

Morphology and morphometrics (μm):		
Source	Dyer (1985)	Vaucher (1987)
Number of intra-uterine eggs		
Genital bulb width	130–160 (133)	
Number of genital spines	8–9	
Genital spine length	25–64 (49)	
Haptoral sucker diameter	330–430 (393)	344–390 × 286–368
Hamulus handle length X		
Hamulus guard length Y		
Hamulus hook length		
Marginal hooklet length	23–24	28
Haptoral L/Body L	0.33	0.58
No of anastomoses	0–2	

Remarks *Riojatrema ecuadorense* was originally described as *Riojatrema ecuadorensis*. This species has been reported from the same host as that of *Wetapolystoma almae* Gray, 1993 in Peru (Gray 1993).

4.4.15 *Sundapolystoma* Lim and Du Preez, 2001

4.4.15.1 *Sundapolystoma chalconotae* Lim and Du Preez, 2001 (Fig. 4.132)

Collection Parasitic Worms Collection, Natural History Museum, London (NHML). Parasitic Worm Collection, National Museum, Bloemfontein, South Africa (NMBP).

Holotype NHML 2000.11.29.1.

Paratypes NHML 2000.11.29.2–3; NMBP 246–249.

Original description Lim and Du Preez (2001).

Other taxonomic contributions None.

Etymology Named after the host.

Type locality University of Malaya field station, Gombak, Selangor, Malaysia.

Other localities None.

Type host *Chalcorana chalconota* (Schlegel, 1837), Schlegel's Frog. In the original description of the polystome, the host was known as *Rana chalconota* Boulenger, 1882.

Additional hosts None.

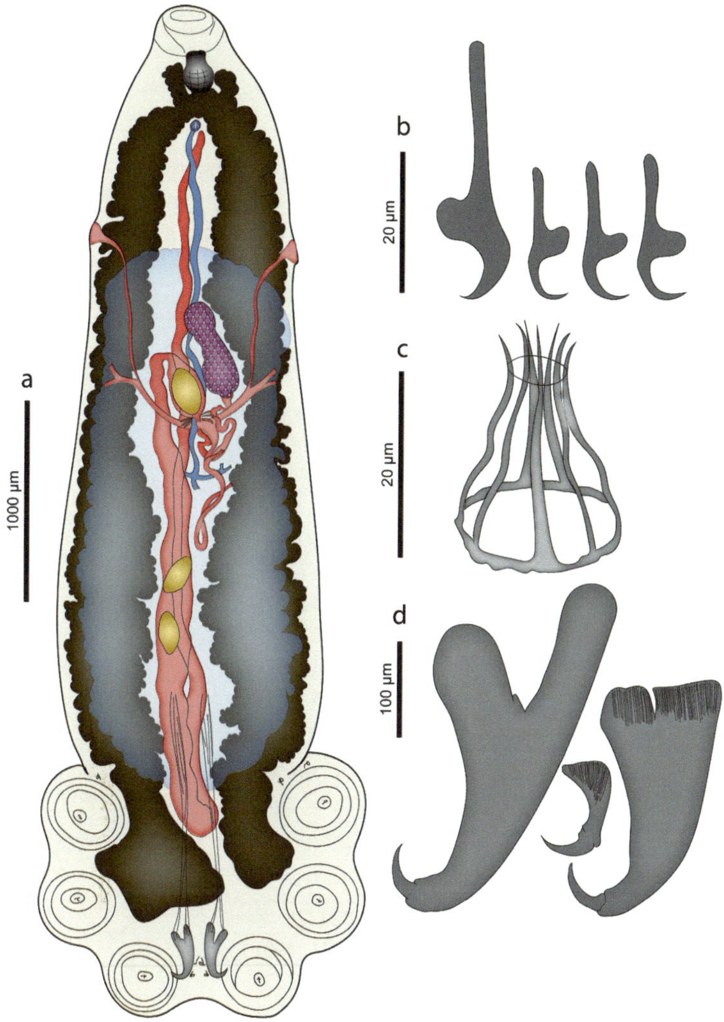

Fig. 4.132 (**a**) *Sundapolystoma chalconotae*, (**b**) marginal hooklets, (**c**) genital spines, (d) hamuli. Redrawn from Lim and Du Preez (2001)

Host geographical distribution *Chalcorana chalconota* occurs in Indonesia, i.e., Java, Bali and Sumatra where it can be found on vegetation and rocks in lowland forest streams and highland areas (Van Dijk et al. 2009).

Host conservation status The Schlegel's Frog has a large population, a wide distribution and is tolerant for habitat modification. It is listed as Least Concern by The IUCN Red List of Threatened Species 2018: e.T78934311A78934546. https://

doi.org/10.2305/IUCN.UK.2018-2.RLTS.T78934311A78934546.en. Accessed on 01 February 2022.

Site on host Urinary bladder.

DNA sequence

- 18S rRNA gene, 2093 bp. GenBank Accession: AM051080
- 18S rRNA gene, 625 bp. GenBank Accession: AJ297778
- 28S rRNA gene, 1453 bp. GenBank Accession: KR856161
- ITS1, 768 bp. GenBank Accession: AJ301696

Morphology and morphometrics (μm):	
Source	Lim and Du Preez (2001)
Sample size	11
Body length	4900–7868 (6459)
Greatest width	1260–2156 (1727)
Width at vagina	
Haptor length	1232–2268 (1815)
Haptor width	1568–2660 (2235)
False oral sucker width	238–406 (327) × 190–280 (239)
Pharynx length	168–224 (217)
Pharynx width	142–202 (179)
Ovary length	294–454 (402)
Ovary width	117–185 (151)
Egg length	197–269 (237)
Egg width	117–128 (120)
Egg incubation time	
Number of intra-uterine eggs	150
Genital bulb width	
Number of genital spines	7–8
Genital spine length	19–22 (27)
Haptoral sucker diameter	308–671 (490)
Hamulus handle length X	275–396 (337)
Hamulus guard length Y	231–330 (284)
Hamulus hook length	50–56 (54)
Marginal hooklet length	C1: 36–42 (39); C2: 21–23 (22); C3–8:19–20 (20)
Haptoral L/Body L	0.28
No of anastomoses	0

Remarks None.

4.4.15.2 *Sundapolystoma crooki* (Vande Vusse, 1976) (Fig. 4.133)

Collection USNM Helminthological Collection.

Holotype 73504.

Paratypes 73505.

Original description Vande Vusse (1976).

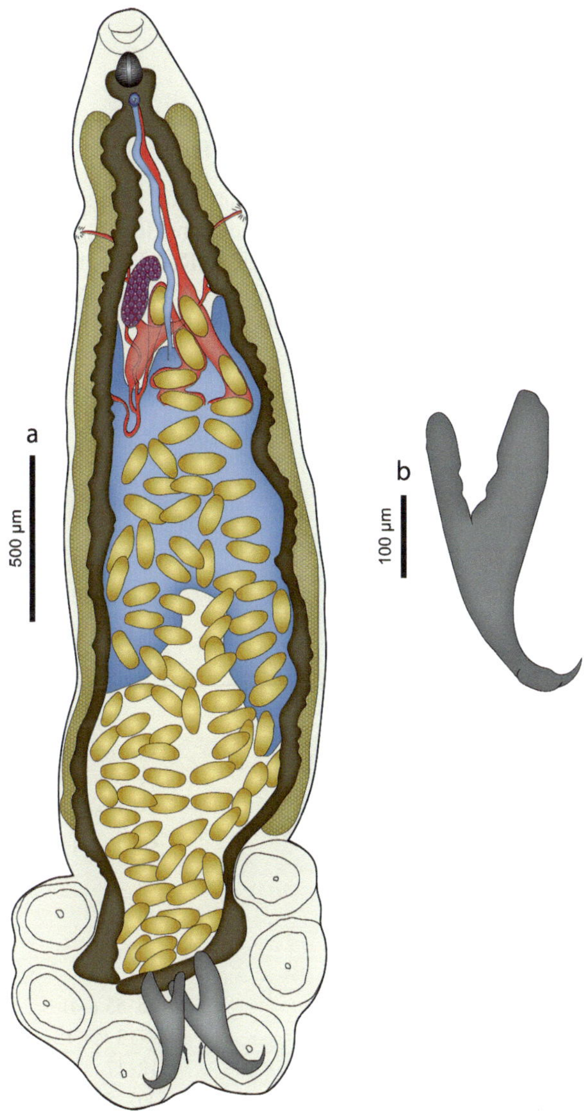

Fig. 4.133 (**a**) *Sundapolystoma crooki*, (**b**) hamuli. Redrawn from Vande Vusse (1976)

Other taxonomic contributions None.

Etymology Named after Dr Philip G. Crook.

Type locality Maite Creek, Valencia, Negros Island, Philippines.

Other localities Ocoy, Tejero and Maloh rivers and Lake Balinsasayao, Negros Oriental, Philippines.

Type host *Limnonectes magnus* (Stejneger, 1910), Mount Arpo Wart Frog. In the original description of the polystome, the host was known as *Rana magna* Stejneger, 1910.

Additional hosts None.

Host geographical distribution *Limnonectes magnus* occurs in the Philippines and in Indonesia where it inhabits streams and rivers (Frost 2021).

Host conservation status The Mount Arpo Wart Frog is in serious decline due to over-harvesting for food and habitat loss, agriculture, logging and pollution. It is currently listed as Near Threatened by The IUCN Red List of Threatened Species 2020: e.T58353A176620668. https://doi.org/10.2305/IUCN.UK.2020-3.RLTS.T58353A176620668.en. Accessed on 01 February 2022.

Site on host Urinary bladder.

DNA sequence None.

Morphology and morphometrics (μm):	
Source	Vande Vusse (1976)
Sample size	20
Body length	1700–7500 (3700)
Greatest width	500–2200 (1100)
Width at vagina	
Haptor length	600–1485 (953)
Haptor width	679–1943 (1127)
False oral sucker width	117–211 (156) × 156–350 (203)
Pharynx length	85–163 (109)
Pharynx width	75–177 (113)
Ovary length	130–422 (260)
Ovary width	50–206 (114)
Egg length	60
Egg width	38
Egg incubation time	
Number of intra-uterine eggs	
Genital bulb width	26–55 (44)
Number of genital spines	9–13 (11,12)
Genital spine length	12–17
Haptoral sucker diameter	207–559 (265)

(continued)

Morphology and morphometrics (μm):	
Source	Vande Vusse (1976)
Hamulus handle length X	329–512 (401)
Hamulus guard length Y	
Hamulus hook length	
Marginal hooklet length	
Haptoral L/Body L	0.257
No of anastomoses	0

Remarks *Sundapolystoma crooki* was originally described as *Parapolystoma crooki* Vande Vusse, 1976.

4.4.16 Wetapolystoma Gray, 1993

4.4.16.1 *Wetapolystoma almae* Gray, 1993 (Fig. 4.134)

Collection United States National Museum (USNM).

Holotype USNM 82387 (specimen lost).

Paratypes Unknown.

Original description Gray (1993).

Other taxonomic contributions None.

Etymology Named after Mrs. Alma (Weta) Gray, USA.

Type locality Peru Departamento de Madre de Dios, Provincia Tam-bopata, Cusco Amazonico, Rio Madre de Dios, ca. 15 km E Puerto Mal-donado.

Other localities None.

Type host *Rhinella margaritifera*, South American Common Toad. In the original description of the polystome, the host was known as *Bufo typhonius*.

Additional hosts None.

Host geographical distribution *Rhinella margaritifera* occurs in Colombia, Panama, Bolivia, Venezuela, Peru, Ecuador, Brazil, Guyana, French Guiana and Suriname. This species inhabits open regions and forest habitats (Solís et al. 2010).

Host conservation status The South American Common Toad has a large population, a wide distribution and is tolerant for various habitat ranges. It is listed as Least Concern by The IUCN Red List of Threatened Species 2010: e. T54701A11185916. https://doi.org/10.2305/IUCN.UK.2010-2.RLTS.T54701A11185916.en. Accessed on 01 February 2022.

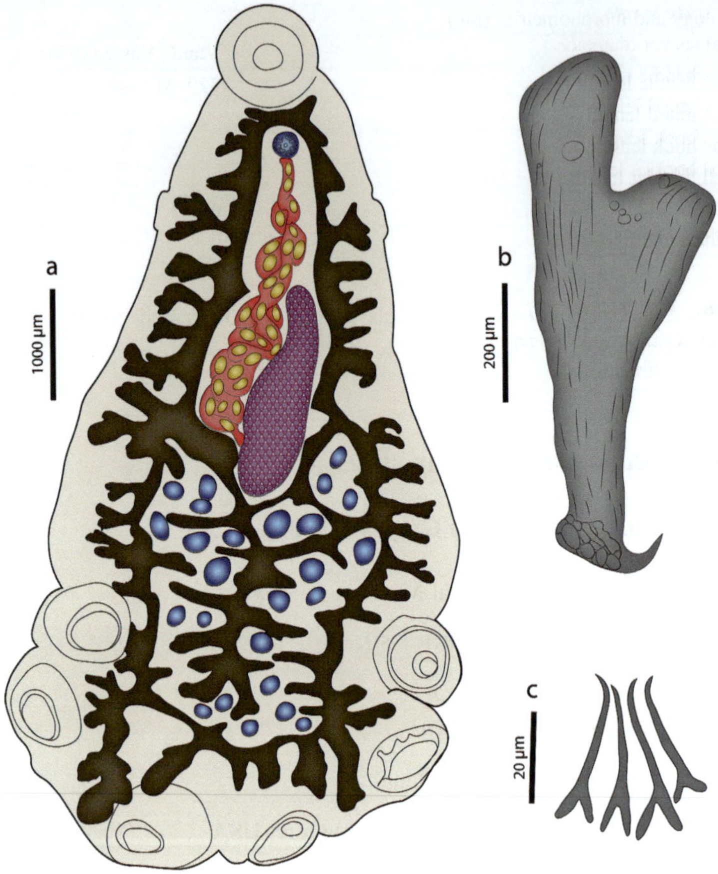

Fig. 4.134 (**a**) *Wetapolystoma almae*, (**b**) hamuli, (**c**) genital spines. Redrawn from Gray (1993)

Site on host Urinary bladder.

DNA sequence

- 18S rRNA gene, 2092 bp. GenBank Accession: AM051081
- 18S rRNA gene, 626 bp. GenBank Accession: AJ301698
- 28S rRNA gene, 1453 bp. GenBank Accession: AM157220
- 28S rRNA gene, 1450 bp. GenBank Accession: AM913876
- 28S rRNA gene, 1449 bp. GenBank Accession: AM913877
- 12S rRNA gene, 491 bp. GenBank Accession: KR856099
- Cytochrome c Oxidase I, 381 bp. GenBank Accessions: AM913866
- Cytochrome c Oxidase I, 396 bp. GenBank Accession: AM913867 and AM913868
- ITS1, 769 bp. GenBank Accession: AJ301693

Morphology and morphometrics (μm):	
Source	Gray (1993)
Sample size	1
Body length	8000
Greatest width	4000
Width at vagina	
Haptor length	2500
Haptor width	4340
False oral sucker width	
Pharynx length	
Pharynx width	276
Ovary length	1972
Ovary width	260–500
Egg length	130–182 (151.45)
Egg width	
Egg incubation time	
Number of intra-uterine eggs	>100
Genital bulb width	325
Number of genital spines	8
Genital spine length	68.25
Haptoral sucker diameter	580–840 (700)
Hamulus handle length X	1056
Hamulus guard length Y	
Hamulus hook length	
Marginal hooklet length	
Haptoral L/Body L	0.31
No of anastomoses	Network

Remarks This species has been reported from the same host as that of *Riojatrema ecuadorense* in Ecuador (Dyer 1985). The validity of this genus has been questioned in Bentz et al. (2006).

4.5 Polystomes of caecilians (Fig. 4.135)

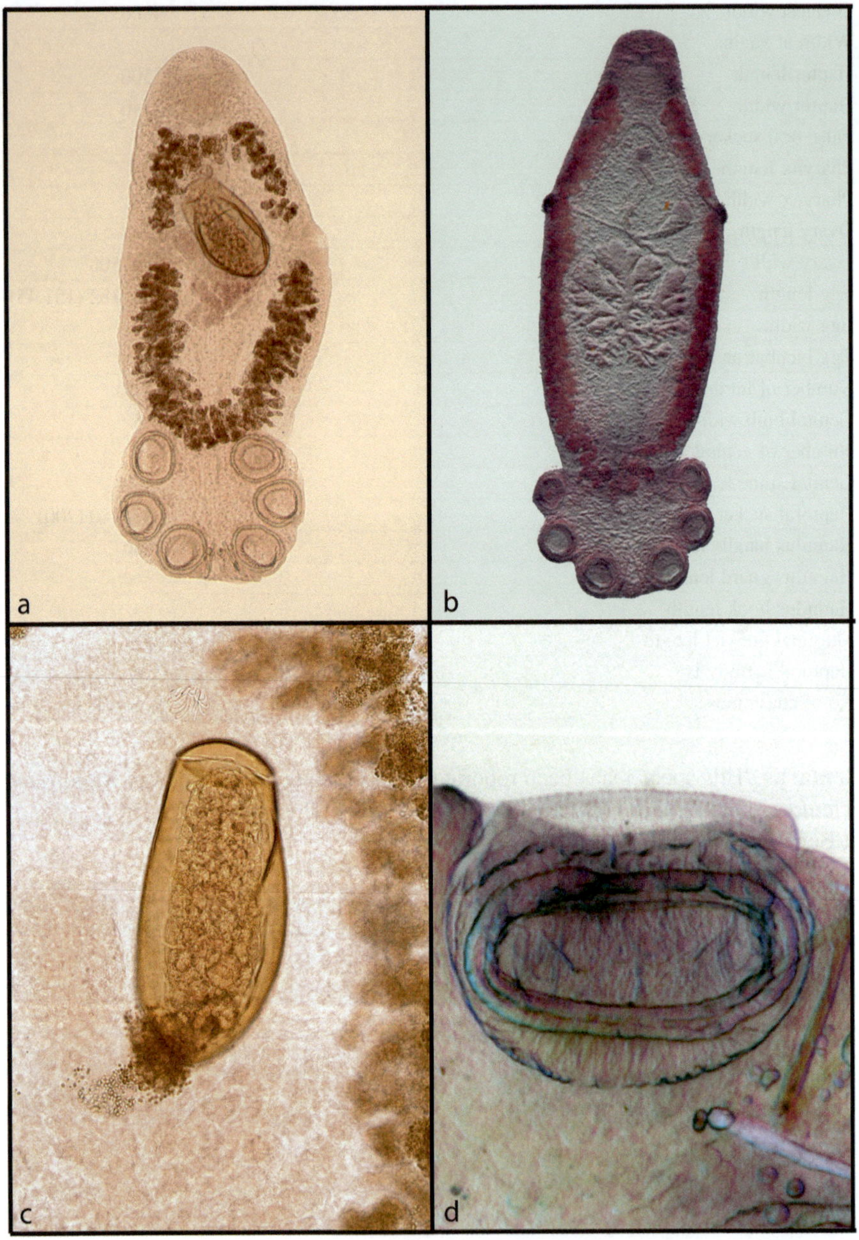

Fig. 4.135 Micrographs of (**a**) *Nanopolystoma brayi*, (**b**) *Nanopolystoma tinsleyi*, (**c**) Operculated egg of *N. brayi*, (**d**) Skeletal ring inside haptoral sucker of *N. tinsleyi*

4.5.1 Nanopolystoma Du Preez, Wilkinson and Huyse, 2008

4.5.1.1 *Nanopolystoma brayi* Du Preez, Wilkinson and Huyse, 2008 (Fig. 4.136)

Collection Parasitic Worms Collection, Natural History Museum, London (NHML).

Holotype NHML 2007.8.15.9.

Paratypes NHML 2007.8.15.10–13.

Original description Du Preez et al. (2008).

Other taxonomic contributions None.

Etymology Named after Dr Rodney Bray. Natural History Museum, London, UK.

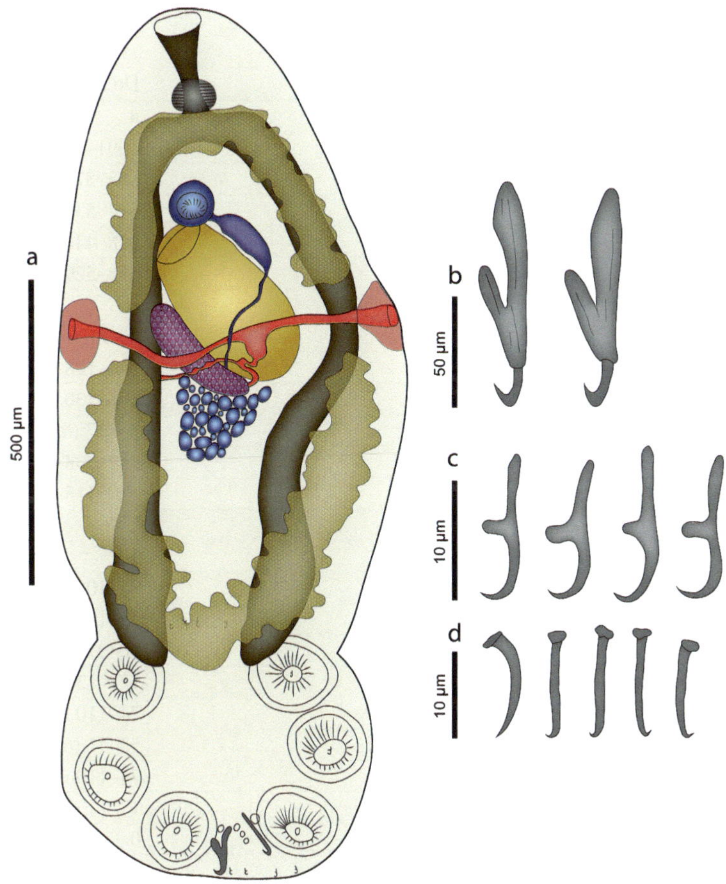

Fig. 4.136 (**a**) *Nanopolystoma brayi*, (**b**) hamuli, (**c**) marginal hooklets, (**d**) genital spines. Redrawn from Du Preez et al. (2008)

Type locality Demerara, Guyana.

Other localities None.

Type host *Caecilia gracilis* Shaw, 1802, Wormlike Caecilia.

Additional hosts None.

Site on host Urinary bladder.

Host geographical distribution *Caecilia gracilis* occurs in Brazil, northern Suriname, northern French Guiana and northeastern Peru, where it is commonly found in old growth forests and savannah (Frost 2021).

Host conservation status The Wormlike Caecilia has a large population and wide distribution. It is considered as Least Concern by The IUCN Red List of Threatened Species 2004: e.T59514A11953373. https://doi.org/10.2305/IUCN.UK.2004.RLTS.T59514A11953373.en. Accessed on 01 February 2022.

DNA sequence None.

Morphology and morphometrics:	
Source	Du Preez et al. (2008)
Sample size	4
Body length	1091–1407 (1237)
Greatest width	543–572 (555)
Width at vagina	514–534 (529)
Haptor length	398–446 (415)
Haptor width	388–529 (457)
False oral sucker width	
Pharynx length	60–86 (68)
Pharynx width	101–108 (106)
Ovary length	106–130 (119)
Ovary width	55–67 (62)
Egg length	182–238 (221)
Egg width	110–115 (114)
Egg incubation time	
Number of intra-uterine eggs	1
Genital bulb width	91–108 (97)
Number of genital spines	16–19
Genital spine length	20–22 (21)
Haptoral sucker diameter	122–139 (129)
Hamulus handle length X	74–110 (95)
Hamulus guard length Y	53–71 (63)
Hamulus hook length	74–110 (95)
Marginal hooklet length	17–21 (19.5)
Haptoral L/Body L	0.335
No of anastomoses	0

Remarks None.

4.5.1.2 *Nanopolystoma lynchi* Du Preez, Wilkinson and Huyse, 2008 (Fig. 4.137)

Collection The Parasitic Worms Collection, Natural History Museum, London (NHML).

Holotype NHML 2007.8.15.1.

Paratypes NHML 2007.8.15.2–8.

Original description Du Preez et al. (2008).

Other taxonomic contributions None.

Etymology Named after Dr John D. Lynch.

Type locality South America.

Other localities None.

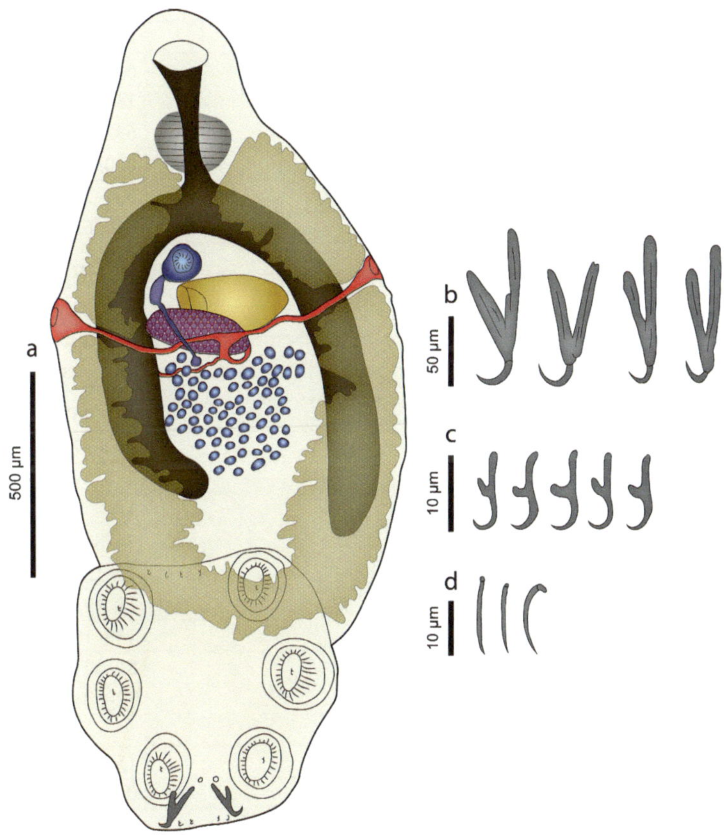

Fig. 4.137 (**a**) *Nanopolystoma lynchi*, (**b**) hamuli, (**c**) marginal hooklets, (**d**) genital spines. Redrawn from Du Preez et al. (2008)

Type host *Caecilia pachynema* Günther, 1859, Intac Caecilian.

Additional hosts None.

Taxonomy None.

Host geographical distribution *Caecilia pachynema* occurs in Colombia and Ecuador in both terrestrial and freshwater systems. It can be found in the wet tropical regions, more precisely in rivers, forests and old growth (Du Preez et al. 2008; Coloma et al. 2004b).

Host conservation status The Intac Caecilian has a fairly large distribution. It is listed as Least Concern by The IUCN Red List of Threatened Species 2019: e. T59524A49355719. https://doi.org/10.2305/IUCN.UK.2019-3.RLTS.T59524A49355719.en. Accessed on 01 February 2022.

Site on host Urinary bladder and phallodeum.

DNA sequence None.

Morphology and morphometrics (μm):	
Source	Du Preez et al. (2008)
Sample size	6
Body length	1358–2012 (1706)
Greatest width	631–737 (690)
Width at vagina	630–698 (661)
Haptor length	504–592 (561)
Haptor width	552–630 (590)
False oral sucker width	
Pharynx length	137–175 (152)
Pharynx width	156–199 (184)
Ovary length	132–204 (173)
Ovary width	65–96 (83)
Egg length	151–262 (218)
Egg width	110–173 (123)
Egg incubation time	
Number of intra-uterine eggs	1
Genital bulb width	74–84 (78)
Number of genital spines	10–12
Genital spine length	18–19 (19)
Haptoral sucker diameter	139–182 (162)
Hamulus handle length X	81–121 (104)
Hamulus guard length Y	72–82 (80)
Hamulus hook length	
Marginal hooklet length	15–19 (17)
Haptoral L/Body L	0.328
No of anastomoses	0

Remarks None.

4.5.1.3 *Nanopolystoma tinsleyi* Du Preez, Badets and Verneau, 2014 (Fig. 4.138)

Collection Parasitic Worm Collection, National Museum, Bloemfontein, South Africa (NMBP); Institute of Parasitology, Biology Centre of ASCR, Ceske Budejovice, Czech Republic (IPCAS).

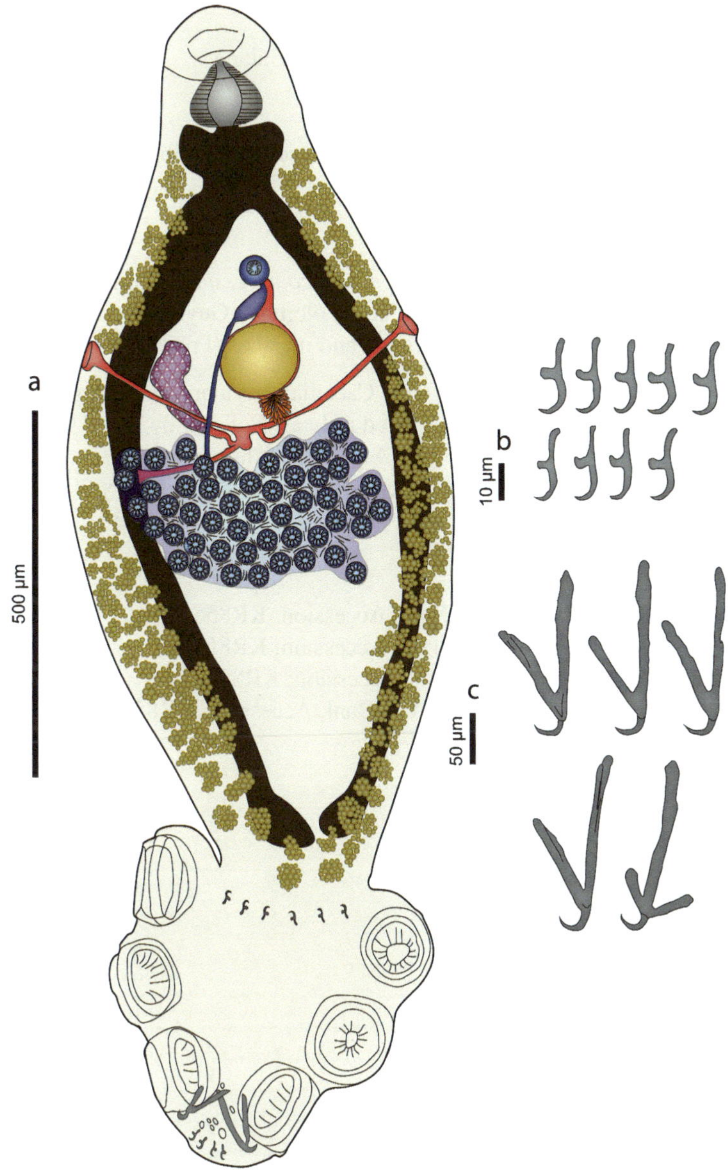

Fig. 4.138 (**a**) *Nanopolystoma tinsleyi*, (**b**) marginal hooklets, (**c**) hamuli. Redrawn from Du Preez et al. (2014)

Holotype NMBP 357.

Paratypes NMBP 358–360; IPCAS M-556.

Original description Du Preez et al. (2014).

Other taxonomic contributions None.

Etymology Named after Professor Richard Tinsley, University of Bristol, Bristol, UK.

Type locality Swamp close to Cayenne, French Guiana.

Other localities None.

Type host *Typhlonectes compressicauda* (Duméril and Bibron, 1841), Cayenne Caecilian.

Additional hosts None.

Host geographical distribution *Typhlonectes compressicauda* is an aquatic (freshwater) species which occurs in Brazil, Peru, Colombia, Guyana, French Guiana and Venezuela. It is found in permanent rivers and marshes (Frost 2021).

Host conservation status The Cayenne Caecilian has a presumed large population size and wide distribution range. It is listed as Least Concern by The IUCN Red List of Threatened Species 2010: e.T59599A11959503. https://doi.org/10.2305/IUCN.UK.2010-2.RLTS.T59599A11959503.en. Accessed on 01 February 2022.

Site on host Urinary bladder.

DNA sequence

- 18S rRNA gene, 1985 bp. GenBank Accession: KR856124
- 28S rRNA gene, 1352 bp. GenBank Accession: KR856142
- 12S rRNA gene, 480 bp. GenBank Accession: KR856077
- Cytochrome c Oxidase I, 346 bp. GenBank Accession: KR856164

Morphology and morphometrics (μm):	
Source	Du Preez et al. (2014)
Sample size	11
Body length	2,8–3,1 (2,9)
Greatest width	1076–1153 (1123)
Width at vagina	971–1060 (1013)
Haptor length	684–866 (774)
Haptor width	934–948 (941)
False oral sucker width	167–180 (175)
Pharynx length	161–190 (174)
Pharynx width	197–248 (218)
Testis length	385–577 (492)
Testis width	674–861 (764)
Ovary length	275–376 (328)

(continued)

Morphology and morphometrics (μm):	
Source	Du Preez et al. (2014)
Ovary width	81–101 (88)
Egg length	236–260 (249)
Egg width	168–193 (183)
Egg incubation time	
Number of intra-uterine eggs	1
Genital bulb width	68–71 (70)
Number of genital spines	16
Genital spine length	19–20 (20)
Haptoral sucker diameter	164–187 (176)
Hamulus handle length X	133–175 (164)
Hamulus guard length Y	
Hamulus hook length	24–33 (28)
Marginal hooklet length	18.4–20.3 (19.6)
Haptoral L/Body L	0.27
No of anastomoses	0

Remarks None.

References

Aisien MSO, Du Preez LH (2009) A redescription of *Polystoma africanum* Szidat, 1932 (Monogenea: Polystomatidae). Zootaxa 2095:37–46

Aisien SO, Salami LA, Obaro FE, Erakpoweri SO (2004) The influence of climate on the distribution of monogeneans of anurans in Nigeria. J Helminthol 78:101–104

Aisien MSO, Du Preez LH, Imasuen AA (2011) *Polystoma okomuensis* n. sp. (Monogenea: Polystomatidae) from Boulenger's striped frog, *Phlyctimantis boulengeri* (Perret, 1986) in Nigeria. J Helminthol 85:153–159

Al-Sorkhy MK, Amr Z (2003) Platyhelminth parasites of some amphibians in Jordan. Turk J Zool 27:89–93

Alvey CH (1933) *Sphyranura oligorchis* n. sp. from *Necturus maculosus*. J Parasitol 20:140

Alvey CH (1936) The morphology and development of the monogenetic trematode *Sphyranura oligorchis* (Alvey 1933) and the description of *Sphyranura polyorchis* n. sp. Parasitology 28: 229–253

Avery RA (1971) A preliminary list of parasites collected from reptiles and amphibians in northern Nigeria. Brit J Herpetol 4:217–219

Badets M, Boissier J, Brémond P, Verneau O (2009) *Polystoma gallieni*: Experimental evidence for chemical cues for developmental plasticity. Exp Parasitol 121:163–166

Badets M, Whittington I, Lalubin F, Allienne JF, Maspimby JL, Bentz S, Du Preez LH, Barton D, Hasegawa H, Tandon V, Imkongwapang R, Ohler A, Combes C, Verneau O (2011) Correlating early evolution of parasitic platyhelminths to Gondwana breakup. Syst Biol 60:762–781

Bain R (2004) *Odorrana livida*. The IUCN Red List of Threatened Species. Version 2014.2. www.iucnredlist.org.

Barrera GS, Hammerson G (2004) *Spea bombifrons*. In:IUCN 2013. IUCN Red List of Threatened Species. Version 2013.2. www.iucnredlist.org.

Barrera GS, Hammerson G, Chaves G, Wilson LD, Bolanos F, Campos PP (2010) *Smilisca baudinii*. In:IUCN 2013. IUCN Red List of Threatened Species. Version 2013.2. www.iucnredlist.org.

Barroso DD, Beebee T, Beja P, Andreone F, Bosch J, Tejedo M, Lizana M, Solano IM, Salvador A, Paris MG, Gil ER, Slimani T, Mouden EHE, Marquez R (2009) *Hyla meridionalis*. In:IUCN 2013. IUCN Red List of Threatened Species. Version 2013.2. www.iucnredlist.org.

Barton DB, Richards SJ (1996) Helminth infracommunities in *Litoria genimaculata* (Amphibia: Anura) from Birthday Creek, an upland rainforest stream in northern Queensland, Australia. Int J Parasitol 26:1381–1385

Batchvarov GK (1982) *Polystoma combesi* n. sp. (Polystomatidae Gambl, 1896) a parasite on *Rana graeca* Boulanger, 1897, in Bulgaria. C R Acad Bulg Sci 35:249–251

Beja P, Bosch J, Tejedo M, Lizana M, Martínez-Solano I, Salvador A, García-París M, Gil ER, Pérez-Mellado V, Paniagua CD, Cheylan M, Marquez R, Geniez P (2009) *Pelobates cultripes*. In:IUCN 2013. IUCN Red List of Threatened Species. Version 2013.2. www.iucnredlist.org.

Bentz S, Leroy S, du Preez L, Mariaux J, Vaucher C, Verneau O (2001) Origin and evolution of African *Polystoma* (Monogenea: Polystomatidae) assessed by molecular methods. Int J Parasitol 31:697–705

Bentz S, Sinnappah-Kang ND, Lim LH, Lebedev B, Combes C, Verneau O (2006) Historical biogeography of amphibian parasites, genus *Polystoma* (Monogenea: Polystomatidae). J Biogeogr 33:742–749

Berthier P, Du Preez LH, Raharivololoniaina L, Vences M, Verneau O (2014) Two new species of polystomes (Monogenea: Polystomatidae) from the anuran host *Guibemantis liber*. Parasitol Int 63:108–119

Beverley-Burton M (1962) Some monogenetic trematodes from Amphibia in Southern Rhodesia including a new species, *Polystoma mashoni*, sp. n., from *Bufo regularis* (Reuss). J Parasitol 1962:752–757

Biserkov VV, Hadjinikolova RI (1993) *Polystoma fuscus* n. sp. (Polystomatidae) from the common spadefoot *Pelobates fuscus* (Pelobatidae) in Bulgaria. Syst Parasitol 25:145–151

Biserkov VY, Yildirimhan HS, Buchvarov G, Ugurtas JH (2001) *Polystoma macrocnemis* n. sp. (Monogenea: Polystomatidae) from the Iranian longlegged wood frog *Rana macrocnemis* (Ranidae) in Turkey. Syst Parasitol 48:61–66

Bourgat R (1975) *Polystoma assoulinei* n. sp. (Monogenea, Polystomatidae) parasite vésical de *Ptychadaena huguettae* (Anura) au Togo. Vie Milieu C 25:197–202

Bourgat R (1977) Etude comparative des polystomes (Monogènes) de ranidés (Anoures) du Sud Togo. Description de *Polystoma togoensis* n. sp. Bull Mus Natl Hist Nat 312:447–463

Bourgat R, Combes C (1976) Présence de *Protopolystoma xenopi* (Price, 1943) au Togo, considérations sur l'aire et les hôtes de ce monogène Polystomatidae. Ann Univ Bénin Togo 2:55–59

Bourgat R, Murith D (1980) *Polystoma lamottei* n. sp. et *P. aeschlimanni* n. sp. deux polystomes (Monogènes) de la même espèce d'Amphibien: *Ptychadena pumilio* (Boulenger, 1920). Z Parasitenkd 62:293–301

Bourgat R, Salami M-L (1973) Contribution à l'étude de *Eupolystoma alluaudi* (de Beauchamp, 1913) (Monogenea, Polystomatidae). Ann Parasitol Hum Comp 48:755–760

Bourgat R, Morere J-J, Kulo S-D (1983) Nouvelles récoltes de *Eupolystoma alluaudi* (de Beauchamp, 1913) en Afrique. Considérations sur l'aire et les hôtes de ce Monogenea. Rev Zool Afric 97:567–580

Brooks DR (1976) Platyhelminthes of amphibians in Nebraska. Bull Univ Nebr State Mus 10:64–92

Buchvarov GK (1980) *Polystoma mazurmovici* n. sp. (Polystomatidae Gambl, 1896), a parasite of *Rana dalmatina* Bonaparte, 1939 in Bulgaria. Trav. Sci. Univ. Plovdiv – Biol 18:183–190

Buchvarov GK (1984a) *Polystoma skrjabini* n. sp. (Polystomatidae Gambl, 1896) - a parasite on *Hyla arborea* (Linné, 1758), in Bulgaria. C R Acad Bulgare Sci 37:697–699

Buchvarov GK (1984b) *Polystoma skuratovitchi* n. sp. (Polystomatidae Gambl, 1986) - a parasite on *Rana arvalis* Nicolsson, 1842 from west Poland. Nauchni Trudove. Trav. Sci. Univ. Plovdiv – Biol 22:137–142

Bwathondi POJ (1977) Short notes of the infection of the African toad, *Bufo regularis* by the monogenean parasite, *Beauchampia alluaudi* (Beauchamp, 1913). Univ Sci J (Dar. Univ) 3:41–44

Caballero C, Cerecero EC (1941) Una nueva especie de *Polystoma* (Trematoda: Polystomatidae) parasito de la vejiga urinaria de *Hyla baudinii* (Dum. y Bibr.) (1). An Inst Biol Univ Nac Auton Mexico 12:615–621

Campbell RA (1967) A comparative study of the parasites of certain Salientia from Pocahontas state park, Virginia. Virginia J Sci 19:13–20

Campbell RA (1969) An unusual infection of *Polystoma nearcticum* (Paul, 1938). Virginia J Sci 20: 174–175

Chaabane A, Verneau O, Du Preez LH (2019) *Indopolystoma* (n. gen.) Monogenea: Polystomatidae with the description of three new species and reassignment of eight known *Polystoma* species from Asian frogs (Anura, Rhacophoridae). Parasite 26:67. https://doi.org/10.1051/parasite/2019067

Chaabane A, Du Preez L, Johnston G, Verneau O (2022) Revision of the systematics of the Polystomoidinae (Platyhelminthes, Monogenea, Polystomatidae) with redefinition of *Polystomoides* Ward, 1917 and *Uteropolystomoides* Tinsley, 2017. Parasite 29:56. https://doi.org/10.1051/parasite/2022056

Coggins JR, Sajdak RA (1982) A survey of helminth parasites in the salamanders and certain anurans from Wisconsin. Proc Helminthol Soc Wash 49:99–102

Coloma LA, Ron S, Jungfer KH (2004a) *Gastrotheca riobambae*. In:IUCN 2013. IUCN Red List of Threatened Species. Version 2013.2. www.iucnredlist.org.

Coloma LS, Ron S, Castro F, Munoñoz-Yánez M, Lynch J, Wilkinson M (2004b) *Caecilia pachynema*. In:IUCN 2013. IUCN Red List of Threatened Species. Version 2013.2. www.iucnredlist.org.

Combes C (1967) Recherches sur les formes néoténiques de Polystomatidae (Monogenea). Présence d'une forme néoténique chez *Polystoma gallieni* Price, 1938. Bull Soc Neuchâtel Sci Natl 90:205–214

Combes C (1968) Biologie, écologie des cycles et biogéographie de digènes et monogènes d'amphibiens dans l'est des Pyrénées. Mém Mus Natl Hist Nat Sér A Zool Fascicule unique:1–195

Combes C (1976) World biogeography of polystomatid monogeneans. Acad Sci USSR Proc 34:55–63

Combes C, Channing A (1978) Polystomatidae (Monogenea) d'amphibiens d'Afrique du Sud: *Polystoma natalensis* n. sp., parasite de *Strongylopus grayi* (Smith, 1849). Vie Milieu C 28–29: 61–68

Combes C, Laurent RF (1974) *Polystoma borellii* n. sp. (Monogenea, Polystomatidae) parasite of *Pleurodema borellii* Peracca (Anura, Leptodactylidae) en République Argentine. Acta Zool Lilloana 31:57–64

Combes C, Laurent RF (1978) Deux nouveaux Polystomatidae (Monogenea) de l'Argentine. Acta Zool Lilloana 33:85–92

Combes C, Laurent RF (1979) Les monogènes Polystomatidae de République Argentine: description de deux nouvelles espèces et essai de synthese. Rev Iber Parasitol 79:545–557

Combes C, Bourgat R, Salami-Cadoux M-L (1973) Découverte d'un cycle interne direct intervenant comme mode habituel dans la reproduction d'un plathelminthe. C R Acad Sci Paris 276:2005–2006

Combes C, Lambert A, Bourgat R, Salami-Cadoux M-L (1978) L'oncomiracidium de *Eupolystoma alluaudi* (de Beauchamp, 1913) monogène parasite de *Bufo regularis* Reuss en Afrique. Bull Mus Natl Hist Nat Série 3(514):211–216

Cortez C, Reichle S, De La Riva I, Köhler J, Lavilla E (2004) *Pleurodema borellii*. In:IUCN 2013. IUCN Red List of Threatened Species. Version 2013.2. www.iucnredlist.org.

Cortez C, Reichle S, De La Riva I, Köhler J, Lavilla E (2010) *Melanophryniscus rubriventris*. In: IUCN 2013. IUCN Red List of Threatened Species. Version 2013.2. www.iucnredlist.org.

Crusz H, Ching CC (1975) Parasites of the relict fauna of Ceylon. VI. More new helminths from amphibians and reptiles with a new host-record and redescription of *Acanthocephalus serendibensis* Crusz and Mills, 1970. Ann Parasitol Hum Comp 50:531–558

Datong Y, Shunqing L (2004a) *Babina pleuraden*. The IUCN Red List of Threatened Species. Version 2014.2. www.iucnredlist.org.

Datong Y, Shunqing L (2004b) *Rana chaochiaoensis*. The IUCN Red List of Threatened Species. Version 2014.2. www.iucnredlist.org.

Datong Y, Shunqing L, Ohler A, Swan S (2004) *Odorrana grahami*. The IUCN Red List of Threatened Species. Version 2014.2. www.iucnredlist.org.

De Beauchamp PM (1913) *Polystomum alluaudi* n. sp. Voyage de Ch. Alluaud et R. Jeannel en Afrique Orientale (1911-1912). Resultats scientifiques, Turbellariés, Trématodes et Gordiacés 17–19.

Diengdoh CR, Tandon V (1991) A new species of *Polystoma* (Monogenea) parasitic in rhacophorid amphibians in Meghalaya, India. Helminthologia 28:173–178

Diesmos A, Alcala A, Brown R, Afuang L, Gee G, Sukumaran J, Yaakob N, Tzi Ming L, Chuaynkern Y, Thirakhupt K, Das I, Iskandar D, Inger R, Stuebing R, Yambun P, Lakim M (2004) *Polypedates leucomystax*. In:IUCN 2013. IUCN Red List of Threatened Species. Version 2013.2. www.iucnredlist.org.

Dollfus R-PH, Euzet L, Combes C (1965) Trématodes de batraciens (fin). Description d'un polystome. Ann Parasitol Hum Comp 40:55–60

Du Preez LH (2011) *Polystoma vernoni* n. sp. (Monogenea: Polystomatidae) from the sharp nosed grass frog *Ptychadena oxyrhynchus* (Smith, 1849) in South Africa. J Helminthol 85:294–299

Du Preez LH (2013) Polystomatidae (Monogenea) of southern African Anura: *Polystoma channingi* n.sp. parasitic in two closely related *Cacosternum* species. Afr Zool 48:64–71

Du Preez LH (2015) *Eupolystoma namibiensis* n. sp. (Monogenea: Polystomatidae) parasitic in *Poyntonophrynus hoeschi* (Ahl) of Namibia. Afr Zool 50:141–145

Du Preez LH, Carruthers VC (2017) Frogs of Southern Africa: A complete guide. Penguin Randomhouse, Cape Town

Du Preez LH, Domingues MV (2019) *Polystoma knoffi* n. sp. and *Polystoma travassosi* n. sp. (Monogenea: Polystomatidae): naming museum-archived specimens from Brazil. Syst Parasitol 96:755–765

Du Preez LH, Kok DJ (1992) Syntopic occurrence of new species of *Polystoma* and *Metapolystoma* (Monogenea: Polystomatidae) in *Ptychadena porosissima* in South Africa. Syst Parasitol 22: 141–150

Du Preez LH, Kok DJ (1993) Polystomatidae (Monogenea) of Anura in southern Africa: *Polystoma testimagna* n. sp. parasitic in *Strongylopus f. fasciatus* (Smith, 1849). Syst Parasitol 25:213–219

Du Preez LH, Kok DJ (1995) Polystomatidae (Monogenea) of southern African Anura: *Polystoma claudecombesi* n. sp. parasitic in *Rana angolensis* Bocage. Syst Parasitol 30:223–231

Du Preez LH, Maritz MF (2006) Demonstrating morphometric protocols using polystome marginal hooklet measurements. Syst Parasitol 63:1–15

Du Preez LH, Vaucher C, Mariaux J (2002) Polystomatidae (Monogenea) of African Anura. *Polystoma dawiekoki* n. sp. parasitic in *Ptychadena anchietae* (Bocage). Syst Parasitol 52:35–41

Du Preez LH, Tinsley RC, De Sa R (2003) Polystomatidae (Monogenea) of Southern African Anura: *Eupolystoma vanasi* n. sp. parasitic in *Schismaderma carens* (Smith). Syst Parasitol 54: 71–79

Du Preez LH, Verneau O, Gross TS (2007) Polystomatidae (Monogenea) of North American Anura: *Polystoma floridana* n. sp. parasitic in the Green Tree Frog *Hyla cinerea* (Schneider, 1799). Zootaxa 1663:33–45

Du Preez LH, Wilkinson M, Huyse T (2008) The first record of polystomes (Monogenea: Polystomatidae) from caecilian hosts (Amphibia: Gymnophiona), with the description of a new genus and two new species. Syst Parasitol 69:201–209

Du Preez LH, Raharivololoniaina L, Verneau O, Vences M (2010) A new genus of polystomatid parasitic flatworm (Monogenea: Polystomatidae) without free-swimming life stage from the Malagasy poison frogs. Zootaxa 2722:54–68

Du Preez LH, Badets M, Verneau O (2014) Assessment of platyhelminth diversity within amphibians of French Guiana revealed a new species of *Nanopolystoma* (Monogenea: Polystomatidae) in the caecilian *Typhlonectes compressicauda*. Folia Parasitol 61:537–542

Du Preez LH, Badets M, Héritier L, Verneau O (2017) Tracking platyhelminth parasite diversity from freshwater turtles in French Guiana: First report of *Neopolystoma* Price, 1939 (Monogenea: Polystomatidae) with the description of three new species. Parasit Vectors 10: 53. https://doi.org/10.1186/s13071-017-1986-y

Dupouy J (1978) *Polystoma ebriensis* Maeder, 1973 (ver monogène) parasite de l'amphibien anoure *Ptychadena aequiplicata* au Cameroun. Description de l'adulte. Rev Zool Afric 923: 775–785

Dupouy J, Combes C (1977) Existence d'un cycle interne de reproduction chez *Polystoma grassei* Euzet, Combes et Knoepffler, 1966 (Monogenea, Polystomatidae) en Afrique Equatoriale. Ann Sci Nat Zool Biol Anim 19:397–400

Dupouy J, Knoepffler LP (1978) *Polystoma chiromantis* n. sp. (Monogenea) chez *Chiromantis rufescens* (Günther, 1868) du Cameroun, du Gabon et de Centrafrique. Bull Mus Natl Hist Nat 514:217–220

Dutta M, Tandon V (1998) Morphology and surface fine topography of *Parapolystoma bulliense* (Johnston, 1912) Ozaki, 1935 (Polyopisthocotylea: Monogenea), a new record from the Indian subcontinent. Acta Parasitol:1–14

Dyer WG (1985) *Riojatrema ecuadorensis* n. sp. (Trematoda: Polystomatidae) from the urinary bladder of *Bufo typhonius* (Linnaeus, 1758). J Parasitol 71:215–217

Euzet L, Combes C (1964) Sur un Polystomatidae (Monogenea) recolté à Madagascar chez *Rana mascareniensis* Duméril et Bibron. Bul Soc Zool Fr:392–401

Euzet L, Combes C (1966) *Polystoma integerrimum pelobatis* n. subsp. (Monogenea) parasite de *Pelobates cultripes* (Cuvier, 1829). Ann Parasitol Hum Comp 41:109–118

Euzet L, Combes C (1967) Présence au Nord-Tchad de *Eupolystoma alluaudi* (de Beauchamp, 1913) (Monogenea, Polystomatidae). Ann Parasitol Hum Comp 4:403–406

Euzet L, Combes C (1975) Présence au Maroc de *Polystoma gallieni* Price, 1939 (Monogenea) chez *Hyla meridionalis* Boettger, 1874. Misc Helminthol Maroccana 41:655–657

Euzet L, Combes C, Knoepffler L-P (1966) Parasites d'amphibiens du Gabon Polystomatidae (Monogenea). Considérations sur la répartition géographique des Polystomatidae africains et malgaches actuellement connus. Biol Gabon 2:215–233

Euzet L, Combes C, Knoepffler LP (1974a) Parasites d'amphibiens de la République Centrafricaine. Polystomatidae (Monogenea). Vie Milieu C 24:141–150

Euzet L, Bourgat R, Salami-Cadoux ML (1974b) *Polystoma galamensis* (Monogenea) parasite de *Rana galamensis* Duméril et Bibron, 1841, au Togo. Ann Parasitol Hum Comp 49:63–68

Euzet L, Combes C, Batchvarov G (1974c) Sur un nouveau Polystomatidae européen, parasite de l'amphibien *Bufo viridis* Laur. Vie Milieu C 24:129–139

Fan LX, Wang ZL, Li JH (2004) A new species of *Polystoma* (Polystomatidae: Monogenea) parasitic in *Polypedates dugritei*. Anim Classific J 29:451–454

Fan XL, Pan H, Wang G (2006) A new species of the genus *Polystoma* (Polystomatidae: Monogenea) parasitic in *Rana pleuraden* Boulenger. Acta Zoot Sin 30:710–713

Fan LX, Wang ZL, Xu HJ (2007) A new species of the genus *Polystoma* (Polystomatidae: Monogenea) parasitic in *Rana grahami* Boulenger. Acta Zoot Sin 32:182–185

Fan LX, Li JH, He ZY (2008) A new species of the genus *Polystoma* (Polystomatidae: Monogenea) parasitic in the host *Rhacophorus carvirostris* Guenther. Acta Zoot Sin 33:340–343

Fan L, Xu W, Jia T, Netherlands EC, Du Preez LH (2020) *Polystoma luohetong* n. sp. (Monogenea: Polystomatidae) from *Rana chaochiaoensis* Liu (Amphibia: Ranidae) in China. Syst Parasitol 97:639–647. https://doi.org/10.1007/s11230-020-09937-1

Feng X, Ohler A. 2004 *Xenophrys glandulosa*. In:IUCN 2013. IUCN Red List of Threatened Species. Version 2013.2. www.iucnredlist.org.

Frandsen F (1974) A study of Danish amphibians parasite fauna. Acta Parasitol Pol 22:49–66

Fröhlich JA (1791) Beitrage zur Naturgeschichte der Eingeweidewürme. Naturforscher, Halle 25: 52–113

Frost DR (2021) Amphibian species of the world: an online reference. Version 6. Electronic database accessible at https://amphibiansoftheworld.amnh.org/index.php. American Museum of Natural History, New York, USA. https://doi.org/10.5531/db.vz.0001

Gallien L (1932) Sur la reproduction néoténique chez *Polystomum integerrimum* Fröl. C R Acad Sci Paris 194:1852

Gallien L (1935) Recherches expérimentales sur le dimorphisme évolutif et la biologie de *Polystomum integerrimum* Fröl. Trav Stat Zool Wimereux 12:1–181

Gallien L (1938) Sur un polystome parasite de *Hyla arborea* L. var. *meridionalis* Boettger. Bull Soc Zool Fr 63:251–254

Gallien L (1947) Description de *Polystomum gallieni* E.-W. Price (Monogenea, Polystomatidae) parasite de *Hyla arborea* var. *meridionalis* Boettger. Ann Parasitol Hum Comp 22:24–29

Gallien L (1956) Deux espèces nouvelles de polystomes africains. Bull Soc Zool Fr 81:369–374

Gao YY, Chen ZM, Fan LX (2012) A new species of the genus *Polystoma* (Monogenea: Polystomatidae) parasitic in *Hyla a. annectans*. Acta Zoot Sin 37:529–534

Gray ME (1993) *Wetapolystoma almae* n. gen., n. sp. (Monogenea: Polystomatidae) parasite of *Bufo typhonius* (Linnaeus, 1758) (Amphibia: Bufonidae) from Tropical Peru. Trans Kans Acad Sci 96:181–185

Hammerson G (2004a) *Eurycea tynerensis*. In:IUCN 2013. IUCN Red List of Threatened Species. Version 2013.2. www.iucnredlist.org.

Hammerson G (2004b) *Necturus maculosus*. In:IUCN 2013. IUCN Red List of Threatened Species. Version 2013.2. www.iucnredlist.org.

Hammerson G (2004c) *Hyla versicolor*. In:IUCN 2013. IUCN Red List of Threatened Species. Version 2013.2. www.iucnredlist.org.

Hammerson G, Hedges B (2004) *Hyla cinerea*. In:IUCN 2013. IUCN Red List of Threatened Species. Version 2013.2. www.iucnredlist.org.

Hedges B, Diaz L, Ibene B, Joglar R, Powell R, Bolanos F, Chaves G (2010) *Osteopilus septentrionalis*. In:IUCN 2013. IUCN Red List of Threatened Species. Version 2013.2. www.iucnredlist.org.

Héritier L, Badets M, Du Preez LH, Aisien MS, Lixian F, Combes C, Verneau O (2015) Evolutionary processes involved in the diversification of chelonian and mammal polystomatid parasites (Platyhelminthes, Monogenea, Polystomatidae) revealed by palaeoecology of their hosts. Mol Phylogenet Evol 92:1–10

Héritier L, Verneau O, Smith KG, Coetzer C, Du Preez LH (2018) Demonstrating the value and importance of combining DNA barcodes and discriminant morphological characters for polystome taxonomy (Platyhelminthes, Monogenea). Parasitol Int 67:38–46

Hero JM, Lemckert F, Robertson P, Mahony M (2004a) *Litoria phyllochroa*. In:IUCN 2013. IUCN Red List of Threatened Species. Version 2013.2. www.iucnredlist.org.

Hero JM, Cunningham M, Alford R, Mcdonald K, Retallick R (2004b) *Litoria nyakalensis*. In: IUCN 2013. IUCN Red List of Threatened Species. Version 2013.2. www.iucnredlist.org.

Hughes RC, Moore GA (1943) *Sphyranura euryceae*, a new polystomatid Monogenean fluke from *Eurycea tynerensis*. Trans Am Micros Soc 62:286–292

Imkongwapang R, Tandon V (2010) *Neoriojatrema mokokchungensis* n. g., n. sp. (Monogenea: Polystomatidae: Polystomatinae) from the urinary bladder of a megophryid frog, *Xenophrys glandulosa*, in Nagaland, north-eastern India. J Helminthol 84:1–6

Johnston SJ (1912) On some trematode parasites of Australian frogs. Proc Linn Soc New South Wales 37:285–362

Kaneko Y, Matsui M. (2004a) *Onychodactylus japonicus*. In:IUCN 2013. IUCN Red List of Threatened Species. Version 2013.2. www.iucnredlist.org.

Kaneko Y, Matsui M (2004b) *Rhacophorus schlegelii*. In:IUCN 2013. IUCN Red List of Threatened Species. Version 2013.2. www.iucnredlist.org.

Kaneko Y, Matsui M (2004c) *Rana japonica*. In:IUCN 2013. IUCN Red List of Threatened Species. Version 2013.2. www.iucnredlist.org.

Kaneko Y, Matsui M (2004d) *Rana ornativentris*. In:IUCN 2013. IUCN Red List of Threatened Species. Version 2013.2. www.iucnredlist.org.

Kaw BL (1950) Studies in helminthology: Helminth parasites of Kashmir. Part I. Trematoda. Indian J Helminthol 2:67–126

Khan MS, Stöck M, Papenfuss T, Anderson S, Kuzmin S, Rastegar-Pouyani N, Sharifi M, Bafti SS, Dutta S, Ohler A (2009) *Pseudepidalea surda*. In:IUCN 2013. IUCN Red List of Threatened Species. Version 2013.2. www.iucnredlist.org.

Kohn A, Combes C, Gomes DC (1978) Représentants du genre *Polystoma* Zeder (Monogenea) au Brésil. Bull Mus Natl Hist Nat 353:227–229

Kok DJ, Du Preez LH (1989) *Polystoma australis* (Monogenea): Development and reproduction in neotenic parasites. S Afr J Zool 24:225–230

Kok DJ, Du Preez LH (1998) The relative importance of bladder versus neotenic stages of *Polystoma marmorati* and *P. umthakathi* in natural frog populations in South Africa. J Helminthol 72:117–121

Kok DJ, Seaman MT (1987) Polystomatidae (Monogenea) parasitic in the anuran genus *Natalobatrachus* in South Africa. S Afr J Zool 22:258–263

Kok DJ, Van Wyk JH (1986) Polystomatidae (Monogenea) parasitic in the anuran genus *Kassina* in South Africa. S Afr J Zool 21:189–195

Kuangyang L, Wenhao C, Baorong G, Huiqing G, Haitao S, Van Dijk PP, Truong NQ (2009) *Hylarana guentheri*. In:IUCN 2013. IUCN Red List of Threatened Species. Version 2013.2. www.iucnredlist.org.

Kulo S-D (1981) Présence au Togo de *Metapolystoma cachani* (Gallien, 1957) Combes, 1976 (Monogenea) chez l'amphibien *Ptychadena longirostris* (Peters, 1870). Bull Soc Zool Fr 106: 177–181

Kuzmin S, Maslova I, Tuniyev B, Matsui M, Pipeng L, Kaneko Y (2004a) *Pelophylax nigromaculatus*. In:IUCN 2013. IUCN Red List of Threatened Species. Version 2013.2. www.iucnredlist.org.

Kuzmin S, Matsui M, Wenge Z, Ishchenko V, Ananjeva, N, Orlov N (2004b) *Rana chensinensis*. In:IUCN 2013. IUCN Red List of Threatened Species. Version 2013.2. www.iucnredlist.org.

Kuzmin S, Tarkhnishvili D, Ishchenko V, Ananjeva N, Orlov N, Tuniyev B, Sparreboom M, Ugurtas I, Rastegar-Pouyani N, Papenfuss T, Anderson S, Eken G, Kiliç T, Gem E, Kaya U (2009a) *Rana macrocnemis*. In: IUCN 2013. IUCN Red List of Threatened Species. Version 2013.2. www.iucnredlist.org.

Kuzmin S, Tarkhnishvili D, Ishchenko V, Tuniyev B, Beebee T, Anthony B, Schmidt, Ogrodowczyk A, Ogielska M, Babik W, Vogrin M, Loman J, Cogalniceanu D, Kovács T, Kiss I (2009b) *Rana arvalis*. In:IUCN 2013. IUCN Red List of Threatened Species. Version 2013.2. www.iucnredlist.org.

Kwet A, Reichle S, Aquino L, Silvano D, Lavilla E, Di Tada I (2010) *Physalaemus biligonigerus*. In:IUCN 2013. IUCN Red List of Threatened Species. Version 2013.2. www.iucnredlist.org.

Kwet A, Aquino L, Lavilla E, Di Tada I (2004) *Hypsiboas pulchellus*. In:IUCN 2013. IUCN Red List of Threatened Species. Version 2013.2. <www.iucnredlist.org>.

La Marca E, Azevedo-Ramos C, Coloma LA, Ron S (2010a) *Osteocephalus taurinus*. In:IUCN 2013. IUCN Red List of Threatened Species. Version 2013.2. www.iucnredlist.org.

La Marca E, Reynolds R, Azevedo-Ramos C, Mijares A, Señaris C (2010b) *Osteocephalus leprieurii*. In:IUCN 2013. IUCN Red List of Threatened Species. Version 2013.2. www.iucnredlist.org.

Lamothe-Argumedo R (1963) Trematodos de los Anfibios de Mexico. I. Sobre un nuevo genero de la familia Polystomatidae Gamble, 1896 hallado en la vejica urinaria de *Tomadactylus amulae* Günther y *Bufo simus* Schmidt. Rev Soc Mex Hist Nat 24:73–88

Lamothe-Argumedo R (1973) Monogéneos de los anfibios de Mexico IV. Redescripcion de *Neodiplorchis scaphiopi* (Rodgers, 1941) Yamaguti 1963. An Inst Biol Univ Nac Auton Mexico 44:1–7

Lamothe-Argumedo R (1976) Monogéneos de los anfibios de Mexico VI. Redescripcion de *Polystoma naevius* Caballero y Cerecero, 1941. An Inst Biol Univ Nac Auton Mexico 47:1–8

Lamothe-Argumedo R (1985) Monogéneos de los anfibios de México VII. Hallazgo de *Pseudodiplorchis americanus* (Rodgers y Kuntz, 1940) Yamaguti, 1963 en baja California sur, México. An Inst Biol Univ Nac Auton Mexico 56:291–300

Landman W, Verneau O, Du Preez LH (2018) First record of viviparity in polystomatid flatworms (Monogenea: Polystomatidae) with the description of two new species of *Madapolystom*a from the Madagascan anuran hosts *Blommersia domerguei* and *Mantella expectata*. Int J Parasitol Parasites Wildl 7:343–354. https://doi.org/10.1016/j.ijppaw.2018.09.004

Landman WJ, Verneau O, Raharivololoniaina L, Du Preez LH (2021) First record of *Metapolystoma* (Monogenea: Polystomatidae) from *Boophis* tree frogs in Madagascar, with the description of five new species. Int J Parasitol Parasites Wild 14:161–178

Landman W, Verneau O, Vences M, Du Preez L (2023) *Metapolystoma ohlerianum* n. sp. (Monogenea: Polystomatidae) from *Aglyptodactylus madagascariensis* (Anura: Mantellidae). Acta Parasitol

Lau MWN, Baorong G, Kuangyang L, Wenhao C (2004a) *Hylarana latouchii*. In:IUCN 2013. IUCN Red List Of Threatened Species. Version 2013.2. Www.IUCNredlist.Org.

Lau MWN, Wenhao C, Datong Y, Swan S, Bain R (2004b) *Rhacophorus dugritei*. The IUCN Red List of Threatened Species. Version 2014.2. www.iucnredlist.org.

Lavilla E (2004) *Telmatobius oxycephalus*. In:IUCN 2013. IUCN Red List of Threatened Species. Version 2013.2. <www.iucnredlist.org>.

Lee DL (1936) On a new and a rare trematode. Contributions from the Institute of Zoology, National Academy of Peiping 3:123–132

Lim LHS, Du Preez LH (2001) *Sundapolystoma chalconotae* n. g., n. sp. (Monogenea: Polystomatidae) from *Rana chalconota* (Schlegel) of Peninsular Malaysia. Syst Parasitol 49: 223–231

López-Neyra CR (1952) *Polystoma palancai* n. sp. (Monogenea: Polystomatidae), parasita de la vejiga urinaria de la rana de San Anton. Rev Iber Parasitol 12:289–295

Lymberakis P, Isailovic JC, Ajtic R, Vogrin M, Haxhiu I (2009) *Rana graeca*. In:IUCN 2013. IUCN Red List of Threatened Species. Version 2013.2. www.iucnredlist.org.

Macé E (1880) Des trématodes parasites des grenouilles. Bull Soc Etudes Scient Finèstere 2:61–89

Macnae W, Rock L, Markowski M (1973) Platyhelminths from the South African clawed toad, or platanna (*Xenopus laevis*). J Helminthol 47:199–235

Maeder A-M (1973) Monogènes et trématodes parasites d'amphibiens en Côte d'Ivoire. Rev Suisse Zool 80:267–322

Maeder AM, Euzet L, Combes C (1970) Espèces nouvelles du genre *Polystoma* (Monogenea) en Afrique Occidentale. Z Parasitenkd 35:140–155

McAllister CT, Trauth SE, Hinck LW (1991) *Sphyranura euryceae* (Monogenea) on *Eurycea* spp. (Amphibia: Caudata), from Northcentral Arkansas. J Helminthol Soc W 58:137–140

Meng YY, Song L, Ding XJ (2010) A new species of the genus *Polystoma* Monogenea, Polystomatidae parasitic in *Polypedates mutus* from Hainan Island, China. Acta Zoot Sin 35: 842–845

Meskal FH (1970) Trematodes of anurans from Ethiopia. Arbok for Universitetet i Bergen Matematik-Naturvitenskapelig Serie 1(1):1–73

Meyer MC, Olsen OW (1972) (1) Class Trematoda. (2) Preparation of specimens for study. In M.C. Brown (ed), Essentials of parasitology. Dubuque, Iowa. 1:37–45; 2:263–293

Mijares A, Rodrigues MT, Baldo D (2010) *Physalaemus cuvieri*. In:IUCN 2013. IUCN Red List of Threatened Species. Version 2013.2. www.iucnredlist.org.

Murith D (1979) Identité des larves de polystomes (Monogenea) parasites du têtard de *Dicroglossus occipitalis* (Günther) en Côte-d'Ivoire. Z Parasitenkd 59:187–194

Murith D (1981) Contribution à l'étude de la systématique des polystomes (Monogènes, Polystomatidae) parasites d'Amphibiens Anoures de basse Côte-d'Ivoire. Rev Suisse Zool 88: 475–533

Murith D, Vaucher C, Combes C (1977) Coexistence de la néoténie et du cycle interne chez un Polystomatidae (Monogenea). C R Acad Sci Paris 284:187–190

Murith D, Miremad-Gassmann M, Vaucher C (1978) Contribution à l'étude des polystomes d'amphibiens du Cameroun. Rev Suisse Zool 85:681–698

Nasir P, Fuentes-Zambrano JL (1983) Algunos trematodos monogeneticos venezolanos. Riv Parassitol 44:335–380

Nussbaum R, Raxworthy C, Andreone F (2008) *Guibemantis liber*. In:IUCN 2008. 2008 IUCN Red List of Threatened Species. www.iucnredlist.org.

Olsen OW (1974) Phylum Platyhelminthes. Animal parasites; their life cycles & ecology, vol 199–213, 3rd edn. University Park Press, London, pp 292–296

Ozaki Y (1931) A new polystome trematode, *Diplorchis ranae*, gen. et. sp. n. Ann Mag Nat Hist Série 10(7):181–184

Ozaki Y (1935) Studies on the frog - trematode *Diplorchis ranae*. I. Morphology of the adult form with review of the family Polystomatidae. J Sci Hiroshima Univ 3:193–223

Ozaki WB (1948) A new trematode, *Polystoma dendriticum* from the urinary bladder of *Onychodactylus japonicus* in Shikoku. Biosphaera 2:33–37

Pandey KC (1969) Studies on monogenetic trematodes of India I. On a new species of the genus *Eupolystoma* Kaw, 1950 from *Bufo* sp. Proc Indian Nat Sci Acad 39:191–193

Park JT (1938) An amphibian trematode, *Polystoma integerrimum* (Frölich, 1791), from Uchikongo, Korea. J Med 11:287–288

Paul AA (1935) *Polystoma integerrimum nearcticum* n. subsp. from the urinary bladder, genital ducts, kidneys and gills of *Hyla versicolor* Le Conte. J Parasitol 21:442

Paul AA (1938) Life history studies of North American freshwater polystomes. J Parasitol 24:489–510

Pérez-Vigueras I (1955) Contribucion al conocimiento de la fauna helmintologica cubana. Mem Soc Cub Hist Nat 22:21–71

Pichelin S (1995) *Parapolystoma johnstoni* n. sp. from *Litoria nyakalensis* (Amphibia) in Australia and taxonomic considerations on the Diplorchiinae (Monogenea). J Parasitol 81:261–265

Price EW (1939) North American monogenetic trematodes. IV. The family Polystomatidae (Polystomatoidea). Proc Helminthol Soc Wash 6:80–92

Price EW (1943) A new trematode of the genus *Polystoma* (Monogenea: Polystomatidae) from *Xenopus laevis* Daud. Proc Helminthol Soc Wash 10:83–85

Pritchard MH (1964) Notes on four helminths from the clawed toad, *Xenopus laevis* (Daudin), in South Africa. Proc Helminthol Soc Wash 10:121–128

Raharivololoniaina L, Verneau O, Berthier P, Vences M, Du Preez L (2011) First monogenean flatworm from a microhylid frog host: *Kankana*, a new polystome genus from Madagascar. Parasitol Int 60:465–473

Rödel MO, Schiøtz A (2004) *Ptychadena superciliaris*. In:IUCN 2013. IUCN Red List of Threatened Species. Version 2013.2. www.iucnredlist.org.

Rödel MO, Largen M, Burger M, Howell K, Salvador A, Lötters S, Joger U (2004) *Hoplobatrachus occipitalis*. In:IUCN 2013. IUCN Red List of Threatened Species. Version 2013.2. www.iucnredlist.org.

Rödel MO, Largen M, Minter L, Howell K, Nussbaum R, Vences M, Baha El Din S (2009a) *Ptychadena mascareniensis*. In:IUCN 2013. IUCN Red List of Threatened Species. Version 2013.2. www.iucnredlist.org.

Rödel MO, Amiet JL, Burger M, Largen M (2009b) *Ptychadena aequiplicata*. In:IUCN 2013. IUCN Red List of Threatened Species. Version 2013.2. www.iucnredlist.org.

Rodgers LO (1941) *Diplorchis scaphiopi*, a new polystomatid monogenean fluke from the spadefoot toad. J Parasitol 27:153–157

Rodgers LO, Kuntz RE (1940) A new polystomatid monogenean fluke from a spadefoot. Wasmann Collect 4:37–40

Rubtsova NY (2018) Redescription and variability of *Polystoma mazurmovici* and *P. skuratovitchi* (Monogenea, Polystomatidae), with a key to *Polystoma* from anurans of Ukraine. Vestn Zool 52:91

Salami-Cadoux M-L (1975) Transmission et développement du monogène réno-vésical *Eupolystoma alluaudi* (de Beauchamp, 1913) Euzet et Combes, 1967. Bull Soc Zool Fr 100: 283–292

Salami-Cadoux M-L (1980) Contribution à l'étude de *Polystoma africanum* SZIDAT, 1932 (Monogenea, Polystomatidae). BIFAN 42:529–544

Sales AN, Du Preez LH, Verneau O, Domingues MV (2023) Morphology and molecular characterization of *Polystoma goeldii* n. sp. (Monogenea, Polystomatidae) parasite from the urinary bladder of *Physalaemus ephippifer* (Steindachner) (Anura, Leptodactylidae). Parasitol Int 92: 102685. https://doi.org/10.1016/j.parint.2022.102685

Santos VGT, Amato SB (2012) *Polystoma cuvieri* (Monogenea: Polystomatidae) in *Physalaemus cuvieri* (Anura, Leiuperidae) in Southern Brazil. Neotrop Helminthol 6:1–7

Saoud MFA (1967) On *Polystoma prudhoei* sp. nov. (Trematoda: Monogenea) from *Rana (Ptychadena) oxyrhynchus* in the Cameroons. J Helminthol 41:229–234

Scott N, Aquino L, Angulo A (2004) *Rhinella diptycha*. In:IUCN 2013. IUCN Red List of Threatened Species. Version 2013.2. www.iucnredlist.org.

Shen J, Wang G, Fan LX (2013) A new species of the genus *Polystoma* (Monogenea: Polystomatidae) parasitic in the tree frog *Philautus palpebralis* Smith. Acta Zoot Sin 38:765–772

Shu-Yi Z, Suo L (1987) Three new monogenetic trematodes of the family Polystomatidae. Acta Zoot Sin 12:230–237

Sinnappah ND, Lim LHS, Rohde K, Tinsley R, Combes C, Verneau O (2001) A paedamorphic parasite associated with a neotenic amphibian host: Phylogenetic evidence suggests a revised systematic position for Sphyranuridae within anuran and turtle Polystomatoineans. Mol Phylogenet Evol 18:189–201. https://doi.org/10.1006/mpev.2000.0877

Solís F, Ibáñez R, Hammerson G, Hedges B, Diesmos A, Matsui M, Hero J, Richards S, Coloma L, Ron S, La Marca E, Hardy J, Powell R, Bolaños F, Chaves G, Ponce P (2009) *Rhinella marina*. In:IUCN 2013. IUCN Red List of Threatened Species. Version 2013.2. www.iucnredlist.org.

Solís S, Ibáñez R, Jaramillo C, Fuenmayor Q, Silvano D, Coloma LA, La Marca E, Ron S, Hoogmoed M (2010) *Rhinella margaritifera*. In:IUCN 2013. IUCN Red List of Threatened Species. Version 2013.2. www.iucnredlist.org.

Song L, Xiao Z, Ding XJ (2008) A new species of the genus *Diplorchis* (Monogenea: Polystomatidae) parasitic in *Rana livida* Blyth, 1855 from Hainan Island, China. Acta Zoot Sin 33:733–736

Stunkard HW (1959) Induced gametogenesis in a monogenetic trematode, *Polystoma stellai* Vigueras, 1955. J Parasitol 45:389–394

Szidat L (1932) Parasiten aus Liberia and Französisch-Guinea. II Teil: Trematoden. Z Parasitenkd 4:506–521

Tandy M, Rödel MO, Largen M, Poynton J, Lotters S, Baha S, Din E, Gerson (2004) *Amietophrynus regularis*. In: IUCN 2013. IUCN Red List of Threatened Species. Version 2013.2. www.iucnredlist.org.

Tinsley RC (1973) Observations on Polystomatidae (Monogenoidea) from East Africa with a description of *Polystoma makereri* n. sp. Z Parasitenkde 42:251–263

Tinsley R (1974a) Observations on *Polystoma africanum* Szidat with a review of the interrelationships of *Polystoma* species in Africa. J Nat Hist 8:355–367

Tinsley RC (1974b) A redescription of *Polystoma pricei* Vercammen-Grandjean, 1960 (Monogenoidea: Polystomatidae). Rev Zool Afric 88:165–174

Tinsley RC (1978a) The morphology and distribution of *Eupolystoma* species (Monogenoidea) in Africa, with a description of *E. anterorchis* sp. n. from *Bufo pardalis* at the Cape. J Helminthol 52:291–302

Tinsley RC (1978b) Oviposition, hatching and the oncomiracidium of *Eupolystoma anterorchis* (Monogenoidea). Parasitology 77:121–132

Tinsley RC, Earle C (1983) Invasion of vertebrate lungs by the polystomatids monogeneans *Pseudodiplorchis americanus* and *Neodiplorchis scaphiopodis*. Parasitology 86:501–517

Tinsley RC, Jackson JA (1998) Speciation of *Protopolystoma* Bychowsky, 1957 (Monogenea: Polystomatidae) in hosts of the genus *Xenopus* (Anura: Pipidae). Syst Parasitol 40:93–141

Tinsley R, Measey J (2004) *Xenopus pygmaeus*. In:IUCN 2013. IUCN Red List of Threatened Species. Version 2013.2. www.iucnredlist.org.

Tinsley RC, Owen RW (1975) Studies on the biology of *Protopolystoma xenopodis* (Monogenoidea): The oncomiracidium and life-cycle. Parasitology 71:445–463

Tinsley R, Measey J, Beier M (2004) *Xenopus fraseri*. In:IUCN 2013. IUCN Red List of Threatened Species. Version 2013.2. www.iucnredlist.org.

Tinsley R, Minter L, Measley J, Howell K, Veloso A, Nunez H, Romano A (2009) *Xenopus laevis*. In:IUCN 2013. IUCN Red List of Threatened Species. Version 2013.2. www.iucnredlist.org.

Uchida A, Itagaki H (1979) Studies on the amphibian helminths in Japan. VI. *Pseudoxyascaris japonicus* n.g. and n.sp. (Oxyascarididae; Nematoda) and *Pseudopolystoma dendriticum* (Monogenea; Trematoda) from a salamander. Jpn J Parasitol 28:43–50

Uchida A, Machida M, Uchida K, Itagaki H (1988) Studies on the amphibian helminths in Japan (X). Two new monogeneans, *Polystoma ozakii* n. sp. and *P. ezoensis* n. sp., and new locality and host of *P. racophori*. Bull Azabu Univ Vet Med 9:7–12

Vaira M (2004) Population morphological variation of the monogenean *Polystoma andinum*, parasitic in *Melanophryniscus rubriventris* (Anura, Bufonidae). Acta Parasitol 49:281–291

Van Dijk PP, Iskandar D, Inger R (2004a) *Rhacophorus nigropalmatus*. In:IUCN 2013. IUCN Red List of Threatened Species. Version 2013.2. www.iucnredlist.org.

Van Dijk PP, Ohler A, Wai Neng Lau M, Haitao S, Datong Y (2004b) *Polypedates mutus*. The IUCN Red List of Threatened Species. Version 2014.2. www.iucnredlist.org.

Van Dijk PP, Truong NQ, Shunqing L, Datong Y (2004c) *Feihyla palpebralis*. The IUCN Red List of Threatened Species. Version 2014.2. www.iucnredlist.org.

Van Dijk PP, Ohler A, Datong Y, Shunqing L, Dutta S, Bordoloi S, Roy D (2004d) *Hyla annectans*. The IUCN Red List of Threatened Species. Version 2014.2. www.iucnredlist.org.

Van Dijk PP, Iskandar D., Inger R, Kusrini M (2009) *Hylarana chalconota*. In:IUCN 2013. IUCN Red List of Threatened Species. Version 2013.2. www.iucnredlist.org.

Van Niekerk S, Kok DJ, Seaman MT (1993) A new species of *Polystoma* (Monogenea: Polystomatidae) parasitic in *Hyperolius marmoratus* (Anura: Hyperoliidae) in South Africa. Syst Parasitol 25:73–80

Vande Vusse FJ (1976) *Parapolystoma crooki* sp. n. (Monogenea: Polystomatidae) from *Rana magna* in the Philippines. J Parasitol 62:552–555

Vaucher C (1981) *Mesopolystoma samiriensis* n.gen, n. sp. (Monogenea: Polystomatidae): parasite de *Osteocephalus taurinus* Steindachner (Amphibia: Hylidae) en Amazonie péruvienne. Rev Suisse Zool 88:797–802

Vaucher C (1983) Helminthes parasites du Paraguay V: *Polystoma lopezromani* Combes et Laurent, 1979 et utilisation d'une technique inédite d'observation des hamuli chez les Polystomatidae (Monogenea). Bull Soc Neuchâtel Sci Nat 106:101–107

Vaucher C (1986) Helminthes parasites du Paraguay XIII: *Polystoma diptychi* n. sp. (Monogenea: Polystomatidae), parasite de *Bufo diptychus* Cope (Amphibia: Bufonidae). Bull Soc Neuchâtel Sci Nat 109:5–10

Vaucher C (1987) Polystomes d'Equateur, avec description de deux nouvelles espèces. Bull Soc Neuchâtel Sci Nat 110:45–56

Vaucher C (1990) *Polystoma cuvieri* n. sp. (Monogenea: Polystomatidae), a parasite of the urinary bladder of the leptodactylid frog *Physalaemus cuvieri* in Paraguay. J Parasitol 76:505–504

Vercammen-Grandjean PH (1960) Les Trématodes du lac Kivu Sud (Vermes.). Ann Mus R Afr Centrl Ser Quat Zool 5:1–171

Verneau O, Bentz S, Sinnappah ND, Du Preez LH, Whittington I, Combes C (2002) A view of early vertebrate evolution inferred from the phylogeny of polystome parasites (Monogenea: Polystomatidae). Proc R Soc Lond Ser B: Biol Sci 269:535–543. https://doi.org/10.1098/rspb.2001.1899

Verneau O, Du Preez LH, Laurent V, Raharivololoniaina L, Glaw F, Vences M (2009) The double odyssey of Madagascan polystome flatworms leads to new insights on the origins of their amphibian hosts. Proc R Soc B Biol Sci 276:1575–1583. https://doi.org/10.1098/rspb.2008.1530

Williams JB (1960a) The dimorphism of *Polystoma integerrimum* (Frölich) Rudolphi and its bearing on relationships within the Polystomatidae: Part I. J Helminthol 34:151–192

Williams JB (1960b) The dimorphism of *Polystoma integerrimum* (Frölich) Rudolphi and its bearing on relationships within the Polystomatidae: Part II. J Helminthol 34:323–346

Williams MO (1969) On *Polystoma africanum* [Trematoda: Monogenea] from *Bufo regularis* in Sierra Leone. BIFAN 31:1071–1073

Wright RR (1879) Contributions to American helminthology. Proc Canadian Inst 1:54–75

Wright RR, Macallum AB (1887) *Sphyranura osleri*: A contribution to American helminthology. J Morphol 1:1–48

Yamaguti S (1936) Studies on the helminth fauna of Japan. P. 14. Amphibian Trematodes. Jpn J Zool 6:551–576

Yamaguti S (1963) Systema Helminthum, IV Monogenea and Aspidocotylea. Interscience publishers, New York, London, p 699

Yildirimhan HS, Ugurtas IH, Altunel FN (1997) An investigation on parasitic helminths of *Rana macrocnemis* Boulenger, 1885 (Uludag Frog). Turk J Zool 21:467–473

Yildirimhan HS, Du Preez LH, Verneau O (2012) *Polystoma nacialtuneli* n. sp. (Monogenea: Polystomatidae) from the eastern spadefoot, *Pelobates syriacus* (Pelobatidae) in Turkey. J Helminthol 86:104–112

Yoshikawa N, Matsui M, Nishikawa K, Kim JB, Kryukov A (2008) Phylogenetic relationships and biogeography of the Japanese clawed salamander, *Onychodactylus japonicus* (Amphibia: Caudata: Hynobiidae), and its congener inferred from the mitochondrial cytochrome b gene. Mol Phylogenet Evol 49:249–259

Zhang S-Y, Long S (1987) Three new monogenetic trematodes of the family Polystomatidae. Acta Zoot Sin 12(3):230–237

Zimkus BM, Lawson LP, Barej MF, Barratt CD, Channing A, Dash KM, Dehling JM, Du Preez L, Gehring PS, Greenbaum E, Gvoždík V (2017) Leapfrogging into new territory: How Mascarene Ridged Frogs diversified across Africa and Madagascar to maintain their ecological niche. Mol Phylogenet Evol 106:254–269

Chapter 5
Polystome Species of the Australian Lungfish, Chelonians and the Common Hippopotamus

Contents

Abstract This chapter brings together all known polystomes of fish, i.e. the Australian lungfish, chelonians and mammal, i.e. the common hippopotamus, namely 67 spp. For each species we provide information on the type material, reference to the description, etymology, type locality, host identity, host geographical distribution, site of infection, morphometrics and a species drawing based on the original species description.

L. H. du Preez et al., *Polystomatid Flatworms*, Zoological Monographs 9,
https://doi.org/10.1007/978-3-031-35887-6_5

5.1 Morphological Features of Taxonomic Importance for Polystomes of the Australian Lungfish, Chelonians and the Common Hippopotamus

They are largely similar as those described for amphibian polystomes under 4.2.

5.2 Taxonomic Corrections for Gender

During the course of history, several polystomes were incorrectly named. Generic names ending in -soma, -stoma, -trema and -nema are neuter and Greek in origin. They have been Latinized and look feminine, but they are not. There was a practice in the nineteenth century of changing them into a true Latin format, e.g. -stomum for -stoma, but the Greek forms have priority if they are older (Gibson pers. comm.). The correct form is thus *Polystoma* Zeder, 1800 (family Polystomatidae) and not *Polystomum* Zeder, 1800 (family Polystomatidae). All other genera based on *Polystoma* are also neuter. A number of species were incorrectly published according to the rules for gender. Therefore, we provide a list with the corrected names (Table 5.1).

5.3 Polystome of the Australian Lungfish

To date only a single species of polystome is known from a fish, namely *Concinnocotyla australensis* (Reichenbach-Klinke, 1966), from Australian lungfish. It is characterized by an elongate body; haptor with three pairs of suckers and one pair of very small hamuli; asymmetrical haptoral suckers, near triangular in shape,

Table 5.1 Polystome taxa of chelonians corrected for gender

Current taxon	New taxon
Aussietrema queenslandensis	*Aussietrema queenslandense*
Fornixtrema guianensis	*Fornixtrema guianense*
Manotrema brasiliensis	*Manotrema brasiliense*
Manotrema uruguayensis	*Manotrema uruguayense*
Polystomoidella oblongum	*Polystomoidella oblonga*
Polystomoides coronatum	*Polystomoides coronatus*
Polystomoides digitatum	*Polystomoides digitatus*
Polystomoides japonicum	*Polystomoides japonicus*
Polystomoides ocellatum	*Polystomoides ocellatus*
Polystomoides opacum	*Polystomoides opacus*
Polystomoides orbiculare	*Polystomoides orbicularis*
Polystomoides rugosa	*Polystomoides rugosus*

with sclerites in shape of a fan; subterminal mouth; bifurcate intestine with caeca extending the full length of the body, not entering haptor, confluent posteriorly; unbranched caeca but with single short one pair of anterior diverticula near the pharynx to form conspicuous pockets; pre-testicular ovary; a single long ellipsoid operculated egg in the oötype, eggs with an abopercular appendage; absence of uterus; absence of vaginae; follicular vitellarium; and numerous intercaecal testes.

5.3.1 *Concinnocotyla* Pichelin, Whittington and Pearson, 1991 (Fig. 5.1)

Fig. 5.1 Micrograph of (**a**) *Concinnocotyla australensis*; (**b**) Haptoral sucker. Images: © Peter Olson

5.3.1.1 *Concinnocotyla australensis* (Reichenbach-Klinke, 1966) (Fig. 5.2)

Collection Type collection unknown; Voucher specimens in Meguro Parasitological Museum, Tokyo, Japan; Muséum National d'Histoire Naturelle, Paris, France (MNHN); Natural History Museum London (NHML); Queensland Museum,

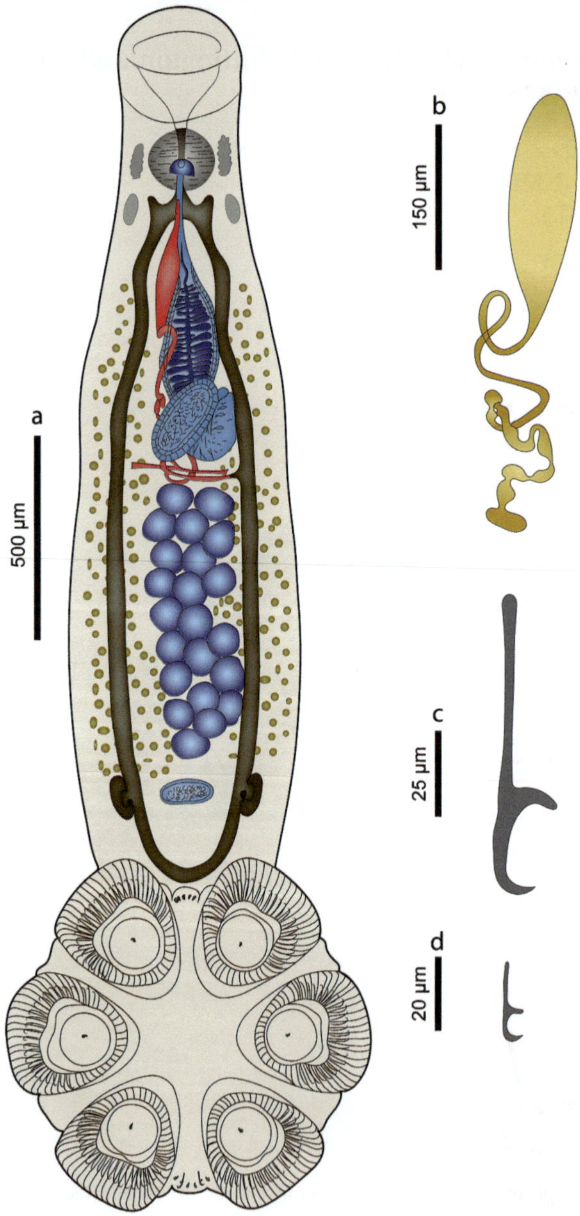

Fig. 5.2 (**a**) *Concinnocotyla australensis*, (**b**) egg, (**c**) hamuli, (**d**) marginal hooklets. Redrawn from Pichelin et al. (1991)

Brisbane, Australia; United States National Museum (USNM); Zoological-Parasitological Institute of Munich University, Munich.

Holotype Unknown.

Paratypes Unknown.

Original description Reichenbach-Klinke (1966).

Other taxonomic contributions Pichelin et al. (1991), Whittington et al. (1991).

Etymology Named after type locality country.

Type locality Aquarium at Hellabrunn Zoo, Munich.

Other localities Brisbane River and Enoggera Creek, S.E. Queensland, Australia.

Type host *Neoceratodus forsteri* (Krefft, 1870), Australian Lungfish.

Additional hosts None.

Host geographical distribution *Neoceratodus forsteri* naturally occurs in south-east Queensland, Australia, where it is present in the Mary and Burnett rivers. However, it can also be found in other rivers including the North Pine River, the Coomera River and the Brisbane River catchment (Arthington 2009).

Host conservation status The habitat range of the Australian Lungfish is threatened by forestry, alien plant and fish species, agriculture, regulation of flows and river impoundment (Arthington 2009). It is listed as Endangered by The IUCN Red List of Threatened Species 2019: e.T122899816A123382021. https://doi.org/10.2305/IUCN.UK.2019-3.RLTS.T122899816A123382021.en. Accessed on 21 January 2022.

Site on host Surfaces of the tooth plates, oral cavity and gills.

DNA sequence

- 18S rRNA gene, 1940 bp. GenBank Accession: AM157183
- 18S rRNA gene, 559 bp. GenBank Accession: AJ313462
- 28S rRNA gene, 1314 bp. GenBank Accession: AM157197
- Abd-A/Lox4 gene, 75 bp. GenBank Accession: FN298164
- Hox2/3 gene, 75 bp. GenBank Accession: FN298152
- Lab/Hox1 gene, 75 bp. GenBank Accession: FN298147

Morphology and morphometrics (μm):

Source	Reichenbach-Klinke (1966)	Whittington and Pichelin (1991)	Pichelin et al. (1991)
Sample size		15–54	43
Body length			1818–3357 (2678)
Greatest width			482–1010 (676)
Width at vagina			
Haptor length			699–948 (824)
Haptor width			777–1041 (936)

(continued)

Morphology and morphometrics (μm):			
Source	Reichenbach-Klinke (1966)	Whittington and Pichelin (1991)	Pichelin et al. (1991)
False oral sucker width			
Pharynx length			128–188 (150)
Pharynx width			156–222 (185)
Testis length			
Testis width			
Ovary length			
Ovary width			
Egg length		269–356 (305)	203–275 (245)
Egg width		79–107 (95)	69–97 (78)
Egg incubation time			
Number of intra-uterine eggs		1	1 (n = 20)
Genital bulb length/width			222–366 (321) × 128–244 (177)
Number of genital spines			
Genital spine length			
Haptoral sucker length/width			225–363 (304) × 213–338 (277)
Hamulus handle length X			90–102 (97)
Hamulus guard length Y			
Hamulus hook length			
Marginal hooklet length			20–25 (22)
Haptoral L/Body L			0.307
No of anastomoses			0

Remarks *Concinnocotyla australensis* was originally described as *Eupolystoma australensis* Reichenbach-Klinke, 1966. It is the only polystomatid infecting fish.

5.4 Chelonian Polystomes (Fig. 5.3)

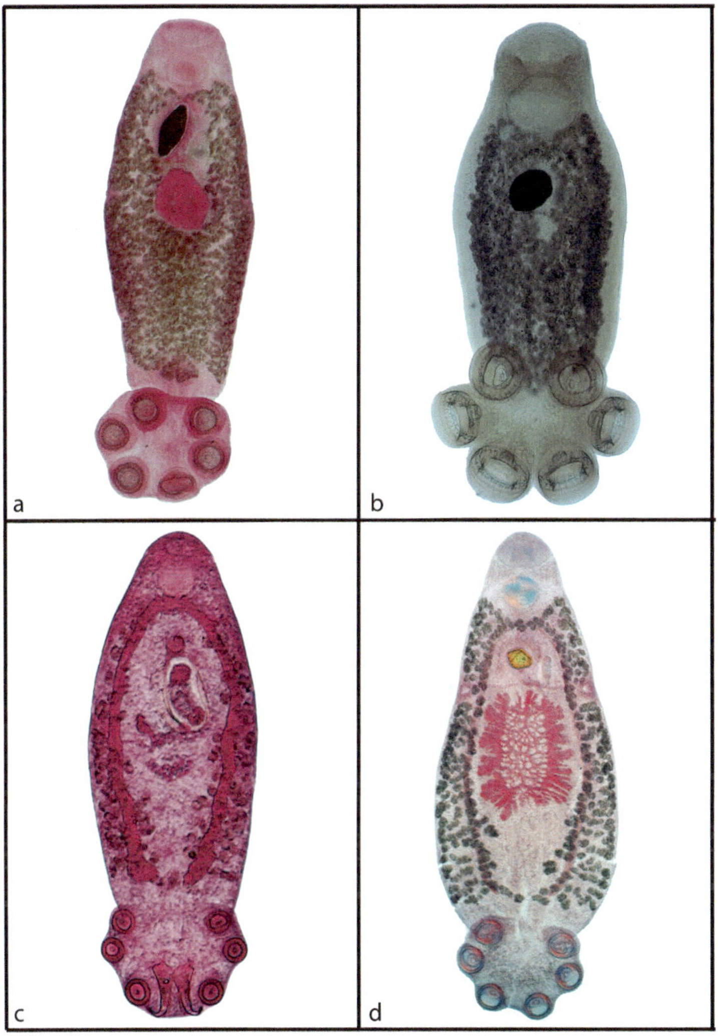

Fig. 5.3 Micrographs of (**a**) *Fornixtrema guianensis;* (**b**) *Pleurodirotrema kreffti*; (**c**) *Uropolystomoides chabaudi*; (**d**) *Polystomoides cayensis*

Nine genera are currently recognized, namely *Apaloneotrema* Du Preez and Verneau, 2020, *Aussietrema* Du Preez and Verneau, 2020, *Fornixtrema* Du Preez and Verneau, 2020, *Manotrema* Du Preez, Domingues and Verneau, 2022, *Pleurodirotrema* Du Preez, Domingues and Verneau, 2022, *Polystomoidella* Price, 1939, *Polystomoides* Ward, 1917 sensu Chaabane et al. (2022), *Uropolystomoides* Tinsley and Tinsley, 2016 and *Uteropolystomoides* Tinsley, 2017.

5.4.1 *Apaloneotrema* Du Preez and Verneau, 2020

5.4.1.1 *Apaloneotrema moleri* (Du Preez and Morrison, 2012) (Fig. 5.4)

Collection Parasitic Worm Collection, National Museum, Bloemfontein, South Africa (NMBP).

Holotype NMBP 335.

Paratypes NMBP 336–340.

Original description Du Preez and Morrison (2012).

Other taxonomic contributions None.

Etymology Named after Paul Moler, Gainesville, Florida, USA.

Type locality Gainesville Florida, USA.

Other localities None.

Type host *Apalone ferox* (Schneider, 1783), Florida Softshell Turtle. In the original description of the polystome, the host was known as *Trionyx ferox* Schweigger, 1812.

Additional hosts None.

Host geographical distribution *Apalone ferox* occurs in the USA (SW South Carolina, S Georgia, Florida, S Alabama) (Uetz et al. 2022).

Host conservation status The predators of the Florida Softshell Turtle include large fish, birds, certain terrapins, and many kinds of mammals. That species is also affected by pollution, habitat desiccation, and roads. It is considered rare in many areas. It is listed as Least Concern by The IUCN Red List of Threatened Species 2011: e.T165597A97397831. https://doi.org/10.2305/IUCN.UK.2011-1.RLTS.T165597A6065209.en. Accessed on 21 January 2022.

Site on host Conjunctival sacs of the eye.

DNA sequence
- 18S rRNA gene, 1957 bp. GenBank Accession: MW029406
- 28S rRNA gene, 1366 bp. GenBank Accession: MW029412
- Cytochrome c Oxidase I, 396 bp. GenBank Accession: MW029424
- 12S rRNA gene, 474 bp. GenBank Accession: MW029418

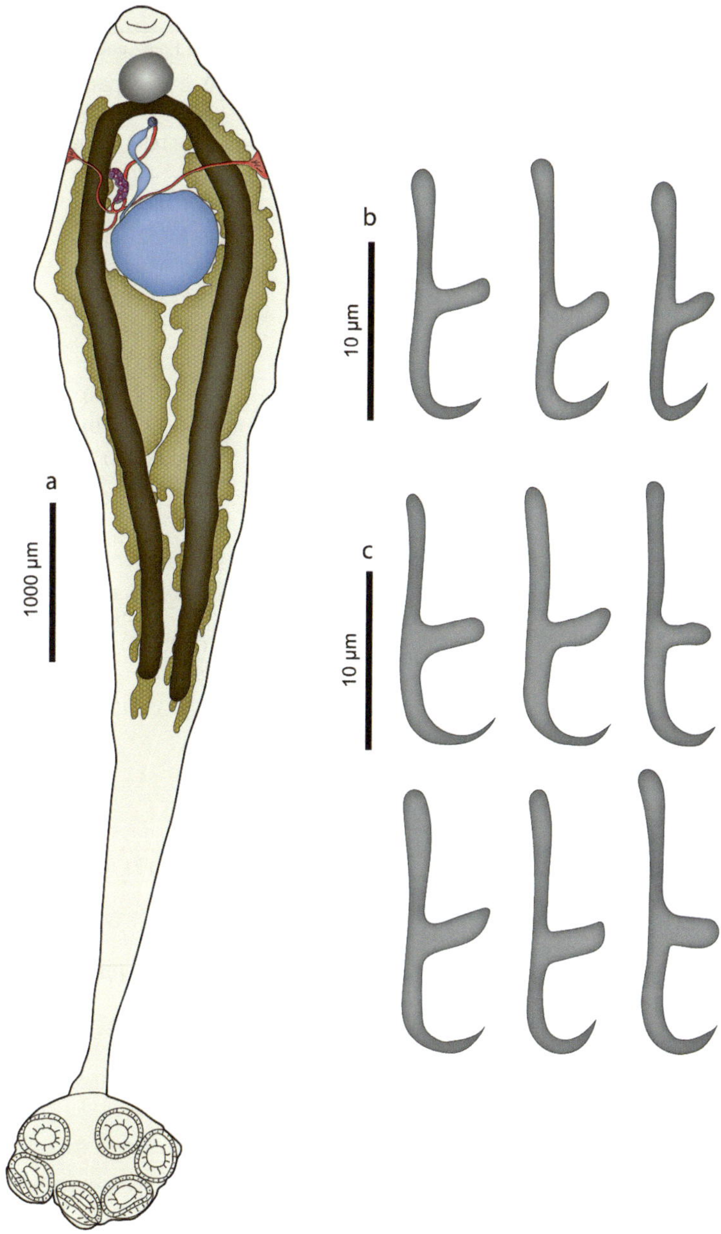

Fig. 5.4 (**a**) *Apaloneotrema moleri*, (**b**) marginal hooks (1), (**c**) marginal hooks (2–8). Redrawn from Du Preez and Morrison (2012)

Morphology and morphometrics (µm):	
Source	Du Preez and Morrison (2012)
Sample size	5
Body length	3249–7944 (5975)
Greatest width	926–1694 (1269)
Width at vagina	599–1426 (984)
Haptor length	601–983 (779)
Haptor width	784–1522 (1027)
False oral sucker width	180–358 (288)
Pharynx length	243–415 (309)
Pharynx width	258–392 (331)
Testis length	449–753 (633)
Testis width	328–719 (548)
Ovary length	124–312 (210)
Ovary width	73–138 (92)
Egg length	358
Egg width	168
Egg incubation time	
Number of intra-uterine eggs	1
Genital bulb width	36–79 (65)
Number of genital spines	12–13
Genital spine length	8–10 (9)
Haptoral sucker diameter	225–351 (280)
Hamulus handle length X	
Hamulus guard length Y	
Hamulus hook length	
Marginal hooklet length	
Haptoral L/Body L	0.11–0.21 (0.16)
No of anastomoses	0

Remarks *Apaloneotrema moleri* was originally described as *Neopolystoma moleri* Du Preez and Morrison, 2012.

5.4.2 *Aussietrema* Du Preez and Verneau, 2020

5.4.2.1 *Aussietrema cribbi* (Pichelin, 1995) (Fig. 5.5)

Collection Queensland Museum, Australia (QM).

Holotype QM GL 18004.

Paratypes QM GL 18005, 18006.

Original description Pichelin (1995).

Other taxonomic contributions None.

Etymology Named after Dr. Tom H Cribb, Brisbane, Australia.

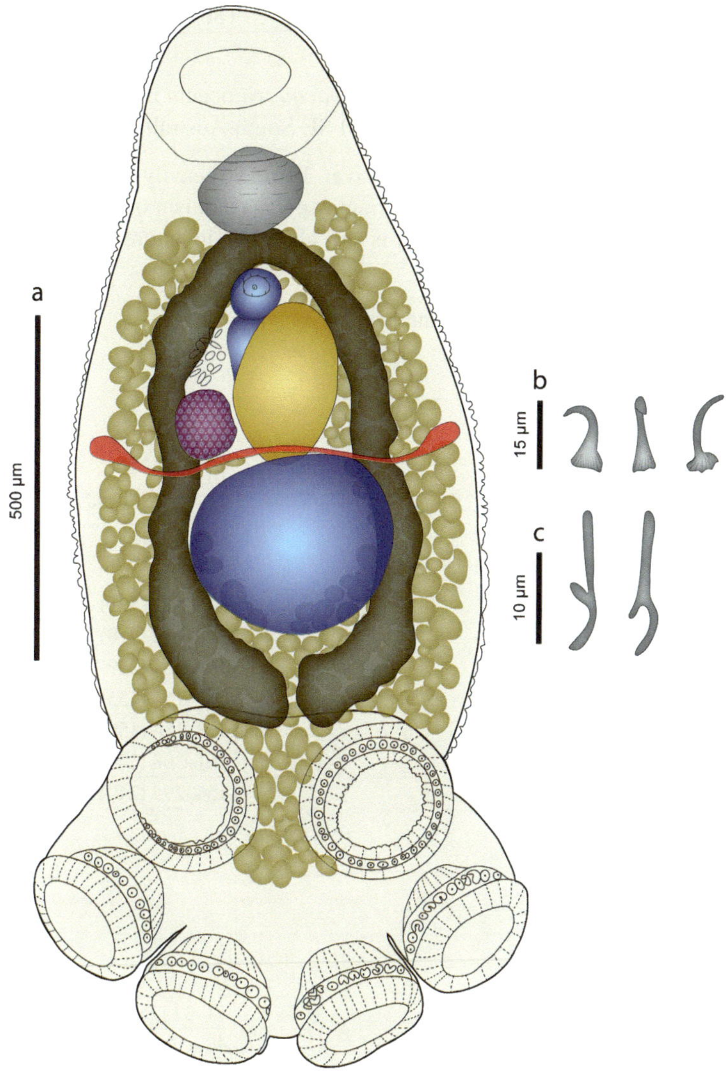

Fig. 5.5 (**a**) *Aussietrema cribbi*, (**b**) genital spines, (**c**) marginal hooklets. Redrawn from Pichelin (1995)

Type locality University of Queensland Veterinary Farm dam (27°32′S, 152°54′E) Pinjarra Hills, Queensland, Australia.

Other localities Darling Downs, Queensland, Australia: Canal Creek (28°01′S, 151°35′E) and Leslie Dam (28°01′S, 152°55′E).

Type host *Emydura macquarii* (Gray, 1830), Murray River Turtle. In the original description of the polystome, the host was known as *Emydura signata* Ahl, 1932.

Additional hosts *Myuchelys latisternum* (Gray, 1867); *Chelodina expansa* Gray, 1857.

Host geographical distribution *Emydura macquarii* occurs in SE Australia, from SC Queensland southward to N Victoria and SE South Australia (Uetz et al. 2022).

Host conservation status The Murray River Turtle is heavily collected for parks and zoos, and is one of the Australian terrapin species that is most frequently found in Western collections. Nevertheless, it is still abundant in nature. It is not listed in The IUCN Red List.

Site on host Conjunctival sacs of the eye.

DNA sequence None.

Morphology and morphometrics (μm):	
Source	Pichelin (1995)
Sample size	10
Body length	1433–1688 (1581)
Greatest width	509–669 (619)
Width at vagina	
Haptor length	446–541 (505)
Haptor width	700–812 (770)
False oral sucker length/width	193–257 (229) × 270–315 (295)
Pharynx length	116–154 (135)
Pharynx width	128–180 (159)
Testis length	161–308 (249)
Testis width	154–295 (240)
Ovary length	
Ovary width	
Egg length	
Egg width	
Egg incubation time	
Number of intra-uterine eggs	0
Genital bulb length/width	71–83 (75) × 64–96 (78)
Number of genital spines	40 (curved); 13 (straight)
Genital spine length	6–10 (8), 11.2–22.4 (17.8)
Haptoral sucker diameter	190–231 (215)
Hamulus handle length X	
Hamulus guard length Y	
Hamulus hook length	
Marginal hooklet length	15.2–17.6 (16.5)
Haptoral L/Body L	0.3194

Remarks *Aussietrema cribbi* was originally described as *Neopolystoma cribbi* Pichelin, 1995.

5.4.2.2 *Aussietrema queenslandense* (Pichelin, 1995) (Fig. 5.6)

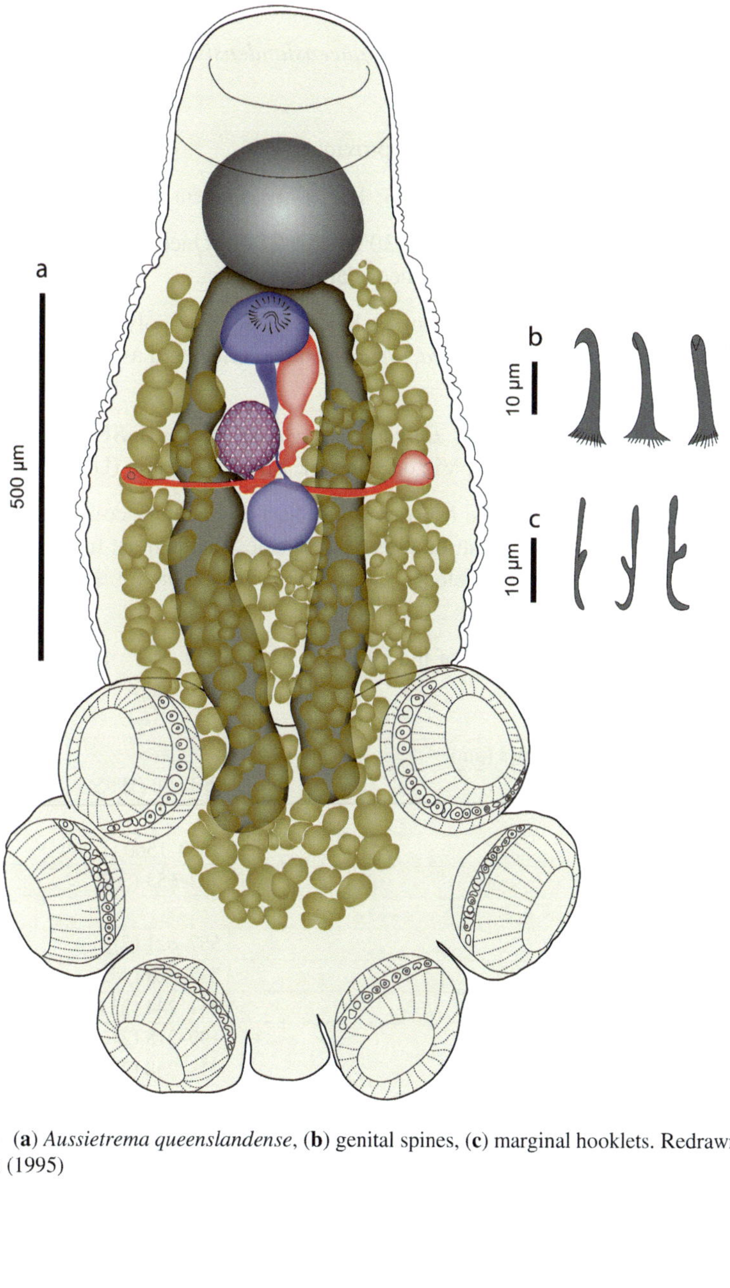

Fig. 5.6 (**a**) *Aussietrema queenslandense*, (**b**) genital spines, (**c**) marginal hooklets. Redrawn from Pichelin (1995)

Collection Queensland Museum, Australia (QM).

Holotype QM GL 18010.

Paratypes QM GL 18011.

Original description Pichelin (1995).

Taxonomic note The name *Aussietrema queenslandensis* was corrected for gender.

Other taxonomic contributions None.

Etymology Named after type locality province.

Type locality Lockyer Creek (27°35′S; 152°07′E), Queensland, Australia.

Other localities Lake Clarendon (27°30′S; 152°21′E), Queensland, Australia; and Canal Creek (28°01′S; 151°35′E), Queensland, Australia.

Type host *Emydura macquarii*, Murray River Turtle. The polystome species was also reported from *Emydura signata*, which is now considered as a junior synonym of *E. macquarii*.

Host geographical distribution *Emydura macquarii* occurs in SE Australia, from SC Queensland southward to N Victoria and SE South Australia (Uetz et al. 2022).

Host conservation status The Murray River Turtle is heavily collected for parks and zoos, and is one of the Australian terrapin species that is most frequently found in Western collections. Nevertheless, it is still abundant in nature. It is not listed in The IUCN Red List.

Site on host Conjunctival sacs of the eye.

DNA sequence None.

Morphology and morphometrics (μm):	
Source	Pichelin (1995)
Sample size	3
Body length	1369–1560 (1480)
Greatest width	462–653 (563)
Width at vagina	
Haptor length	557–621 (578)
Haptor width	732–828 (785)
False oral sucker length/width	231–244 (233) × 263–308 (289)
Pharynx length	199–218 (210)
Pharynx width	212–238 (225)
Testis length	77–135 (103)
Testis width	77–135 (103)
Ovary length	
Ovary width	
Egg length	

(continued)

Morphology and morphometrics (μm):	
Source	Pichelin (1995)
Egg width	
Egg incubation time	
Number of intra-uterine eggs	0
Genital bulb length/width	96–135 (111) × 103–128 (116)
Number of genital spines	22–28 or 29 Straight with hooked tip
Genital spine length	19.2–25.6 (21.6)
Haptoral sucker diameter	207–215 (211)
Hamulus handle length X	
Hamulus guard length Y	
Hamulus hook length	
Marginal hooklet length	16.8–19.2 (17.8)
Haptoral L/Body L	0.3905

Remarks *Aussietrema queenslandense* was originally described as *Neopolystoma queenslandensis* Pichelin 1995.

5.4.2.3 *Aussietrema spratti* (Pichelin, 1995) (Fig. 5.7)

Collection Queensland Museum, Australia (QM).

Holotype QM GL 18001.

Paratypes QM GL 18002, 18003.

Original description Pichelin (1995).

Other taxonomic contributions None.

Etymology Named after Dr. DM Spratt.

Type locality Macquarie Marshes, New South Wales, Australia.

Other localities None.

Type host *Chelodina longicollis* (Shaw, 1794), Common Snake-Necked Turtle.

Additional hosts None.

Host geographical distribution *Chelodina longicollis* occurs in Australia New South Wales, Queensland, South Australia and Victoria (Uetz et al. 2022).

Host conservation status The Common Snake-Necked Turtle is a popular pet and zoo animal, and there is a definite drain upon wild populations. It is not listed in The IUCN Red List.

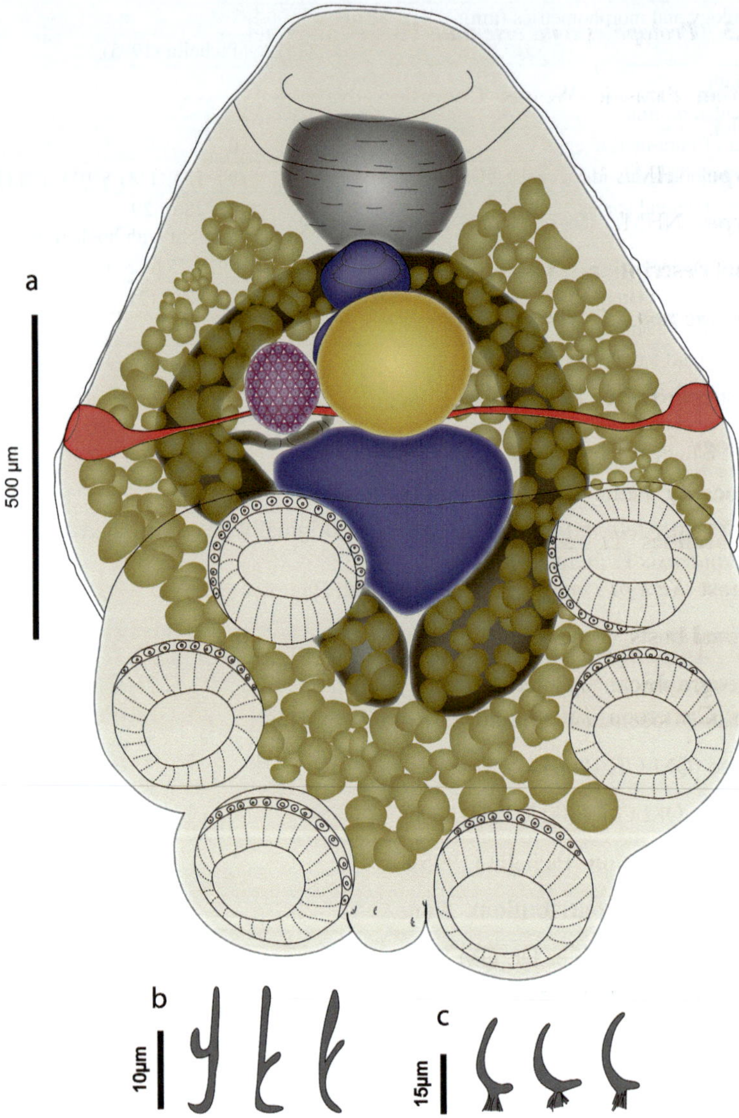

Fig. 5.7 (**a**) *Aussietrema spratti*, (**b**) genital spines, (**c**) marginal hooklets. Redrawn from Pichelin (1995)

Site on host Conjunctival sacs of the eye.

DNA sequence

- 18S rRNA gene, 2015 bp. GenBank Accession: AJ228788
- 18S rRNA gene, 561 bp. GenBank Accession: AJ297782
- 28S rRNA gene, 1368 bp. GenBank Accession: FM992702

- 28S rRNA gene, 957 bp. GenBank Accession: Z83006
- Cytochrome c Oxidase I, 396 bp. GenBank Accession: Z83007
- 12S rRNA gene, 373 bp. GenBank Accession: KR856105

Morphology and morphometrics (μm):	
Source	Pichelin (1995)
Sample size	5
Body length	1146–1305 (1239)
Greatest width	828–923 (866)
Width at vagina	
Haptor length	589–716 (665)
Haptor width	828–955 (907)
False oral sucker length/width	180–257 (229) × 302–385 (351)
Pharynx length	167–238 (209)
Pharynx width	193–231 (217)
Testis length	205–321 (287)
Testis width	289–360 (315)
Ovary length	
Ovary width	
Egg length	
Egg width	
Egg incubation time	
Number of intra-uterine eggs	1
Genital bulb length/width	77–109 (94) × 97–135 (114)
Number of genital spines	20–26 (23)
Genital spine length	19.2–27.2 (23.9)
Haptoral sucker diameter	178–212 (202)
Hamulus handle length X	
Hamulus guard length Y	
Hamulus hook length	
Marginal hooklet length	14.4–19.2 (17.3)
Haptoral L/Body L	0.5367

Remarks *Aussietrema spratti* was originally described as *Neopolystoma spratti* Pichelin, 1995.

5.4.2.4 *Aussietrema tinsleyi* (Pichelin, 1995) (Fig. 5.8)

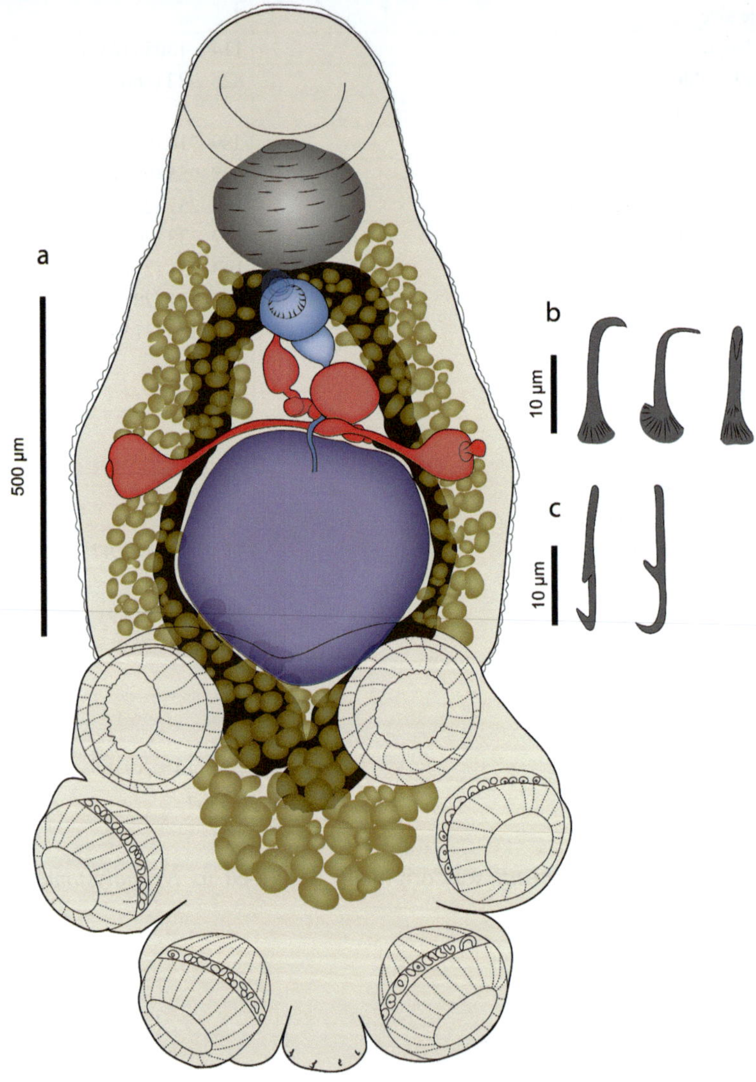

Fig. 5.8 (**a**) *Aussietrema tinsleyi*, (**b**) genital spines, (**c**) marginal hooklets. Redrawn from Pichelin (1995)

Collection Queensland Museum, Australia (QM).

Holotype QM GL 18007.

Paratypes QM GL 18008, 18009.

Original description Pichelin (1995).

Other taxonomic contributions None.

Etymology Named after Prof Richard C Tinsley, University of Bristol, Bristol, UK.

Type locality Canal Creek (28°01′S, 151°35′E), Darling Downs, Queensland, Australia.

Other localities Queensland, Australia–The University of Queensland Veterinary Farm dam (37°32′S, 152°54′E).

Type host *Chelodina expansa* Gray, 1857, Giant Snake-Necked Turtle.

Additional hosts None.

Host geographical distribution *Chelodina expansa* occurs in Australia—Queensland: from Rockhampton southwest through the coastal and interior areas to W New South Wales and the Murray River drainage, and on Fraser Island off Queensland; South Australia, Victoria (Uetz et al. 2022).

Host conservation status The Giant Snake-Necked Turtle makes popular pets. Precise counts of populations in the wild are extremely difficult because it hides very effectively and it is almost impossible to find. It is not listed in The IUCN Red List.

Site on host Conjunctival sacs of the eye.

DNA sequence None.

Morphology and morphometrics (μm):	
Source	Pichelin (1995)
Sample size	9
Body length	1305–1656 (1500)
Greatest width	557–732 (640)
Width at vagina	
Haptor length	509–685 (601)
Haptor width	700–877 (804)
False oral sucker length/width	218–321 (246) × 244–321 (276)
Pharynx length	167–205 (188)
Pharynx width	199–238 (219)
Testis length	315–482 (383)
Testis width	289–398 (332)
Ovary length	
Ovary width	

(continued)

Morphology and morphometrics (μm):	
Source	Pichelin (1995)
Egg length	
Egg width	
Egg incubation time	
Number of intra-uterine eggs	1
Genital bulb length/width	83–96 (92) × 83–96 (87)
Number of genital spines	21–27 (23)
Genital spine length	14.4–20.8 (16.8)
Haptoral sucker diameter	180–212 (193)
Hamulus handle length X	
Hamulus guard length Y	
Hamulus hook length	
Marginal hooklet length	14.4–19.2 (17.1)
Haptoral L/Body L	0.2651–0.4006

Remarks *Aussietrema tinsleyi* was originally described as *Neopolystoma tinsleyi* Pichelin, 1995.

5.4.3 *Fornixtrema* Du Preez and Verneau, 2020

5.4.3.1 *Fornixtrema elizabethae* (Platt, 2000) (Fig. 5.9)

Collection United States National Museum (USNM).

Holotype USNM 1384593.

Paratypes USNM 1384594–5.

Original description Platt (2000a).

Other taxonomic contributions None.

Etymology Named after Dr. Elizabeth Scarborough.

Type locality Kinwamakwad Lake (46′14″N, 89′30″W), UNDERC, Gogebic County, Michigan.

Other localities Plum Lake (46′13″N, 89′30″W), UNDERC, Gogebic County, Michigan and Vilas County, Wisconsin, Unnamed pond (41′37″N, 86′17″W), Centre Township, St Joseph County, Indiana.

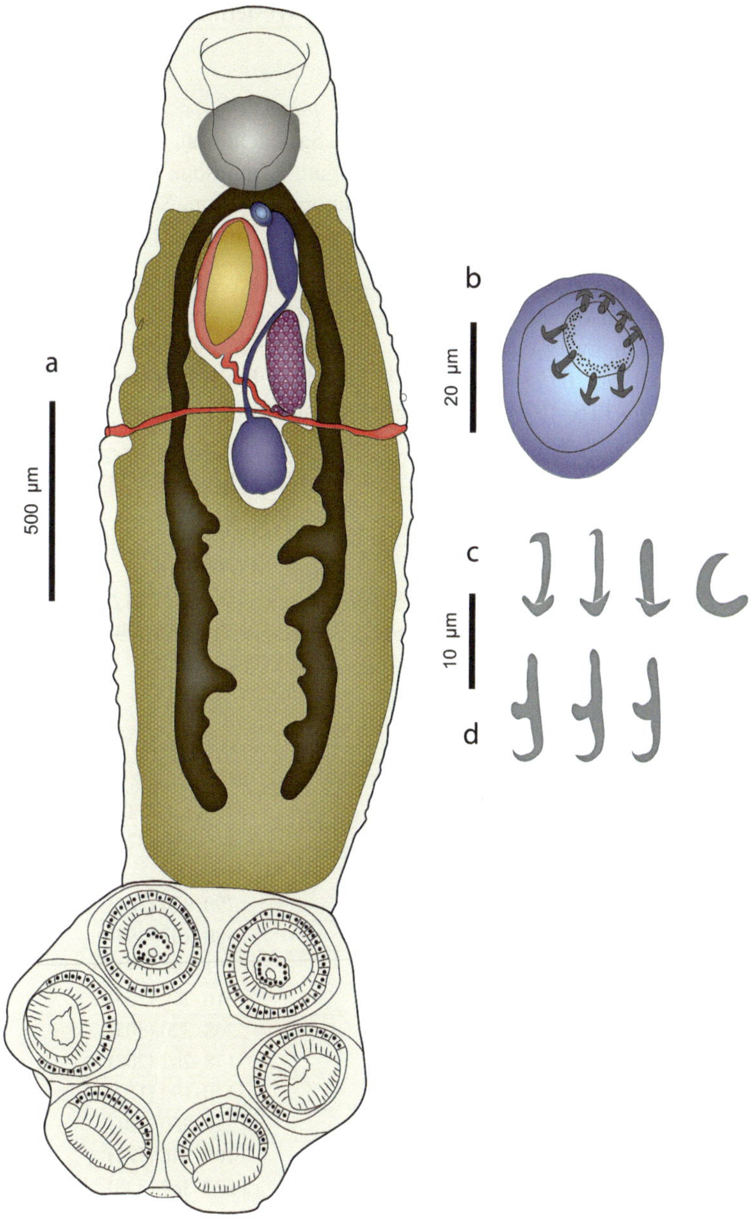

Fig. 5.9 (**a**) *Fornixtrema elizabethae*, (**b**) genital bulb, (**c**) genital spines, (**d**) marginal hooklets. Redrawn from Platt (2000a)

Type host *Chrysemys picta bellii* (Gray, 1831), Western Painted Turtle.

Additional hosts None.

Host geographical distribution *Chrysemys picta bellii* occurs in S Canada—Alberta, British Columbia, Manitoba, Ontario, Saskatchewan, USA—Arizona, Colorado, Idaho, Illinois, Iowa, Kansas, Michigan, Minnesota, Missouri, Montana, Nebraska, New Mexico, North Dakota, Oklahoma, Oregon, South Dakota, Texas, Utah, Washington, Wisconsin, Wyoming, Mexico (Chihuahua) (Uetz et al. 2022).

Host conservation status The worst predator for The Western Painted Turtle is the raccoon, which can detect and excavate eggs. Other threats also include habitat degradation and highways. It is listed as Least Concern by The IUCN Red List of Threatened Species 2011: e.T163467A97410447. https://doi.org/10.2305/IUCN.UK.2011-1.RLTS.T163467A5608383.en. Accessed on 21 January 2022.

Site on host Conjunctival sacs of the eye.

DNA sequence
- 18S rRNA gene, 1981 bp. GenBank Accession: MW029402
- 28S rRNA gene, 1366 bp. GenBank Accession: MW029408
- Cytochrome c Oxidase I, 396 bp. GenBank Accession: KY744297
- Cytochrome c Oxidase I, 372 bp. GenBank Accession: MW029420
- 12S rRNA gene, 465 bp. GenBank Accession: MW029414

Morphology and morphometrics (μm):	
Source	Platt (2000a)
Sample size	4
Body length	2550–3675 (3125)
Greatest width	640–990 (823)
Width at vagina	
Haptor length	790–970 (865)
Haptor width	880–1070 (975)
False oral sucker length/width	251–292 (271) × 449–540 (473)
Pharynx length	216–269 (255)
Pharynx width	268–320 (305)
Testis length	178–262 (208)
Testis width	140–192 (155)
Ovary length	218–350 (301)
Ovary width	100–146 (122)
Egg length	322–367 (348)
Egg width	117–122 (120)
Egg incubation time	
Number of intra-uterine eggs	3
Genital bulb length/width	50–63 (58) × 45–63 (57)
Number of genital spines	8
Genital spine length	10

(continued)

Morphology and morphometrics (μm):	
Source	Platt (2000a)
Haptoral sucker diameter	344–408 (372)
Hamulus handle length X	
Hamulus guard length Y	
Hamulus hook length	
Marginal hooklet length	11.8–12.7 (12.3); $n = 8$
Haptoral L/Body L	0.28

Remarks *Fornixtrema elizabethae* was originally described as *Neopolystoma elizabethae* Platt, 2000.

5.4.3.2 *Fornixtrema fentoni* (Platt, 2000) (Fig. 5.10)

Collection United States National Museum (USNM); Helminthological Collection, Oswaldo Cruz Institute (CHIOC).

Holotype USNM 1385135 (89943).

Paratypes USNM 1385136–8 (89944–6); CHIOC 34292, 34293a–b, 34294.

Original description Platt (2000b).

Other taxonomic contributions None.

Etymology Named after Mr. William S Fenton.

Type locality Quebrada Costa Rica (10°49.666′N; 85°38.216′W), Guanacaste Conservation Area, Santa Rosa, Guanacaste, Costa Rica, South America.

Other localities Quebrada el Duende (10°50.236′N; 85°38.724′W) and San Gerardo (10°52.55′N; 85°23.27′W), Guanacaste Conservation Area, Santa Rosa, Guanacaste, Costa Rica.

Type host *Kinosternon leucostomum* (Duméril, Bibron and Duméril, 1851), White-Lipped Mud Turtle.

Additional hosts *Rhinoclemmys pulcherrima* (Gray, 1855).

Host geographical distribution *Kinosternon leucostomum* occurs in Mexico—Campeche, Quintana Roo, CS Veracruz, Oaxaca, Tabasco, Chiapas, Yucatan, Belize, Guatemala, El Salvador, Honduras, Nicaragua, Costa Rica, Panama, Colombia, Ecuador and N Peru (Uetz et al. 2022).

Host conservation status The White-Lipped Mud Turtle is regularly captured and consumed, but still seems to be abundant in nature. It is not listed in the IUCN Red List.

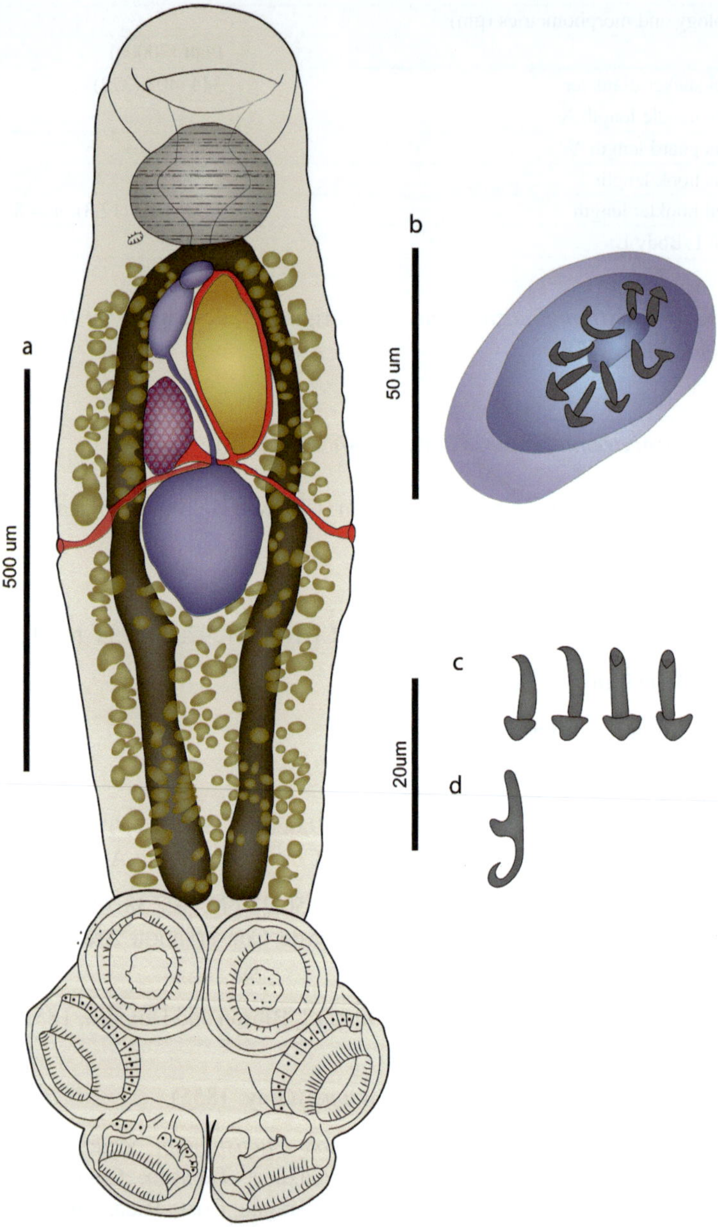

Fig. 5.10 (**a**) *Fornixtrema fentoni*, (**b**) genital bulb, (**c**) genital spines, (**d**) marginal hooklets. Redrawn from Platt (2000b)

Site on host Conjunctival sacs of the eye.

DNA sequence from polystomes of *K. leucostomum*

- 18S rRNA gene, 1982 bp. GenBank Accession: KR856133
- 28S rRNA gene, 1148 bp. GenBank Accession: KR856152
- Cytochrome c Oxidase I, 395 bp. GenBank Accession: KR856175
- 12S rRNA gene, 478 bp. GenBank Accession: KR856109

DNA sequence of polystomes of *R. pulcherrima*

- 18S rRNA gene, 1994 bp. GenBank Accession: KR856134
- 28S rRNA gene, 1147 bp. GenBank Accession: KR856153
- Cytochrome c Oxidase I, 396 bp. GenBank Accession: FR822555
- 12S rRNA gene, 440 bp. GenBank Accession: KR856110

Morphology and morphometrics (μm):	
Source	Platt (2000b)
Sample size	10
Body length	1500–2450 (1985)
Greatest width	426–760 (568)
Width at vagina	
Haptor length	449–690 (571)
Haptor width	550–850 (683)
False oral sucker length/width	150–303 (230) × 240–496 (370)
Pharynx length	156–257 (216)
Pharynx width	185–367 (278)
Testis length	98–367 (225)
Testis width	78–251 (181)
Ovary length	80–245 (103)
Ovary width	55–169 (105)
Egg length	245–332 (286); $n = 7,3$
Egg width	122–146 (136)
Egg incubation time	
Number of intra-uterine eggs	1
Genital bulb length/width	43–70 (55) × 30–83 (60)
Number of genital spines	8
Genital spine length	11
Haptoral sucker diameter	210–326 (265); $n = 60, 24$
Hamulus handle length X	
Hamulus guard length Y	
Hamulus hook length	
Marginal hooklet length	11.8–12.7 (12.5); $n = 4$
Haptoral L/Body L	0.2877

Remarks *Fornixtrema fentoni* was originally described as *Neopolystoma fentoni* Platt, 2000. According to Héritier et al. (2015), the polystomes from both reported host species may be regarded as two distinct species following genetic analyses. Further morphological investigation is required to conclude.

5.4.3.3 *Fornixtrema grossi* (Du Preez and Morrison, 2012) (Fig. 5.11)

Collection Specimens are deposited in the Parasitic Worm Collection, National Museum, Bloemfontein, South Africa (NMBP).

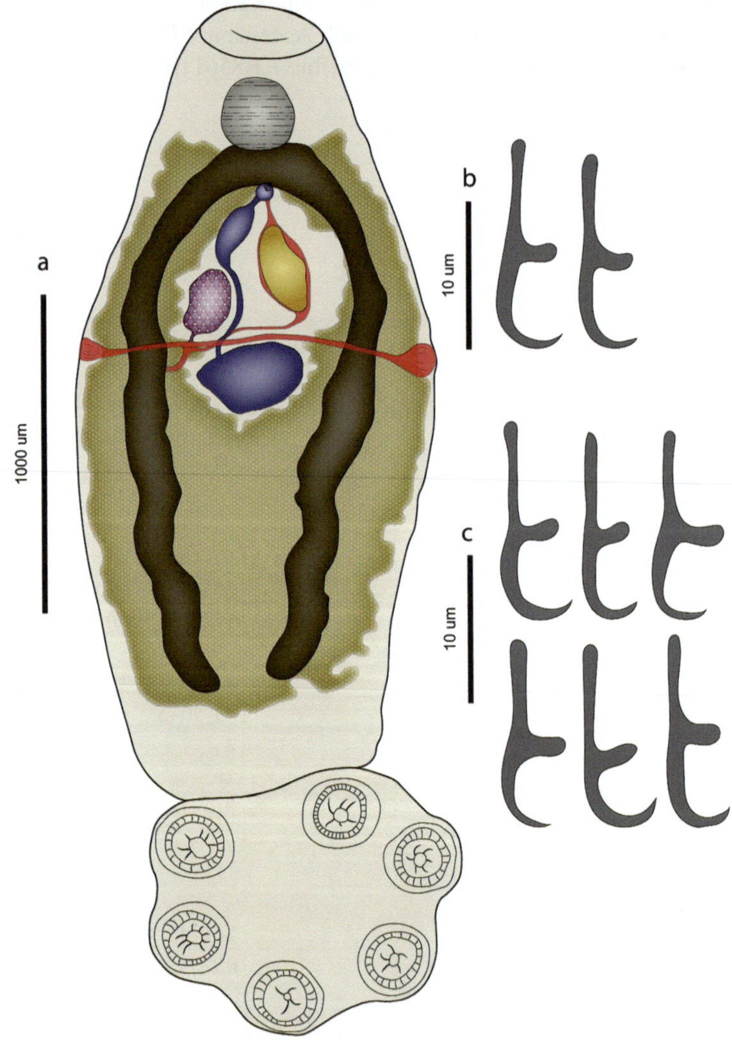

Fig. 5.11 (**a**) *Fornixtrema grossi*, (**b**) marginal hooklets (1), (**c**) marginal hooklets (2–8). Redrawn from Du Preez and Morrison (2012)

Holotype NMBP 341.

Paratypes NMBP 342–343.

Original description Du Preez and Morrison (2012).

Other taxonomic contributions None.

Etymology Named after Dr. Timothy Gross, Gainesville, Florida, USA.

Type locality Pond at the United States Geological Survey USGS-BRD facility, Florida Integrated Science Centres, 7920 NW 71st Street, Gainesville FL 32653, USA (29.725278 N; 82.417778 W).

Other localities None.

Type host *Pseudemys concinna floridana* (Le Conte, 1830), Florida Cooter.

Additional hosts None.

Host geographical distribution *Pseudemys concinna floridana* occurs in the USA—along the Atlantic Coastal Plain in North Carolina, South Carolina, Florida, Georgia, and west across the Gulf Coastal Plain in Alabama, Mississippi, Louisiana, E Texas, Oklahoma, SE Kansas, Arkansas, S Missouri, SW Indiana, W Kentucky, W Tennessee and Virginia (Uetz et al. 2022).

Host conservation status The Florida Cooter is widely distributed but suffers from exploitation for consumption, road mortality, predation by raccoons, foxes, opossums and crows that takes juveniles. Populations seem to be dropping because of habitat destruction and pollution due to dredging, mining, impoundment, industrial and agricultural pollution. It is listed as Least Concern by The IUCN Red List of Threatened Species 2011: e.T163444A97425355. https://doi.org/10.2305/IUCN.UK.2011-1.RLTS.T163444A5606651.en. Accessed on 21 January 2022.

Site on host Conjunctival sacs of the eye.

DNA sequence None.

Morphology and morphometrics (μm):	
Source	Du Preez and Morrison (2012)
Sample size	3
Body length	3298–4873 (4018)
Greatest width	911–1130 (988)
Width at vagina	900–1125 (983)
Haptor length	707–819 (769)
Haptor width	709–989 (891)
False oral sucker width	369–436 (403)
Pharynx length	206–232 (222)
Pharynx width	209–246 (229)
Testis length	186–293 (235)
Testis width	149–245 (236)

(continued)

Morphology and morphometrics (μm):	
Source	Du Preez and Morrison (2012)
Ovary length	210–222 (218)
Ovary width	118–135 (128)
Egg length	288–301 (293)
Egg width	137–140 (139)
Egg incubation time	
Number of intra-uterine eggs	1
Genital bulb width	58–60 (59)
Number of genital spines	6–8
Genital spine length	8–9 (8)
Haptoral sucker diameter	208–244 (241)
Hamulus handle length X	
Hamulus guard length Y	
Hamulus hook length	
Marginal hooklet length	
Haptoral L/Body L	0.15–0.24 (0.20)
No of anastomoses	0

Remarks *Fornixtrema grossi* was originally described as *Neopolystoma grossi* Du Preez and Morrison, 2012.

5.4.3.4 *Fornixtrema guianense* (Du Preez, Badets, Héritier and Verneau, 2017) (Fig. 5.12)

Collection Parasitic Worm Collection, National Museum, Bloemfontein, South Africa (NMBP).

Holotype NMBP 404.

Paratypes NMBP 405–P406.

Original description Du Preez et al. (2017).

Taxonomic note The name *Fornixtrema guianensis* was corrected for gender.

Other taxonomic contributions None.

Etymology Named after the type locality country.

Type locality Pond on the outskirts of the capital Cayenne, French Guiana (4.87082 N, 52.33678 W).

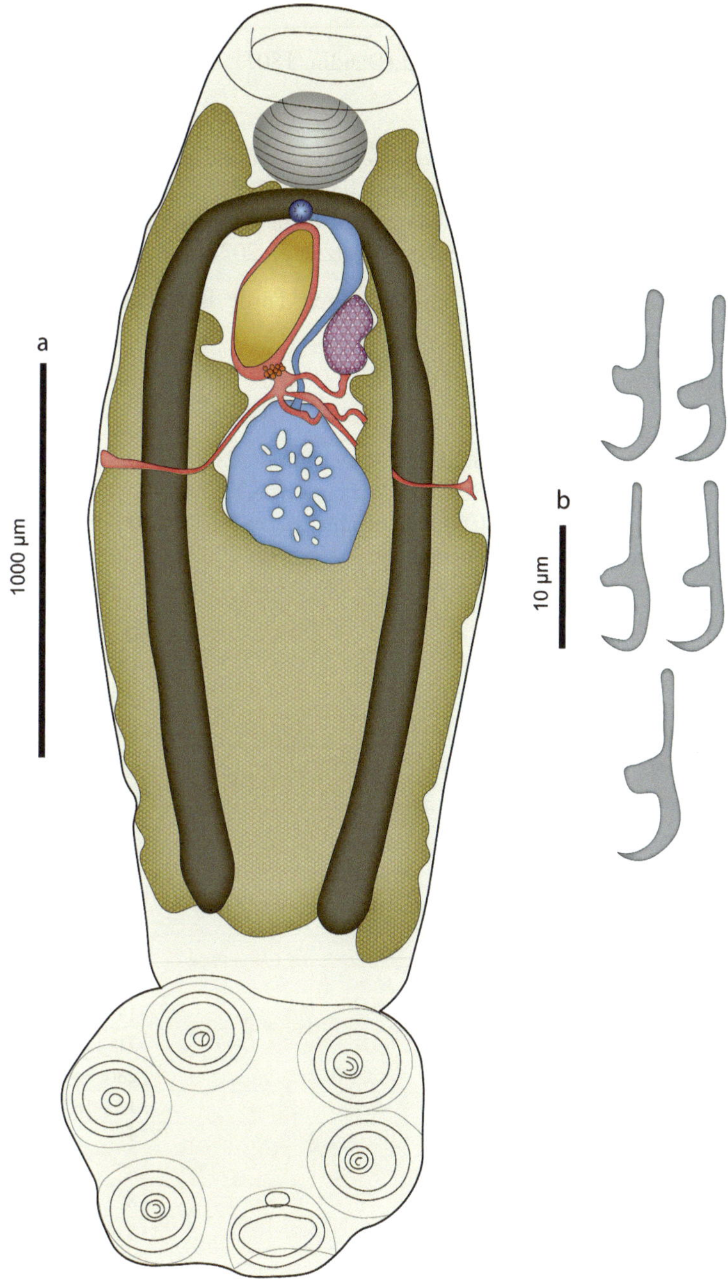

Fig. 5.12 (**a**) *Fornixtrema guianense*, (**b**) marginal hooklets. Redrawn from Du Preez et al. (2017)

Other localities None.

Type host *Rhinoclemmys punctularia* (Daudin, 1801), Spotted-legged Turtle.

Additional hosts None.

Host geographical distribution *Rhinoclemmys punctularia* occurs in E Colombia, Venezuela, Trinidad, Tobago, Guyana, French Guiana, Suriname, NE Brazil (Tocantins, Pará, Amazon, Bahia, Maranhão, as far upstream as the lower reaches of the Tapajós river and Rio de Janeiro) (Uetz et al. 2022).

Host conservation status The Spotted-legged Turtle is not listed in the IUCN Red List.

Site on host Conjunctival sacs of the eye.

DNA sequence

- 18S rRNA gene, 1993 bp. GenBank Accession: KY200987
- 28S rRNA gene, 1361 bp. GenBank Accession: KY200989
- Cytochrome c Oxidase I, 319 bp. GenBank Accession: KY200995
- 12S rRNA gene, 494 bp. GenBank Accession: KY200992

Morphology and morphometrics (μm):	
Source	Du Preez et al. (2017)
Sample size	11
Body length	2284–3342 (2737)
Greatest width	830–1164 (1037)
Width at vagina	792–1091 (989)
Haptor length	634–852 (755)
Haptor width	784–1014 (910)
False oral sucker width	189–391 (290)
Pharynx length	189–260 (213)
Pharynx width	226–283 (252)
Testis length	364–541 (437)
Testis width	380–527 (477)
Ovary length	149–243 (197)
Ovary width	71–120 (97)
Egg length	280–344 (317)
Egg width	127–163 (148)
Number of intra-uterine eggs	1
Genital bulb width	49
Number of genital spines	8
Genital spine length	8.4–9.8 (8.9)
Haptoral sucker diameter	208–242 (226)
Marginal hooklet length	13.3–15.1 (14.3)
Haptoral L/Body L	0.28

Remarks *Fornixtrema guianense* was originally described as *Neopolystoma guianensis* Du Preez, Badets, Héritier and Verneau, 2017.

5.4.3.5 *Fornixtrema liewi* (Du Preez and Lim, 2000) (Fig. 5.13)

Collection Parasitic Worms Collection, Natural History Museum London (NHML); Institute of Parasitology, České Budějovice, Czech Republic (M); Parasitic Worm Collection, National Museum, Bloemfontein, South Africa (NMBP).

Holotype NHML 1999.10.29.1.

Paratypes NHML 1999.10.29.2–4; M 355/1; NMBP 222–227.

Original description Du Preez and Lim (2000).

Other taxonomic contributions Platt et al. (2011).

Etymology Named after Mr. George Liew, Kuala Lumpur, Malaysia.

Type locality Well vegetated earth-walled dam in the Botanical Garden (Hutan Rimba) on the campus of the University of Malaya, Kuala Lumpur, Malaysia.

Other localities Pet shops in Kuala Lumpur, collected in the Perak area; Selangor State, Malaysia.

Type host *Cuora amboinensis* (Daudin, 1801), Southeast Asian Box Turtle.

Additional hosts None.

Host geographical distribution *Cuora amboinensis* occurs in India, Bhutan, Bangladesh, Myanmar, Thailand, Cambodia, Vietnam, West Malaysia, Singapore, and Philippines (Uetz et al. 2022).

Host conservation status The Southeast Asian Box Turtle is the most frequently consumed terrapin species in China restaurants, also captured and released in Buddhist temple ponds, and gathered for exports. These practices may lead to a population collapse in the future. It is listed as Endangered by The IUCN Red List of Threatened Species 2020: e.T5958A3078812. https://doi.org/10.2305/IUCN.UK.2020-2.RLTS.T5958A3078812.en. Accessed on 21 January 2022.

Site on host Conjunctival sacs of the eye.

DNA sequence
- 18S rRNA gene, 1970 bp. GenBank Accession: KR856128
- 18S rRNA gene, 573 bp. GenBank Accession: AJ297785
- 28S rRNA gene, 1367 bp. GenBank Accession: KR856147
- 28S rRNA gene, 1214 bp. GenBank Accession: AF382066
- Cytochrome c Oxidase I, 396 bp. GenBank Accession: FR822530
- 12S rRNA gene, 482 bp. GenBank Accession: KR856102

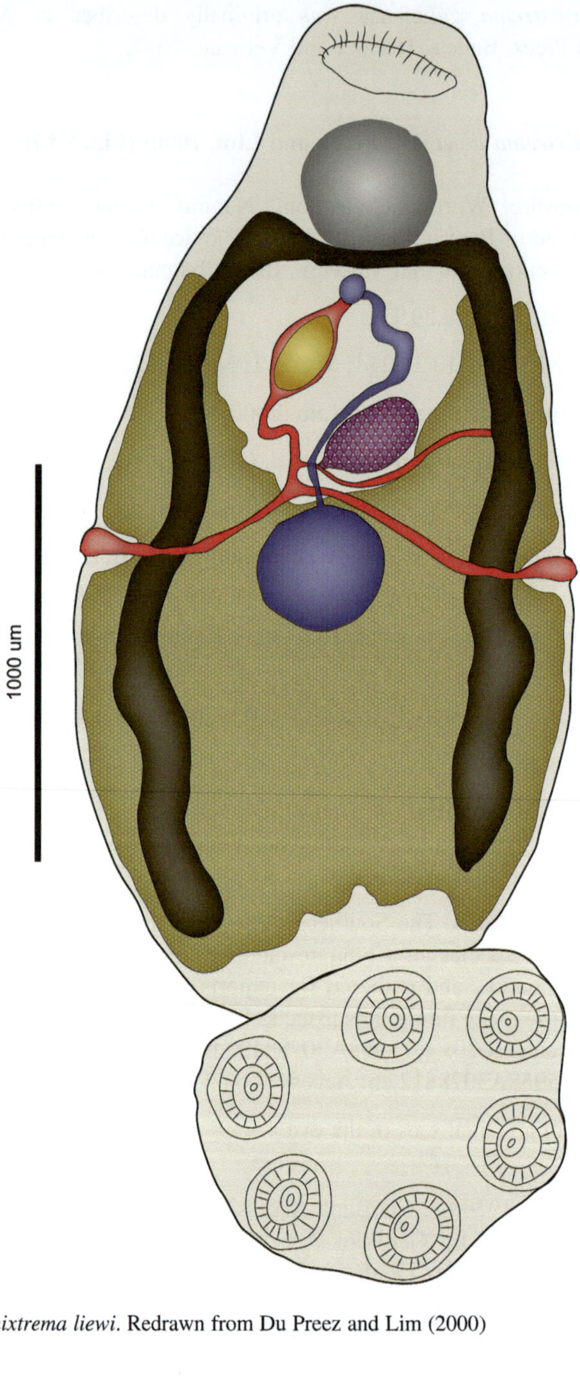

Fig. 5.13 *Fornixtrema liewi*. Redrawn from Du Preez and Lim (2000)

Morphology and morphometrics (μm):		
Source	Du Preez and Lim (2000)	Platt et al. (2011)
Sample size	11	11
Body length	2120–4169 (3474)	1925–2800 (2243)
Greatest width	1012–1566 (1321)	724–938 (848)
Width at vagina	1012–1446 (1246)	704–867 (804)
Haptor length	795–1205 (921)	571–683 (628)
Haptor width	988–1470 (1179)	745–877 (821)
False oral sucker width	252–446 (367)	329–428 (374)
Pharynx length	296–398 (349)	204–251 (234)
Pharynx width	330–417 (382)	256–308 (292)
Testis length	271–446 (341)	120–339 (236)
Testis width	349–436 (363)	120–339 (212)
Ovary length	213–339 (287)	115–245 (181)
Ovary width	116–155 (136)	80–131 (98)
Egg length	265–294 (283)	266–308 (287)
Egg width	109–126 (120)	110–125 (118)
Egg incubation time	1.5 (0–3) eggs per 24 h	
Number of intra-uterine eggs	1	1
Genital bulb length/width	63–87 (78) × 53–92 (72)	68–88 (74) × 60–93 (72)
Number of genital spines	8–11	8–12
Genital spine length	10.8–13.2 (11.8)	10–13
Haptoral sucker diameter	184–252 (217)	210–246 (231) (diameter)
Hamulus handle length X		
Hamulus guard length Y		
Hamulus hook length		
Marginal hooklet length	12.2–12.7 (12.6)	12.2–12.7 (12.6)
Haptoral L/Body L	0.27	0.41

Remarks *Fornixtrema liewi* was originally described as *Neopolystoma liewi* Du Preez and Lim, 2000. Measures reported by Platt et al. (2011) are of unflattened specimens only.

5.4.3.6 *Fornixtrema palpebrae* (Strelkov, 1950) (Fig. 5.14)

Collection Unknown.

Holotype Unknown.

Paratypes Unknown.

Original description Strelkov (1950).

Other taxonomic contributions None.

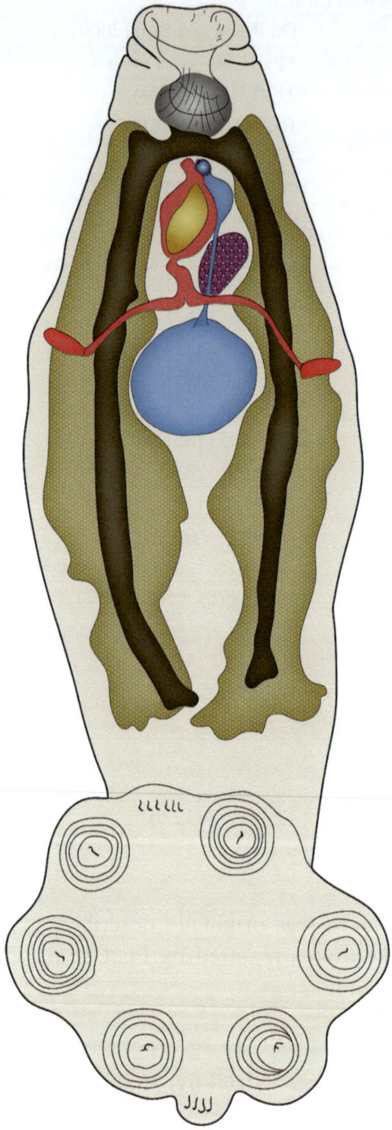

Fig. 5.14 *Fornixtrema palpebrae*. Redrawn from Strelkov (1950)

Etymology Refers to the site on the host.

Type locality Lake Hanka, East Asia.

Other localities None.

Type host *Pelodiscus sinensis* Wiegmann, 1835, Chinese Softshell Turtle. In the original description of the polystome, the host was known as *Amyda sinensis* Schmidt, 1927.

Additional hosts None.

Host geographical distribution *Pelodiscus sinensis* occurs in C/S China (Yunnan, Liaoning, Shaanxi, Anhui, Zhejiang, Guangdong, Fujian), N Vietnam, Hainan, Taiwan, S Far East (Russia), Japan (Bonin Islands), Indonesia (Timor), Japan and Philippines (Panay, Cebu) (Uetz et al. 2022).

Host conservation status The Chinese Softshell Turtle is bred and sold for its meat, thus natural populations have almost disappeared. Another problem is that its habitats have been almost completely altered or cleared. It is listed as Vulnerable by The IUCN Red List of Threatened Species 2000: e.T39620A97401140. https://doi.org/10.2305/IUCN.UK.2000.RLTS.T39620A10251914.en. Accessed on 21 January 2022.

Site on host Conjunctival sacs of the eye.

DNA sequence
- 18S rRNA gene, 1993 bp. GenBank Accession: FM992696
- 28S rRNA gene, 1168 bp. GenBank Accession: AF382065
- Cytochrome c Oxidase I, 374 bp. GenBank Accession: FR822601
- 12S rRNA gene, 449 bp. GenBank Accession: KR856104

Morphology and morphometrics (μm):	
Source	Strelkov (1950)
Sample size	
Body length	2200–5100
Greatest width	1100–1800
Width at vagina	
Haptor length	
Haptor width	1000–1600
False oral sucker width	280–350
Pharynx length	180–330
Pharynx width	250–480
Testis length	
Testis width	150–180
Ovary length	190–310
Ovary width	110–150
Egg length	280–300
Egg width	130–150
Egg incubation time	24 days in 18–25 °C
Number of intra-uterine eggs	1
Genital bulb width	20–60
Number of genital spines	16
Genital spine length	16–14
Haptoral sucker diameter	260–420

(continued)

Morphology and morphometrics (μm):	
Source	Strelkov (1950)
Hamulus handle length X	
Hamulus guard length Y	
Hamulus hook length	
Marginal hooklet length	11–14
Haptoral L/Body L	

Remarks *Fornixtrema palpebrae* was originally described as *Neopolystoma palpebrae* Strelkov, 1950.

5.4.3.7 *Fornixtrema scorpioides* (Du Preez, Badets, Héritier and Verneau, 2017) (Fig. 5.15)

Collection Parasitic Worm Collection, National Museum, Bloemfontein, South Africa (NMBP).

Holotype NMBP 407.

Paratypes NMBP 408–409.

Original description Du Preez et al. (2017).

Other taxonomic contributions None.

Etymology Named after the host.

Type locality Forest pond along the Cayenne-Kaw road south of the town of Roura, French Guiana (4.66997 N, 52.30560 W).

Other localities None.

Type host *Kinosternon scorpioides* (Linnaeus, 1766), Scorpion Mud Turtle.

Additional hosts None.

Host geographical distribution *Kinosternon scorpioides* occurs in Mexico, Guatemala, Belize, El Salvador, Honduras, Nicaragua, Costa Rica, Panama, Colombia, Ecuador, Venezuela, French Guiana, Guyana, Brazil, Argentina, Paraguay, Bolivia, N Peru, Suriname and Trinidad (Uetz et al. 2022).

Host conservation status The Scorpion Mud Turtle is found in freshwater swamps and quiet rivers. It is not listed in the IUCN Red List.

Site on host Conjunctival sacs of the eye.

DNA sequence

- 28S rRNA gene, 931 bp. GenBank Accession: KY200990
- Cytochrome c Oxidase I, 323 bp. GenBank Accession: KY200996
- 12S rRNA gene, 470 bp. GenBank Accession: KY200993

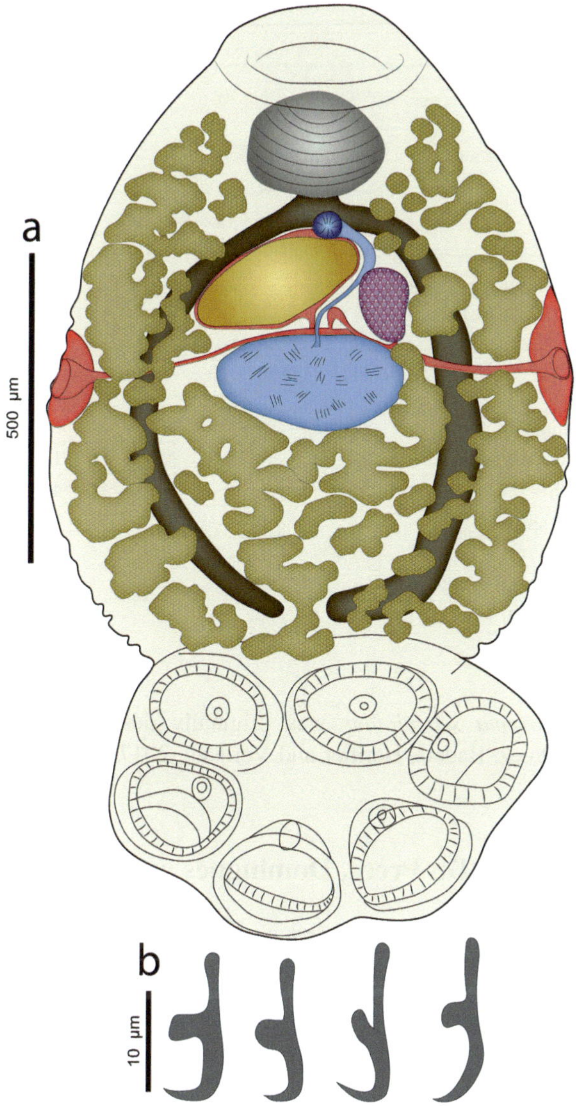

Fig. 5.15 (**a**) *Fornixtrema scorpioides*, (**b**) marginal hooklets. Redrawn from Du Preez et al. (2017)

Morphology and morphometrics (µm):	
Source	Du Preez et al. (2017)
Sample size	
Body length	1512–1770 (1658)
Greatest width	868–907 (894)
Width at vagina	837–876 (863)

(continued)

Morphology and morphometrics (µm):	
Source	Du Preez et al. (2017)
Haptor length	445–511 (485)
Haptor width	719–765 (735)
False oral sucker width	240–245 (242)
Pharynx length	170–184 (177)
Pharynx width	226–231 (228)
Testis length	120–142 (127)
Testis width	246–301 (271)
Ovary length	136–187 (153)
Ovary width	66–71 (69)
Egg length	253–279 (265)
Egg width	133–156 (143)
Egg incubation time	
Number of intra-uterine eggs	1
Genital bulb width	45
Number of genital spines	8
Genital spine length	5.5–8.7 (7.7)
Haptoral sucker diameter	133–279 (220)
Marginal hooklet length	16.4–17.2 (16.8)
Haptoral L/Body L	0.29

Remarks *Fornixtrema scorpioides* was originally described as *Neopolystoma scorpioides* Du Preez, Badets, Héritier and Verneau, 2017.

5.4.4 Manotrema Du Preez, Domingues and Verneau, 2022

5.4.4.1 *Manotrema brasiliense* (Vieira, Novelli, Sousa and de SouzaLima, 2008) (Fig. 5.16)

Collection Helminthological Collection, Instituto Oswaldo Cruz, Rio de Janeiro, Brazil (CHIOC).

Holotype CHIOC 36902.

Paratypes CHIOC 36903a–d.

Original description Vieira et al. (2008).

Taxonomic note The name *Manotrema brasiliensis* was corrected for gender.

Other taxonomic contributions Santana et al. (2019); Du Preez et al. (2022).

Etymology Named after type locality country.

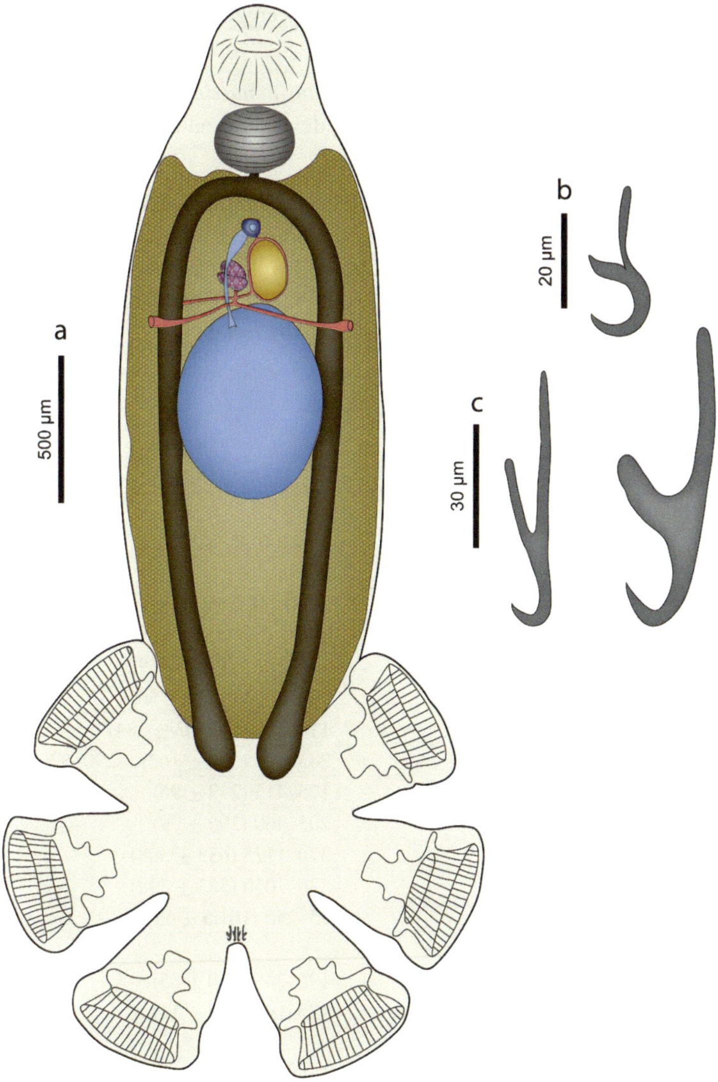

Fig. 5.16 (**a**) *Manotrema brasiliense*, (**b**) marginal hooklets, (**c**) hamuli. Redrawn from Vieira et al. (2008)

Type locality Juiz de For-a, Minas Gerais, Brazil, 21°41′20″S, 43°20′40″W.

Other localities Caatinga and Atlantic forest in an area of northeastern Brazil.

Type host *Hydromedusa maximiliani* (Mikan, 1825), Maximilian's Snake-headed Turtle.

Additional hosts *Phrynops geoffroanus* (Schweigger, 1812) and *Mesoclemmys tuberculate* (Lüderwaldt, 1926).

Host geographical distribution *Hydromedusa maximiliani* occurs in Brazil (S Bahia, Minas Gerais, Espirito Santo, Rio de Janeiro and Sao Paulo) (Uetz et al. 2022).

Host conservation status The range of The Maximilian's Snake-headed Turtle is threatened by human-caused habitat degradation. The species is threatened and vulnerable at national as well as international levels. It is listed as Vulnerable by The IUCN Red List of Threatened Species 1996: e.T10309A97269236. https://doi.org/10.2305/IUCN.UK.1996.RLTS.T10309A3191766.en. Accessed on 21 January 2022.

Site on host Buccal and pharyngeal cavities.

DNA sequence None.

Morphology and morphometrics (μm):	
Source	Vieira et al. (2008)
Sample size	8
Body length	2450–6062 (3600 ± 1479)
Greatest width	
Width at vagina	545–1875 (1100 ± 592)
Haptor length	998–2200 (1325 ± 503)
Haptor width	1100–2325 (1503 ± 484)
False oral sucker length/width	245–490 (312 ± 102) × 290–520 (364 ± 89)
Pharynx length	175–415 (249 ± 96)
Pharynx width	205–460 (290 ± 99)
Testis length	370–1125 (633 ± 308)
Testis width	330–1050 (583 ± 298)
Ovary length	95–250 (161.5 ± 63)
Ovary width	
Egg length	195–245 (220 ± 35)
Egg width	175–195 (185 ± 14)
Egg incubation time	
Number of intra-uterine eggs	1
Genital bulb width	
Number of genital spines	8–9
Genital spine length	11.3–17.5 (12.8 ± 1.6)
Haptoral sucker diameter	300–813 (455 ± 179)
Hamulus handle length X	62.5–80 (72.1 ± 5.5)
Hamulus guard length Y	55–71.3 (65.1 ± 4.9)
Hamulus hook length	
Marginal hooklet length	25–36.3 (31.9 ± 3.4)
Haptoral L/Body L	0.4073

Remarks *Manotrema brasiliense* was originally described as *Polystomoides brasiliensis* Vieira, Novelli, Sousa and de SouzaLima, 2008.

5.4.4.2 *Manotrema fuquesi* (Mañé-Garzón and Gil, 1962) (Fig. 5.17)

Collection Museum of Natural History of Montevideo, S. Alfredo Ximenez collection.

Holotype Unknown.

Paratypes Unknown.

Original description Mañé-Garzón and Gil (1962).

Other taxonomic contributions None.

Etymology Named after the collector, Mr. Calisto Fuques, Uruguay.

Type locality Department of Artigas, Uruguay, South America.

Other localities None.

Type host *Phrynops hilarii* (Duméril and Bibron, 1835); Hilaire's Toadhead Turtle. In the original description of the polystome, the host was known as *Phrynops geoffroana hillarii* Müller, 1939.

Additional hosts None.

Host geographical distribution *Phrynops hilarii* occurs in Brazil (Rio Grande do Sul), Uruguay, Argentina (Corrientes) and Paraguay (Uetz et al. 2022).

Host conservation status Eggs of The Hilaire's Toadhead Turtle are eaten, but some local people consider that species to be poisonous. Despite its habitat being altered and destroyed, it does not appear on any protected list. It is not listed in the IUCN Red List.

Site on host Oral cavity.

DNA sequence None.

Morphology and morphometrics (μm):	
Source	Mañé-Garzón and Gil (1962)
Sample size	28
Body length	7480–7550
Greatest width	2300–2600
Width at vagina	
Haptor length	
Haptor width	2660–2780
False oral sucker length/width	450 × 670–700
Pharynx length	480–510

(continued)

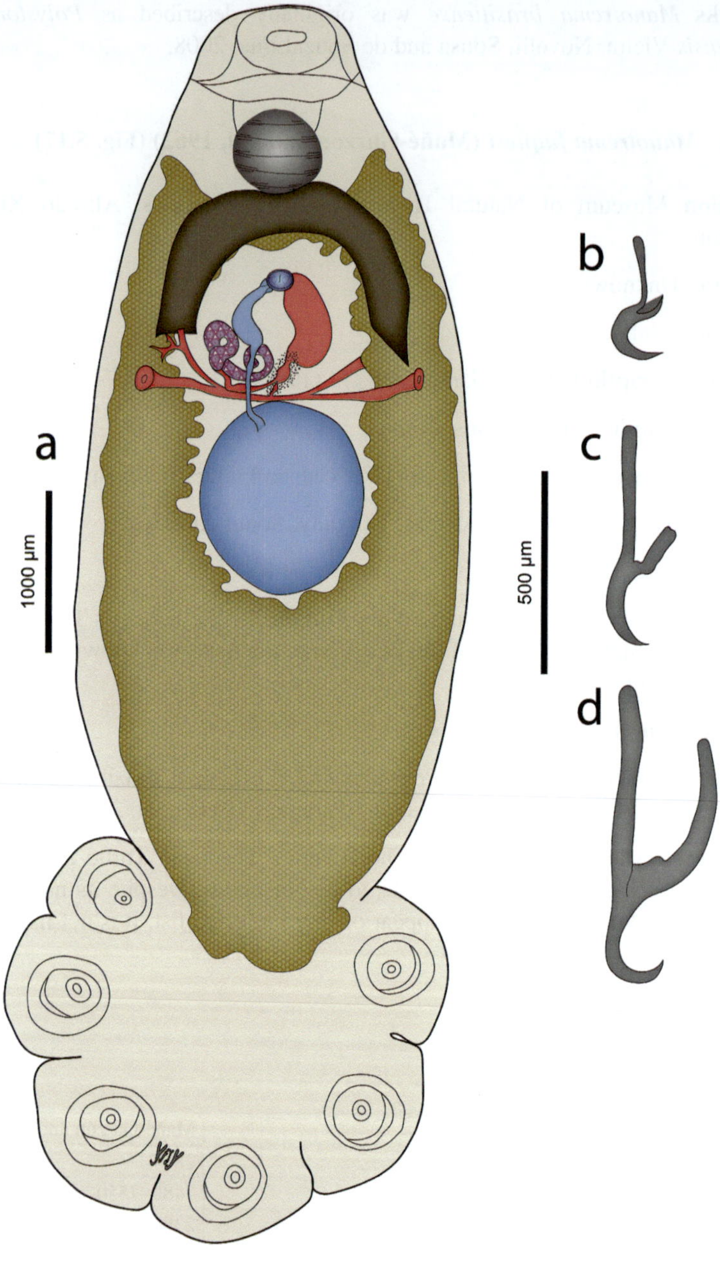

Fig. 5.17 (**a**) *Manotrema fuquesi*, (**b**) marginal hooklets, (**c**) hamulus pair two, (**d**) hamuli. Redrawn from Mané-Garzon and Gil (1962)

Morphology and morphometrics (μm):	
Source	Mañé-Garzón and Gil (1962)
Pharynx width	660–700
Testis length	114–1400
Testis width	1000–1100
Ovary length	1179
Ovary width	900
Egg length	
Egg width	250–360
Egg incubation time	
Number of intra-uterine eggs	1
Genital bulb width	176 × 118
Number of genital spines	2
Genital spine length	56
Haptoral sucker diameter	481–518
Hamulus handle length X	65–68
Hamulus guard length Y	45
Hamulus hook length	
Marginal hooklet length	24–27
Haptoral L/Body L	

Remarks *Manotrema fuquesi* was originally described as *Polystomoides fuquesi* Mañé-Garzón and Gil, 1962. This species could be a junior synonym of *Manotrema uruguayensis* (Mañé-Garzón and Gil, 1961) as it was collected from the same host and locality.

5.4.4.3 *Manotrema uruguayense* (Mañé-Garzón and Gil, 1961) (Fig. 5.18)

Collection Museum of Natural History of Montevideo, S. Alfredo Ximenez collection.

Holotype 10.III.1960.

Paratypes None.

Original description Mañé-Garzón and Gil (1961).

Taxonomic note The name *Manotrema uruguayensis* was corrected for gender.

Other taxonomic contributions None.

Etymology Named after type locality country.

Type locality Cuaró stream, Department of Artigas, Uruguay, South America.

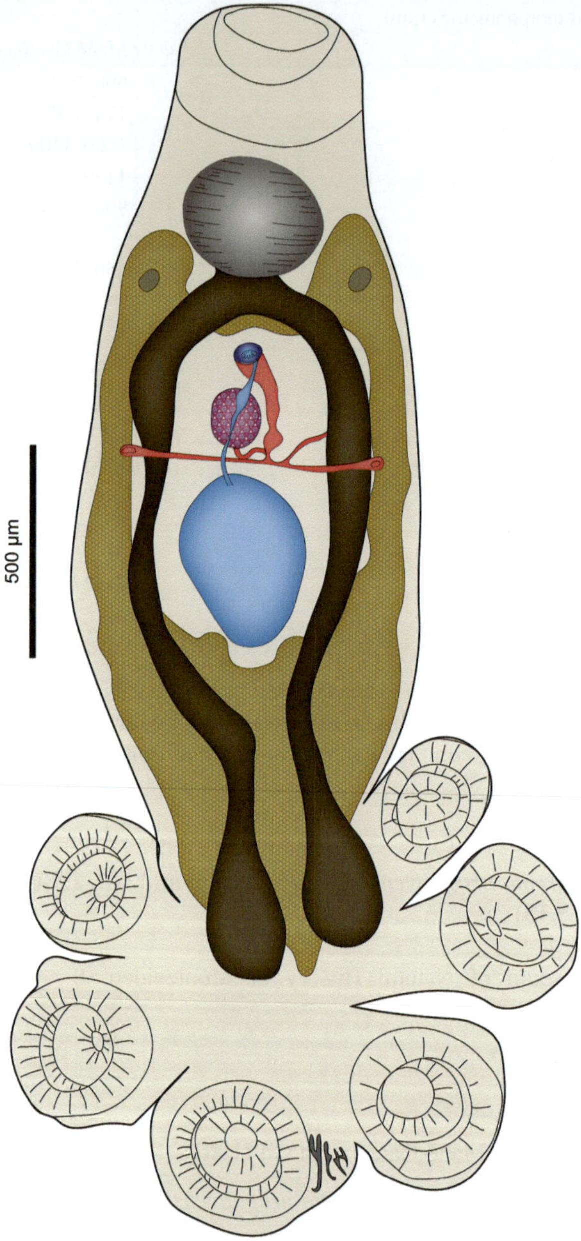

Fig. 5.18 *Manotrema uruguayense*. Redrawn from Mané-Garzon and Gil (1961)

Other localities None.

Type host *Phrynops hilarii*; Hilaire's Toadhead Turtle. In the original description of the polystome, the host was known as *Phrynops geoffroana hillarii*.

Additional hosts None.

Host geographical distribution Brazil (Rio Grande do Sul), Uruguay, Argentina (Corrientes), Paraguay (Uetz et al. 2022).

Host conservation status Eggs of The Hilaire's Toadhead Turtle are eaten, but some local people consider that species to be poisonous. Despite its habitat being altered and destroyed, it does not appear on any protected list. It is not listed in the IUCN Red List.

Site on host Oral cavity.

DNA sequence None.

Morphology and morphometrics (µm):	
Source	Mañé-Garzón and Gil (1961)
Sample size	
Body length	2560–2650
Greatest width	780–800
Width at vagina	
Haptor length	
Haptor width	1250–1310
False oral sucker length/width	350 × 270
Pharynx length	
Pharynx width	210
Testis length	300–390
Testis width	230–290
Ovary length	
Ovary width	100–120
Egg length	
Egg width	
Egg incubation time	
Number of intra-uterine eggs	
Genital bulb width	
Number of genital spines	8–10
Genital spine length	10
Haptoral sucker diameter	310–320
Hamulus handle length X	52
Hamulus guard length Y	38
Hamulus hook length	
Marginal hooklet length	21
Haptoral L/Body L	

Remarks *Manotrema uruguayense* was originally described as *Polystomoides uruguayensis* Mañé-Garzón and Gil, 1961.

5.4.5 *Pleurodirotrema* Du Preez, Domingues and Verneau, 2022

5.4.5.1 *Pleurodirotrema chelodinae* (MacCallum, 1918) (Fig. 5.19)

Collection Type in United States National Museum (USNM). Voucher specimens in Australian Museum, Sydney; Australian Helminthological Collection, South Australian Museum, Adelaide; Queensland museum, Australia (OM).

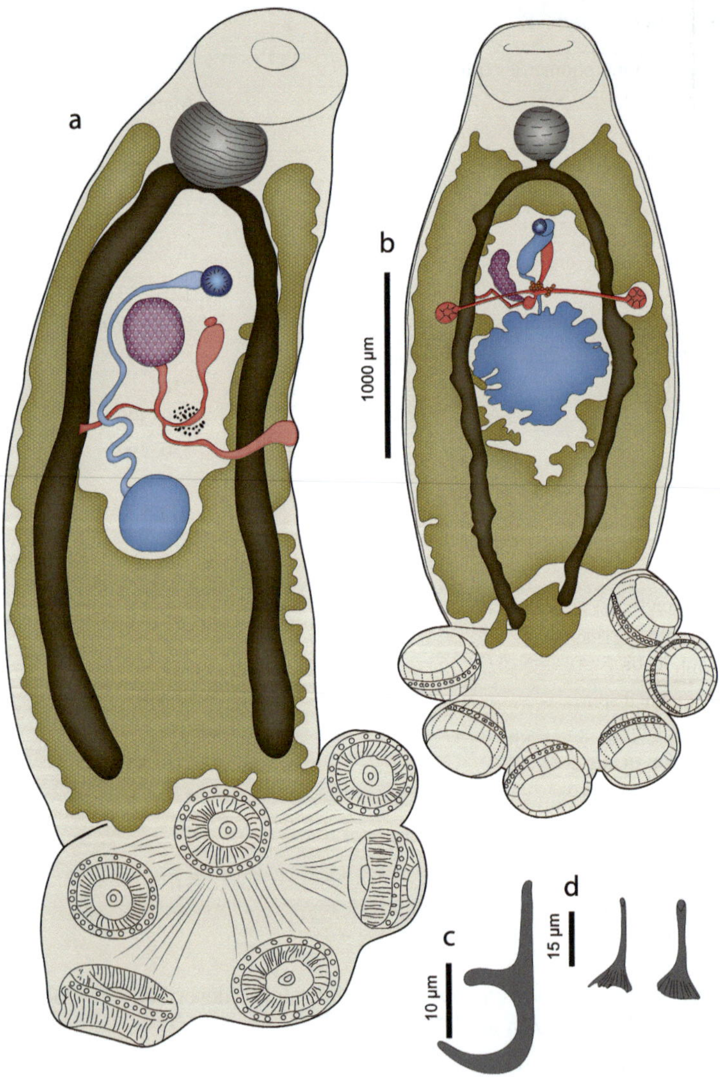

Fig. 5.19 (**a**, **b**) *Pleurodirotrema chelodinae*, (**c**) marginal hooklets, (**d**) genital spines. (**a**): Redrawn from MacCallum (1918); (**b–d**): Redrawn from Pichelin (1995)

Holotype USNM 1336304.

Paratypes None.

Original description MacCallum (1918).

Other taxonomic contributions Price (1939); Rohde and Pearson (1980); Pichelin (1995).

Etymology Named after the host.

Type locality United States (New York Zoological Park), North America.

Other localities Brisbane river, Brisbane; Mt. Cootha, Brisbane; Armidale, New South Wales (Australia) (see Rohde and Pearson 1980); Tailem Bend (South Australia) and Lake Manchester (Queensland) (see Pichelin 1995).

Type host *Chelodina longicollis*, Common Snake-Necked Turtle.

Additional hosts *Chelodina* sp. (see Pichelin 1995).

Host geographical distribution *Chelodina longicollis* occurs in Australia (New South Wales, Queensland, South Australia, Victoria) (Uetz et al. 2022).

Host conservation status The Common Snake-Necked Turtle is a popular pet and zoo animal, and there is a definite drain upon wild populations. It is not listed in The IUCN Red List.

Site on host Urinary bladder and cloaca.

DNA sequence

- 18S rRNA gene, 1994 bp. GenBank Accession: KR856126
- 18S rRNA gene, 561 bp. GenBank Accession: AJ287996
- 28S rRNA gene, 1368 bp. GenBank Accession: KR856145
- 28S rRNA gene, 944 bp. GenBank Accession: Z83004
- Cytochrome c Oxidase I, 388 bp. GenBank Accession: Z83005
- 12S rRNA gene, 491 bp. GenBank Accession: KR856100

Morphology and morphometrics (μm):

Source	MacCallum (1918)	Price (1939)	Rohde and Pearson (1980)	Pichelin (1995)
Sample size	1	1	5	4
Body length	3500	2900	3210–5320 (4658)	3646–4712 (4115)
Greatest width	1000		1110–1990 (1684)	1353–1608 (1508)
Width at vagina		1100		
Haptor length	1750	935	990–1480 (874)	987–1274 (1119)
Haptor width		1230	1490–1670 (1590)	1449–1815 (1648)
False oral sucker length/width	300	265 × 500	470–580 (510) × 610–790 (730)	350–462 (422) × 589–748 (669)
Pharynx length	350	340	320–420 (356)	276–360 (305)
Pharynx width		425	320–490	289–327 (315)

(continued)

Morphology and morphometrics (μm):				
Source	MacCallum (1918)	Price (1939)	Rohde and Pearson (1980)	Pichelin (1995)
Testis length	400		440–790 (680)	509–764 (653)
Testis width			510–700 (472)	669–828 (764)
Ovary length		340		
Ovary width		170		
Egg length				
Egg width				
Egg incubation time				
Number of intra-uterine eggs	1	0		0
Genital bulb length/width		250	94–148 (128) × 104–160 (128)	68–113 (87) × 71–113 (88)
Number of genital spines	15	14	12–16 (14)	13–15 (14)
Genital spine length		15		20.8–27.2 (23.6)
Haptoral sucker diameter	300	340	350–470 (388)	417–499 (449)
Hamulus handle length X				
Hamulus guard length Y				
Hamulus hook length				
Marginal hooklet length				21.6–25.6 (23.8)
Haptoral L/Body L	0.5	0.3224	0.1876	0.2719

Remarks *Pleurodirotrema chelodinae* was originally described as *Polystoma chelodinae* MacCallum, 1918. MacCallum (1918) omitted a 0 from the end of each of the measurements given in μm, e.g. 30 instead of 300.

5.4.5.2 *Pleurodirotrema kreffti* (Rohde, 1984) (Fig. 5.20)

Collection Australian Museum, Sydney A(M); United States National Museum (USNM); Natural History Museum London (NHML).

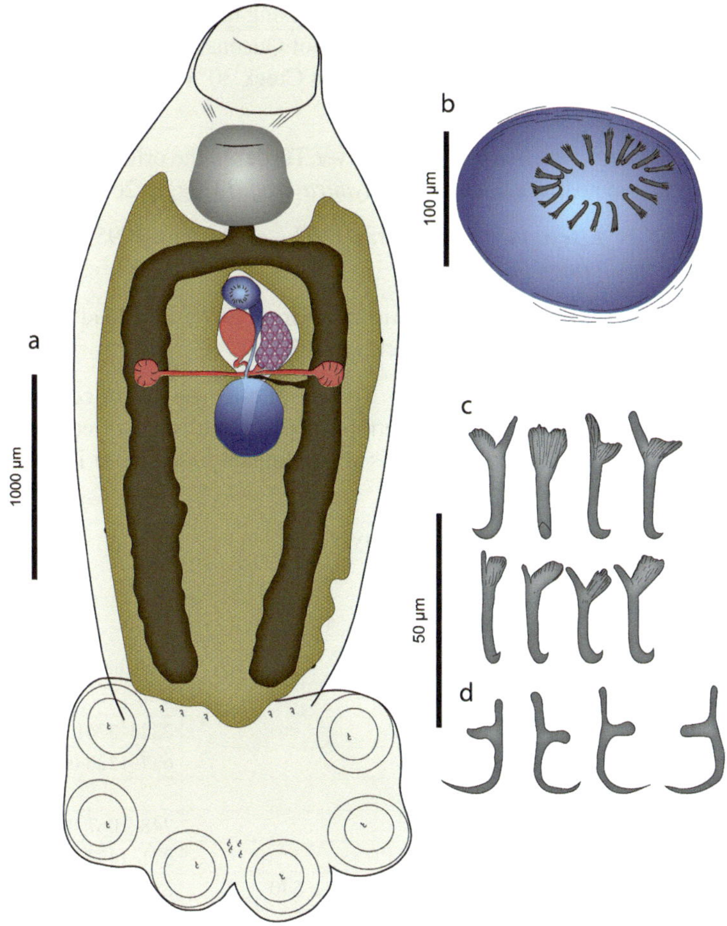

Fig. 5.20 (**a**) *Pleurodirotrema kreffti*, (**b**) genital bulb, (**c**) genital spines, (**d**) marginal hooklets. Redrawn from Rohde (1984)

Holotype AM W19291.

Paratypes USNM 77256; NHML 1982.11.3.22.

Original description Rohde (1984).

Other taxonomic contributions Pichelin (1995).

Etymology Named after the host.

Type locality Frogmore Lagoon 23°26′S; 150°32′E, Fitzroy drainage, Australia.

Other localities Barambah Creek 25°41′S; 150°48′E, Burnett drainage, Queensland, Australia, Mulgrave River (Queensland), Leslie Dam (28°01′S, 152°55′E),

Canal Creek (28°01′S, 151°35′E), College's Crossing (27°33′S 152°48′E), Lockyer Creek (27°35′S, 152°07′E) and The University of Queensland Veterinary Farm dam (27°32′S, 152°54′E) (all Queensland); Georges Creek 30°43′S, 152°11′E, Macleay drainage, N.S.W.

Type host *Emydura macquarii*, Murray River Turtle. In the original description of the polystome, the host was known as *Emydura krefftii* Cogger 2000.

Additional hosts *Elseya latisternum* Gray 1867 (see Pichelin 1995), known today as *Myuchelys latisternum* (Gray, 1867).

Host geographical distribution SE Australia (from SC Queensland southward to N Victoria and SE South Australia) (Uetz et al. 2022).

Host conservation status The Murray River Turtle is heavily collected for parks and zoos, and is one of the Australian terrapin species that is most frequently found in Western collections. Nevertheless, it is still abundant in nature. It is not listed in the IUCN Red List.

Site on host Mouth and pharyngeal cavity.

DNA sequence None.

Morphology and morphometrics (μm):		
Source	Rohde (1984)	Pichelin (1995)
Sample size	3	9
Body length	2900/4100/4600	2324–3741 (2958)
Greatest width	1000/1500/1500	637–1226 (922)
Width at vagina		
Haptor length	890/950/1500	748–1067 (892)
Haptor width	1170/1640/1300	971–1433 (1150)
False oral sucker length/width	290 × 460/380 × 590/510 × 540	289–480 (388) × 360–642 (501)
Pharynx length	320/510/510	270–494 (348)
Pharynx width	440/540/540	295–520 (404)
Testis length	270/290/370	257–404 (301)
Testis width	360/400/370	205–282 (245)
Ovary length		
Ovary width		
Egg length	290/300/240	
Egg width	230/250/210	
Egg incubation time		
Number of intra-uterine eggs	1	1
Genital bulb length/width	1st specimen unmeasurable 150 × 160; 200 × 220	141–199 (165) × 128–218 (169)
Number of genital spines	27/22/18	20–26 (24)

(continued)

Morphology and morphometrics (μm):		
Source	Rohde (1984)	Pichelin (1995)
Genital spine length	17–20	24.0–32.0 (27.5)
Haptoral sucker diameter	300/390/390	279–427 (244)
Hamulus handle length X		
Hamulus guard length Y		
Hamulus hook length		
Marginal hooklet length	20–28	22.4–27.2 (26.1)
Haptoral L/Body L	0.3069/0.2317/0.3261	0.3016

Remarks *Pleurodirotrema kreffti* was originally described as *Neopolystoma kreffti* Rohde, 1984. Rohde (1984) mistakenly stated the length of the marginal hooklets as 200–28 μm.

5.4.5.3 *Pleurodirotrema macleayi* (Rohde, 1984) (Fig. 5.21)

Collection Australian Museum, Sydney (AM); United States National Museum (USNM); Natural History Museum of London (NHML).

Holotype AM 19292.

Paratypes AM 77257; USNM 1372755; NHML 1982.11.3.24.

Original description Rohde (1984).

Etymology Named after the type locality Macleay drainage.

Other taxonomic contributions Pichelin (1995).

Type locality Georges Creek, 30°43′S; 152°11′E, Macleay drainage, New South Wales, Australia.

Other localities Apsley River, 31°4′S; 152°1′E, Macleay drainage, New South Wales, Australia; Australia–the University of Queensland Veterinary Farm dam (27° 32′S, 152°54′E) in south-east Queensland and the Tinaroo Falls Dam (17°10′S, 145°33′E) in northern Queensland.

Type host *Emydura* sp.

Additional hosts *Emydura macquarii*; *Myuchelys latisternum* (see Pichelin, 1995).

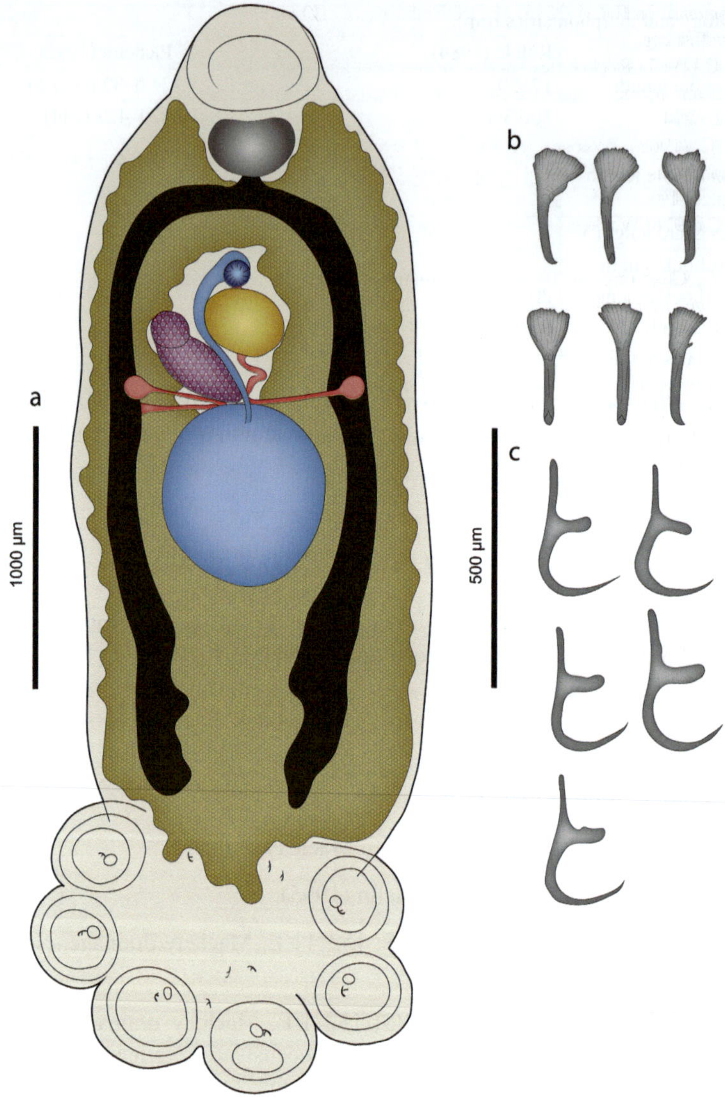

Fig. 5.21 (**a**) *Pleurodirotrema macleayi*, (**b**) genital spines, (**c**) marginal hooklets. Redrawn from Rohde (1984)

Host geographical distribution Not available due to uncertainty of host.

Host conservation status Not available.

Site on host Cloacal bursa; Urinary and accessory bladders.

DNA sequence None.

Morphology and morphometrics (μm):		
Source	Rohde (1984)	Pichelin (1995)
Sample size	4	7
Body length	2800/3500/3700/4000	2690–3184 (2907)
Greatest width	760/1400/1600/1300	669–971 (855)
Width at vagina		
Haptor length	630/860/1000/950	844–923 (885)
Haptor width	1000/1200/1600/1400	987–1067 (1019)
False oral sucker length/width	290 × 330/390 × 460/290 × 460/380 × 490	350–411 (380) × 366–475 (440)
Pharynx length	200/190/220/220	207–263 (231)
Pharynx width	210/340/330/330	255–287 (271)
Testis length	270/?/730/700	462–571 (507)
Testis width	190/?/820/660	327–379 (354)
Ovary length		
Ovary width		
Egg length	0/360/360/320	
Egg width	0/290/260/260	
Egg incubation time		
Number of intra-uterine eggs	1	1
Genital bulb length/width	70 × 80/100 × 100/?/140 × 120	74–104 (93) × 64–115 (90)
Number of genital spines	12 (or 13)/12/ca. 15/12	11–13 (12)
Genital spine length	13–15 (blade); 22–23 (blade + long root)	20.8–27.4 (23.6) Straight with hooked tip, including roots
Haptoral sucker diameter	240/320/320/340	302–326 (306)
Hamulus handle length X		
Hamulus guard length Y		
Hamulus hook length		
Marginal hooklet length	22–28	25.6–28.8 (26.4)
Haptoral L/Body L	0.225/0.2457/0.2703/0.2375	

Remarks *Pleurodirotrema macleayi* was originally described as *Neopolystoma macleayi* Rohde, 1984.

5.4.5.4 *Pleurodirotrema novaeguineae* (Fairfax, 1990) (Fig. 5.22)

Collection Australian Museum, Sydney, Australia (AM); Queensland Museum, Australia (QM).

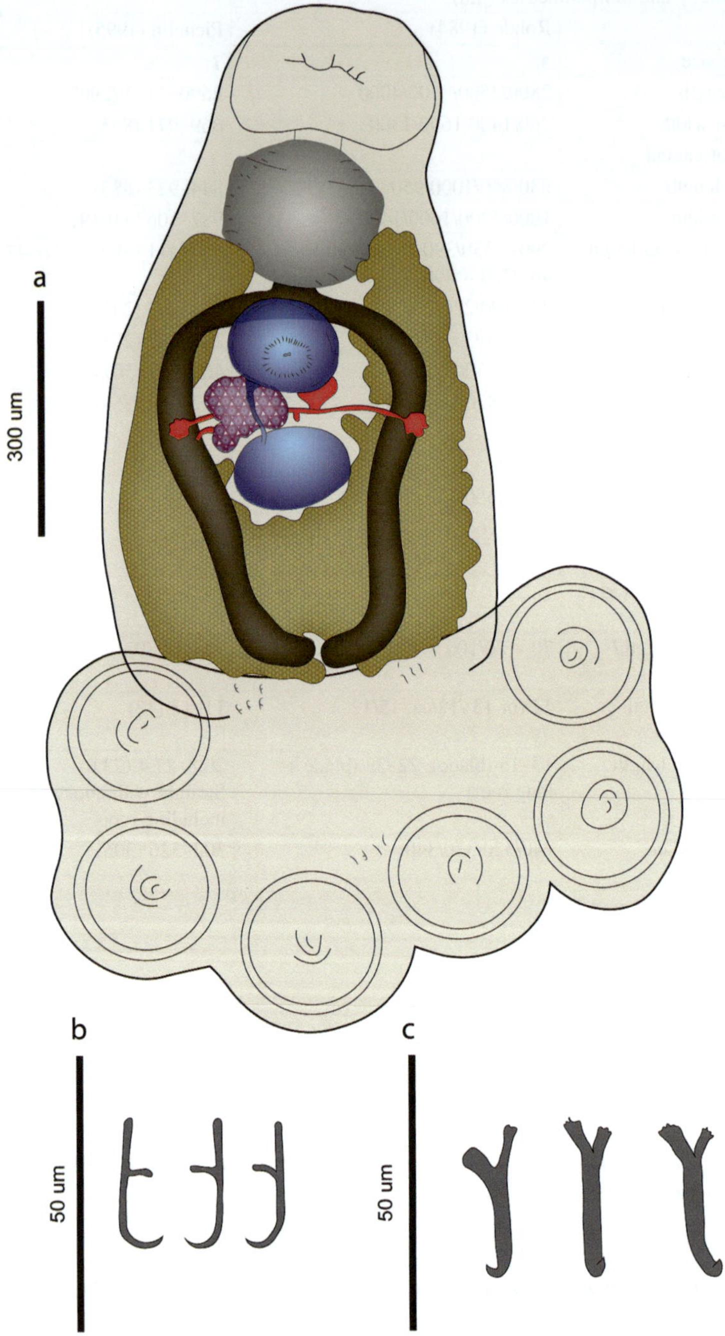

Fig. 5.22 (**a**) *Pleurodirotrema novaeguineae*, (**b**) marginal hooklets, (**c**) genital spines. Redrawn from Fairfax (1990)

Syntypes QM GL13052–8.

Paratypes AM R 125031–9.

Original description Fairfax (1990).

Other taxonomic contributions None.

Etymology Named after the host.

Type locality Mount Garagos near Gabensis Village, Morobe Province, Papua New Guinea. Watul River drainage, 6°47′S; 146°46′E, elevation 400 m.

Other localities None.

Type host *Elseya novaeguineae* (Meyer, 1874), Western New Guinea Stream Turtle.

Additional hosts None.

Host geographical distribution *Elseya novaeguineae* occurs in Indonesia and Papua New Guinea (West Papua, Waigeo Island, Aru Islands) (Uetz et al. 2022).

Host conservation status The Western New Guinea Stream Turtle is regularly caught for food, but despite this, populations in nature are in sound condition. It is listed as Least Concern by The IUCN Red List of Threatened Species 2000: e. T46581A97268667. https://doi.org/10.2305/IUCN.UK.2000.RLTS.T46581A11061765.en. Accessed on 21 January 2022.

Site on host Mouth and pharyngeal cavity.

DNA sequence None.

Morphology and morphometrics (μm):	
Source	Fairfax (1990)
Sample size	9
Body length	1300–3530 (2190)
Greatest width	510–1230 (840)
Width at vagina	
Haptor length	510–920 (690)
Haptor width	720–1300 (950)
False oral sucker length/width	150–510 (250) × 200–470 (380)
Pharynx length	170–410 (270)
Pharynx width	220–470 (320)
Testis length	90–290 (220)
Testis width	150–350 (220)
Ovary length	
Ovary width	
Egg length	180–240 (210)
Egg width	180–220 (200)
Egg incubation time	

(continued)

Morphology and morphometrics (μm):	
Source	Fairfax (1990)
Number of intra-uterine eggs	1
Genital bulb length/width	90–230 (170) × 100–240 (190)
Number of genital spines	32 (usually)/33
Genital spine length	23–29 (26)
Haptoral sucker diameter	200–320 (270)
Hamulus handle length X	
Hamulus guard length Y	
Hamulus hook length	
Marginal hooklet length	18–23 (23)
Haptoral L/Body L	0.3151

Remarks *Pleurodirotrema novaeguineae* was originally described as *Neopolystoma novaeguineae* Fairfax, 1990.

5.4.6 *Polystomoidella* Price, 1939

5.4.6.1 *Polystomoidella mayesi* Richardson and Brooks, 1987 (Fig. 5.23)

Collection Invertebrate Collection (Parasites), Canadian Museum of Nature (CMNPA).

Holotype CMNPA 1987-1609.

Paratypes CMNPA 1987-1610 to 1614.

Original description Richardson and Brooks (1987).

Other taxonomic contributions None.

Etymology Named after Dr. Monte A. Mayes, Dow Chemical, USA.

Type locality Telok Anson, Malaysia.

Other localities None.

Type host *Cuora amboinensis*, Southeast Asian Box Turtle.

Host geographical distribution *Cuora amboinensis* occurs in India, Bhutan, Bangladesh, Myanmar, Thailand, Cambodia, Vietnam, West Malaysia, Singapore and Philippines (Uetz et al. 2022).

Host conservation status The Southeast Asian Box Turtle is the most frequently consumed terrapin species in China restaurants, also captured and released in Buddhist temple ponds, and gathered for exports. These practices may lead to a population collapse in the future. It is listed as Endangered by The IUCN Red List of

Threatened Species 2020: e.T5958A3078812. https://doi.org/10.2305/IUCN.UK.2020-2.RLTS.T5958A3078812.en. Accessed on 21 January 2022.

Site on host Urinary bladder.

DNA sequence None.

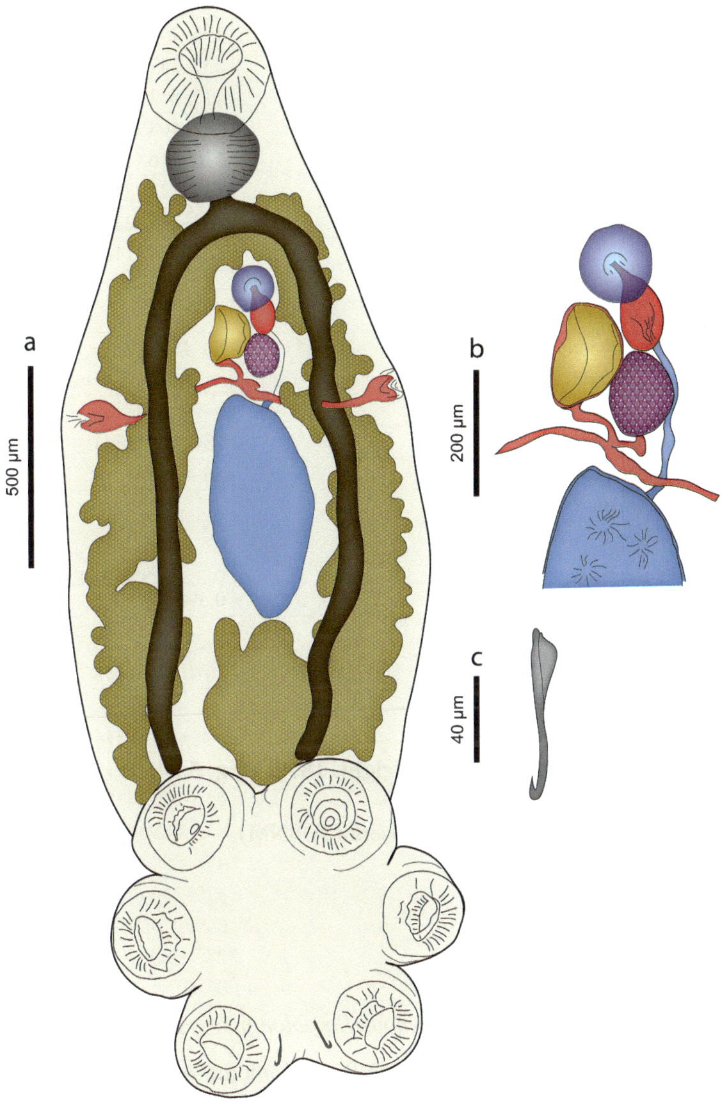

Fig. 5.23 (**a**) *Polystomoidella mayesi*, (**b**) reproductive system, (**c**) hamuli. Redrawn from Richardson and Brooks (1987)

Morphology and morphometrics (µm):	
Source	Richardson and Brooks (1987)
Sample size	7
Body length	1652–2734 (2136)
Greatest width	612–1142 (863)
Width at vagina	
Haptor length	530–1428 (835)
Haptor width	669–1061 (859)
False oral sucker length/width	171–377 (293) × 282–439 (352)
Pharynx length	196–296 (231)
Pharynx width	235–347 (276)
Testis length	135–612 (392)
Testis width	122–408 (250)
Ovary length	110–216 (163)
Ovary width	82–112 (98)
Egg length	153–285 (233)
Egg width	102–138 (116)
Egg incubation time	
Number of intra-uterine eggs	1
Genital bulb length/width	86–173 (126) × 47–112 (71)
Number of genital spines	
Genital spine length	
Haptoral sucker diameter	184–296 (234)
Hamulus handle length X	58–82 (74)
Hamulus guard length Y	
Hamulus hook length	
Marginal hooklet length	
Haptoral L/Body L	0.3909

Remarks None.

5.4.6.2 *Polystomoidella oblonga* (Wright, 1879) (Fig. 5.24)

Collection United States National Museum (USNM).

Holotype USNM 1316681.

Paratypes USNM 1340029.

Original description Wright (1879).

Taxonomic note The name *Polystomoidella oblongum* was corrected for gender.

Other taxonomic contributions Price (1939); Lamothe-Argumedo (1972).

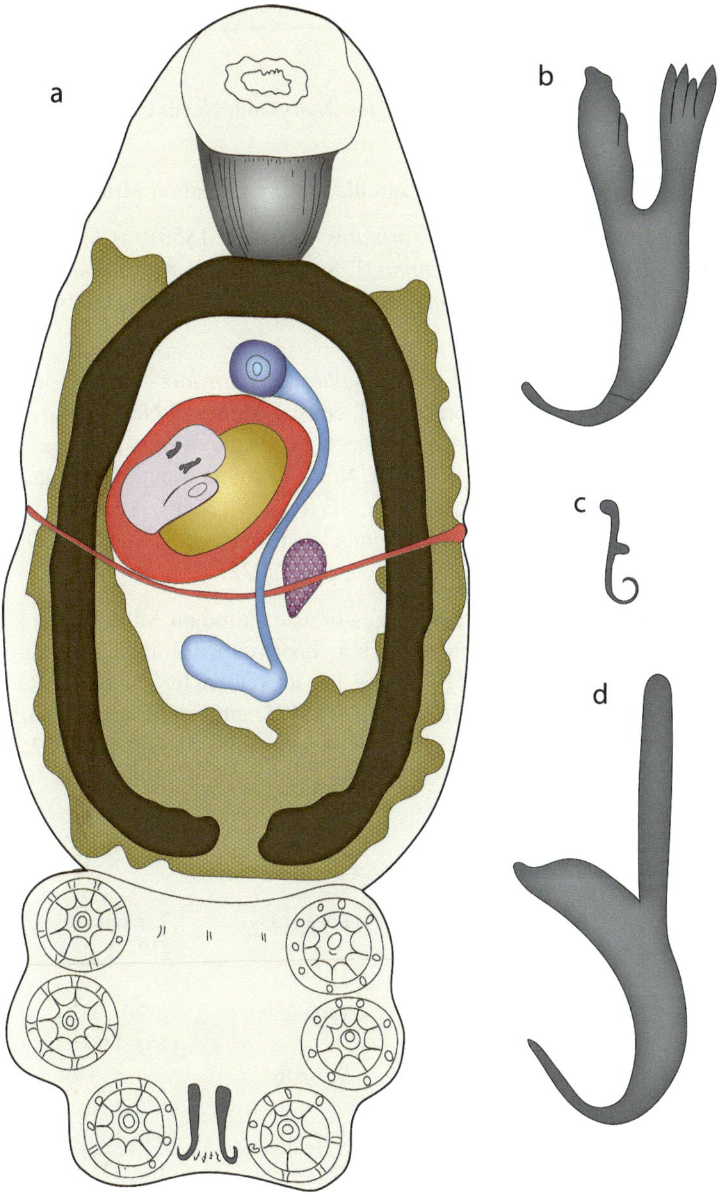

Fig. 5.24 (**a**) *Polystomoidella oblonga*, (**b**) hamuli, (**c**) marginal hooklets, (**d**) hamulus pair two. Redrawn from Wright (1879)

Etymology Refers to the elongated body.

Type locality Toronto, Canada.

Other localities Canada and United States (Maryland, North Carolina, Texas, Iowa and Virginia).

Type host *Sternotherus odoratus* (Latreille, 1802), Common Musk Turtle.

Additional hosts *Sternotherus carinatus* (Gray, 1856); *Chrysemys picta* (Schneider, 1783); *Chelydra serpentina* (Linnaeus, 1758); *Kinosternon subrubrum* Bonnaterre, 1789; *Kinosternon integrum* (Le Conte, 1854); *Kinosternon hirtipes* (Wagler, 1830).

Host geographical distribution *Sternotherus odoratus* occurs in Canada (S Ontario, Quebec), USA (Rhode Island, coastal Maine, S New Hampshire, Vermont, S New York, Massachusetts, Connecticut, Pennsylvania, New Jersey, Delaware, Maryland, West Virginia, Virginia, North Carolina, South Carolina, Georgia, Florida, Alabama, Mississippi, Louisiana, SE Iowa, Missouri, Arkansas, S Wisconsin, Illinois, Indiana, Michigan, Ohio, Kentucky, Tennessee, E Kansas, E Oklahoma, C/E Texas) and Mexico (Chihuahua) (Uetz et al. 2022).

Host conservation status Although eggs of The Common Musk Turtle are eaten by predators and that human threats such as boat propeller impacts, pollution and drying exist, it is not very rare. It is listed as Least Concern by The IUCN Red List of Threatened Species 2015: e.T163450A97384475. https://doi.org/10.2305/IUCN.UK.2015-3.RLTS.T163450A79816811.en. Accessed on 21 January 2022.

Site on host Urinary bladder.

DNA sequence None.

Morphology and morphometrics (μm):			
Source	Wright (1879)	Price (1939)	Lamothe-Argumedo (1972)
Sample size	4		
Body length	2500	1300–2300	2720–3332
Greatest width	1500		885–1181
Width at vagina		510–616	
Haptor length			676–724
Haptor width		460–715	998–1014
False oral sucker length/width		95–190 × 210–360	161–322 × 513 × 531
Pharynx length		114–190	198–232
Pharynx width		95–190	198–225
Testis length		250	305–445
Testis width		340	515–611
Ovary length			210–257
Ovary width		76	116–128
Egg length	235	235	289–354

(continued)

Morphology and morphometrics (μm):			
Source	Wright (1879)	Price (1939)	Lamothe-Argumedo (1972)
Egg width	195	195	183–241
Egg incubation time			
Number of intra-uterine eggs	1	1	1
Genital bulb width			
Number of genital spines	16	16	16
Genital spine length 1	20	20	18–22
Gen spine length 2	15	15	11–15
Haptoral sucker length/width	200	133–190	177–209 × 177–209
Hamulus handle length X/width	150	121–152	131–161 × 71–75
Hamulus guard length Y			
Hamulus hook length			
Marginal hooklet length	15		18
Haptoral L/Body L		0.23	

Remarks *Polystomoidella oblonga* was originally described as *Polystoma oblongum* Wright, 1879. Wright (1879) observed in two specimens that an oncomiracidium hatched inside the ootype.

5.4.6.3 *Polystomoidella whartoni* Price, 1939 (Fig. 5.25)

Collection United States National Museum (USNM).

Holotype USNM 1341286.

Paratypes USNM 1341287.

Original description Price (1939).

Other taxonomic contributions Lamothe-Argumedo (1972).

Etymology Unknown.

Type locality Canada, United States (Florida and Texas), and Mexico.

Other localities None.

Type host *Kinosternon baurii* (Garman, 1891), Striped Mud Turtle; *Kinosternon hirtipes*, Mexican Mud Turtle; *Kinosternon steindachneri* Siebenrock, 1906, Florida Mud Turtle; *Kinosternon subrubrum*, Eastern Mud Turtle.

Additional hosts *Kinosternon integrum*; *Kinosternon leucostomum*.

Site on host Urinary bladder.

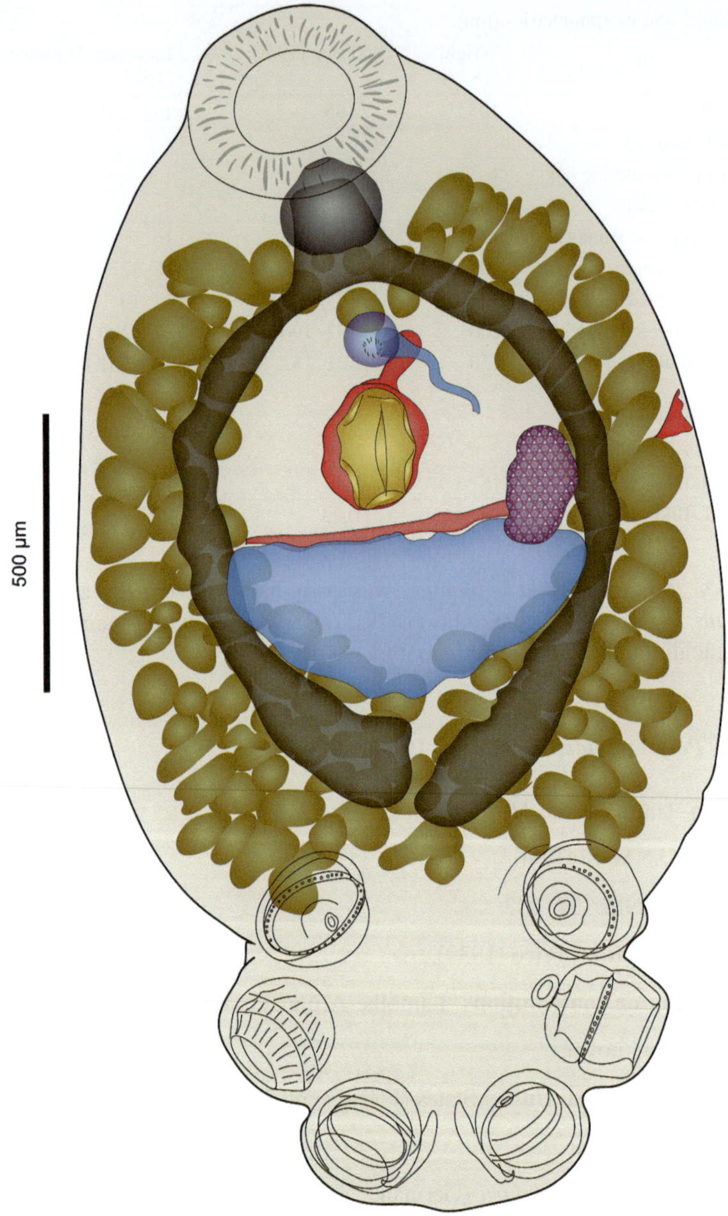

Fig. 5.25 *Polystomoidella whartoni*. Redrawn from micrograph of holotype supplied by Yolanda Villacampa, Department of Invertebrate Zoology, National Museum of Natural History, Smithsonian Institution, Washington DC, USA

Host geographical distribution *Kinosternon baurii* occurs in SE USA (coastal Maryland, Virginia, South Carolina, North Carolina, along the Atlantic Coastal Plain of Georgia southward through peninsular Florida and the Keys) (Uetz et al. 2022).

Host conservation status The Striped Mud Turtle, the Mexican Mud Turtle and the Eastern Mud Turtle are listed as Least Concern by The IUCN Red List of Threatened Species 2011: e.T163435A97382608. https://doi.org/10.2305/IUCN.UK.2011-1.RLTS.T163435A5606303.en. Accessed on 21 January 2022. The Florida Mud Turtle is not listed in IUCN Red List.

DNA sequence Note these sequences were obtained from a polystome extracted from *Kinosternon bauri* in the USA.

- 18S rRNA gene, 1981bp. GenBank Accession: MW029405
- 28S rRNA gene, 1368bp. GenBank Accession: MW029411
- Cytochrome c Oxydase I, 372bp. GenBank Accession: MW029423
- 12S rRNA gene, 469bp. GenBank Accession: MW029417

Morphology and morphometrics (μm):

Source	Price (1939)	Lamothe-Argumedo (1972)	Louis Du Preez collection
Sample size		3	3
Body length	2700	2189–2640	1917–3008 (2582)
Greatest width	1000	660–821	749.5–1182 (994
Width at vagina			729–1149 (968)
Haptor length		483–563	473–662 (576)
Haptor width	680	756–933	734–938 (869)
False oral sucker width	375	209–273 × 289–370	314–389 (358)
Pharynx length		128–193	159–254 (222)
Pharynx width	100	161–177	176–236 (201)
Testis length		414–322	295–347 (329)
Testis width		289–450	215–624 (409)
Ovary length		144–193	134–238 (202)
Ovary width		96–112	81–119 (99)
Egg length	290	322–418	283 and 213 (248)
Egg width	190	225–241	149 and 169 (159)
Egg incubation time			
Number of intra-uterine eggs	1	1	1
Genital bulb width			58–90 (72)
Number of genital spines 1		8	8
Number of genital spines 2		8	6–8 (7)
Genital spine length 1		15–18	11.2–19.4 (15.4)
Genital spine length 2		11	7.1–10.7 (8.4)
Haptoral sucker length/width	200	161–193 × 161–193	136.3–176.2 (161.4)
Hamulus handle length X	148–185		127.7–146.5 (137.8)
Hamulus guard length Y			

(continued)

Morphology and morphometrics (μm):			
Source	Price (1939)	Lamothe-Argumedo (1972)	Louis Du Preez collection
Hamulus hook length			27.9–29.0 (28.4)
Marginal hooklet length	20	18 × 7	18.3–20.1 (19.2)
Haptoral L/Body L			0.21 –0.25 (0.23)

Remarks None.

5.4.7 *Polystomoides* Ward, 1917

5.4.7.1 *Polystomoides albicollis* (MacCallum, 1918) (Fig. 5.26)

Collection Unknown.

Holotype Unknown.

Paratypes Unknown.

Original description MacCallum (1918).

Other taxonomic contributions None.

Etymology Unknown.

Type locality New York Aquarium, U.S.A.

Other localities None.

Type host *Trachemys scripta elegans* (Wied, 1838), Red-eared Slider.

Additional hosts *Kinosternon integrum*; *Kinosternon leucostomum*.

Site on host Urinary bladder.

Host geographical distribution *Trachemys scripta* (Pond slider) occurs in south-central and eastern United States, from south-eastern Virginia to northern Florida. The subspecies *T. s. elegans* has been introduced in many countries worldwide. In the Nearctic Realm this terrapin occurs in Canada (southern Quebec to Ontario) and southern United States (Tennessee, northern Alabama, Pennsylvania, Florida and Virginia). Also introduced to South America, Africa, Australia, Europe, Asia and several islands (Uetz et al. 2022).

Host conservation status The Pond Slider is not threatened and is a popular pet, but is regarded an invasive in several countries where it competes with native species. It is listed as Least Concern by The IUCN Red List of Threatened Species 2011: e.T22028A97429935. https://doi.org/10.2305/IUCN.UK.2011-1.RLTS.T22028A9347395.en. Accessed on 21 January 2022.

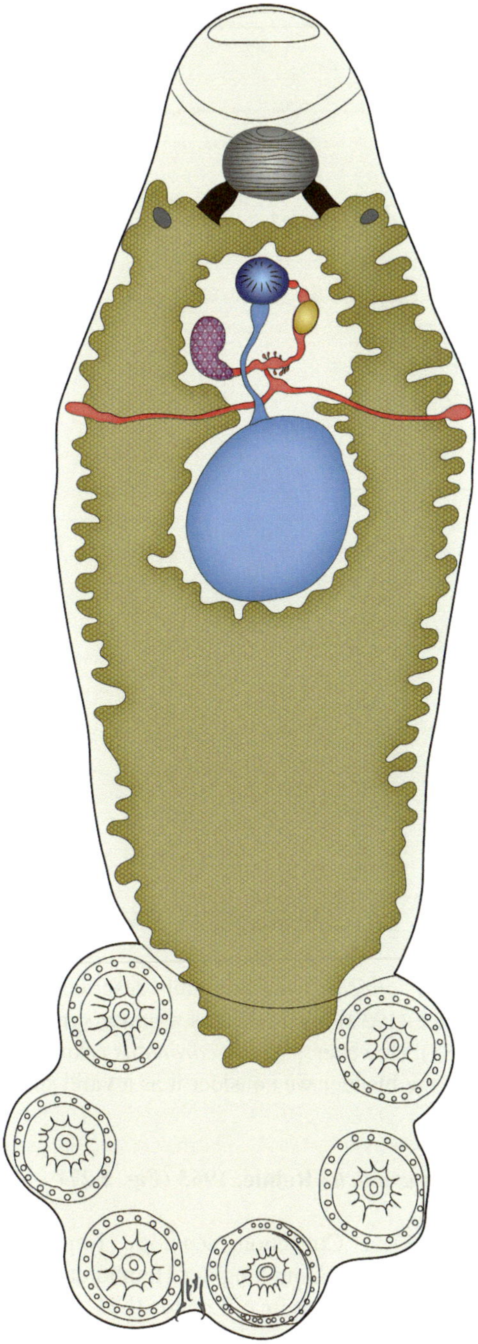

Fig. 5.26 *Polystomoides albicollis*. Redrawn from MacCallum (1918)

DNA sequence None.

Morphology and morphometrics (μm):	
Source	MacCallum (1918)
Sample size	3
Body length	5000–5400
Greatest width	1800
Width at vagina	
Haptor length	
Haptor width	2000
False oral sucker width	400
Pharynx length	
Pharynx width	
Testis length	400
Testis width	
Ovary length	
Ovary width	
Egg length	
Egg width	
Egg incubation time	
Number of intra-uterine eggs	1
Genital bulb width	
Number of genital spines	33; 34; 17
Genital spine length	80
Haptoral sucker diameter	400–450
Hamulus handle length X	
Hamulus guard length Y	
Hamulus hook length	
Marginal hooklet length	
Haptoral L/Body L	

Remarks *Polystomoides albicollis* was originally described al *Polystoma albicollis* MacCallum, 1918. Price (1939) regarded it as a junior synonym of *Polystomoides coronatus* (Leidy, 1888). However, since *P. coronatus* is known from the oral cavity and *P. albicollis* from the bladder we consider it as a valid species.

5.4.7.2 *Polystomoides asiaticus* Rohde, 1965 (Fig. 5.27)

Collection Helminthological Collection, Zoology Department, University of Malaya (UM). Helmithologishe Sammlung der Humboldt Universität, Berlin, Germany (Berlin); Natural History Museum of London (NHML); United States National Museum (USNM).

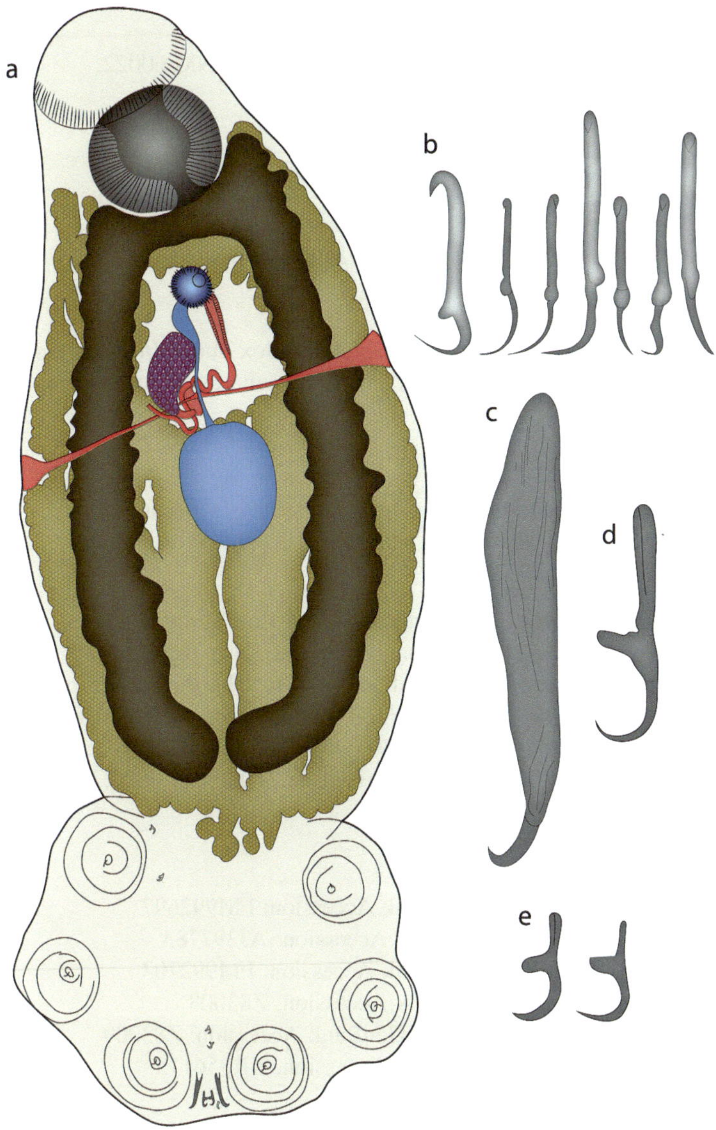

Fig. 5.27 (**a**) *Polystomoides asiaticus*, (**b**) genital spines, (**c**) hamuli, (**d**) hamulus pair two, (**e**) marginal hooklets. Redrawn from Rohde (1965)

Syntypes (UM) R129-135.

Topotype nos Berlin–6533; NHML 1963.10.23.3; USNM 60022.

Original description Rohde (1965).

Other taxonomic contributions None.

Etymology Unknown.

Type locality Malaysia (Chinese shop in Kuala Lumpur).

Other localities None.

Type host *Cuora amboinensis*, Southeast Asian Box Turtle. In the original description of the polystome, the host was known as *Cyclemys amboinensis* Boulenger, 1889.

Additional hosts None.

Host geographical distribution *Cuora amboinensis* occurs in India, Bhutan, Bangladesh, Myanmar, Thailand, Cambodia, Vietnam, West Malaysia, Singapore, Philippines (Uetz et al. 2022).

Host conservation status The Southeast Asian Box Turtle is the most frequently consumed terrapin species in China restaurants, also captured and released in Buddhist temple ponds, and gathered for exports. These practices may lead to a population collapse in the future. It is listed as Endangered by The IUCN Red List of Threatened Species 2020: e.T5958A3078812. https://doi.org/10.2305/IUCN.UK.2020-2.RLTS.T5958A3078812.en. Accessed on 21 January 2022.

Site on host Pharyngeal cavity.

DNA sequence
- 18S rRNA gene, 1964 bp. GenBank Accession: FM992697
- 18S rRNA gene, 572 bp. GenBank Accession: AJ297783
- 28S rRNA gene, 1361 bp. GenBank Accession: FM992703
- 28S rRNA gene, 388 bp. GenBank Accession: Z83008
- Cytochrome c Oxidase I, 388 bp. GenBank Accession: Z83009
- 12S rRNA gene, 498 bp. GenBank Accession: KR856113

Morphology and morphometrics (μm):	
Source	Rohde (1965)
Sample size	10
Body length	3700–5600 (4600)
Greatest width	1100–1700 (1400)
Width at vagina	
Haptor length	1000–1000 (1100)
Haptor width	1100–2500 (1700)
False oral sucker length/width	300–530 (380) × 550–620 (630)
Pharynx length	400–510 (450)

(continued)

Morphology and morphometrics (μm):	
Source	Rohde (1965)
Pharynx width	470–610 (570)
Testis length	240–710 (450)
Testis width	240–520 (360)
Ovary length	90–150 (140)
Ovary width	150–390 (280)
Egg length	240 × 180; 270 × 230; 290 × 210
Egg width	
Egg incubation time	
Number of intra-uterine eggs	
Genital bulb length/width	150–210 (180) × 150–220 (190)
Number of genital spines	34–40 (37)
Genital spine length	Short: 30–45 (36); long: 39–60 (50)
Haptoral sucker diameter	290–410 (340)
Hamulus handle length X	110–160 (140)
Hamulus guard length Y	40–70 (60)
Hamulus hook length	
Marginal hooklet length	
Haptoral L/Body L	0.2391

Remarks None.

5.4.7.3 *Polystomoides aspidonectis* (MacCallum, 1918) (Fig. 5.28)

Collection Unknown.

Holotype Unknown.

Paratypes Unknown.

Original description MacCallum (1918).

Taxonomic note None.

Other taxonomic contributions None.

Etymology Unknown.

Type locality New York Aquarium, United States.

Other localities None.

Type host *Apalone ferox*, Florida Softshell Turtle.

Additional hosts None.

Host geographical distribution *Apalone ferox* occurs in the USA (SW South Carolina, S Georgia, Florida, S Alabama) (Uetz et al. 2022).

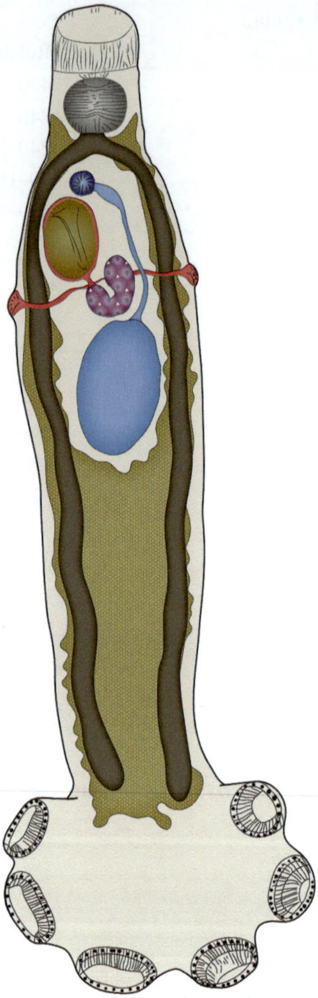

Fig. 5.28 *Polystomoides aspidonectis*. Redrawn from MacCallum (1918)

Host conservation status The predators of the Florida Softshell Turtle include large fish, birds, certain terrapins and many kinds of mammals. That species is also affected by pollution, habitat desiccation and roads. It is considered rare in many areas. It is listed as Least Concern by The IUCN Red List of Threatened Species 2011: e.T165597A97397831. https://doi.org/10.2305/IUCN.UK.2011-1.RLTS.T165597A6065209.en. Accessed on 21 January 2022.

Site on host Nasal cavities, lungs and intestines.

DNA sequence None.

Morphology and morphometrics (μm):	
Source	MacCallum (1918)
Sample size	
Body length	2600
Greatest width	400
Width at vagina	
Haptor length	
Haptor width	800
False oral sucker width	
Pharynx length	
Pharynx width	
Testis length	
Testis width	600
Ovary length	200
Ovary width	
Egg length	
Egg width	200
Number of intra-uterine eggs	
Genital bulb width	
Number of genital spines	17
Genital spine length	
Haptoral sucker diameter	
Marginal hooklet length	
Haptoral L/Body L	

Remarks *Polystomoides aspidonectis* was originally described as *Polystoma aspidonectis* MacCallum, 1918. It was synonymized with *Polystomoides orbicularus* (Price 1939). However based on the host identity and site of infection, we conclude that this species should be regarded as a valid species until more material can be examined.

5.4.7.4 *Polystomoides cayensis* (Du Preez, Badets, Héritier and Verneau, 2017) (Fig. 5.29)

Collection Parasitic Worm Collection, National Museum, Bloemfontein, South Africa (NMBP).

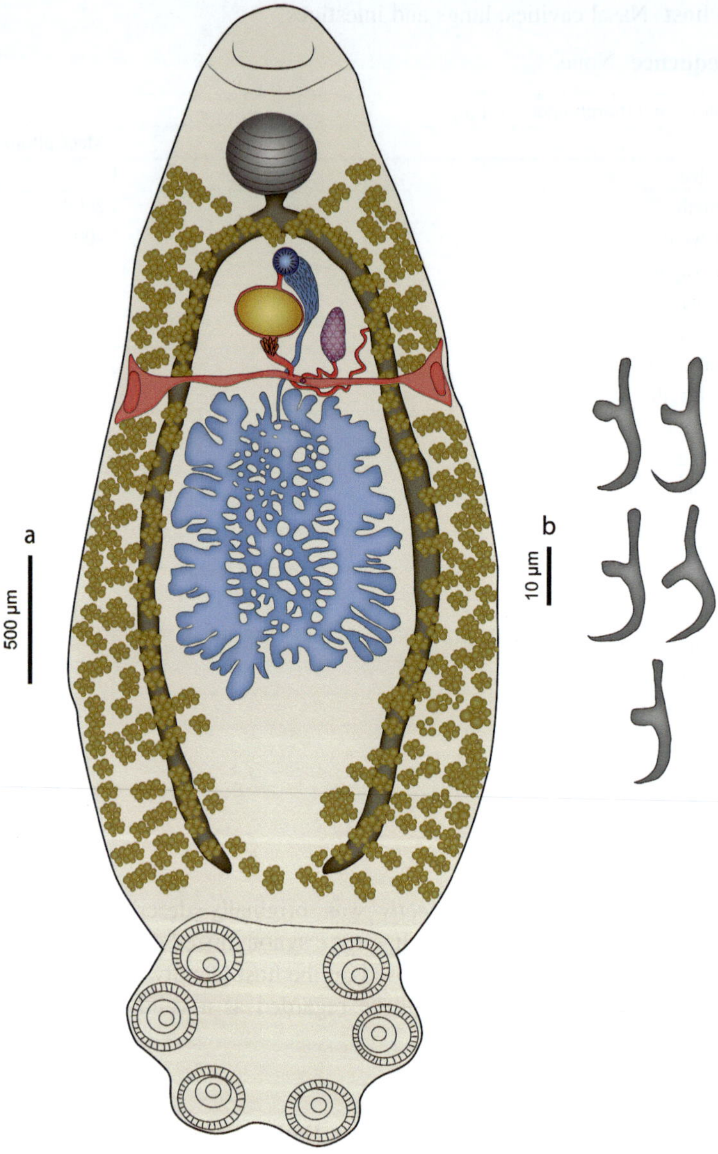

Fig. 5.29 (**a**) *Polystomoides cayensis*, (**b**) marginal hooklets. Redrawn from Du Preez et al. (2017)

Holotype NMBP 394.

Paratypes NMBP 395–403.

Original description Du Preez et al. (2017).

Other taxonomic contributions None.

Etymology Named after the type locality.

Type locality Pond on the outskirts of Cayenne, French Guiana (4.87082 N, 52.33678 W).

Other localities Swampy area near Cayenne, French Guiana (4.90149 N, 52.35767 W).

Type host *Rhinoclemmys punctularia*, Spotted-legged Turtle.

Additional hosts None.

Host geographical distribution *Rhinoclemmys punctularia* occurs in E Colombia, Venezuela, Trinidad, Tobago, Guyana, French Guiana, Suriname and NE Brazil (Tocantins, Pará, Amazon, Bahia, Maranhão, as far upstream as the lower reaches of the Tapajós river, Rio de Janeiro) (Uetz et al. 2022).

Host conservation status The Spotted-legged Turtle is not listed in the IUCN Red List.

Site on host Urinary Bladder.

DNA sequence

- 18S rRNA gene, 1994 bp. GenBank Accession: KY200986
- 28S rRNA gene, 1344 bp. GenBank Accession: KY200988
- Cytochrome c Oxidase I, 372 bp. GenBank Accession: KY200994
- 12S rRNA gene, 471 bp. GenBank Accession: KY200991

Morphology and morphometrics (μm):	
Source	Du Preez et al. (2017)
Sample size	22
Body length	3342–5575 (4139)
Greatest width	1353–1961 (1753)
Width at vagina	1015–1504 (1281)
Haptor length	759–1139 (912)
Haptor width	1095–1459 (1249)
False oral sucker width	130–356 (189)
Pharynx length	293–401 (330)
Pharynx width	276–457 (357)
Testis length	560–983 (762)
Testis width	468–1283 (933)
Ovary length	140–357 (195)

(continued)

Morphology and morphometrics (μm):	
Source	Du Preez et al. (2017)
Ovary width	62–118 (88)
Egg length	209–289 (246)
Egg width	155–187 (165)
Number of intra-uterine eggs	1
Genital bulb width	73–129 (99)
Number of genital spines	16–17
Genital spine length	19–26 (23)
Haptoral sucker diameter	227–297 (261)
Marginal hooklet length	19.6–21.8 (21.2)
Haptoral L/Body L	0.22

Remarks *Polystomoides cayensis* was originally described as *Neopolystoma cayensis* Du Preez, Badets, Héritier and Verneau, 2017.

5.4.7.5 *Polystomoides coronatus* (Leidy, 1888) (Fig. 5.30)

Collection United States National Museum (USNM).

Holotype USNM 1350595.

Syntypes USNM 1315426, 1336024, 1336290, 1336292.

Paratypes USNM 1350596, 1350597.

Original description Leidy (1888).

Taxonomic note The name *Polystomoides coronatum* was corrected for gender.

Other taxonomic contributions Stunkard (1916); Ozaki (1935); Mané-Garzon (1958).

Etymology Unknown.

Type locality North and South America (Mañé-Garzón and Gil 1962).

Other localities None.

Type host Leidy (1888) only states that the host is a common terrapin. Stunkard (1916) gives three possibilities: *Trachemys terrapin* (Lacépède, 1788), Jamaican Slider, *Trachemys decussata decussata* (Gray, 1831), North Antillean Slider and *Terrapene carolina carolina* (Linnaeus, 1758), Carolina Box Turtle.

Additional hosts *Chrysemys picta*; *Chelydra serpentina*; *Trachemys scripta scripta*; *Apalone ferox*; *Apalone spinifera*; *Trachemys scripta elegans*; *Graptemys pseudogeographica pseudogeographica*; *Trachemys dorbigni*; *Actinemys marmorata*.

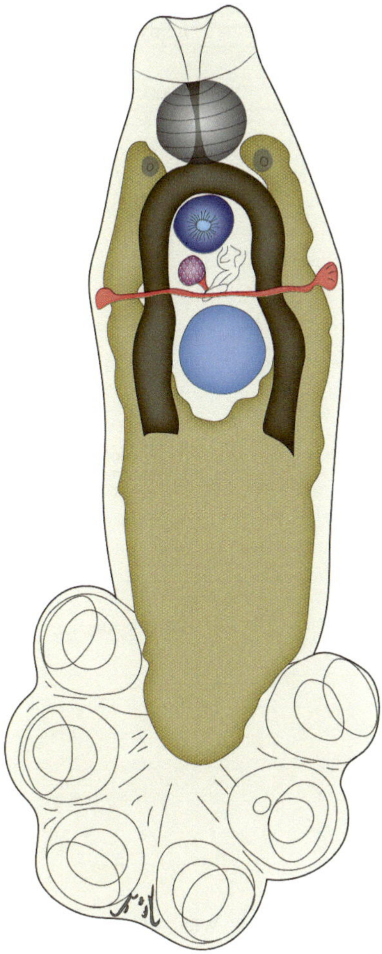

Fig. 5.30 *Polystomoides coronatus*. Redrawn from Leidy (1888)

Host geographical distribution Not available due to uncertainty of true host.

Host conservation status Not available.

Site on host Oral region and nasal passages.

DNA sequence None.

Morphology and morphometrics (μm):		
Source	Leidy (1888)	Stunkard (1916)
Sample size		1
Body length	4000–6000	3150
Greatest width		
Width at vagina		830
Haptor length		
Haptor width		1240
False oral sucker length/width		160 × 400
Pharynx length		
Pharynx width		300 (circular)
Testis length		
Testis width		300 (circular)
Ovary length		
Ovary width		94 (circular)
Egg length		
Egg width		
Egg incubation time		
Number of intra-uterine eggs		θ
Genital bulb width		190
Number of genital spines	32	32
Genital spine length		
Haptoral sucker diameter		370
Hamulus handle length X		132
Hamulus guard length Y		51
Hamulus hook length		
Marginal hooklet length		20
Haptoral L/Body L		

Remarks *Polystomoides coronatus* was originally described as *Polystomum coronatum* Leidy, 1888. Mañé-Garzón and Gil (1962) reported the species in the nose and mouth of *Apalone ferox*, *Apalone spinifera* (LeSueur, 1827), *Trachemys scripta elegans*, *Trachemys scripta scripta* (Schoepff, 1792), *Chelydra serpentina*, *Graptemys pseudogeographica pseudogeographica* (Gray, 1831) from North America and *Trachemys dorbigni* (Duméril and Bibron, 1835) from South America. Baruš and Moravec (1967) found the species in *Actinemys marmorata* (Baird and Girard, 1852) from Oregon, North America.

5.4.7.6 *Polystomoides cyclemydis* Fischthal and Kuntz, 1964 (Fig. 5.31)

Collection United States National Museum (USNM).

Holotype and Paratype USNM 1355972.

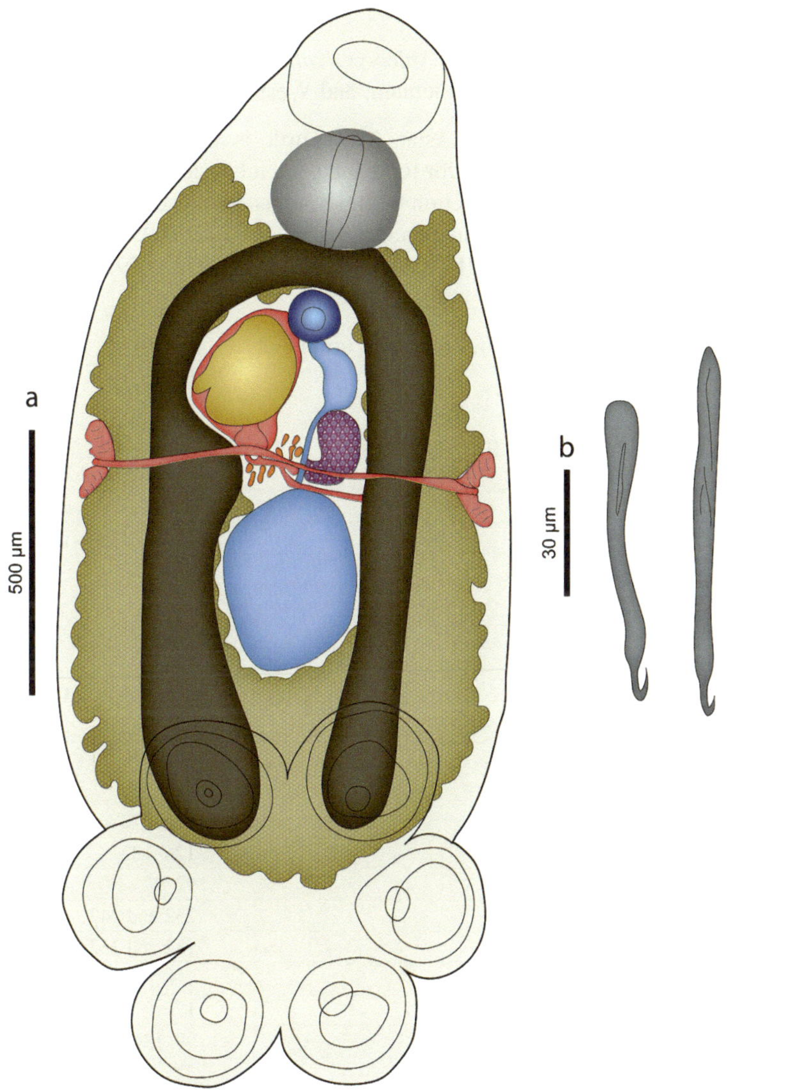

Fig. 5.31 (**a**) *Polystomoides cyclemydis*, (**b**) hamuli. Redrawn from Fischthal and Kuntz (1964)

Original description Fischthal and Kuntz (1964).

Other taxonomic contributions None.

Etymology Named after the host.

Type locality Puerto Princesa, Palawan Island, Philippines.

Other localities None.

Type host *Cyclemys dentata* (Gray, 1831), Asian Leaf Turtle.

Additional hosts None.

Host geographical distribution *Cyclemys dentata* occurs in N India, Bangladesh, Myanmar, Thailand, Cambodia, Vietnam, and West Malaysia (Uetz et al. 2022).

Host conservation status The Asian Leaf Turtle is widespread and reasonably abundant although it is exploited for food and pets. It has no special protection apart from the inclusive national protection of all terrapins. It is listed as Near Threatened by The IUCN Red List of Threatened Species 2021: e.T195849722A2929066. https://doi.org/10.2305/IUCN.UK.2021-1.RLTS.T195849722A2929066.en. Accessed on 21 January 2022.

Site on host Large intestine but probably oral cavity.

DNA sequence None.

Morphology and morphometrics (μm):	
Source	Fischthal and Kuntz (1964)
Sample size	2
Body length	1733–2178 (without haptor)
Greatest width	859–882
Width at vagina	
Haptor length	706–851
Haptor width	815–994
False oral sucker length/width	242–295 × 324–390
Pharynx length	222–265
Pharynx width	242–310
Testis length	340–350
Testis width	210–270
Ovary length	126–182
Ovary width	92–97
Egg length	232 and 242
Egg width	182 and 162
Egg incubation time	
Number of intra-uterine eggs	1
Genital bulb width	99–136 × 92–106
Number of genital spines	32
Genital spine length	30–37 (large); 23–24 (small)

(continued)

Morphology and morphometrics (μm):	
Source	Fischthal and Kuntz (1964)
Haptoral sucker length/width	215–295 × 225–295
Hamulus handle length X	72–91
Hamulus guard length Y	32–47
Hamulus hook length	
Marginal hooklet length	17–23 (larval hooklets)
Haptoral L/Body L	0.39–0.41

Remarks None.

5.4.7.7 *Polystomoides cyclovitellum* (Caballero, Zerecero and Grocott, 1956) (Fig. 5.32)

Collection Helminthological collection, Institute of Biology, National University of Mexico.

Holotype 214–4.

Paratypes None.

Original description Caballero et al. (1956).

Other taxonomic contributions None.

Etymology Refers to the large ovoid vitelline follicles.

Type locality City of Panama, Panama, Central America.

Other localities None.

Type host *Rhinoclemmys melanosterna* (Gray, 1861), Colombian Wood Turtle. In the original description of the polystome, the host was known as *Geoemyda melanosterna* Siebenrock, 1909

Additional hosts None.

Host geographical distribution *Rhinoclemmys melanosterna* occurs in Caribbean drainages of SE Panama and N Colombia, and Pacific drainages of W Colombia and NW Ecuador (Uetz et al. 2022)

Host conservation status The Colombian Wood Turtle is consumed by local people. It is not listed in The IUCN Red List.

Site on host Urinary bladder.

DNA sequence None.

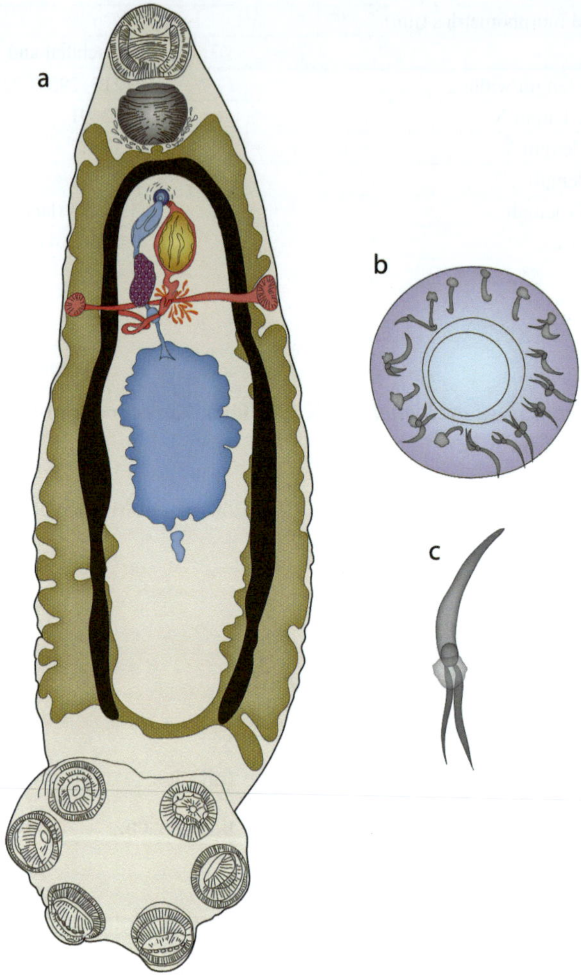

Fig. 5.32 (**a**) *Polystomoides cyclovitellum*, (**b**) genital bulb, (**c**) genital spine. Redrawn from Caballero et al. (1956)

Morphology and morphometrics (μm):	
Source	Caballero et al. (1956)
Sample size	5
Body length	3486–3652
Greatest width	1112–1610
Width at vagina	
Haptor length	747
Haptor width	996–1046
False oral sucker length/width	171–183 × 334–349
Pharynx length	213–232

(continued)

Morphology and morphometrics (μm):	
Source	Caballero et al. (1956)
Pharynx width	277–282
Testis length	730–747
Testis width	498–664
Ovary length	201–239
Ovary width	80–84
Egg length	232
Egg width	149
Egg incubation time	
Number of intra-uterine eggs	1
Genital bulb width	87 × 95
Number of genital spines	16
Genital spine length	19–23
Haptoral sucker length/width	217–232 × 239–243
Hamulus handle length X	
Hamulus guard length Y	
Hamulus hook length	
Marginal hooklet length	
Haptoral L/Body L	0.21

Remarks *Polystomoides cyclovitellum* was originally described as *Neopolystoma cyclovitellum* Caballero, Zerecero and Grocott, 1956.

5.4.7.8 *Polystomoides digitatus* (MacCallum, 1918) (Fig. 5.33)

Collection Unknown. Voucher specimens in the United States National Museum (USNM).

Holotype Unknown.

Paratypes Unknown.

Original description MacCallum (1918).

Taxonomic note The name *Polystomoides digitatum* was corrected for gender.

Other taxonomic contributions None.

Etymology Refers to the finger like shape of the suckers.

Type locality New York Aquarium, U.S.A.

Other localities None.

Type host *Apalone spinifera*, Spiny Softshell Turtle. In the original description of the polystome, the host was known as *Aspidonectes spinifer* Baur, 1888.

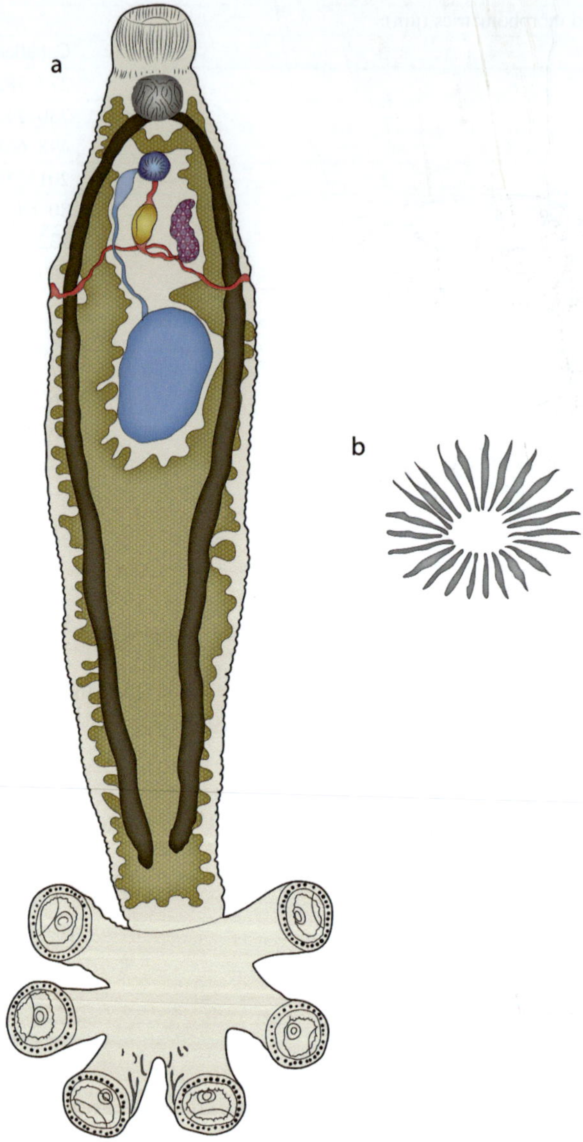

Fig. 5.33 (**a**) *Polystomoides digitatus*, (**b**) genital spines. Redrawn from MacCallum (1918)

Additional hosts *Apalone ferox*.

Host geographical distribution *Apalone spinifera* has a wide distribution in the USA, Canada (S Ontario, Quebec) and Mexico. *Apalone ferox* occurs in the USA (SW South Carolina, S Georgia, Florida, S Alabama) (Uetz et al. 2022).

Host conservation status The Spiny Softshell Turtle is listed as Least Concern by The IUCN Red List of Threatened Species 2011: e.T163451A97398618. https://doi.

org/10.2305/IUCN.UK.2011-1.RLTS.T163451A5607536.en. Accessed on 27 August 2022.

Site on host Nasal cavities.

DNA sequence None.

Morphology and morphometrics (μm):	
Source	MacCallum (1918)
Sample size	3
Body length	4000
Greatest width	800
Width at vagina	
Haptor length	
Haptor width	1400
False oral sucker width	
Pharynx length	200
Pharynx width	
Testis length	600
Testis width	
Ovary length	300
Ovary width	
Egg length	
Egg width	
Egg incubation time	
Number of intra-uterine eggs	1
Genital bulb width	
Number of genital spines	31–35 or 36
Genital spine length	
Haptoral sucker diameter	300 or 400
Hamulus handle length X	
Hamulus guard length Y	
Hamulus hook length	
Marginal hooklet length	
Haptoral L/Body L	

Remarks *Polystomoides digitatus* was originally described as *Polystoma digitatum* MacCallum, 1918. According to Price (1939), it may be regarded as *Polystomoides coronatus*. However, based on unique morphological features and the fact that uncertainty exists as to the true host for *P. coronatus,* we regard it as valid for the moment.

5.4.7.9 *Polystomoides domitilae* (Caballero, 1938) (Fig. 5.34)

Collection Helminthological collection of the Institute of Biology (University of Mexico).

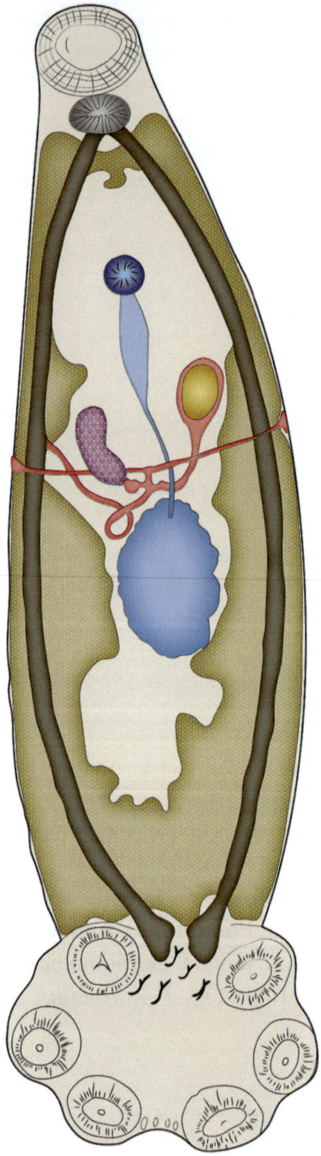

Fig. 5.34 *Polystomoides domitilae*. Redrawn from Caballero (1938)

Holotype 19–7 (Specimen is lost).

Paratypes 225–19.

Neotype no 225–18.

Original description Caballero (1938).

Other taxonomic contributions Price (1939); Lamothe-Argumedo (1972).

Type locality Alvarado, Veracruz (Mexico).

Etymology Unknown.

Other localities None.

Type host *Trachemys ornata* (Gray, 1831), Ornate Slider. In the original description of the polystome, the host was known as *Chrysemys ornata* Savage and Bolanos, 2009.

Additional hosts *Chelydra serpentina.*

Host geographical distribution *Trachemys ornata* occurs in El Salvador, Nicaragua, Costa Rica, Panama and the USA (introduced to Florida) (Uetz et al. 2022).

Host conservation status The Ornate Slider is collected for the pet trade as well as for collections, and with the added threats of habitat degradation and pollution, some populations have become almost depleted. It is listed as Vulnerable by The IUCN Red List of Threatened Species 2007: e.T63661A97430544. https://doi.org/10.2305/IUCN.UK.2007.RLTS.T63661A12704799.en. Accessed on 21 January 2022.

Site on host Urinary bladder.

DNA sequence None.

Morphology and morphometrics (μm):		
Source	Caballero (1938)	Lamothe-Argumedo (1972)
Sample size	2	2
Body length		4039–4057
Greatest width		1320–1722
Width at vagina		
Haptor length	624–780	1046–1067
Haptor width	877–975	1416–1851
False oral sucker length/width	273–312 × 351–429	370–450 × 547
Pharynx length	205–225	305–322
Pharynx width	295–303	328
Testis length	585–224	450–644
Testis width	390–526	289–338
Ovary length	253	322–402
Ovary width	117	112–127
Egg length	155.8	305

(continued)

Morphology and morphometrics (μm):		
Source	Caballero (1938)	Lamothe-Argumedo (1972)
Egg width	104.5	177
Egg incubation time		
Number of intra-uterine eggs	1	1
Genital bulb length/width	164	209–273 × 241–273
Number of genital spines	19–20	20–21
Genital spine length	45	55–71 × 7
Haptoral sucker length/width	273–292	305–322 × 305–483
Hamulus handle length X		
Hamulus guard length Y		
Hamulus hook length		
Marginal hooklet length	24.6	
Haptoral L/Body L		0.258–0.263

Remarks *Polystomoides domitilae* was originally described as *Polystoma domitilae* Caballero, 1938. Lamothe-Argumedo (1972) stated that the original type and paratypes were lost. He assigned a neotype and paratype provided by Dr. Venon. The neotype was found in *Trachemys ornata* and the paratype in *Chelydra serpentina*, both from Tabasco, Mexico.

5.4.7.10 *Polystomoides euzeti* (Combes and Ktari, 1976) (Fig. 5.35)

Collection Unknown.

Holotype Unknown.

Paratypes Unknown.

Original description Combes and Ktari (1976).

Other taxonomic contributions None.

Etymology Named after Prof Louis Euzet, Montpellier, France.

Type locality Streams in the West of Tunis, Tunisia (Africa).

Other localities None.

Type host *Mauremys leprosa* (Schweigger, 1812), Mediterranean Pond Turtle. In the original description of the polystome, the host was known as *Clemmys caspica var. leprosa* Loveridge and Williams, 1957.

Additional hosts None.

Host geographical distribution *Mauremys leprosa* occurs in Spain, Portugal, France, W Libya, Tunisia, Algeria, Morocco, Western Sahara, Senegal, Benin, Niger, Mauritania and Gambia (Uetz et al. 2022).

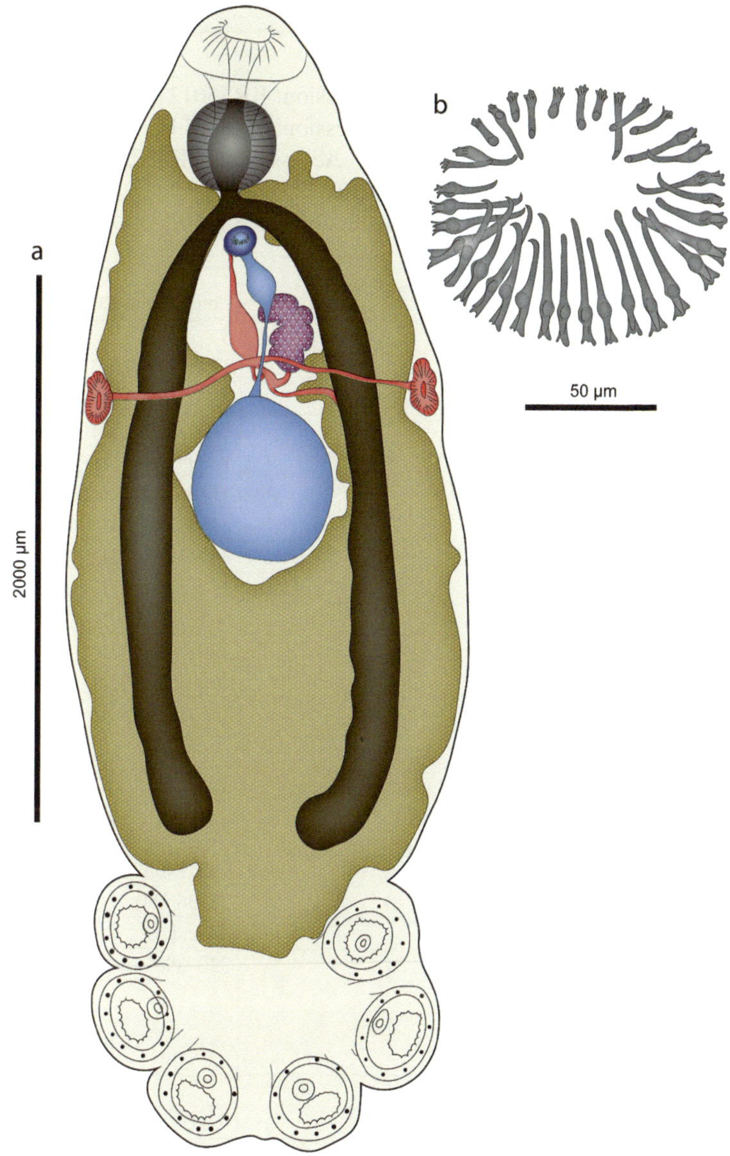

Fig. 5.35 (**a**) *Polystomoides euzeti*, (**b**) genital spines. Redrawn from Combes and Ktari (1976)

Host conservation status The Mediterranean Pond Turtle is very resistant to pollution, not heavily collected and not really threatened by habitat destruction. It is not listed in The IUCN Red List.

Site on host Urinary bladder and cloaca.

DNA sequence

- 18S rRNA gene, 1993 bp. GenBank Accession: KR856127
- 28S rRNA gene, 1366 bp. GenBank Accession: KR856146
- Cytochrome c Oxidase I, GenBank Accessions: FR822587, KM258887, KM258894 and KY704804 to KY704813
- 12S rRNA gene, 497 bp. GenBank Accession: KR856101

Morphology and morphometrics (μm):

Source	Combes and Ktari (1976)
Sample size	6
Body length	3790–5780 (4548)
Greatest width	1570–1880 (1683)
Width at vagina	1420–1600 (1510)
Haptor length	1000–1380 (1190)
Haptor width	1340–2180 (1668)
False oral sucker length/width	286–411 (364) × 457–640 (527)
Pharynx length	340–423 (363)
Pharynx width	350–570 (411)
Testis length	605–730 (680)
Testis width	420–710 (557)
Ovary length	260–350 (302)
Ovary width	115–180 (148)
Egg length	
Egg width	
Egg incubation time	
Number of intra-uterine eggs	0
Genital bulb width	183–240 (209)
Number of genital spines	33–36 (34)
Genital spine length	48 and 57
Haptoral sucker diameter	320–390
Hamulus handle length X	
Hamulus guard length Y	
Hamulus hook length	
Marginal hooklet length	
Haptoral L/Body L	0,263

Remarks *Polystomoides euzeti* was originally known as *Neopolystoma euzeti* Combes and Ktari, 1976.

5.4.7.11 *Polystomoides exhamatum* (Ozaki, 1935) (Fig. 5.36)

Collection Zoological Laboratory of Hiroshima University.

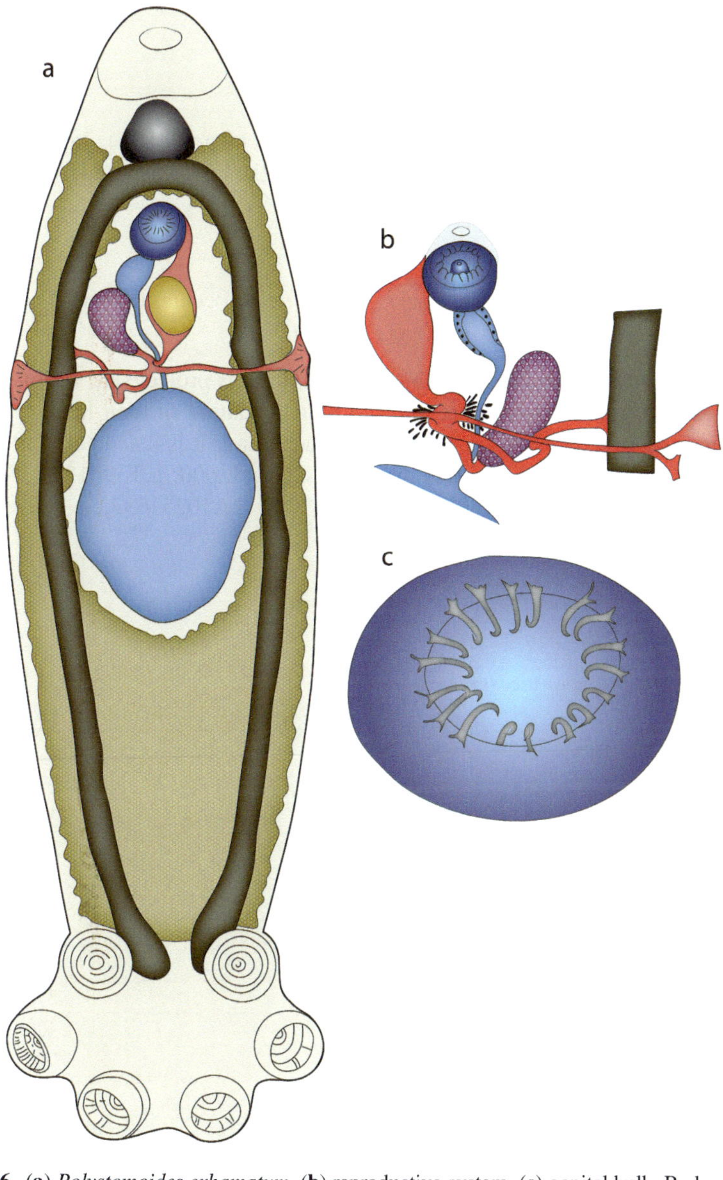

Fig. 5.36 (**a**) *Polystomoides exhamatum*, (**b**) reproductive system, (**c**) genital bulb. Redrawn from Ozaki (1935)

Holotype Unknown.

Paratypes Unknown.

Original description Ozaki (1935).

Other taxonomic contributions Price (1939).

Etymology Unknown.

Type locality Hiroshima, Japan.

Other localities None.

Type host *Mauremys japonica* (Temminck and Schlegel, 1835), Japanese Pond Turtle. In the original description of the polystome, the host was known as *Clemmys japonica* Strauch, 1862.

Additional hosts None.

Host geographical distribution *Mauremys japonica* occurs in Central and South Japan (Honshu, Kyushu, Shikoku) (Uetz et al. 2022).

Host conservation status The Japanese Pond Turtle is protected at national level, and population numbers are being reduced by degradation of their habitat, vehicle deaths and the wildlife trade. Populations are declining. It is listed as Near Threatened by The IUCN Red List of Threatened Species 2000: e.T39612A97370705. https://doi.org/10.2305/IUCN.UK.2000.RLTS.T39612A10251032.en. Accessed on 21 January 2022.

Site on host Urinary bladder.

DNA sequence None.

Morphology and morphometrics (μm):	
Source	Ozaki (1935)
Sample size	
Body length	4600–5400
Greatest width	1500–2000
Width at vagina	
Haptor length	850–1000
Haptor width	1400–2000
False oral sucker width	
Pharynx length	
Pharynx width (diameter)	370–400
Testis length	1200–1600
Testis width	700–1300
Ovary length	370–500
Ovary width	170–200
Egg length	260–300
Egg width	190–210

(continued)

Morphology and morphometrics (μm):	
Source	Ozaki (1935)
Egg incubation time	
Number of intra-uterine eggs	1
Genital bulb width	240
Number of genital spines	16–18
Genital spine length	
Haptoral sucker diameter	300–350
Hamulus handle length X	
Hamulus guard length Y	
Hamulus hook length	
Marginal hooklet length	
Haptoral L/Body L	0.18 – 0.19

Remarks *Polystomoides exhamatum* was originally described as *Neopolystoma exhamatum* Ozaki, 1935.

5.4.7.12 *Polystomoides japonicus* Ozaki, 1935 (Fig. 5.37)

Collection Zoological Laboratory of Hiroshima University, Japan.

Holotype Unknown.

Paratypes Unknown.

Original description Ozaki (1935).

Taxonomic note The original name *Polystomoides japonicum* was corrected for gender.

Other taxonomic contributions None.

Etymology Named after the host.

Type locality Saijo, Hiroshima Prefecture, Japan.

Other localities Author noted that he has collected the species in other localities as well, but provided no other details.

Type host *Mauremys japonica*, Japanese Pond Turtle. In the original description of the polystome, the host was known as *Clemmys japonica*.

Additional hosts None.

Host geographical distribution *Mauremys japonica* occurs in central and southern Japan, on the Honshu, Kyushu and Shikoku islands (Uetz et al. 2022).

Host conservation status The Japanese Pond Turtle is protected at national level, and population numbers are being reduced by degradation of their habitat, vehicle deaths and the wildlife trade. Populations are declining. It is listed as Near Threatened by The IUCN Red List of Threatened Species 2000: e.T39612A97370705. https://doi.org/10.2305/IUCN.UK.2000.RLTS.T39612A10251032.en. Accessed on 21 January 2022.

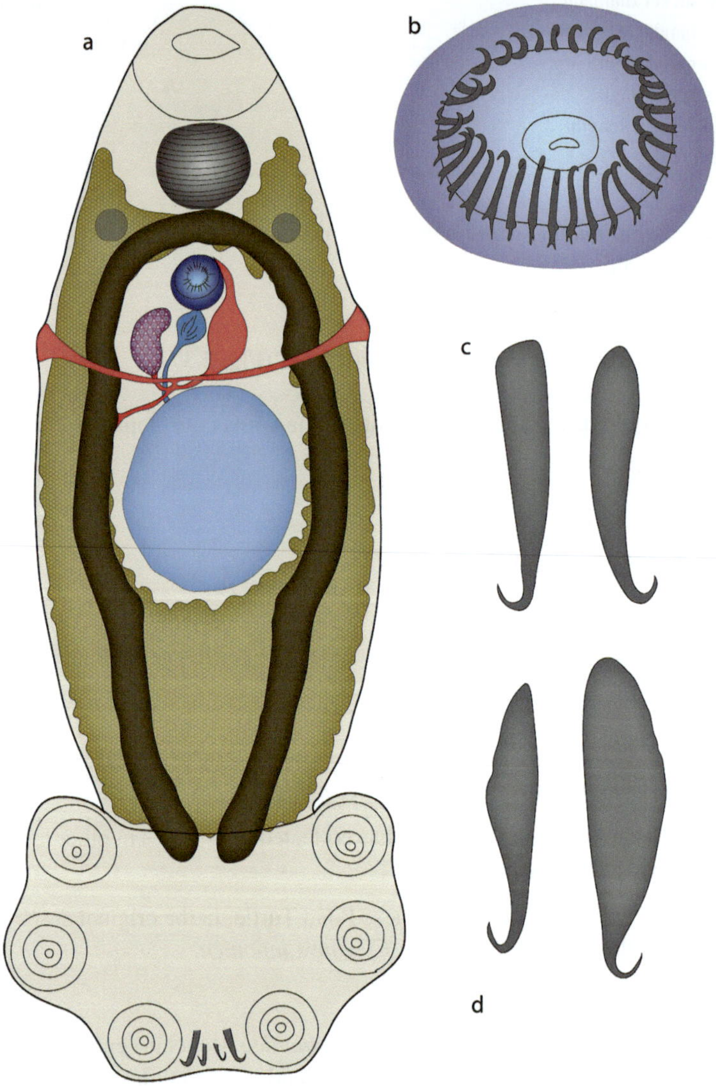

Fig. 5.37 (**a**) *Polystomoides japonicus*, (**b**) genital bulb, (**c**) hamuli, (**d**) hamulus pair two. Redrawn from Ozaki (1935)

Site on host Mouth and oesophagus.

DNA sequence None.

Morphology and morphometrics (μm):	
Source	Ozaki (1935)
Sample size	
Body length	3000–3400
Greatest width	1250–1700
Width at vagina	
Haptor length	1100–1450
Haptor width	1500–1800
False oral sucker width	
Pharynx length	
Pharynx width	350–400
Testis length	850–900
Testis width	650–700
Ovary length	300
Ovary width	150
Egg length	245–265
Egg width	155–160
Egg incubation time	
Number of intra-uterine eggs	1
Genital bulb width	
Number of genital spines	31–39
Genital spine length	
Haptoral sucker diameter	300–350
Hamulus handle length X	100–120
Hamulus guard length Y	50
Hamulus hook length	
Marginal hooklet length	
Haptoral L/Body L	0.39 (0.36–0.42)

Remarks *Polystomoides japonicus* was originally described as *Polystomoides japonicum.*

5.4.7.13 *Polystomoides magdalenensis* Lenis and García-Prieto, 2009 (Fig. 5.38)

Collection Collección de Trematodos de la Universidad de Antioquia (CTUA); Colección Nacional de Helmintos, Instituto de Biología, Universidad Nacional Autónoma de México, Mexico City (CNHE).

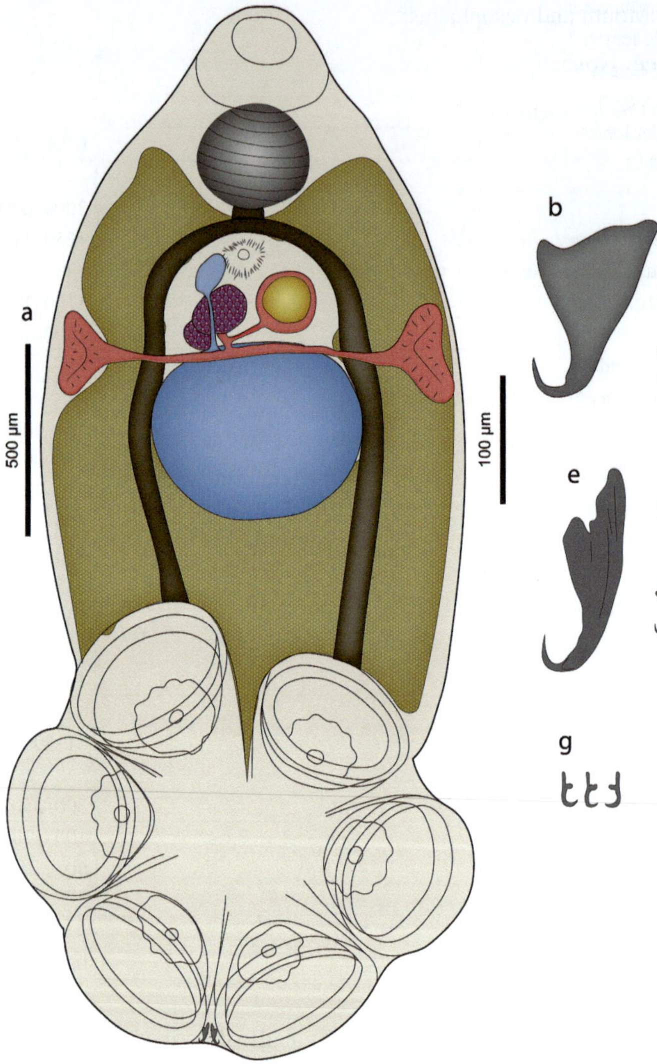

Fig. 5.38 (**a**) *Polystomoides magdalenensis*, (**b**) hamuli, (**c**) hamulus pair two, (**d**) marginal hooklets, (**e**) 2 from photo in type, (**f**) 3 from photo in type, (**g**) 1 from photo in type. Redrawn from Lenis and Garcia-Prieto (2009)

Holotype CTUA (116.10.001).

Paratypes CNHE (6470–71); CTUA (116.02–116.09).

Original description Lenis and García-Prieto (2009).

Other taxonomic contributions None.

Etymology Named after the type locality.

Type locality Coyongal (08°55′11″N, 74°29′25″W), Río Magdalena, Departamento de Bolivar, Colombia.

Other localities Mompox (09°14′40″N, 74°25′29″W) (Departemento de Bolívar); Ricaurte (09°7′7″N, 74°11′40″W) (Departemento del Magdalena), Colombia.

Type host *Trachemys callirostris* (Gray, 1855), Colombian Slider. In the original description of the polystome, the host was known as *Trachemys callirostris callirostris* Martinez Silvestre and Arias Bernal 2005.

Additional hosts None.

Host geographical distribution *Trachemys callirostris* occurs in Belize, Guatemala, Honduras and Mexico (Campeche, Chiapas, Oaxaca, Quintana Roo, Tabasco, Veracruz, Yucatán) (Uetz et al. 2022).

Host conservation status The Colombian Slider is heavily exploited for meat and eggs. Harvesting is especially heavy during Holy Week (Semana Santa), even though it is protected by law in both Colombia and Venezuela, the tradition of egg-harvesting is hard to break. It is not listed in the IUCN Red List.

Site on host Buccal cavity.

DNA sequence None.

Morphology and morphometrics (μm):	
Source	Lenis and García-Prieto (2009)
Sample size	13
Body length	1600–3100 (2400)
Greatest width	
Width at vagina	775–1157 (1114)
Haptor length	840–1469 (1159)
Haptor width	1020–1570 (1290)
False oral sucker length/width	166–428 (220) × 317–483 (392)
Pharynx length	158–333 (249)
Pharynx width	238–428 (333)
Testis length	261–554 (454)
Testis width	388–768 (554)
Ovary length	79–238 (131)
Ovary width	
Egg length	166–246 (219)
Egg width	150–214 (176)
Egg incubation time	
Number of intra-uterine eggs	1
Genital bulb length/width	143–238 (188) × 158–309 (215)

(continued)

Morphology and morphometrics (μm):	
Source	Lenis and García-Prieto (2009)
Number of genital spines	29–35
Genital spine length	45–51 (48)
Haptoral sucker length/width	301–451 (362) × 316–554 (437)
Hamulus handle length X	123–150 (137); $n = 6$
Hamulus guard length Y	54–63 (57); $n = 6$
Hamulus hook length	
Marginal hooklet length	21–27 (25); $n = 10$
Haptoral L/Body L	0.4829

Remarks Authors refer to one polystome incidentally found in cloaca. This parasite may be part of another polystome species. Type drawings were corrected based on a micrograph that appeared in the type description.

5.4.7.14 *Polystomoides megacotyle* (Stunkard, 1916) (Fig. 5.39)

Collection Helminthological Collection of the University of Illinois.

Holotype Unknown.

Paratypes Unknown.

Original description Stunkard (1916).

Other taxonomic contributions Ozaki (1935).

Etymology Unknown.

Type locality Creston, Iowa, North America.

Other localities None.

Type host *Chrysemys picta* (Schneider, 1783), Painted Turtle. In the original description of the polystome, the host was known as *Chrysemys marginata* Williams, 1959.

Additional hosts *Trachemys scripta elegans.*

Host geographical distribution *Chrysemys picta* has a very wide range in most of the United States, in Southern Canada, from Nova Scotia to British Columbia, and south to Georgia, Alabama, Mississippi, Louisiana, Oklahoma, Colorado, Wyoming, Idaho and Oregon. The species can also be found in scattered localities in Texas, New Mexico, Arizona, Utah, and Chihuahua, Mexico. It has also been introduced in California (Uetz et al. 2022).

Host conservation status The worst predator of The Painted Turtle is the raccoon, which can detect and excavate eggs. Other threats also include habitat degradation and highways. That species is listed as Least Concern by The IUCN Red List of

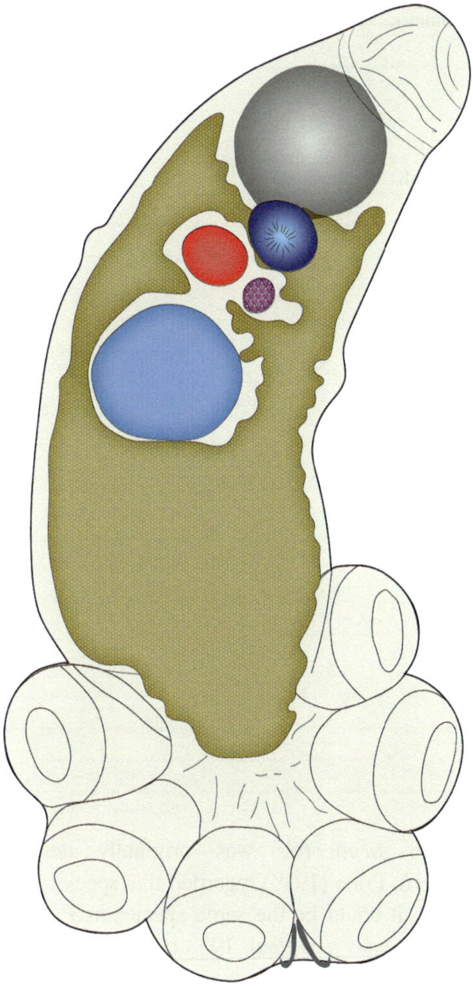

Fig. 5.39 *Polystomoides megacotyle*. Redrawn from Stunkard (1916)

Threatened Species 2011: e.T163467A97410447. https://doi.org/10.2305/IUCN.UK.2011-1.RLTS.T163467A5608383.en. Accessed on 21 January 2022.

Site on host Oral cavity.

DNA sequence None.

Morphology and morphometrics (μm):	
Source	Stunkard (1916)
Sample size	
Body length	2500–2700
Greatest width	710–780

(continued)

Morphology and morphometrics (μm):	
Source	Stunkard (1916)
Width at vagina	
Haptor length	
Haptor width	
False oral sucker length/width	280 × 350–420
Pharynx length	350–380
Pharynx width	380–440
Testis length	280–330
Testis width	330–380
Ovary length	100
Ovary width	75
Egg length	
Egg width	
Egg incubation time	
Number of intra-uterine eggs	
Genital bulb width	
Number of genital spines	36; 42
Genital spine length	
Haptoral sucker diameter	
Hamulus handle length X	
Hamulus guard length Y	
Hamulus hook length	
Marginal hooklet length	
Haptoral L/Body L	

Remarks *Polystomoides megacotyle* was originally described as *Polystoma megacotyle* Stunkard, 1916. Price (1939) regarded this species as a junior synonym of *Polystomoides coronatus*. It could be the same species as *Polystomoides microcotyle* (Stunkard, 1916), *Polystomoides oris* Paul, 1938 or *Polystomoides pauli* Timmers and Lewis, 1979. Until further material can be examined we keep all four species.

5.4.7.15 *Polystomoides microcotyle* (Stunkard, 1916) (Fig. 5.40)

Collection Helminthological Collection of the University of Illinois.

Holotype Unknown.

Paratypes Unknown.

Original description Stunkard (1916).

Other taxonomic contributions Ozaki (1935).

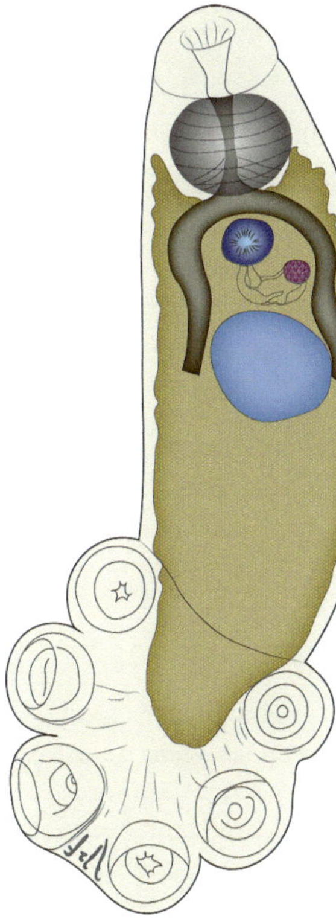

Fig. 5.40 *Polystomoides microcotyle*. Redrawn from Stunkard (1916)

Etymology Unknown.

Type locality Creston, Iowa, North America.

Other localities None.

Type host *Chrysemys picta*, Painted Turtle.

Additional hosts None.

Host geographical distribution *Chrysemys picta* has a very wide range in most of the United States, in Southern Canada, from Nova Scotia to British Columbia, and south to Georgia, Alabama, Mississippi, Louisiana, Oklahoma, Colorado, Wyoming, Idaho and Oregon. The species can also be found in scattered localities in Texas, New Mexico, Arizona, Utah, and Chihuahua, Mexico. It has also been introduced in California (Uetz et al. 2022).

Host conservation status The worst predator of The Painted Turtle is the raccoon, which can detect and excavate eggs. Other threats also include habitat degradation and highways. That species is listed as Least Concern by The IUCN Red List of Threatened Species 2011: e.T163467A97410447. https://doi.org/10.2305/IUCN.UK.2011-1.RLTS.T163467A5608383.en. Accessed on 21 January 2022.

Site on host Oral cavity

DNA sequence None.

Morphology and morphometrics (μm):	
Source	Stunkard (1916)
Sample size	
Body length	3000
Greatest width	780
Width at vagina	
Haptor length	
Haptor width	
False oral sucker length/width	200 × 420
Pharynx length	370
Pharynx width	400
Testis length	360
Testis width	420
Ovary length	75
Ovary width	100
Egg length	
Egg width	
Egg incubation time	
Number of intra-uterine eggs	
Genital bulb width	
Number of genital spines	32
Genital spine length	
Haptoral sucker diameter	
Hamulus handle length X	
Hamulus guard length Y	
Hamulus hook length	
Marginal hooklet length	
Haptoral L/Body L	

Remarks *Polystomoides microcotyle* was originally described as *Polystoma microcotyle* Stunkard, 1916. Price (1939) regarded this species as a junior synonym of *P. coronatus*. It could be the same species as *P. megacotyle*, *P. oris* or *P. pauli*. Until further material can be examined we keep all four species.

5.4.7.16 *Polystomoides microrchis* Fukui and Ogata, 1936 (Fig. 5.41)

Collection Unknown.

Holotype Unknown.

Paratypes Unknown.

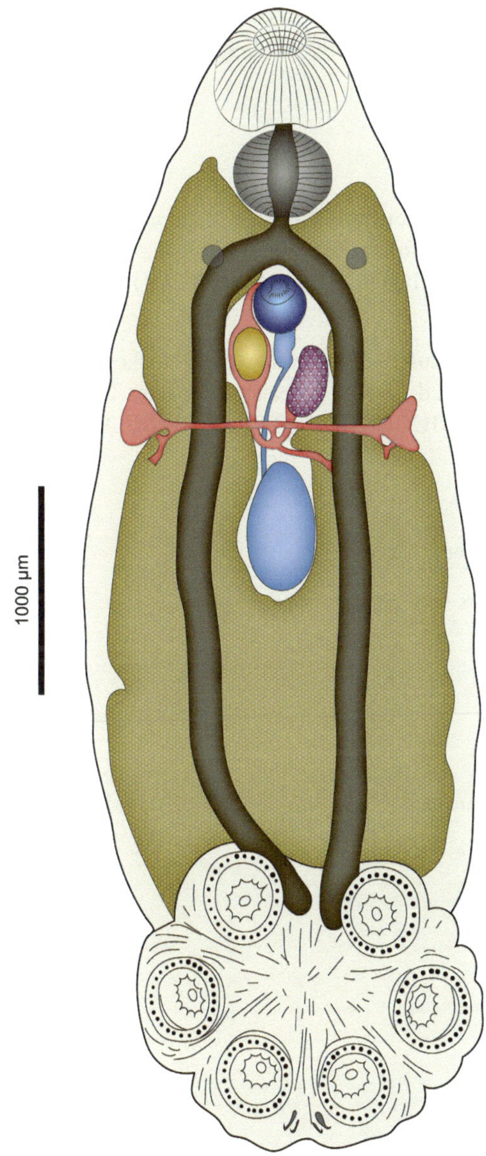

Fig. 5.41 *Polystomoides microrchis*. Redrawn from Fukui and Ogata (1936)

Original description Fukui and Ogata (1936).

Other taxonomic contributions None.

Etymology Unknown.

Type locality Taipei area, Formosa, China.

Other localities None.

Type host *Mauremys sinensis* (Gray, 1834), Chinese Stripe-Necked Turtle. In the original description of the polystome, the host was known as *Ocadia sinensis* Boulenger, 1889.

Additional hosts None.

Site on host Mouth.

Host geographical distribution *Mauremys sinensis* occurs in Taiwan, S China (Guangdong, Guangxi, Fujian, Zhejiang, Jiangsu, Fukien, Hanchow, Soochow, Kwangtung, Shanghai, Hainan Island, Hong Kong), Laos and N Vietnam (Uetz et al. 2022).

Host conservation status The Chinese Stripe-Necked Turtle is listed as Endangered by The IUCN Red List of Threatened Species 2000: e.T15026A97372848. https://doi.org/10.2305/IUCN.UK.2000.RLTS.T15026A4488909.en. Accessed on 21 January 2022.

DNA sequence None.

Morphology and morphometrics (μm):	
Source	Fukui and Ogata (1936)
Sample size	
Body length	5300
Greatest width	2035
Width at vagina	
Haptor length	1270
Haptor width	1900
False oral sucker length/width	730 × 400
Pharynx length	430
Pharynx width	540
Testis length	540
Testis width	430
Ovary length	330
Ovary width	190
Egg length	270
Egg width	190
Egg incubation time	
Number of intra-uterine eggs	
Genital bulb length/width	270 × 250

(continued)

Morphology and morphometrics (μm):	
Source	Fukui and Ogata (1936)
Number of genital spines	42–47
Genital spine length	75–88
Haptoral sucker diameter	390–430
Hamulus handle length/width X	97–110 × 17–22
Hamulus guard length Y	47–55
Hamulus hook length	
Marginal hooklet length	22–25
Haptoral L/Body L	0.24

Remarks None.

5.4.7.17 *Polystomoides ocellatus* (Rudolphi, 1819) (Fig. 5.42)

Collection Unknown.

Holotype Unknown.

Paratypes Unknown.

Original description Rudolphi (1819).

Taxonomic note The name *Polystomoides ocellatum* was corrected for gender by Sproston (1946).

Other taxonomic contributions Ozaki (1935); Knoepffler and Combes (1977).

Etymology Unknown.

Type locality Europe, France according to Knoepffler and Combes (1977).

Other localities Corsica, Morocco.

Type host *Emys orbicularis* (Linnaeus, 1758), European Pond Turtle.

Additional hosts *Mauremys caspica* (Gmelin, 1774).

Host geographical distribution *Emys orbicularis* occurs in Germany, Austria, Switzerland, Poland, Hungary, Albania, Yugoslavia, Czech Republic, Slovakia, Italy, Sardinia, France, Corsica, Spain, Balearic Islands: (Mallorca, Menorca), Portugal, Greece, Turkey, Bulgaria, Romania, Iran, Soviet Union, Latvia, Lithuania, Morocco, Algeria and Tunisia (Uetz et al. 2022).

Host conservation status The European Pond Turtle is completely protected by law throughout its range. Its habitats are being degraded, and American terrapins that are released or escaped are encroaching on habitat and food. More and more protection measurements are being taken, and this is now one of the most studied and best protected species. It is listed as Near Threatened by The IUCN Red List of

Threatened Species 1996: e.T7717A97292665. https://doi.org/10.2305/IUCN.UK.1996.RLTS.T7717A12844431.en. Accessed on 21 January 2022.

Site on host Oral and nasal cavities.

DNA sequence

- 28S rRNA gene. GenBank Accession: OP795805
  Cytochrome c Oxydase I, GenBank Accessions: KY704624 to KY704672

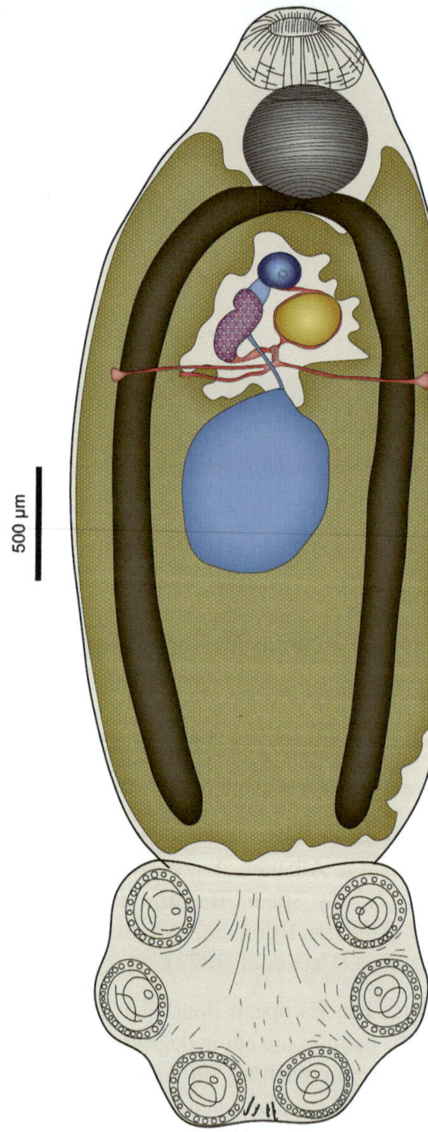

Fig. 5.42 *Polystomoides ocellatus*. Redrawn from Rudolphi (1819)

Morphology and morphometrics (μm):	
Source	Knoepffler and Combes (1977)
Sample size	4
Body length	3130–4060 (3463)
Greatest width	1100–1170 (1143)
Width at vagina	
Haptor length	1000–1200 (1125)
Haptor width	1230–1370 (1323)
False oral sucker width	318–350 (333)
Pharynx length	193–248 (221)
Pharynx width	276–331 (311)
Testis length	400–660 (536)
Testis width	415–590 (501)
Ovary length	219–260 (235)
Ovary width	114–130 (122)
Egg length	230–260 (248)
Egg width	172–200 (193)
Egg incubation time	
Number of intra-uterine eggs	1
Genital bulb width	115–170 (139)
Number of genital spines	29–34
Genital spine length	Short = 27–28; long = 25–37
Haptoral sucker diameter	
Hamulus handle length X	55–62 (58)
Hamulus guard length Y	43–68 (55)
Hamulus hook length	
Marginal hooklet length	
Haptoral L/Body L	0.32

Remarks *Polystomoides ocellatus* was originally described as *Polystoma ocellatum* Rudolphi, 1819. Rudolphi (1819) provided no measurements or sketches of the species.

5.4.7.18 *Polystomoides opacus* (Stunkard, 1916) (Fig. 5.43)

Collection American Museum of National History (AMNH).

Syntypes AMNH1469.1-4.

Original description Stunkard (1916).

Taxonomic note The name *Polystomoides opacum* was corrected for gender.

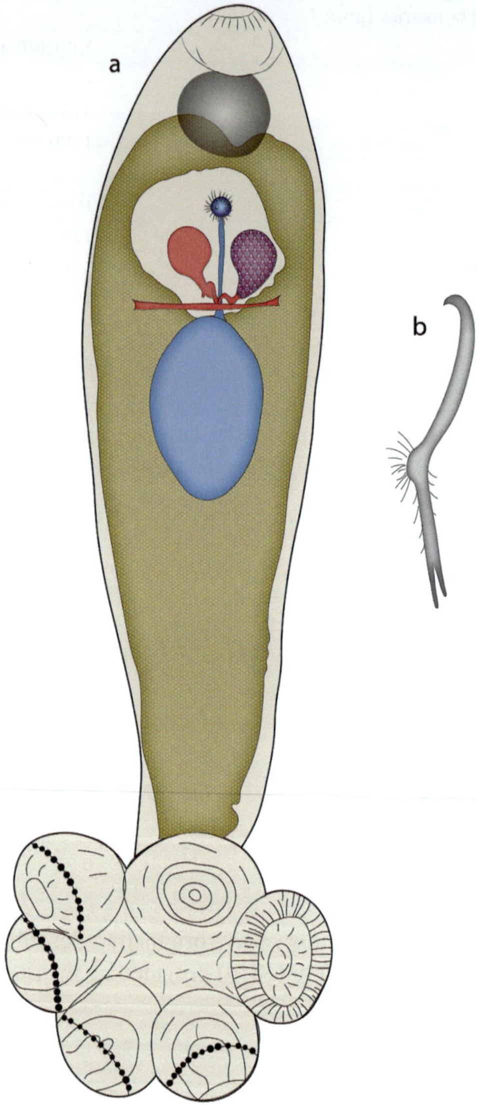

Fig. 5.43 (**a**) *Polystomoides opacus*, (**b**) genital spines. Redrawn from Stunkard (1916)

Other taxonomic contributions Ozaki (1935), Price (1939).

Etymology Unknown.

Type locality Newton, Texas, North America.

Other localities None.

Type host *Apalone ferox*, Florida Softshell Turtle.

Additional hosts None.

Host geographical distribution *Apalone ferox* occurs in the USA (SW South Carolina, S Georgia, Florida, S Alabama) (Uetz et al. 2022).

Host conservation status The predators of The Florida Softshell Turtle include large fish, birds, certain terrapins and many kinds of mammals. That species is also affected by pollution, habitat desiccation and roads. It is considered rare in many areas. It is listed as Least Concern by The IUCN Red List of Threatened Species 2011: e.T165597A97397831. https://doi.org/10.2305/IUCN.UK.2011-1.RLTS.T165597A6065209.en. Accessed on 21 January 2022.

Site on host Oral region and oesophagus.

DNA sequence None.

Morphology and morphometrics (μm):	
Source	Stunkard (1916)
Sample size	
Body length	3250–4000
Greatest width	800–1000
Width at vagina	
Haptor length	
Haptor width	
False oral sucker width	200–220 × 230
Pharynx length	
Pharynx width	300 (spherical)
Testis length	
Testis width	400–500
Ovary length	160–200
Ovary width	800–1200
Egg length	
Egg width	
Egg incubation time	
Number of intra-uterine eggs	
Genital bulb width	
Number of genital spines	32
Genital spine length	
Haptoral sucker diameter	
Hamulus handle length X	
Hamulus guard length Y	
Hamulus hook length	
Marginal hooklet length	
Haptoral L/Body L	

Remarks *Polystomoides opacus* was originally described as *Polystoma opacum* Stunkard, 1916. According to Price (1939), this species should be regarded as *P. coronatus*. In the light that there is uncertainty as to what is true host of *P. coronatus* we keep this species for now.

5.4.7.19 *Polystomoides orbicularis* (Stunkard, 1916) (Fig. 5.44)

Collection American Museum of Natural History (AMNH).

Syntypes AMNH 1434.1, 1434.3, 1467, 1468.1–2.

Original description Stunkard (1916).

Taxonomic note The name *Polystomoides orbiculare* was corrected for gender.

Other taxonomic contributions Price (1939); Lamothe-Argumedo (1972); Dutton et al. (2021).

Etymology Unknown.

Type locality Raleigh, N.C., and Chicago, Illinois and Creston, Iowa, North America.

Other localities United States (North Carolina, Illinois, Iowa, New York, Minnesota, Oklahoma, Florida and Texas).

Type host *Trachemys scripta* (Thunberg, 1792), Pond Slider and *Chrysemys picta*, Painted Turtle. In the original description of the polystome, the hosts were known as *Pseudemys scripta* Stebbins, 1985 and *Chrysemys marginata*, respectively.

Additional hosts *Apalone ferox*; *Kinosternon leucostomum*; *Malaclemys terrapin terrapin* (Schoepff, 1793); *Pseudemys alabamensis* Baur, 1893; *Pseudemys concinna concinna* (Le Conte, 1830); *Trachemys decussata decussata*; *Trachemys scripta elegans*; *Trachemys scripta troostii* (Holbrook, 1836); *Trachemys venusta cataspila* (Gray, 1855); *Macrochelys temminckii* (Troost in Harland, 1835).

Host geographical distribution *Trachemys scripta* occurs in south-central and eastern United States, from south-eastern Virginia to northern Florida. The subspecies *T. s. elegans* has been introduced in many countries worldwide. In the Nearctic Realm this terrapin occurs in Canada (southern Quebec to Ontario) and southern United States (Tennessee, northern Alabama, Pennsylvania, Florida and Virginia). Also introduced to South America, Africa, Australia, Europe, Asia and several islands (Uetz et al. 2022). *Chrysemys picta* has a very wide range in most of the United States, in Southern Canada, from Nova Scotia to British Columbia, and south to Georgia, Alabama, Mississippi, Louisiana, Oklahoma, Colorado, Wyoming, Idaho and Oregon. The species can also be found in scattered localities in Texas, New Mexico, Arizona, Utah, and Chihuahua, Mexico. It has also been introduced in California (Uetz et al. 2022).

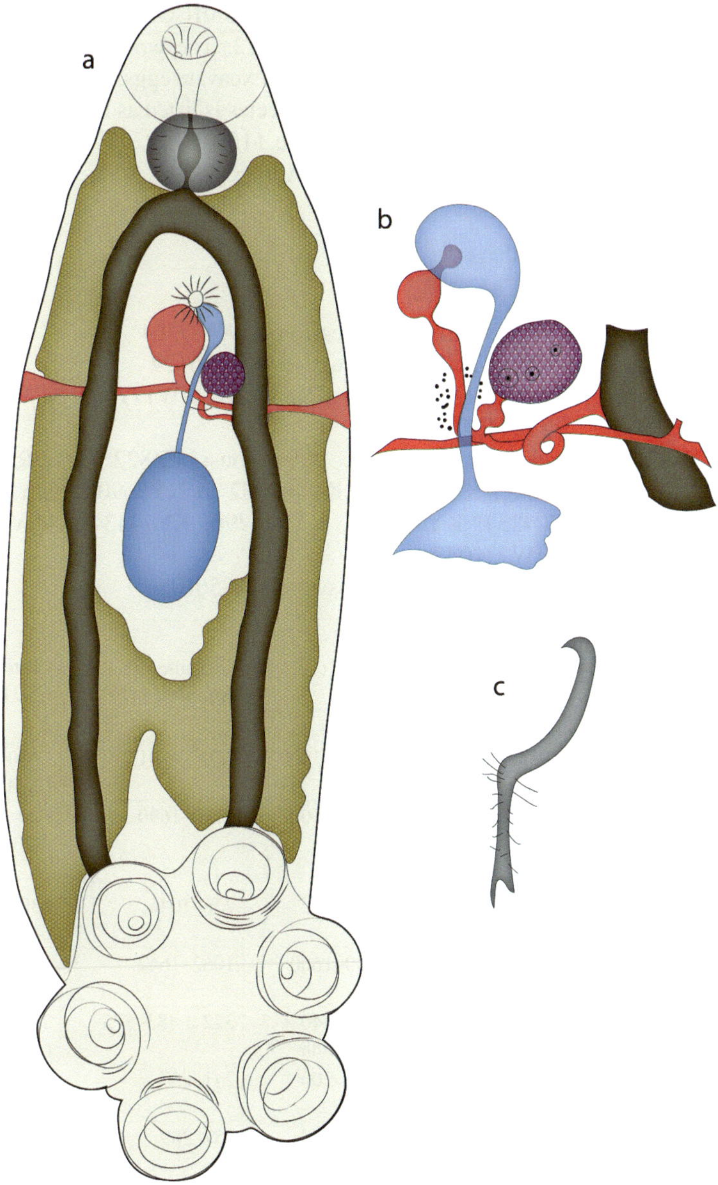

Fig. 5.44 (**a**) *Polystomoides orbicularis*, (**b**) reproductive system, (**c**) genital spines singular. Redrawn from Stunkard (1916)

Host conservation status The Pond Slider is not threatened and is a popular pet, but is regarded an invasive in several countries where it competes with native species. It is listed as Least Concern by The IUCN Red List of Threatened Species

2011: e.T22028A97429935. https://doi.org/10.2305/IUCN.UK.2011-1.RLTS.T22028A9347395.en. Accessed on 21 January 2022. The worst predator of The Painted Turtle is the raccoon, which can detect and excavate eggs. Other threats also include habitat degradation and highways. That species is listed as Least Concern by The IUCN Red List of Threatened Species 2011: e.T163467A97410447. https://doi.org/10.2305/IUCN.UK.2011-1.RLTS.T163467A5608383.en. Accessed on 21 January 2022.

Site on host Urinary bladder.

DNA sequence

- 18S rRNA gene, 1932 bp GenBank Accession: MW177564
- 18S rRNA gene, 1992 bp GenBank Accession: KR856129
- 28S rRNA gene, GenBank Accessions: KR856148, MW177565, MW219725, MW219726 and MW264484
- Cytochrome c Oxidase I, GenBank Accessions: FR822531, FR822532, FR822538 to FR822542, KM258896, KY569087 to KY569091, KY704697, KY704831 to KY704834, KY744295, KY744296, KY744342, MW174247 and MW221029 to MW221033
- 12S rRNA gene, 489 bp. GenBank Accession: KR856103

Morphology and morphometrics (μm):				
Source	Stunkard (1916)	Price (1939)	Lamothe-Argumedo (1972)	Dutton et al. (2021)
Sample size	7		2	4
Body length	2700–3750	2400–5800	3703–4701	2400–2750 (2600)
Greatest width	900–1200	318–1600	1110–1690	700–950 (823)
Width at vagina				
Haptor length			756–1046	630–910 (748)
Haptor width	800–1070 (diameter)	700–1600	1062–1642	780–1100 (871)
False oral sucker length/width	250–270 × 370–420	170–340 × 272–588	322 × 483–575	245–300 (273)
Pharynx length	240–280 (spherical)	187–300	241–257	175–270 (220)
Pharynx width	240–280 (spherical) (diameter)	204–390	257–289	
Testis length	360–500	425–1000	563	290–390 (360)
Testis width	290–390	340–680	322–402	215–275 (253)
Ovary length	140–185	120–375	257–289	75–115 (85)
Ovary width	100–140	65–170	80	60–85 (69)
Egg length		228–272	281	

(continued)

Morphology and morphometrics (μm):				
Source	Stunkard (1916)	Price (1939)	Lamothe-Argumedo (1972)	Dutton et al. (2021)
Egg width	210–224	153–170	206	
Egg incubation time				
Number of intra-uterine eggs	1	1	1	
Genital bulb width		76–148	150–193	158–200 (174)
Number of genital spines	16	16	16–17	16
Genital spine length		20	33–37	45–58 (52)
Haptoral sucker diameter	250–270 × 370–420	170–425	177–402 × 305–402	215–375 (286)
Hamulus handle length X				
Hamulus guard length Y				
Hamulus hook length				
Marginal hooklet length		20		25–25 (25)
Haptoral L/Body L			0.19–0.22	

Remarks *Polystomoides orbicularis* was originally described as *Polystoma orbiculare* Stunkard, 1916. *Polystoma floridanum* Stunkard, 1924 later was later accepted as *Neopolystoma orbiculare* (Stunkard, 1916). Price, 1939 regarded it as as junior synonym of *Polystomoides orbicularis*.

5.4.7.20 *Polystomoides oris* Paul, 1938 (Fig. 5.45)

Collection American Museum of Natural History (AMNH), United States National Museum (USNM).

Holotype AMNH 280.

Syntype USNM 1341297.

Original description Paul (1938).

Other taxonomic contributions None.

Etymology Unknown.

Type locality Cold Spring, New York, U.S.A.

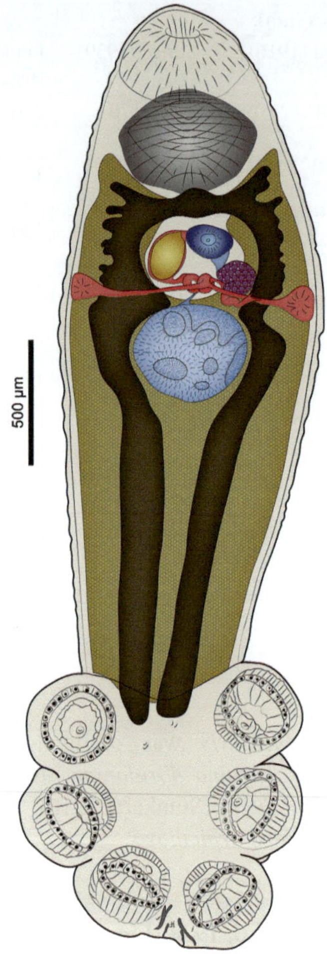

Fig. 5.45 *Polystomoides oris*. Redrawn from Paul (1938)

Other localities None.

Type host *Chrysemys picta*, Painted Turtle.

Additional hosts None.

Host geographical distribution *Chrysemys picta* has a very wide range in most of the United States, in Southern Canada, from Nova Scotia to British Columbia, and south to Georgia, Alabama, Mississippi, Louisiana, Oklahoma, Colorado, Wyoming, Idaho and Oregon. The species can also be found in scattered localities in Texas, New Mexico, Arizona, Utah, and Chihuahua, Mexico. It has also been introduced in California (Uetz et al. 2022).

Host conservation status The worst predator of the Painted Turtle is the raccoon, which can detect and excavate eggs. Other threats also include habitat degradation and highways. That species is listed as Least Concern by The IUCN Red List of Threatened Species 2011: e.T163467A97410447. https://doi.org/10.2305/IUCN.UK.2011-1.RLTS.T163467A5608383.en. Accessed on 21 January 2022.

Site on host Oral cavity.

DNA sequence

- 18S rRNA gene, 1993 bp. GenBank Accession: FM992698
- 28S rRNA gene, 1340 bp. GenBank Accession: FM992705
- Cytochrome c Oxidase I, GenBank Accessions: FR822533, FR822534, KM258893, KY704511 to KY704565, KY704842, KY704843 and KY744282 to KY744294, OP793161 to O793165.
- 12S rRNA gene, 490 bp. GenBank Accession: KR856115

Morphology and morphometrics (μm):	
Source	Paul (1938)
Sample size	12
Body length	2900–4200 (3600)
Greatest width	800–1600 (1100)
Width at vagina	
Haptor length	720–1100
Haptor width	950–1300
False oral sucker width	570
Pharynx length	
Pharynx width	500–650 (555)
Testis length	200–450
Testis width	240–500
Ovary length	90–260
Ovary width	
Egg length	250
Egg width	180
Egg incubation time	2/3 per 24 h
Number of intra-uterine eggs	1
Genital bulb width	
Number of genital spines	24–28 (usually 27)
Genital spine length	58
Haptoral sucker diameter	300–360 (315)
Hamulus handle length X	120
Hamulus guard length Y	65
Hamulus hook length	
Marginal hooklet length	30
Haptoral L/Body L	0.25

Remarks *Polystomoides oris* could be the same species as *P. microcotyle*, *P. megacotyle* or *P. pauli*. Until further material can be examined we keep all four species.

5.4.7.21 *Polystomoides pauli* Timmers and Lewis, 1979 (Fig. 5.46)

Collection United States National Museum (USNM); Canadian Museum of Nature (CMNPA).

Holotype USNM 1370295.

Paratypes USNM 1370296; CMNPA 1978-187, 1978-188.

Original description Timmers and Lewis (1979).

Other taxonomic contributions None.

Etymology Named after Anthony Allard Paul, USA.

Type locality Brokenhead River and Whitemouth River, Manitoba, Canada.

Other localities None.

Type host *Chrysemys picta bellii*, Western Painted Turtle.

Additional hosts *Chrysemys dorsalis* Agassiz, 1857.

Host geographical distribution *Chrysemys picta bellii* occurs in S Canada—Alberta, British Columbia, Manitoba, Ontario, Saskatchewan, USA—Arizona, Colorado, Idaho, Illinois, Iowa, Kansas, Michigan, Minnesota, Missouri, Montana, Nebraska, New Mexico, North Dakota, Oklahoma, Oregon, South Dakota, Texas, Utah, Washington, Wisconsin, Wyoming, Mexico (Chihuahua) (Uetz et al. 2022).

Host conservation status The worst predator for The Western Painted Turtle is the raccoon, which can detect and excavate eggs. Other threats also include habitat degradation and highways. It is listed as Least Concern by The IUCN Red List of Threatened Species 2011: e.T163467A97410447. https://doi.org/10.2305/IUCN.UK.2011-1.RLTS.T163467A5608383.en. Accessed on 21 January 2022.

Site on host Oral mucosa.

DNA sequence None.

Morphology and morphometrics (μm):	
Source	Timmers and Lewis (1979)
Sample size	10
Body length	4950–7830 (6120)
Greatest width	
Width at vagina	1410–1990 (1660)
Haptor length	1480–2150 (1790)
Haptor width	1480–2130 (1700)
False oral sucker length/width	430–722 (537) × 612–951 (833)
Pharynx length	503–777 (642)
Pharynx width	631–941 (795)
Testis length	402–740 (543)

(continued)

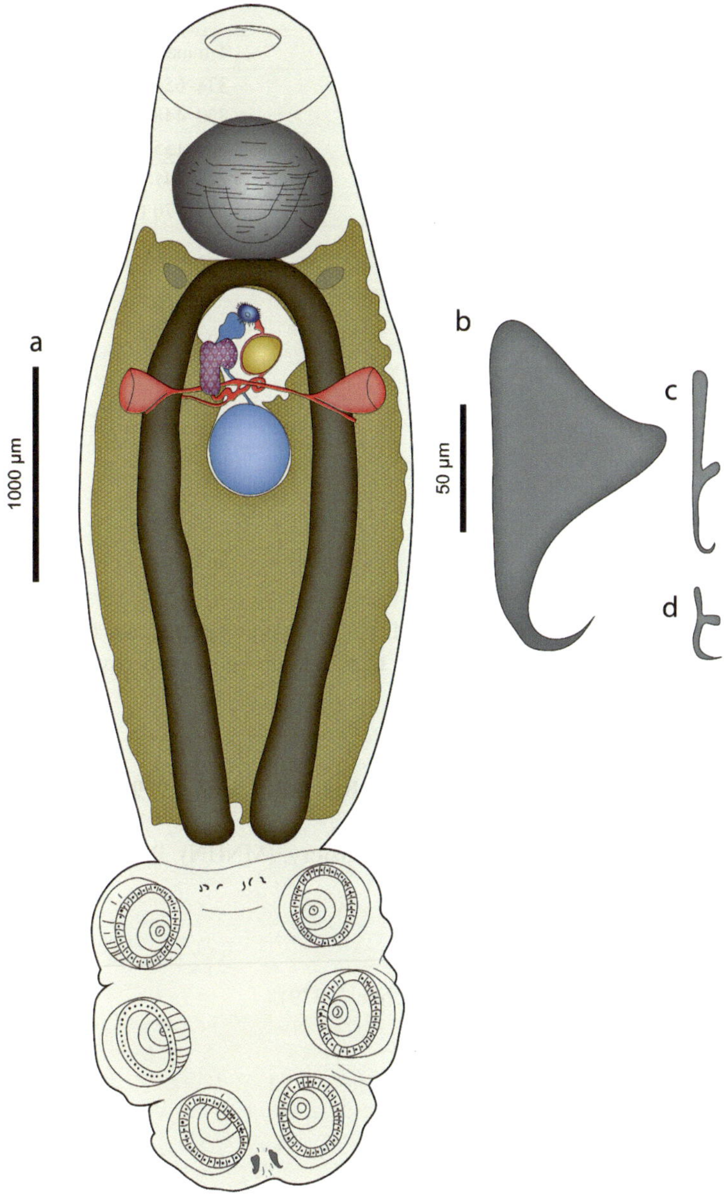

Fig. 5.46 (**a**) *Polystomoides pauli*, (**b**) hamuli, (**c**) hamulus pair two, (**d**) marginal hooklets. Redrawn from Timmers and Lewis (1979)

Morphology and morphometrics (μm):	
Source	Timmers and Lewis (1979)
Testis width	338–658 (513)
Ovary length	329–448 (405)
Ovary width	100–155 (129)
Egg length	213–241 (228)
Egg width	143–170 (157)
Egg incubation time	
Number of intra-uterine eggs	1
Genital bulb width	
Number of genital spines	36–49 (39)
Genital spine length	
Haptoral sucker diameter	366–466 (408)
Hamulus handle length X	116–146 (135)
Hamulus guard length Y	61–77 (70)
Hamulus hook length	
Marginal hooklet length	25–27 (27)
Haptoral L/Body L	0.29

Remarks *Polystomoides pauli* could be the same species as *P. microcotyle*, *P. megacotyle* or *P. oris*. Until further material can be examined we keep all four species.

5.4.7.22 *Polystomoides platynotae* Combes and Rohde, 1979 (Fig. 5.47)

Collection National Museum of Natural History, (MNHN), Paris, France.

Holotype MNHN 238 PE.

Paratypes MNHN 239 PE.

Original description Combes and Rohde (1979).

Other taxonomic contributions None.

Etymology Named after the host.

Type locality Malaysia.

Other localities None.

Type host *Notochelys platynota* (Gray, 1834), Malayan Flat-Shelled Turtle.

Additional hosts None.

Host geographical distribution *Notochelys platynota* occurs in Myanmar, Thailand, Cambodia, West Malaysia, Indonesia (Sumatra, Java, Borneo, Bangka), Vietnam, Singapore (Uetz et al. 2022).

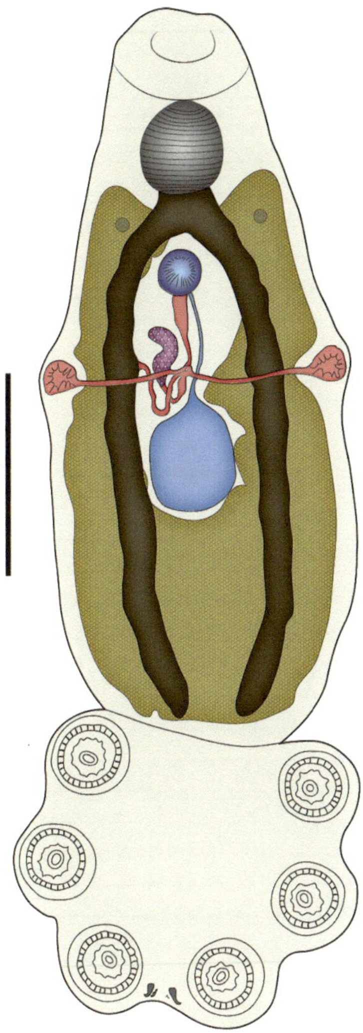

Fig. 5.47 *Polystomoides platynotae*. Redrawn from Combes and Rohde (1979)

Ecology The Malayan Flat-Shelled Turtle is scarce and poorly known, and for lack of information receives no legal protection. However, it is sometimes eaten, kept in ponds and captured for export. It is listed as Vulnerable by The IUCN Red List of Threatened Species 2021: e.T14856A546301. https://doi.org/10.2305/IUCN.UK.2021-1.RLTS.T14856A546301.en. Accessed on 21 January 2022.

Site on host Oral cavity.

DNA sequence None.

Morphology and morphometrics (µm):	
Source	Combes and Rohde (1979)
Sample size	2
Body length	4920–5220 (5070)
Greatest width	1610–2060 (1835)
Width at vagina	
Haptor length	1440–1470 (1455)
Haptor width	1820–1890 (1855)
False oral sucker length/width	440–470 (455) × 732–790 (761)
Pharynx length	455–485 (470)
Pharynx width	470–550 (510)
Testis length	470–510 (490)
Testis width	360–415 (388)
Ovary length	285–300 (293)
Ovary width	95–115 (105)
Egg length	
Egg width	
Egg incubation time	
Number of intra-uterine eggs	0
Genital bulb width	225–240 (233)
Number of genital spines	27–30
Genital spine length	60–70
Haptoral sucker diameter	373–414 (394)
Hamulus handle length X	100–108 (104)
Hamulus guard length Y	60–62 (61)
Hamulus hook length	
Marginal hooklet length	
Haptoral L/Body L	0.29

Remarks None.

5.4.7.23 *Polystomoides renschi* Rohde, 1965 (Fig. 5.48)

Collection Syntypes are in the Helminthological Collection, Zoology Department, University of Malaya (UM). Voucher specimens in Helminthologische Sammlung der Humboldt University, Berlin, Natural History Museum of London and the United States National Museum.

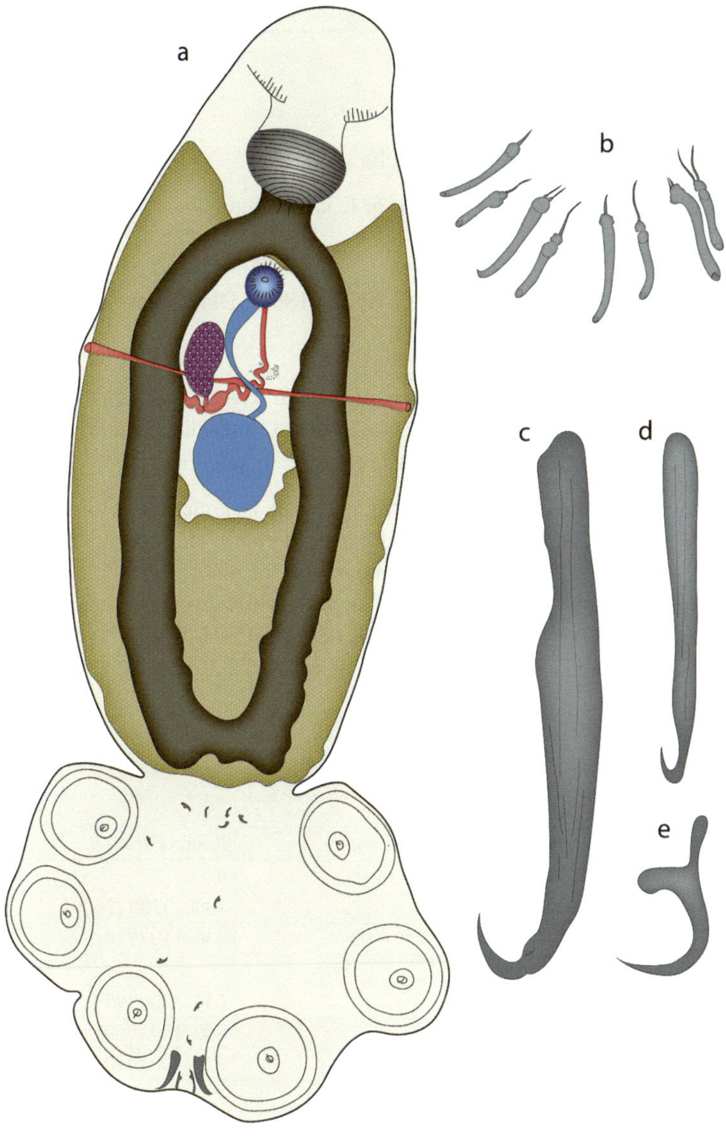

Fig. 5.48 (**a**) *Polystomoides renschi*, (**b**) genital spines, (**c**) hamuli, (**d**) hamulus pair two, (**e**) marginal hooklets. Redrawn from Rohde (1965)

Syntypes R220–222.

Original description Rohde (1965).

Other taxonomic contributions None.

Etymology Named after Prof Dr. Renche.

Type locality Malaysia (Chinese shop in Kuala Lumpur).

Other localities None.

Type host *Siebenrockiella crassicollis* (Gray, 1831), Black Marsh Turtle.

Additional hosts None.

Host geographical distribution *Siebenrockiella crassicollis* has a fragmented range in south-east Asia including S Vietnam, Cambodia, Thailand, Myanmar, Laos, Malaysia, Indonesia and Singapore (Uetz et al. 2022).

Host conservation status The Black Marsh Turtle is sometimes consumed or sold in markets, but the flesh is little sought after. Despite this, populations may be dropping because of progressive fragmentation of their habitat. It is listed as Vulnerable by The IUCN Red List of Threatened Species 2000: e. T39616A97377799. https://doi.org/10.2305/IUCN.UK.2000.RLTS. T39616A10251374.en. Accessed on 21 January 2022.

Site on host Oral cavity.

DNA sequence None.

Morphology and morphometrics (μm):	
Source	Rohde (1965)
Sample size	9
Body length	1800–3700 (2600)
Greatest width	430–110 (770)
Width at vagina	
Haptor length	650–1200 (950)
Haptor width	800–150 (960)
False oral sucker length/width	220–300 (240) × 250–460 (340)
Pharynx length	120–300 (200)
Pharynx width	160–330 (270)
Testis length	120–450 (220)
Testis width	100–280 (200)
Ovary length	40–100 (90)
Ovary width	60–220 (150)
Egg length	
Egg width	
Egg incubation time	
Number of intra-uterine eggs	1
Genital bulb length/width	93–140 (100) × 105–140 (120)

(continued)

Morphology and morphometrics (μm):	
Source	Rohde (1965)
Number of genital spines	20–25 (24)
Genital spine length	15 (short); 17–20 (long)
Haptoral sucker diameter	180–360 (250)
Hamulus handle length X	70–100 (88)
Hamulus guard length Y	42–60 (54)
Hamulus hook length	
Marginal hooklet length	
Haptoral L/Body L	0.37

Remarks None.

5.4.7.24 *Polystomoides rohdei* Mañé-Garzón and Holcman-Spector, 1968 (Fig. 5.49)

Collection Unknown.

Holotype Unknown.

Paratypes Unknown.

Original description Mañé-Garzón and Holcman-Spector (1968).

Other taxonomic contributions None.

Etymology Named after Prof Klaus Rohde, University of New England (UNE), Australia.

Type locality Uruguay, South America.

Other localities None.

Type host *Trachemys dorbigni*, Black Bellied Slider. In the original description of the polystome, the host was known as *Pseudemys dorbigni* Mertens et al., 1934.

Additional hosts None.

Host geographical distribution *Trachemys dorbigni* occurs in Brazil (Rio Grande do Sul, Santa Catarina, Rio de Janeiro), Uruguay and N Argentina (Buenos Aires, Chaco, Corrientes, Entre Rios, Santa Fe) (Uetz et al. 2022).

Host conservation status The Black Bellied Slider is not listed in the IUCN Red List.

Site on host Oral cavity.

DNA sequence

- 28S rRNA gene. GenBank Accession: OP795806
- Cytochrome c Oxydase I, 381 bp. GenBank Accession: FR828364

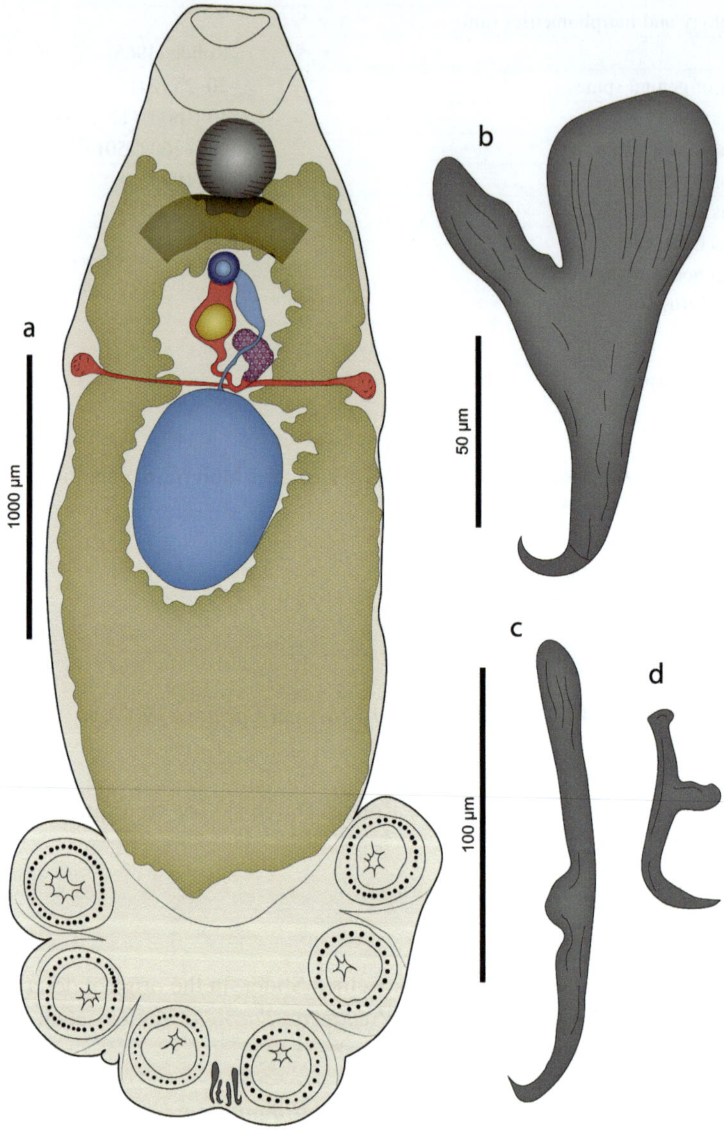

Fig. 5.49 (**a**) *Polystomoides rohdei*, (**b**) hamuli, (**c**) hamulus pair two, (**d**) marginal hooklets. Redrawn from Mané-Garzon and Holcman-Spector (1968)

- Dfd/Hox4 gene, 75 bp. GenBank Accession: FN298161
- Hox2/3 gene, 75. GenBank Accession: FN298155
- Lab/Hox1 gene, 75. GenBank Accession: FN298150

Morphology and morphometrics (μm):	
Source	Mañé-Garzón and Holcman-Spector (1968)
Sample size	
Body length	3000–4020
Greatest width	930–1300
Width at vagina	
Haptor length	
Haptor width	1260–1960
False oral sucker length/width	250–300 × 330–440
Pharynx length	200–290
Pharynx width	210–290
Testis length	540–810
Testis width	510–740
Ovary length	290–370
Ovary width	
Egg length	230–270
Egg width	200
Egg incubation time	
Number of intra-uterine eggs	1
Genital bulb width	
Number of genital spines	29–32
Genital spine length	34–52
Haptoral sucker diameter	340–470
Hamulus handle length X	160
Hamulus guard length Y	70
Hamulus hook length	
Marginal hooklet length	30
Haptoral L/Body L	

Remarks None.

5.4.7.25 *Polystomoides rugosus* (MacCallum, 1918) (Fig. 5.50)

Collection United States National Museum (USNM).

Syntypes USNM 1336302.

Original description MacCallum (1918).

Taxonomic note The name *Polystomoides rugosa* was corrected for gender.

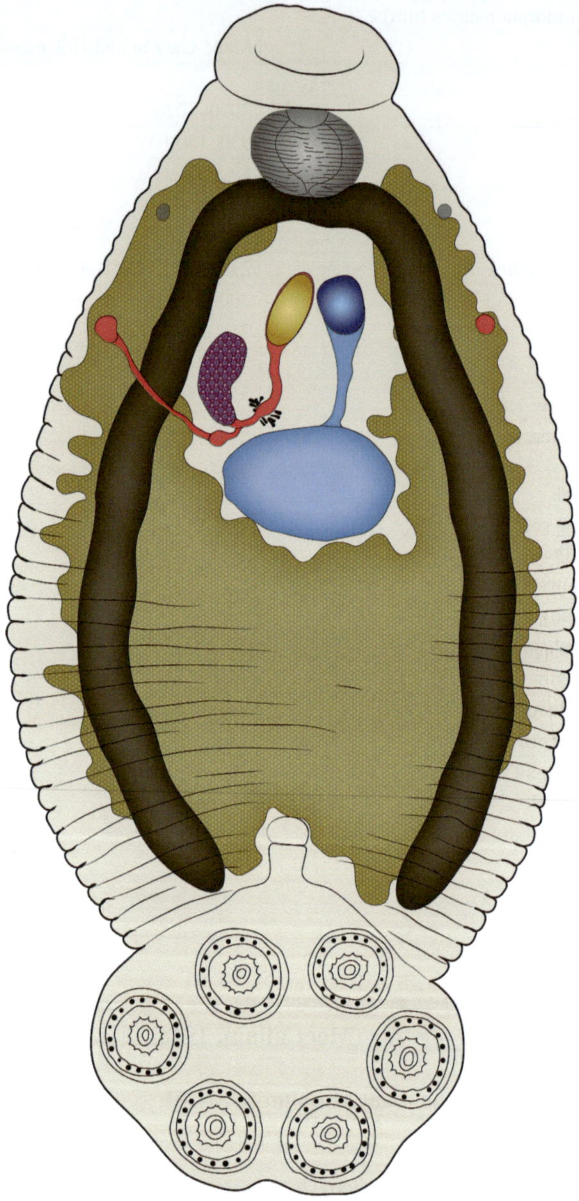

Fig. 5.50 *Polystomoides rugosus*. Redrawn from MacCallum (1918)

Other taxonomic contributions Price (1939).

Etymology Unknown.

Type locality New York Aquarium (United States).

Other localities None.

Type host *Apalone ferox*, Florida Softshell Turtle. In the original description of the polystome, the host was known as *Trionyx ferox*.

Additional hosts None.

Host geographical distribution *Apalone ferox* occurs in the USA (SW South Carolina, S Georgia, Florida, S Alabama) (Uetz et al. 2022).

Host conservation status The predators of the Florida Softshell Turtle include large fish, birds, certain terrapins and many kinds of mammals. That species is also affected by pollution, habitat desiccation and roads. It is considered rare in many areas. It is listed as Least Concern by The IUCN Red List of Threatened Species 2011: e.T165597A97397831. https://doi.org/10.2305/IUCN.UK.2011-1.RLTS.T165597A6065209.en. Accessed on 21 January 2022.

Site on host Nostrils.

DNA sequence Cytochrome c Oxydase I, GenBank Accessions: OP793460 and OP793461

Morphology and morphometrics (μm):		
Source	MacCallum (1918)	Price (1939)
Sample size	4	4
Body length	4000	2960–3710
Greatest width	2000	1920–1960
Width at vagina		
Haptor length		
Haptor width	1400	1260–1370
False oral sucker width	300	425 × 510
Pharynx length		290–340
Pharynx width	450	340–400
Testis length	600	340–680
Testis width	300	510–765
Ovary length	200	340
Ovary width	150	136
Egg length	400	360
Egg width	200	150
Egg incubation time		
Number of intra-uterine eggs	1	1
Genital bulb width		40
Number of genital spines		14
Genital spine length		9

(continued)

Morphology and morphometrics (μm):		
Source	MacCallum (1918)	Price (1939)
Haptoral sucker diameter	400	306–340
Hamulus handle length X		
Hamulus guard length Y		
Hamulus hook length		
Marginal hooklet length	12.2–12.7 (12.6)	
Haptoral L/Body L		

Remarks *Polystomoides rugosus* was originally described as *Polystoma rugosa* MacCallum, 1918. The validity of this species is in question. We retain the species till further evidence becomes available.

5.4.7.26 *Polystomoides scriptanus* Héritier, Verneau, Smith, Coetzer and Du Preez, 2018 (Fig. 5.51)

Collections National Museum of Natural History (MNHN), Paris, France. Parasitic Worm Collection, National Museum, Bloemfontein, South Africa (NMBP).

Holotype MNHN HEL622.

Paratypes MNHN HEL623_624; NMBP 435–436.

Original description Héritier et al. (2018).

Other taxonomic contributions None.

Etymology Named after the host.

Type locality Gainesville, Florida (USA).

Other localities American aquatic ecosystems (North Carolina) and outdoor turtle enclosures (France) for *Trachemys scripta*; outdoor turtle enclosures (France) for *Mauremys leprosa*.

Type host *Trachemys scripta elegans*, Red-eared Slider Turtle.

Additional hosts *Mauremys leprosa*.

Host geographical distribution *Trachemys scripta* occurs in south-central and eastern United States, from south-eastern Virginia to northern Florida. The subspecies *T. s. elegans* has been introduced in many countries worldwide. In the Nearctic Realm this terrapin occurs in Canada (southern Quebec to Ontario) and southern United States (Tennessee, northern Alabama, Pennsylvania, Florida and Virginia). Also Introduced to South America, Africa, Australia, Europe, Asia and several islands (Uetz et al. 2022).

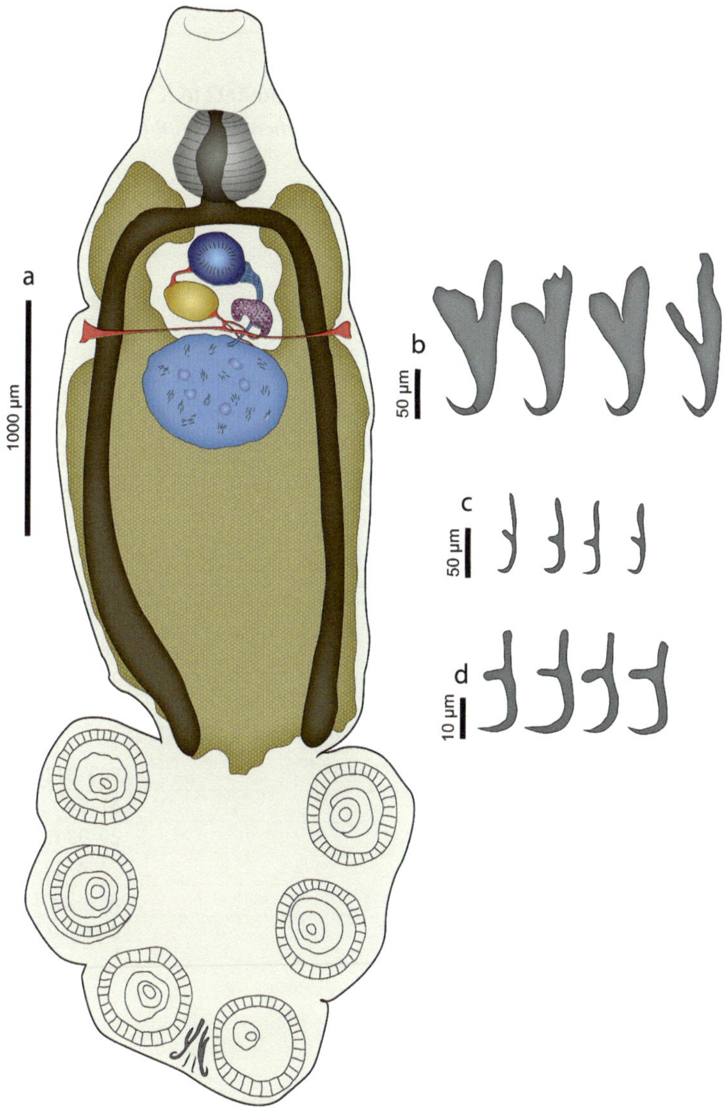

Fig. 5.51 (**a**) *Polystomoides scriptanus*, (**b**) hamuli, (**c**) hamulus pair two, (**d**) marginal hooklets. Redrawn from Héritier et al. (2018)

Host conservation status The Pond Slider is not threatened and is a popular pet, but is regarded an invasive in several countries where it competes with native species. It is listed as Least Concern by The IUCN Red List of Threatened Species 2011: e.T22028A97429935. https://doi.org/10.2305/IUCN.UK.2011-1.RLTS.T22028A9347395.en. Accessed on 21 January 2022.

Site on host Pharyngeal cavity.

DNA sequence
Cytochrome c Oxidase I, GenBank Accessions: KY569092 to KY569099, OP793434, OP793435, OP793438 to OP793442 (from *Trachemys scripta*) and OP793443 to OP793445 (from *Pseudemys peninsularis*)

Morphology and morphometrics (μm):	
Source	Héritier et al. (2018)
Sample size	5
Body length	2789–4954 (3745)
Greatest width	633–1491 (993)
Width at vagina	608–1420 (912)
Haptor length	876–884 (880)
Haptor width	1198–1250 (1224)
False oral sucker width	214–512 (315)
Pharynx length	214–390 (302)
Pharynx width	232–420 (326)
Testis length	197–276 (236)
Testis width	292–334 (313)
Ovary length	333–376 (354)
Ovary width	98–114 (106)
Egg length	218–249 (232)
Egg width	167–184 (172)
Number of intra-uterine eggs	1
Genital bulb width	303–312 (307)
Number of genital spines	31–34
Genital spine length	29–33 (31)
Haptoral sucker diameter	74–418 (207)
Marginal hooklet length	26.4–28.8 (27.3)
Haptoral L/Body L	0.22 – 0.38 (0.29).

Remarks None.

5.4.7.27 *Polystomoides soredensis* Héritier, Verneau, Smith, Coetzer and Du Preez, 2018 (Fig. 5.52)

Collections National Museum of Natural History (MNHN), Paris, France; Parasitic Worm Collection, National Museum, Bloemfontein, South Africa (NMBP).

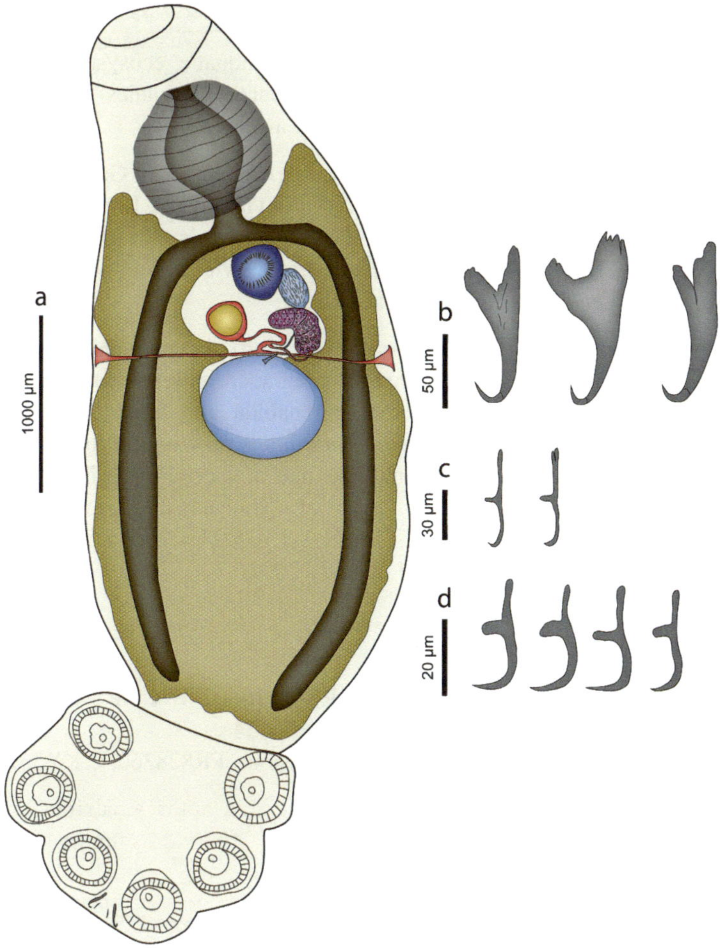

Fig. 5.52 (**a**) *Polystomoides soredensis*, (**b**) hamuli, (**c**) hamulus pair two, (**d**) marginal hooklets. Redrawn from Héritier et al. (2018)

Holotype MNHN HEL620.

Paratypes MNHN HEL621; NMBP 430–434.

Original description Héritier et al. (2018).

Other taxonomic contributions None.

Etymology Named after the type locality.

Type locality Outdoor turtle enclosures of the "Vallée des Tortues" at Sorède (France).

Other localities Pyrénées Orientales in France; Indiana, Maine and North Carolina in USA.

Type host *Emys orbicularis*; European Pond Turtle.

Additional hosts *Mauremys leprosa* in French aquatic ecosystems (Pyrénées Orientales); *Trachemys scripta* in French (Pyrénées Orientales) and American aquatic ecosystems (Indiana, Maine and North Carolina).

Host geographical distribution *Emys orbicularis* occurs in Germany, Austria, Switzerland, Poland, Hungary, Albania, Yugoslavia, Czech Republic, Slovakia, Italy, Sardinia, France, Corsica, Spain, Balearic Islands: (Mallorca, Menorca), Portugal, Greece, Turkey, Bulgaria, Romania, Iran, Soviet Union, Latvia, Lithuania, Morocco, Algeria and Tunisia (Uetz et al. 2022).

Host conservation status The European Pond Turtle is completely protected by law throughout its range. Its habitats are being degraded, and American terrapins that are released or escaped are encroaching on habitat and food. More and more protection measurements are being taken, and this is now one of the most studied and best protected species due to the mobilization of specialists. It is listed as Near Threatened by The IUCN Red List of Threatened Species 1996: e. T7717A97292665. https://doi.org/10.2305/IUCN.UK.1996.RLTS.T7717A12844431.en. Accessed on 21 January 2022.

Site on host Pharyngeal cavity.

DNA sequence

- 18S rRNA gene. GenBank Accession: KR856135
- 28S rRNA gene. GenBank Accessions: KR856154
- Cytochrome c Oxidase I, GenBank Accession: FR828360, KY569080 to KY569086
- 12S rRNA gene. GenBank Accession: KR856111

Morphology and morphometrics (μm):	
Source	Héritier et al. (2018)
Sample size	7
Body length	3033–5286 (4390)
Greatest width	1051–1875 (1412)
Width at vagina	1019–1754 (1352)
Haptor length	756–1119 (931)
Haptor width	1133–1706 (1459)
False oral sucker width	433–687 (545)
Pharynx length	415–757 (532)
Pharynx width	474–815 (604)
Testis length	438–667 (574)
Testis width	419–690 (598)
Ovary length	354–544 (439)
Ovary width	90–148 (123)
Egg length	214–238 (226)
Egg width	148–151 (149)
Number of intra-uterine eggs	1

(continued)

Morphology and morphometrics (μm):	
Source	Héritier et al. (2018)
Genital bulb width	329–408 (374)
Number of genital spines	34–36
Genital spine length	39–43 (41)
Haptoral sucker diameter	277–380 (312)
Marginal hooklet length	27.7–30.4 (28.8)
Haptoral L/Body L	0.19–0.25 (0.22)

Remarks None.

5.4.7.28 *Polystomoides terrapenis* (Harwood, 1932) (Fig. 5.53)

Collection United States National Museum (USNM).

Holotype USNM 1332517.

Paratypes None.

Original description Harwood (1932).

Other taxonomic contributions Price (1939).

Etymology Named after the host.

Type locality Houston, Texas, North America.

Other localities None.

Type host *Terrapene carolina triunguis* (Agassiz, 1857), Three-toed Box Turtle.

Additional host None.

Host geographical distribution *Terrapene carolina triunguis* occurs in the USA (Alabama, Arkansas, Illinois, Kansas, Louisiana, Mississippi, Missouri, Oklahoma, Texas) (Uetz et al. 2022).

Host conservation status The Three-toed Box Turtle is in decline in some areas. It suffers from drying of wetlands, urban development, agricultural proliferation and overgrazing. It is listed as Vulnerable by The IUCN Red List of Threatened Species 2011: e.T21641A97428179. https://doi.org/10.2305/IUCN.UK.2011-1.RLTS.T21641A9303747.en. Accessed on 21 January 2022

Site on host Urinary bladder.

DNA sequence None.

Morphology and morphometrics (μm):	
Source	Harwood (1932)
Sample size	
Body length	1900–2500

(continued)

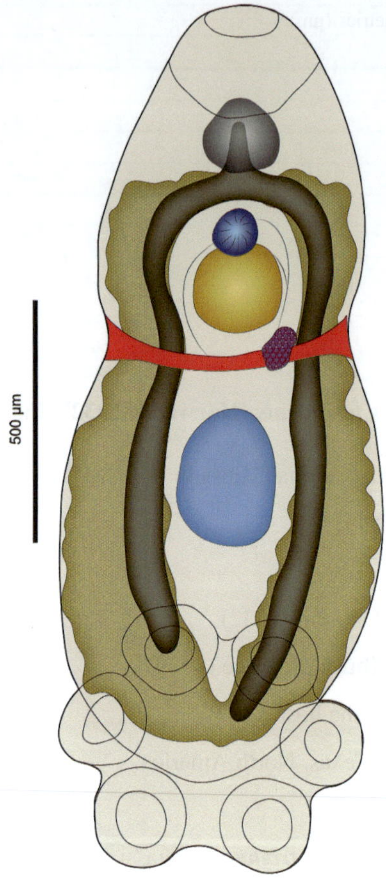

Fig. 5.53 *Polystomoides terrapenis*. Redrawn from Harwood (1932)

Morphology and morphometrics (µm):	
Source	Harwood (1932)
Greatest width	720–820
Width at vagina	
Haptor length	
Haptor width	640–800
False oral sucker length/width	260–280 × 290–360
Pharynx length	130–170
Pharynx width	190–220
Testis length	300–330
Testis width	230–280
Ovary length	
Ovary width	67–85 (spherical)
Egg length	

(continued)

Morphology and morphometrics (μm):	
Source	Harwood (1932)
Egg width	180–220
Egg incubation time	
Number of intra-uterine eggs	1
Genital bulb width	82–90
Number of genital spines	16
Genital spine length	
Haptoral sucker diameter	180–200
Hamulus handle length X	
Hamulus guard length Y	
Hamulus hook length	
Marginal hooklet length	20
Haptoral L/Body L	

Remarks *Polystomoides terrapenis* was originally described as *Polystoma* (*Polystomoides*) *terrapenis* Harwood, 1932.

5.4.7.29 *Polystomoides tunisiensis* Gonzales and Mishra, 1977 (Fig. 5.54)

Collection Unknown.

Holotype Unknown.

Paratypes Unknown.

Original description Gonzales and Mishra (1977).

Other taxonomic contributions None.

Etymology Named after type locality country.

Type locality Tunisia, Africa.

Other localities None.

Type host *Mauremys leprosa*, Mediterranean Pond Turtle. In the original description of the polystome, the host was known as *Clemmys caspica leprosa*.

Additional hosts None.

Host geographical distribution *Mauremys leprosa* occurs in Spain, Portugal, France, W Libya, Tunisia, Algeria, Morocco, Western Sahara, Senegal, Benin, Niger, Mauritania and Gambia (Uetz et al. 2022).

Host conservation status The Mediterranean Pond Turtle is very resistant to pollution, not heavily collected and not really threatened by habitat destruction. It is not listed in The IUCN Red List.

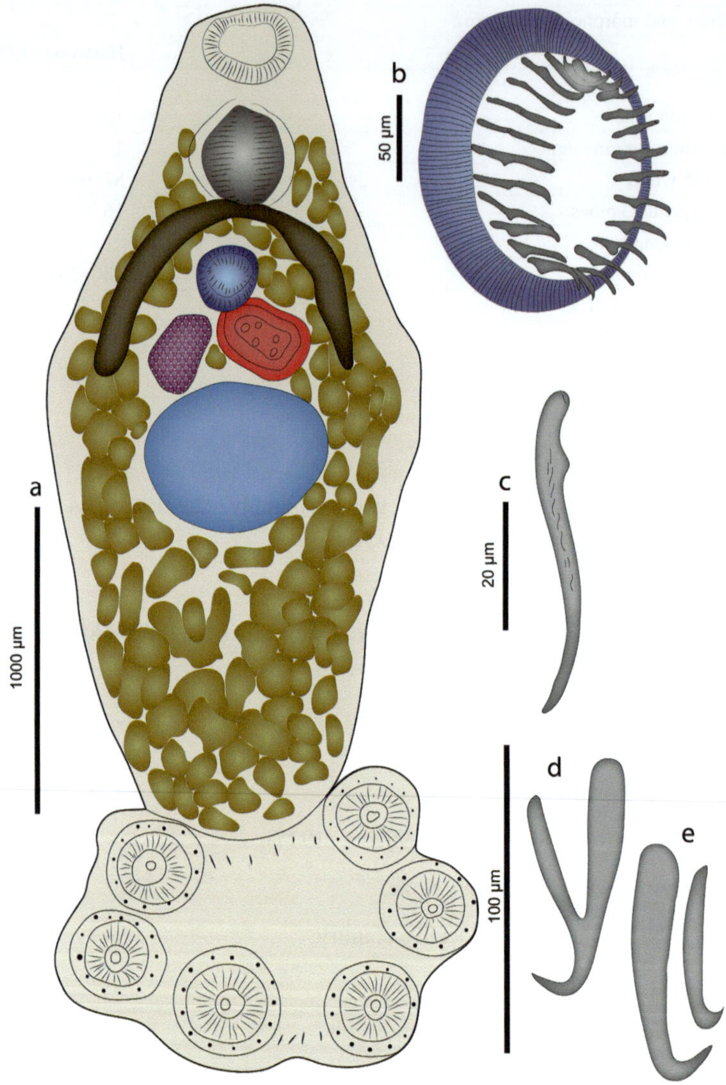

Fig. 5.54 (**a**) *Polystomoides tunisiensis*, (**b**) genital bulb, (**c**) genital spines, (**d**) hamuli, (**e**) genital spines. Redrawn from Gonzales and Mishra (1977)

Site on host Oral cavity, pharynx, oesophagus.

DNA sequence

- 18S rRNA gene, 1993 bp. GenBank Accession: KR856136
- 28S rRNA gene, 1366 bp. GenBank Accession: KR856155

- Cytochrome c Oxidase I, GenBank Accessions: FR822570 to FR822575, KM258885, KM258886, KM258888, KM258890, KM258892 and KY704814 to KY704830
- 12S rRNA gene, 497 bp. GenBank Accession: KR856116

Morphology and morphometrics (μm):	
Source	Gonzales and Mishra (1977)
Sample size	
Body length	1500–5700
Greatest width	700–1700
Width at vagina	
Haptor length	900–2200
Haptor width	500–1700
False oral sucker length/width	100–400 × 200–500
Pharynx length	200–500
Pharynx width	150–600
Testis length	200–700
Testis width	300–500
Ovary length	
Ovary width	100–400
Egg length	300–600
Egg width	200–400
Egg incubation time	
Number of intra-uterine eggs	1
Genital bulb width	
Number of genital spines	26–28
Genital spine length	40–50
Haptoral sucker diameter	300–500
Hamulus handle length/width X	90–120 × 10
Hamulus guard length Y	50–70
Hamulus hook length	
Marginal hooklet length	10–40
Haptoral L/Body L	0.4 (0.4–0.6)

Remarks None.

5.4.8 *Uropolystomoides* Tinsley and Tinsley, 2016

5.4.8.1 *Uropolystomoides australiensis* (Rohde and Pearson, 1980) (Fig. 5.55)

Collection Australian Museum, Sydney (AM); United States National Museum (USNM).

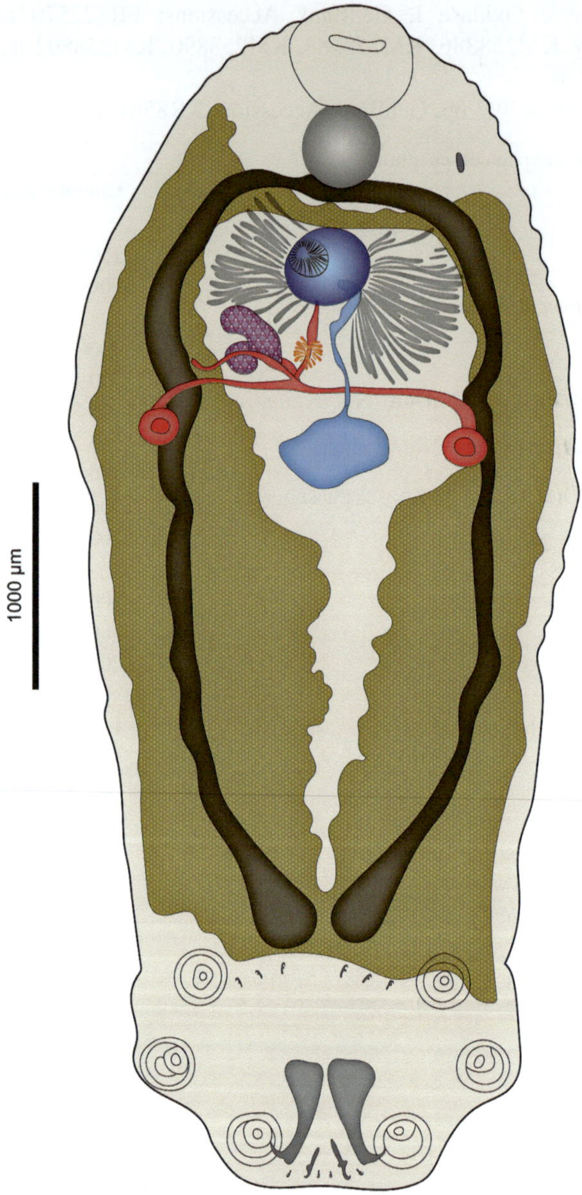

Fig. 5.55 *Uropolystomoides australiensis*. Redrawn from Rohde and Pearson (1980)

Holotype AU W 15050.

Paratypes AU W 15051; 14856; USNM 1370414.

Original description Rohde and Pearson (1980).

Other taxonomic contributions None.

Etymology Named after type locality country.

Type locality Mt. Crosby Weir and Brisbane River, Queensland, Australia.

Other localities Ugly Gully, Pullenvale, a branch of the Brisbane River (26.2.1968), Queensland, and male Fitzroy River, 149°55′E 23°11′S, 25 km east and 63 km north of Qualinga, Queensland, Australia.

Type host *Emydura macquarii*, Murray River Turtle. In the original description of the polystome, the host was known as *Emydura krefftii*.

Additional hosts *Myuchelys latisternum* (Gray, 1867).

Host geographical distribution *Emydura macquarii* occurs in SE Australia (from SC Queensland southward to N Victoria and SE South Australia) (Uetz et al. 2022).

Host conservation status The Murray River Turtle has legal protection and it also constitutes the most abundant terrapin species on Fraser Island. It is not listed in the IUCN Red List.

Site on host Bladder and cloacal bursa.

DNA sequence
- 28S rRNA gene, 945 bp. GenBank Accession: Z83012
- Cytochrome c Oxidase I, 401 bp. GenBank Accession: Z83013

Morphology and morphometrics (μm):	
Source	Rohde and Pearson (1980)
Sample size	4
Body length	4680–9290 (6193)
Greatest width	1370–3030 (2190)
Width at vagina	
Haptor length	1110–1900 (1353)
Haptor width	1260–2970 (1920)
False oral sucker length/width	510–880 × 600–1300
Pharynx length	290–590 (410)
Pharynx width	360–450 (398)
Testis length	300–560 (420)
Testis width	430–830 (653)
Ovary length	
Ovary width	
Egg length	ca. 298
Egg width	ca. 195

(continued)

Morphology and morphometrics (μm):	
Source	Rohde and Pearson (1980)
Egg incubation time	
Number of intra-uterine eggs	1
Genital bulb length/width	ca. 360–980 × 360–850 (ca. 598 × 558)
Number of genital spines	74–95 (78)
Genital spine length	78–ca. 106 (ca. 93)
Haptoral sucker diameter	310–620 (380)
Hamulus handle length X	ca. 490–770 (593); 500–800 (605)
Hamulus guard length Y	220–270 (227); 160–210 (185)
Hamulus hook length	
Marginal hooklet length	
Haptoral L/Body L	0.22

Remarks *Uropolystomoides australiensis* was originally described as *Polystomoides australiensis* Rohde and Pearson, 1980.

5.4.8.2 *Uropolystomoides bourgati* (Combes and Kulo, 1978) (Fig. 5.56)

Collection Royal Museum of Central Africa, Tervuren, Belgium (MRAC).

Holotype MRAC 36.521.

Paratypes MRAC 36.522.

Original description Combes and Kulo (1978).

Other taxonomic contributions None.

Etymology Named after Dr. Robert Bourgat, Perpignan, France.

Type locality Vicinity of Lomé (Togo), Africa.

Other localities Senegal.

Type host *Pelusios castaneus* (Schweigger, 1812), West African Mud Turtle. In the original description of the polystome, the host was known as *Pelusios castaneus derbianus*.

Additional hosts *Pelusios adansonii* (Schweiger, 1812).

Host geographical distribution *Pelusios castaneus* occurs in Guinea (Conakry), Senegal, Nigeria, W Democratic Republic of the Congo; Burkina Faso, Mali, Cameroon, Gabon, São Tomé, Central African Republic, Gambia, Benin and Angola (Uetz et al. 2022).

Host conservation status The West African Mud Turtle is heavily harvested for food, suffers from the drying of its habitats and is also offered as stuffed specimens to tourists. It is not listed in the IUCN Red List.

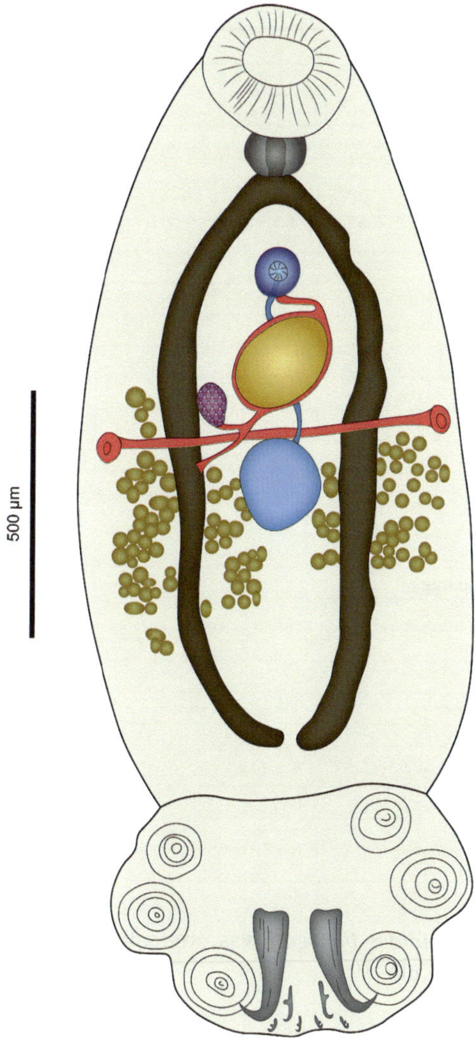

Fig. 5.56 *Uropolystomoides bourgati*. Redrawn from Combes and Kulo (1978)

Site on host Urinary bladder.

DNA sequence

- 18S rRNA gene, 560 bp. GenBank Accession: AJ297781
- 28S rRNA gene, 1168 bp. GenBank Accession: AF382068
- Cytochrome c Oxidase I, 396 bp. GenBank Accession: FR822602

Morphology and morphometrics (μm):	
Source	Combes and Kulo (1978)
Sample size	6
Body length	2340–2900 (2700)
Greatest width	800–1060 (900)
Width at vagina	
Haptor length	540–680 (620)
Haptor width	720–920 (840)
False oral sucker width	340–390 (370)
Pharynx length	137–190 (168) 140–190 (170)
Pharynx width	131–177 (154) 130–180 (150)
Testis length	154–290 (216) 150–290 (220)
Testis width	148–263 (210) 140–260 (210)
Ovary length	74–131 (104) 70–130 (100)
Ovary width	50–86 (69) 50–90 (70)
Egg length	240–300 (270) 240–300 (270)
Egg width	183–212 (196) 180–210 (200)
Egg incubation time	
Number of intra-uterine eggs	1
Genital bulb width	103–120 (110) 100–120 (110)
Number of genital spines	
Genital spine length	
Haptoral sucker diameter	154–171 (163) 150–170 (160)
Hamulus handle length X	257–314 (279) 260–310 (280)
Hamulus guard length Y	80–90 (85)
Hamulus hook length	
Marginal hooklet length	
Haptoral L/Body L	0.23

Remarks *Uropolystomoides bourgati* was originally described as *Polystomoides bourgati* Combes and Kulo, 1978.

5.4.8.3 *Uropolystomoides chabaudi* (Euzet and Combes, 1965) (Fig. 5.57)

Collection Unknown.

Holotype Unknown.

Paratypes Unknown.

Original description Euzet and Combes (1965).

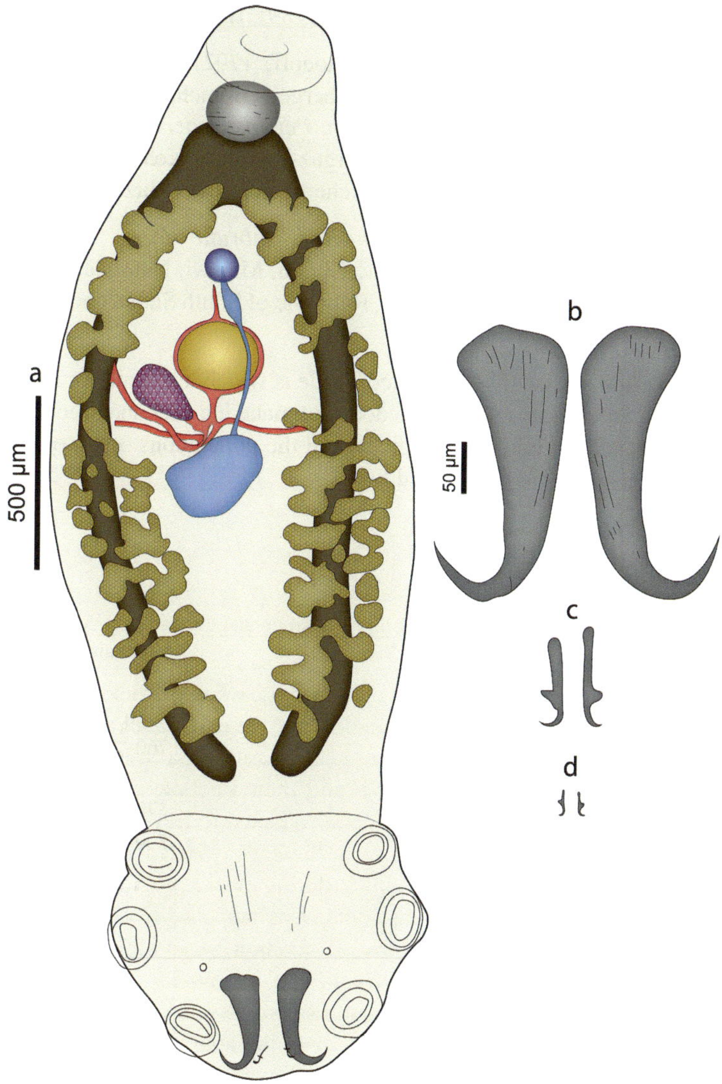

Fig. 5.57 (**a**) *Uropolystomoides chabaudi*, (**b**) hamuli, (**c**) hamulus pair two, (**d**) marginal hooklets. Redrawn from Euzet and Combes (1965)

Other taxonomic contributions None.

Etymology Named after Prof Alain Chabaud, National Museum of Natural History, Paris, France.

Type locality Betioky, Madagascar.

Other localities Uganda.

Type host *Pelomedusa subrufa* (Bonnaterre, 1789), Helmeted Turtle.

Additional hosts *Pelomedusa galeata* (Schoepff, 1792); *Pelomedusa neumanni* Petzold, Vargaz-Ramirez, Kehlmaier, Vamberger, Branch, Du Preez, Hofmeyr, Meyer, Schleicher, Siroky and Fritz, 2014; *Pelomedusa olivacea* (Schweigger, 1812); *Pelomedusa variabilis* Petzold, Vargaz-Ramirez, Kehlmaier, Vamberger, Branch, Du Preez, Hofmeyr, Meyer, Schleicher, Siroky and Fritz, 2014.

Host geographical distribution *Pelomedusa subrufa* occurs in S Angola, Botswana, SE Democratic Republic of the Congo, Malawi, Namibia, South Africa and Tanzania. Introduced to Madagascar, Republic of South Sudan and Sudan (Uetz et al. 2022).

Host conservation status The Helmeted Turtle is widely captured for food as well as sold to motorists, but its modest size does not make it a cherished food item, thus collection does not lead to drastic declines, and the populations seem to be in good shape. It is not listed in the IUCN Red List.

Site on host Urinary bladder.

DNA sequence None.

Morphology and morphometrics (μm):	
Source	Euzet and Combes (1965)
Sample size	9
Body length	2280–3290 (2910)
Greatest width	760–1070 (850)
Width at vagina	
Haptor length	570–720 (620)
Haptor width	800–1070 (880)
False oral sucker width	
Pharynx length	
Pharynx width	
Testis length	
Testis width	
Ovary length	
Ovary width	
Egg length	260–280 (270)
Egg width	170–200 (180)
Egg incubation time	
Number of intra-uterine eggs	1
Genital bulb width	105–162 (120)
Number of genital spines	31–36
Genital spine length	22–31 (27)
Haptoral sucker diameter	108–183 (150)
Hamulus handle length X	251–314 (268)
Hamulus guard length Y	80–103 (91)

(continued)

Morphology and morphometrics (μm):	
Source	Euzet and Combes (1965)
Hamulus hook length	
Marginal hooklet length	22–25 (24)
Haptoral L/Body L	0.21

Remarks *Uropolystomoides chabaudi* was originally described as *Polystomoides chabaudi* Euzet and Combes, 1965. *Uropolystomoides chabaudi* was also reported from *Pelomedusa subrufa* in Uganda (Tinsley 1973). Because African helmeted terrapins were further splitted into several species (see Petzold et al. 2014), we should now consider that *U. chabaudi* also infects several host species, as *P. neumanni* in Uganda (see Tinsley 1973), *P. galeata* in South Africa (Du Preez and Verneau, unpublished results), *P. olivacea* and *P. variabilis* in Nigeria (Aisien and Verneau, unpublished results).

5.4.8.4 *Uropolystomoides chauhani* (Pandey and Agarwal, 1978) (Fig. 5.58)

Collection Unknown.

Holotype Unknown.

Paratypes Unknown.

Original description Pandey and Agarwal (1978).

Other taxonomic contributions None.

Etymology Named after BS Cauhan, India.

Type locality Ghaghra river, district Faizabad, India.

Other localities None.

Type host *Hardella thurjii* (Gray, 1831), Crowned River Turtle.

Additional hosts None.

Host geographical distribution *Hardella thurjii* occurs in N India, Nepal, Bangladesh (watersheds of the Ganges, Brahmaputra and Indus rivers), and Pakistan (Uetz et al. 2022).

Host conservation status The Crowned River Turtle is found in markets in eastern India and in Bangladesh, having a long history of human consumption, and is sometimes collected for export to China. Thus populations are becoming increasingly decimated. It is listed as Endangered by The IUCN Red List of Threatened Species 2021: e.T9696A3152073. https://doi.org/10.2305/IUCN.UK.2021-1.RLTS.T9696A3152073.en. Accessed on 21 January 2022.

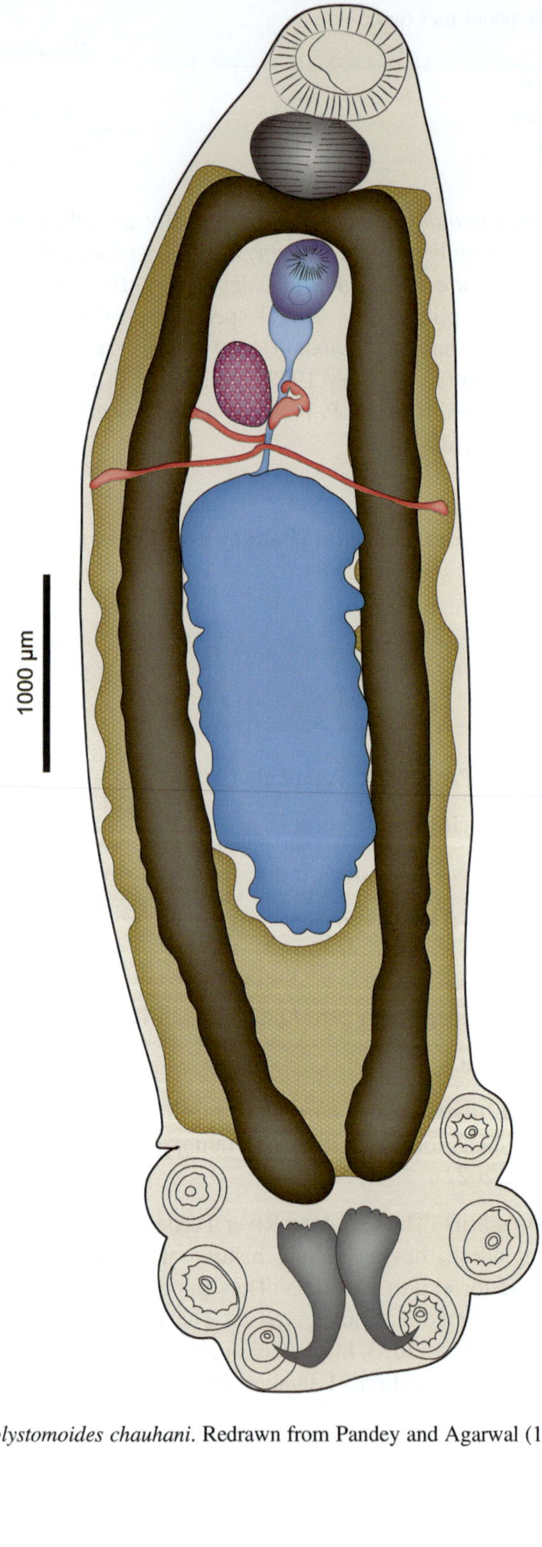

Fig. 5.58 *Uropolystomoides chauhani*. Redrawn from Pandey and Agarwal (1978)

Site on host Urinary bladder.

DNA sequence None.

Morphology and morphometrics (μm):	
Source	Pandey and Agarwal (1978)
Sample size	2
Body length	3640–3840
Greatest width	1120–1450
Width at vagina	
Haptor length	
Haptor width	
False oral sucker width	
Pharynx length	210–240
Pharynx width	280–330
Testis length	920–960
Testis width	480–500
Ovary length	80–96
Ovary width	
Egg length	
Egg width	
Egg incubation time	
Number of intra-uterine eggs	0
Genital bulb width	
Number of genital spines	40
Genital spine length	48
Haptoral sucker diameter	170–240
Hamulus handle length X	330–400 (base: 96–190)
Hamulus guard length Y	140–160
Hamulus hook length	
Marginal hooklet length	
Haptoral L/Body L	

Remarks *Uropolystomoides chauhani* was originally described as *Polystomoides chauhani* Pandey and Agarwal, 1978.

5.4.8.5 *Uropolystomoides kachugae* (Stewart, 1914) (Fig. 5.59)

Collection Unknown.

Holotype Unknown.

Paratypes Unknown.

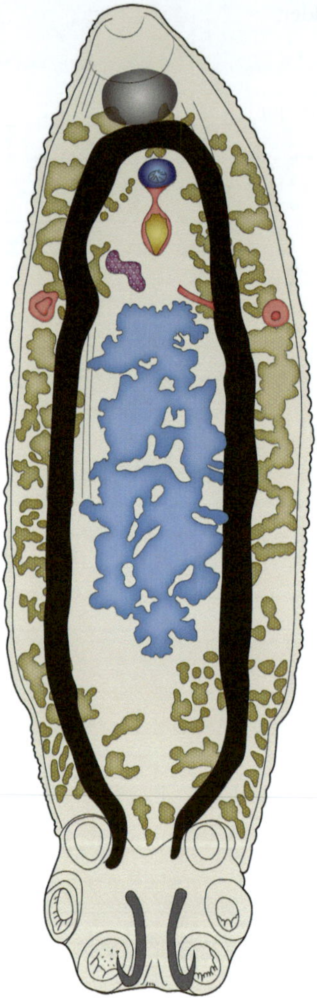

Fig. 5.59 *Uropolystomoides kachugae*. Redrawn from Stewart (1914)

Original description Stewart (1914).

Other taxonomic contributions Ozaki (1935).

Etymology Named after the host.

Type locality Lucknow, India.

Other localities None.

Type host *Batagur kachuga* (Gray, 1831), Red-Crowned Roofed Turtle. In the original description of the polystome, the host was known as *Kachuga lineata*.

Additional hosts None.

Host geographical distribution *Batagur kachuga* occurs in C Nepal, NE India and Bangladesh (Uetz et al. 2022).

Host conservation status The Red-Crowned Roofed Turtle is heavily fished for its meat. Reintroduction programmes are necessary to restock some of the former colonies, because degradation of habitats, barrages, sand mining and water pollution. It is listed as Critically Endangered by The IUCN Red List of Threatened Species 2019: e.T10949A152043133. https://doi.org/10.2305/IUCN.UK.2019-1.RLTS.T10949A152043133.en. Accessed on 21 January 2022.

Site on host Urinary bladder.

DNA sequence None.

Morphology and morphometrics (μm):	
Source	Stewart (1914)
Sample size	2
Body length	6500
Greatest width	2000
Width at vagina	
Haptor length	
Haptor width	1330
False oral sucker width	
Pharynx length	
Pharynx width	
Testis length	2200
Testis width	1000
Ovary length	
Ovary width	
Egg length	
Egg width	
Egg incubation time	
Number of intra-uterine eggs	1
Genital bulb width	
Number of genital spines	40
Genital spine length	
Haptoral sucker diameter	400
Hamulus handle length X	900
Hamulus guard length Y	166
Hamulus hook length	
Marginal hooklet length	
Haptoral L/Body L	Stewart (1914)

Remarks *Uropolystomoides kachugae* was originally described as *Polystomum kachugae* Stewart, 1914.

5.4.8.6 *Uropolystomoides ludhianae* (Gupta and Randev, 1974) (Fig. 5.60)

Collection Unknown.

Holotype Unknown.

Paratypes Unknown.

Original description Gupta and Randev (1974).

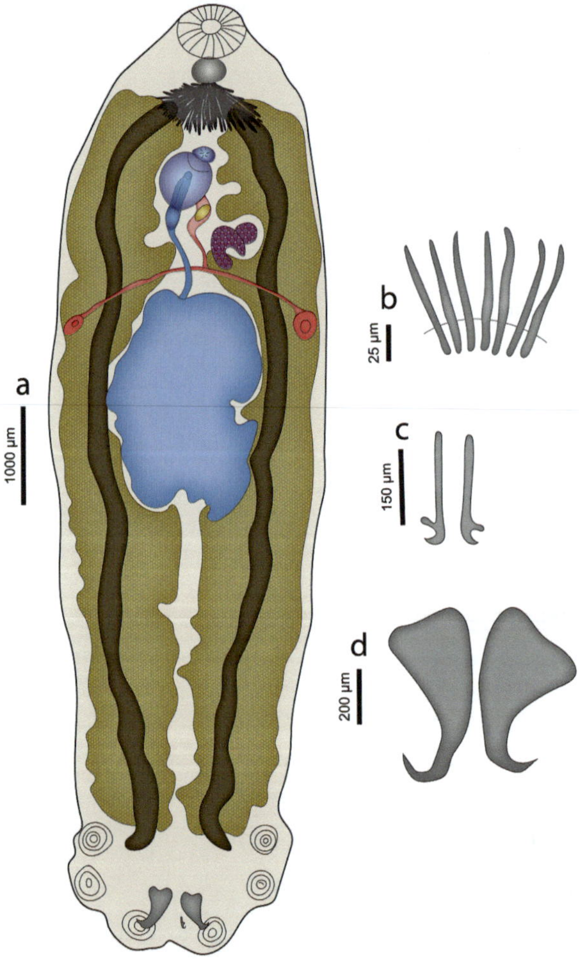

Fig. 5.60 (**a**) *Uropolystomoides ludhianae*, (**b**) genital spines, (**c**) hamulus pair two, (**d**) hamuli. Redrawn from Gupta and Randev (1974)

Other taxonomic contributions None.

Etymology Named after the type locality.

Type locality Ludhiana (Punjab), India.

Other localities None.

Type host *Pangshura tecta* (Gray, 1831), Indian Roofed Turtle and *Pangshura smithii* (Gray, 1863), Brown Roofed Turtle. In the original description of the polystome, the hosts were known as *Kachuga tectum* Boulenger, 1889 and *Kachuga smithi* Baig et al., 2008, respectively.

Additional hosts None.

Host geographical distribution *Pangshura tecta* and *Pangshura smithii* occur in Pakistan, India (Ganges, Brahmaputra, Indus drainages, Assam, Madhya Pradesh, Gujarat, Jammu and Kashmir), Bangladesh and Nepal (Uetz et al. 2022).

Host conservation status The Indian Roofed Turtle is listed in Appendix 1 of CITES, but populations are reasonably abundant. It is collected for consumption and sales, and the collection of eggs, habitat destruction, dams and barrages, and the destruction of nesting sites have made the species highly vulnerable; It is listed as Vulnerable by The IUCN Red List of Threatened Species 2021: e.T46370A3005714. https://doi.org/10.2305/IUCN.UK.2021-1.RLTS.T46370A3005714.en. Accessed on 21 January 2022. The Brown Roofed Turtle is listed as Near Threatened by The IUCN Red List of Threatened Species 2021: e.T39554A2929235. https://doi.org/10.2305/IUCN.UK.2021-1.RLTS.T39554A2929235.en. Accessed on 21 January 2022.

Site on host Urinary bladder.

DNA sequence None.

Morphology and morphometrics (μm):	
Source	Gupta and Randev (1974)
Sample size	
Body length	6640–10,060
Greatest width	2430–3050
Width at vagina	
Haptor length	
Haptor width	1742–2396
False oral sucker length/width	399–726 × 577–799
Pharynx length	218–290
Pharynx width	254–363
Testis length	1488–2214
Testis width	1416–2033
Ovary length	
Ovary width	109–182

(continued)

Morphology and morphometrics (μm):	
Source	Gupta and Randev (1974)
Egg length	182–290
Egg width	145–218
Egg incubation time	
Number of intra-uterine eggs	1
Genital bulb length/width	49–53 × 504–748
Number of genital spines	54–64
Genital spine length	
Haptoral sucker diameter	254–363
Hamulus handle length X	545–617
Hamulus guard length Y	136–154
Hamulus hook length	
Marginal hooklet length	
Haptoral L/Body L	Gupta and Randev (1974)

Remarks *Uropolystomoides ludhianae* was originally described as *Polystomoides ludhianae* Gupta and Randev, 1974.

5.4.8.7 *Uropolystomoides malayi* (Rohde, 1963) (Fig. 5.61)

Collection Helminthological Collection, Zoology department, University of Malaysia, Kuala Lumpur (UM); Helminthological collection of the Humboldt University of Berlin (Berlin); Natural History Museum of London (NHML); United States National Museum (USNM).

Syntype no (UM) R. 129-135.

Topotype nos Berlin 6533; NHML 1963.10.23.3; USNM 60022.

Original description Rohde (1963).

Other taxonomic contributions None.

Etymology Named after the type locality country.

Type locality Different localities in Selangor, Malaysia.

Other localities None.

Type host *Cuora amboinensis*, Southeast Asian Box Turtle.

Additional hosts None.

Host geographical distribution *Cuora amboinensis* occurs in India, Bhutan, Bangladesh, Myanmar, Thailand, Cambodia, Vietnam, West Malaysia, Singapore and Philippines (Uetz et al. 2022).

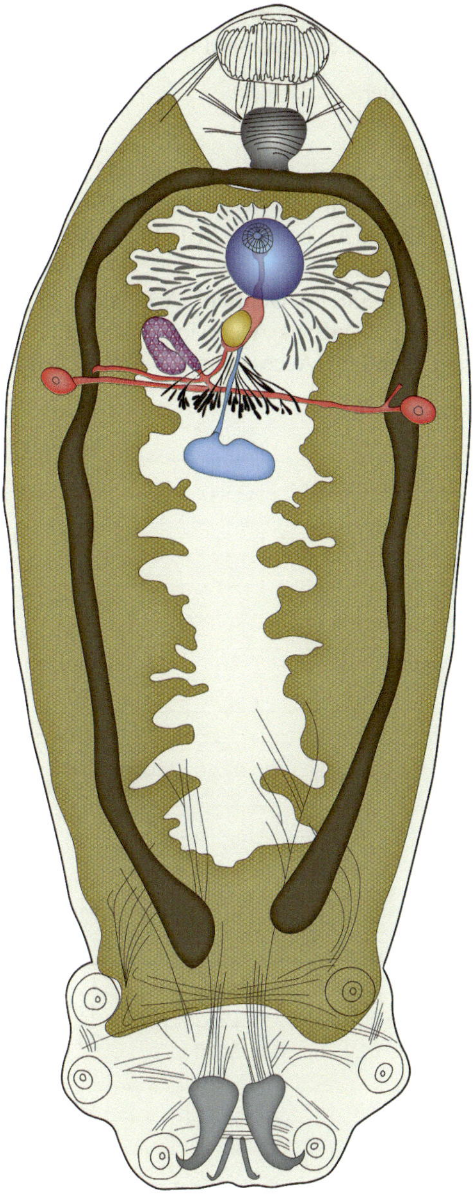

Fig. 5.61 *Uropolystomoides malayi*. Redrawn from Rohde (1963)

Host conservation status The Southeast Asian Box Turtle is the most frequently consumed terrapin species in China restaurants, also captured and released in Buddhist temple ponds, and gathered for exports. These practices may lead to a population collapse in the future. It is listed as Endangered by The IUCN Red List of

Threatened Species 2020: e.T5958A3078812. https://doi.org/10.2305/IUCN.UK.2020-2.RLTS.T5958A3078812.en. Accessed on 21 January 2022.

Site on host Urinary bladder.

DNA sequence

- 18S rRNA gene, 2014 bp. GenBank Accession: AJ228792
- 18S rRNA gene, 560 bp. GenBank Accession: AJ287995
- 28S rRNA gene. GenBank Accessions: AF131718, AY157170, FM992704 and Z83010
- Cytochrome c Oxidase I, 338 bp. GenBank Accession: Z83011
- 12S rRNA gene, 506 bp. GenBank Accession: KR856112
- Hox2/3 gene, 75 bp. GenBank Accessions: FN298156 and FN298157

Morphology and morphometrics (μm):	
Source	Rohde (1963)
Sample size	5
Body length	5900–8200 (7200)
Greatest width	2300–3300 (2900)
Width at vagina	
Haptor length	1200–1700 (1500)
Haptor width	1500–2400 (2000)
False oral sucker length/width	460–500 (480) × 690–870 (790)
Pharynx length	370–420 (380)
Pharynx width	460–520 (480)
Testis length	320–430 (370)
Testis width	530–960 (690)
Ovary length	
Ovary width	
Egg length	270–300 (280)
Egg width	220–230 (230)
Egg incubation time	
Number of intra-uterine eggs	1
Genital bulb length/width	490–630 (570) × 440–540 (520)
Number of genital spines	70–80 (77)
Genital spine length	
Haptoral sucker length/width	270–390 (320) × 260–390 (320)
Hamulus handle length X	570–680 (630)
Hamulus guard length Y	120–320 (220)
Hamulus hook length	
Marginal hooklet length	21–31 (26)
Haptoral L/Body L	0.2083

Remarks *Uropolystomoides malayi* was originally described as *Polystomoides malayi* Rohde, 1963.

5.4.8.8 *Uropolystomoides megaovum* (Ozaki, 1936) (Fig. 5.62)

Collection Unknown.

Holotype Unknown.

Paratypes Unknown.

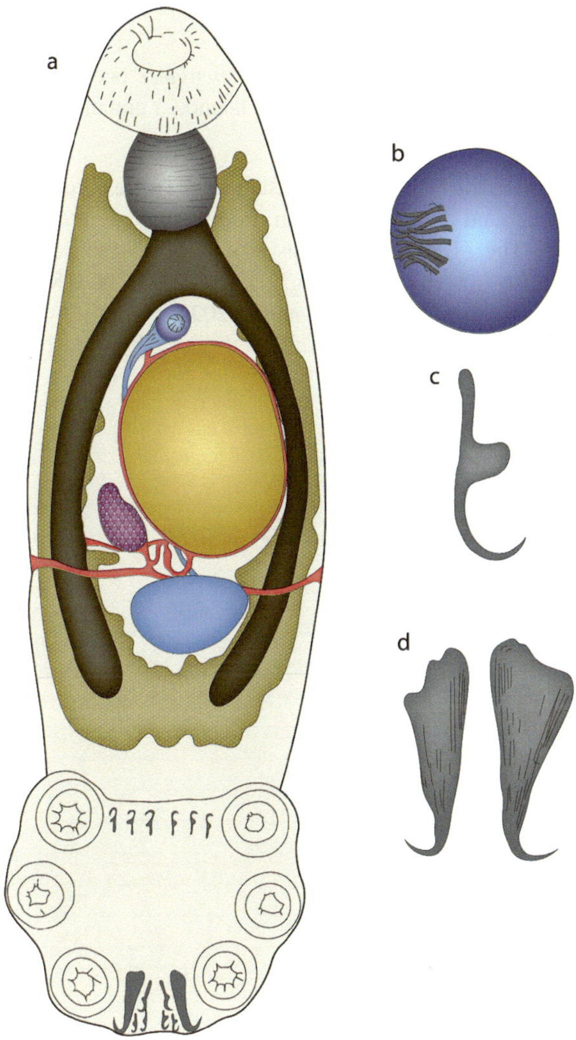

Fig. 5.62 (**a**) *Uropolystomoides megaovum*, (**b**) genital bulb, (**c**) marginal hooklets, (**d**) hamuli. Redrawn from Ozaki (1936)

Original description Ozaki (1936).

Other taxonomic contributions None.

Etymology Refers to big eggs.

Type locality Kunigami district of the Loochoo Island proper, China.

Other localities None.

Type host *Geoemyda spengleri* (Gmelin, 1789), Black-Breasted Leaf Turtle.

Additional hosts None.

Host geographical distribution *Geoemyda spengleri* occurs in S China (from Kwangsi, Kwangtung, Guangdong, Hainan Island, Yunnan), Vietnam and Laos (Uetz et al. 2022).

Host conservation status The Black-Breasted Leaf Turtle is collected by hobbyists because it is a small and attractive terrapin. Populations are reduced. It is listed as Endangered by The IUCN Red List of Threatened Species 2020: e. T39552A2929166. https://doi.org/10.2305/IUCN.UK.2020-2.RLTS.T39552A2929166.en. Accessed on 21 January 2022.

Site on host Urinary bladder.

DNA sequence None.

Morphology and morphometrics (μm):	
Source	Ozaki (1936)
Sample size	21
Body length	1350–2100
Greatest width	420–650
Width at vagina	
Haptor length	
Haptor width	480–720
False oral sucker length/width	140 × 300–370
Pharynx length	130–180
Pharynx width	130–160
Testis length	130–230
Testis width	80–170
Ovary length	100–190
Ovary width	40–100
Egg length	240–400
Egg width	160–320
Egg incubation time	
Number of intra-uterine eggs	1
Genital bulb width	80
Number of genital spines	12–14

(continued)

Morphology and morphometrics (μm):	
Source	Ozaki (1936)
Genital spine length	
Haptoral sucker diameter	150
Hamulus handle length X	
Hamulus guard length Y	
Hamulus hook length	
Marginal hooklet length	25
Haptoral L/Body L	Ozaki (1936)

Remarks *Uropolystomoides megaovum* was originally described as *Polystomoides megaovum* Ozaki, 1936.

5.4.8.9 *Uropolystomoides nabedei* (Kulo, 1980) (Fig. 5.63)

Collection National Museum of Natural History (MNHN), Paris, France.

Holotype MNHN TJ 63.

Paratypes None.

Original description Kulo (1980).

Other taxonomic contributions None.

Etymology Unknown.

Type locality Temporary pond in Siborototi (Dapaong, Togo), Africa.

Other localities None.

Type host *Pelomedusa subrufa*, Helmeted Turtle.

Additional hosts None.

Host geographical distribution *Pelomedusa subrufa* occurs in S Angola, Botswana, SE Democratic Republic of the Congo, Malawi, Namibia, South Africa and Tanzania. Introduced to Madagascar, Republic of South Sudan and Sudan (Uetz et al. 2022).

Host conservation status The Helmeted Turtle is widely captured for food as well as sold to motorists, but its modest size does not make it a cherished food item, thus collection does not lead to drastic declines, and the populations seem to be in good shape. It is not listed in the IUCN Red List.

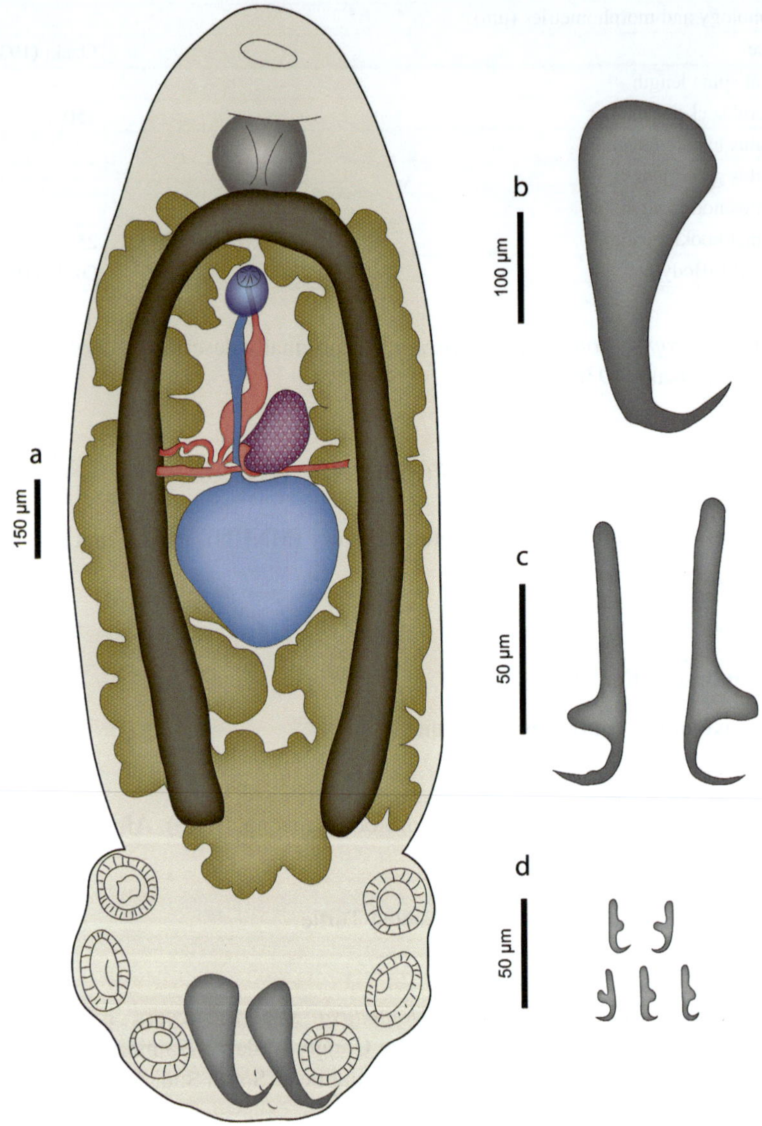

Fig. 5.63 (**a**) *Uropolystomoides nabedei*, (**b**) hamuli, (**c**) hamulus pair two, (**d**) marginal hooklets. Redrawn from Kulo (1980)

Site on host Urinary bladder.

DNA sequence None.

Morphology and morphometrics (µm):	
Source	Kulo (1980)
Sample size	10

(continued)

Morphology and morphometrics (µm):	
Source	Kulo (1980)
Body length	2280–3370 (2940)
Greatest width	860–1210 (1080)
Width at vagina	
Haptor length	460–760 (650)
Haptor width	760–1010 (900)
False oral sucker length/width	268–445 (356) × 210–294 (248)
Pharynx length	142–285 (223)
Pharynx width	134–226 (185)
Testis length	294–588 (413)
Testis width	142–420 (254)
Ovary length	126–252 (167)
Ovary width	58–109 (79)
Egg length	275–332
Egg width	
Egg incubation time	
Number of intra-uterine eggs	
Genital bulb width	126–168 (140)
Number of genital spines	36–39
Genital spine length	42–46
Haptoral sucker diameter	150–163 (156)
Hamulus handle length X	294–344 (324)
Hamulus guard length Y	92–151 (110)
Hamulus hook length	
Marginal hooklet length	23–24
Haptoral L/Body L	0.22

Remarks *Uropolystomoides nabedei* was originally described as *Polystomoides nabedei* Kulo, 1980. It could be a junior synonym of *U. chabaudi* as it was reported from the same host species. This must be further investigated.

5.4.8.10 *Uropolystomoides ocadiae* (Fukui and Ogata, 1936) (Fig. 5.64)

Collection Unknown.

Holotype Unknown.

Paratypes Unknown.

Original description Fukui and Ogata (1936).

Other taxonomic contributions None.

Etymology Named after the host.

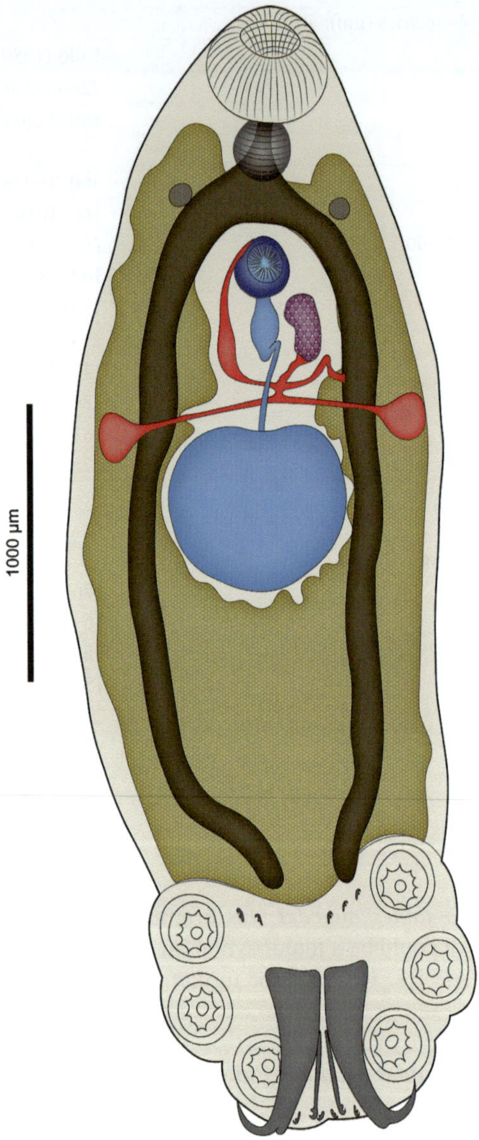

Fig. 5.64 *Uropolystomoides ocadiae*. Redrawn from Fukui and Ogata (1936)

Type locality Taipei, Formosa, China.

Other localities None.

Type host *Mauremys sinensis*, Chinese Stripe-Necked Turtle. In the original description of the polystome, the host was known as *Ocadia sinensis*.

Additional hosts None.

Host geographical distribution *Mauremys sinensis* occurs in Taiwan, S China (Guangdong, Guangxi, Fujian, Zhejiang, Jiangsu, Fukien, Hanchow, Soochow, Kwangtung, Shanghai, Hainan Island, Hong Kong), Laos and N Vietnam (Uetz et al. 2022).

Host conservation status The Chinese Stripe-Necked Turtle is listed as Endangered by The IUCN Red List of Threatened Species 2000: e.T15026A97372848. https://doi.org/10.2305/IUCN.UK.2000.RLTS.T15026A4488909.en. Accessed on 21 January 2022.

Site on host Urinary bladder.

DNA sequence None.

Morphology and morphometrics (μm):	
Source	Fukui and Ogata (1936)
Sample size	
Body length	7300–9000; 3200–5240; 3950
Greatest width	1800–2200; 1140–1850; 1560
Width at vagina	
Haptor length	860
Haptor width	1270
False oral sucker width	390 × 490
Pharynx length	
Pharynx width	250
Testis length	560
Testis width	830
Ovary length	270
Ovary width	130
Egg length	330
Egg width	220
Egg incubation time	
Number of intra-uterine eggs	
Genital bulb width	250
Number of genital spines	46–59
Genital spine length	69
Haptoral sucker diameter	270–320
Hamulus handle length X	640 × 190
Hamulus guard length Y	330
Hamulus hook length	
Marginal hooklet length	25
Haptoral L/Body L	0.22

Remarks *Uropolystomoides ocadiae* was originally described as *Polystomoides ocadiae* Fukui and Ogata, 1936.

5.4.8.11 *Uropolystomoides scottae* (Pichelin, 1995) (Fig. 5.65)

Collection Queensland Museum, Australia (QM).

Holotype QM G211517.

Paratype QM G211518.

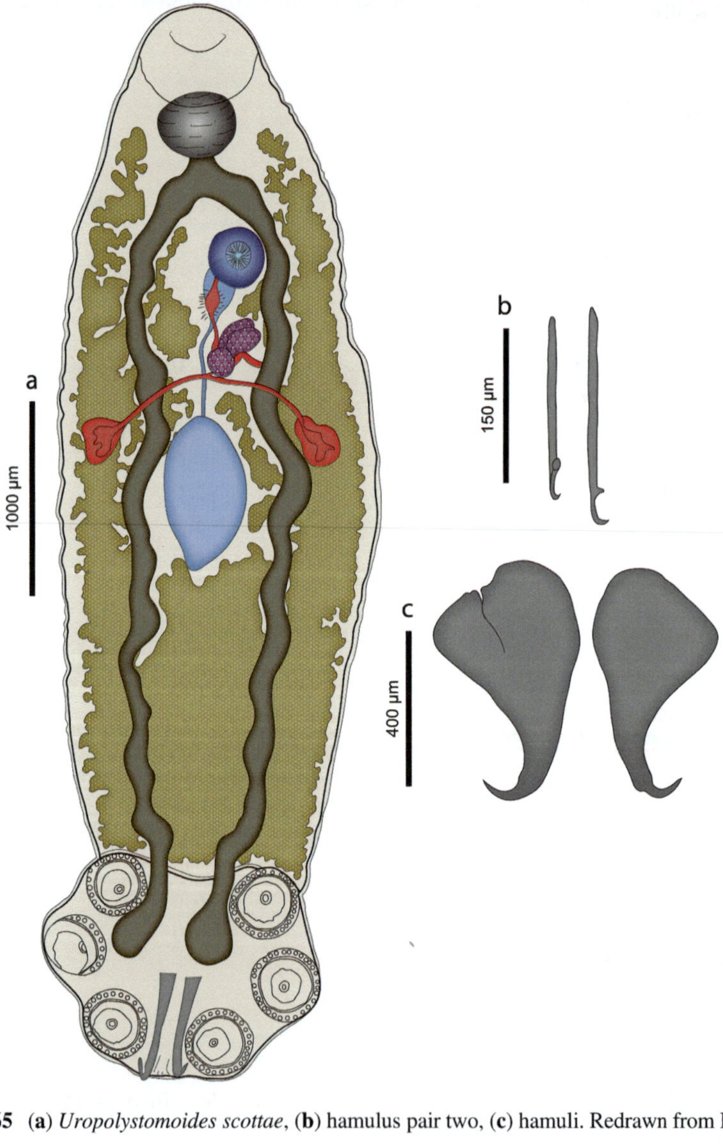

Fig. 5.65 (**a**) *Uropolystomoides scottae*, (**b**) hamulus pair two, (**c**) hamuli. Redrawn from Pichelin (1995)

Original description Pichelin (1995).

Other taxonomic contributions None.

Etymology Named after Mrs. D. Scott, Australia.

Type locality Downfall Cr. (27°40′S, 153°0.′E), Brisbane, Australia.

Other localities Queensland, Australia–Canal Cr. (28°01′S, 151°35′E), Darling Downs and The University of Queensland Veterinary Farm dam (27°32′S, 152° 54′E), Pinjarra Hills.

Type host *Chelodina expansa*, Giant Snake-Necked Turtle.

Additional hosts None.

Host geographical distribution *Chelodina expansa* occurs in Australia—Queensland: from Rockhampton southwest through the coastal and interior areas to W New South Wales and the Murray River drainage, and on Fraser Island off Queensland; South Australia, Victoria (Uetz et al. 2022).

Host conservation status The Giant Snake-Necked Turtle makes popular pets. Precise counts of populations in the wild are extremely difficult because it hides very effectively and it is almost impossible to find. It is not listed in The IUCN Red List.

Site on host Urinary and accessory bladders, cloaca.

DNA sequence None.

Morphology and morphometrics (μm):	
Source	Pichelin (1995)
Sample size	7
Body length	3948–5588 (4978)
Greatest width	1576–2054 (1760)
Width at vagina	
Haptor length	1114–1274 (1201)
Haptor width	1465–1735 (1597)
False oral sucker length/width	469–578 (523) × 591–687 (648)
Pharynx length	347–424 (383)
Pharynx width	379–475 (416)
Testis length	475–854 (652)
Testis width	449–559 (506)
Ovary length	
Ovary width	
Egg length	
Egg width	
Egg incubation time	
Number of intra-uterine eggs	1

(continued)

Morphology and morphometrics (μm):	
Source	Pichelin (1995)
Genital bulb length/width	276–308 (289) × 257–308 (253)
Number of genital spines	68–79 (73)
Genital spine length	54.4–65.6 (60.4)
Haptoral sucker diameter	313–417 (347)
Hamulus handle length X	565–725 (632)
Hamulus guard length Y	148–225 (186)
Hamulus hook length	
Marginal hooklet length	20.8–23.2 (22.1)
Haptoral L/Body L	0.24

Remarks *Uropolystomoides scottae* was originally described as *Polystomoides scottae* Pichelin, 1995.

5.4.8.12 *Uropolystomoides siebenrockiellae* (Rohde, 1965) (Fig. 5.66)

Collection Helminthological Collection, Zoology Department, University of Malaya. Voucher specimens in Helminthologische Sammlung-Humboldt University, Natural History Museum of London and the United States National Museum.

Holotype None.

Paratypes R 228–230.

Original description Rohde (1965)

Other taxonomic contributions None.

Etymology Named after the host.

Type locality Malaysia (Chinese shop in Kuala Lumpur).

Other localities None.

Type host *Siebenrockiella crassicollis*, Black Marsh Turtle.

Additional hosts None.

Host geographical distribution *Siebenrockiella crassicollis* has a fragmented range in south-east Asia including S Vietnam, Cambodia, Thailand, Myanmar, Laos, Malaysia, Indonesia and Singapore (Uetz et al. 2022).

Host conservation status The Black Marsh Turtle is sometimes consumed or sold in markets, but the flesh is little sought after. Despite this, populations may be dropping because of progressive fragmentation of their habitat. It is listed as

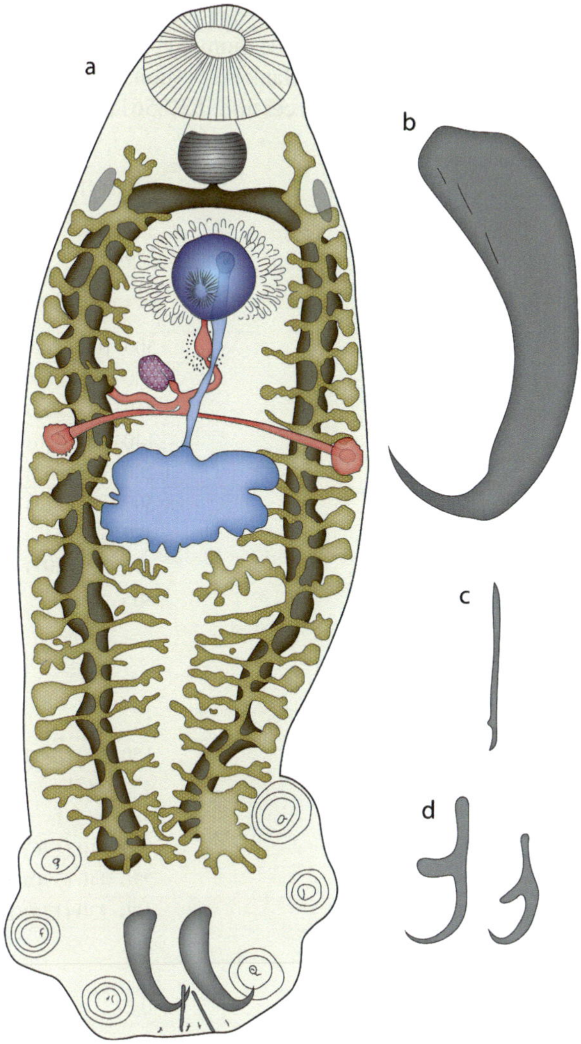

Fig. 5.66 (**a**) *Uropolystomoides siebenrockiellae*, (**b**) hamuli, (**c**) hamulus pair two, (**d**) marginal hooklets. Redrawn from Rohde (1965)

Vulnerable by The IUCN Red List of Threatened Species 2000: e. T39616A97377799. https://doi.org/10.2305/IUCN.UK.2000.RLTS. T39616A10251374.en. Accessed on 21 January 2022.

Site on host Urinary bladder.

DNA sequence

- 18S rRNA gene, 1971 bp. GenBank Accession: FM992699
- 18S rRNA gene, 573 bp. GenBank Accession: AJ297784

- 28S rRNA gene, 1368 bp. GenBank Accession: FM992706
- 28S rRNA gene, 1166 bp. GenBank Accession: AF382067
- Cytochrome c Oxidase I, 391 bp. GenBank Accession: FR822604
- 12S rRNA gene, 498 bp. GenBank Accession: KR856114

Morphology and morphometrics (μm):	
Source	Rohde (1965)
Sample size	5
Body length	1900–4400 (3580)
Greatest width	710–1200 (1082)
Width at vagina	
Haptor length	500–1000 (780)
Haptor width	600–1300 (1060)
False oral sucker length/width	210–400 (354) × 350–590 (510)
Pharynx length	140–280 (216)
Pharynx width	200–310 (262)
Testis length	300–360 (323)
Testis width	480–610 (543)
Ovary length	190–220 (207)
Ovary width	80
Egg length	
Egg width	
Egg incubation time	
Number of intra-uterine eggs	1
Genital bulb length/width	150–360 (272) × 150–300 (252)
Number of genital spines	45–56 (51)
Genital spine length	54–60 (58)
Haptoral sucker diameter	130–210 (176)
Hamulus handle length X	270–440 (390)/250–420 (366)
Hamulus guard length Y	130–230 (174)/150–230 (187)
Hamulus hook length	
Marginal hooklet length	
Haptoral L/Body L	0.22

Remarks *Uropolystomoides siebenrockiellae* was originally described as *Polystomoides siebenrockiellae* Rohde, 1965.

5.4.8.13 *Uropolystomoides stewarti* (Pandey, 1973) (Fig. 5.67)

Collection Unknown.

Holotype Unknown.

Paratypes Unknown.

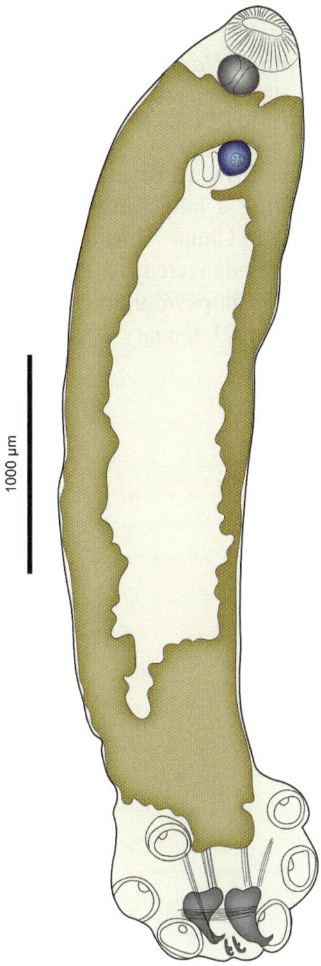

Fig. 5.67 *Uropolystomoides stewarti*. Redrawn from Pandey (1973)

Original description Pandey (1973).

Other taxonomic contributions None.

Etymology Named after Dr. FH Stewart.

Type locality Lucknow, India.

Other localities None.

Type host *Hardella thurjii*, Crowned River Turtle. In the original description of the polystome, the host was known as *Hardella thurjii* Gray, 1870.

Additional hosts None.

Host geographical distribution *Hardella thurjii* occurs in N India, Nepal, Bangladesh (watersheds of the Ganges, Brahmaputra and Indus rivers) and Pakistan (Uetz et al. 2022).

Host conservation status The Crowned River Turtle is found in markets in eastern India and in Bangladesh, having a long history of human consumption, and is sometimes collected for export to China. Thus, populations are becoming increasingly decimated. It is listed as Endangered by The IUCN Red List of Threatened Species 2021: e.T9696A3152073. https://doi.org/10.2305/IUCN.UK.2021-1.RLTS.T9696A3152073.en. Accessed on 21 January 2022.

Site on host Urinary bladder.

DNA sequence None.

Morphology and morphometrics (μm):	
Source	Pandey (1973)
Sample size	1
Body length	4240
Greatest width	1320
Width at vagina	
Haptor length	810
Haptor width	1250
False oral sucker length/width	210 × 360
Pharynx length	120–150
Pharynx width	
Testis length	320
Testis width	210
Ovary length	90–120
Ovary width	
Egg length	
Egg width	
Egg incubation time	
Number of intra-uterine eggs	
Genital bulb width	
Number of genital spines	36
Genital spine length	
Haptoral sucker length/width	220 × 270
Hamulus handle length/width X	120–150 × 180–210
Hamulus guard length Y	80–110
Hamulus hook length	
Marginal hooklet length	
Haptoral L/Body L	0.19

Remarks *Uropolystomoides stewarti* was originally described as *Polystomoides stewarti* Pandey, 1973.

5.4.9 *Uteropolystomoides* Tinsley, 2017

5.4.9.1 *Uteropolystomoides multifalx* (Stunkard, 1924) (Fig. 5.68)

Collection American Museum of Natural History (AMNH); United States National Museum (USNM). Vouchers in Parasite Worm Collection, National Museum, Bloemfontein, South Africa (NMBP).

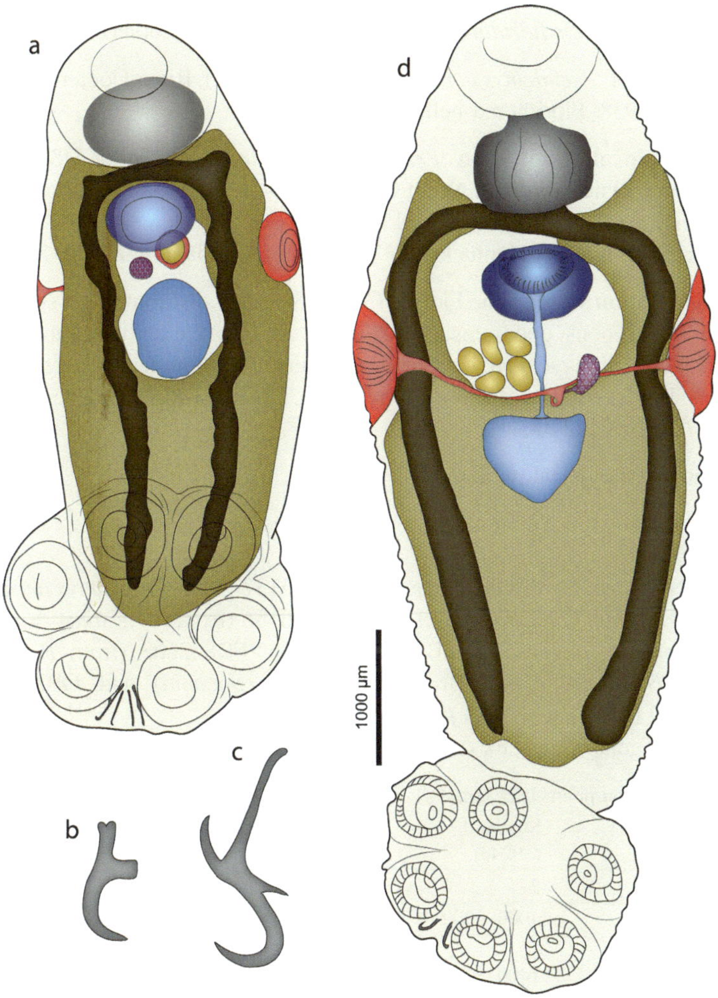

Fig. 5.68 (**a**, **d**) *Uteropolystomoides multifalx*, (**b**) marginal hooklets, (**c**) genital spines. (**a**–**c**): Redrawn from Stunkard (1924a); (**d**): Redrawn from Du Preez and Van Rooyen (2015)

Syntypes AMNH 882; 883.1-883.5; 1466.1-10.

Paratypes USNM 1332519.

Original description Stunkard (1924a).

Other taxonomic contributions Du Preez and Van Rooyen (2015); Chaabane et al. (2022).

Etymology Refers to the large number of sickle shaped genital spines.

Type locality Central Florida, U.S.A.

Other localities Gainesville, Florida, USA.

Type host *Pseudemys floridana* (LeConte, 1830), Florida Cooter.

Additional hosts *Pseudemys concinna* (LeConte, 1830), River Cooter; *Pseudemys nelsoni* Carr, 1938, Florida red-bellied cooter.

Host geographical distribution *Pseudemys floridana* occurs in the USA (Indiana, West Virginia, Ohio, Virginia, North Carolina, South Carolina, Georgia, Alabama, Mississippi, Louisiana, Arkansas, E Oklahoma, Texas, SE New Mexico, SE Kansas, S Missouri, S Illinois, W Kentucky, W Tennessee, Florida) (Uetz et al. 2022).

Host conservation status The Florida Cooter is widely distributed but suffers from predation by raccoons, opossums, certain carnivorous fish, other terrapins, alligators and birds. Populations seem to be dropping because of pollution due to fertilizers and chemicals. It is listed as Least Concern by The IUCN Red List of Threatened Species 2011. T170496A97427004.

Site on host Pharyngeal region.

DNA sequence

- 18S rRNA gene. GenBank Accession: KR856137 (from *Pseudemys nelsoni*)
- 28S rRNA gene. GenBank Accession: KR856156 (from *P. nelsoni*)
- Cytochrome c Oxydase. GenBank Accession: FR822603 (from *P. nelsoni*), OP793446 to OP793456 (from *Pseudemys concinna*) and OP793457 to OP793459 (from *Pseudemys floridana*)
- 12S rRNA gene. GenBank Accession: KR856117 (from *P. nelsoni*)

Morphology and morphometrics (μm):

Source	Stunkard (1924a)	Du Preez and Van Rooyen (2015)	Chaabane et al. (2022)
Sample size	2	10	20
Body length	4000–5000	3052–7378 (5707)	4730–10,691 (6743)
Greatest width	1400–2100	1276–2751 (2278)	1761–3058 (2449)
Haptor length		912–1616 (1310)	1130–2043 (1459)
Haptor width		1232–2182 (1931)	1409–2657 (2026)

(continued)

Morphology and morphometrics (μm):			
Source	Stunkard (1924a)	Du Preez and Van Rooyen (2015)	Chaabane et al. (2022)
False oral sucker width	600	398–1036 (788)	684–1281 (962)
Pharynx length	460	345–917 (539)	417–676 (554)
Pharynx width	620	391–881 (658)	619–959 (783)
Testis length	430–540	108–687 (401)	342–892 (545)
Testis width	400–520	148–781 (564)	425–778 (632)
Ovary length	90–100	102–330 (251)	131–350 (233)
Ovary width		27–124 (86)	70–192 (124)
Egg length	210	182–274 (227)	137–269 (232)
Egg width	200–180	118–194 (144)	137–193 (169)
Egg incubation time			
Number of intra-uterine eggs	1	8	0–12 (3.6)
Genital bulb length/ width	370–460 × 460–530	302–816 (586)	438–847 (650)
Number of genital spines	120–124	108–132	118–136 (125)
Genital spine length		101 (93–106) (123)	83–98 (93)
Haptoral sucker diameter	400–460	564 (148–781)	343–477 (419)
Hamulus handle length X	200	104–173 (138)	105–175 (137)
Hamulus guard length Y	100	48–95 (69)	86–167 (121)
Hamulus hook length		19–26 (22)	59–86 (70)
Marginal hooklet C1 length	30	25–30 (28)	25–30 (28)
Marginal hooklet C2–8 length		25–29 (27)	24–29 (27)
Haptoral L/Body L		0.23	

Remarks *Uteropolystomoides multifalx* was originally described as *Polystoma multifalx* Stunkard, 1924.

5.5 Mammalian Polystomes

There is a single polystome species that parasitizes a mammal. The generic diagnosis for the genus *Oculotrema* Stunkard, 1924 is a reduction of the cirrus sac, the absence of the lateral vaginae, the absence of the coronet of the genital hooks, the absence of vitellaria in the posterior portion of the body as well as short, unequal digestive caeca (Stunkard 1924b).

5.5.1 *Oculotrema* Stunkard, 1924 (Fig. 5.69)

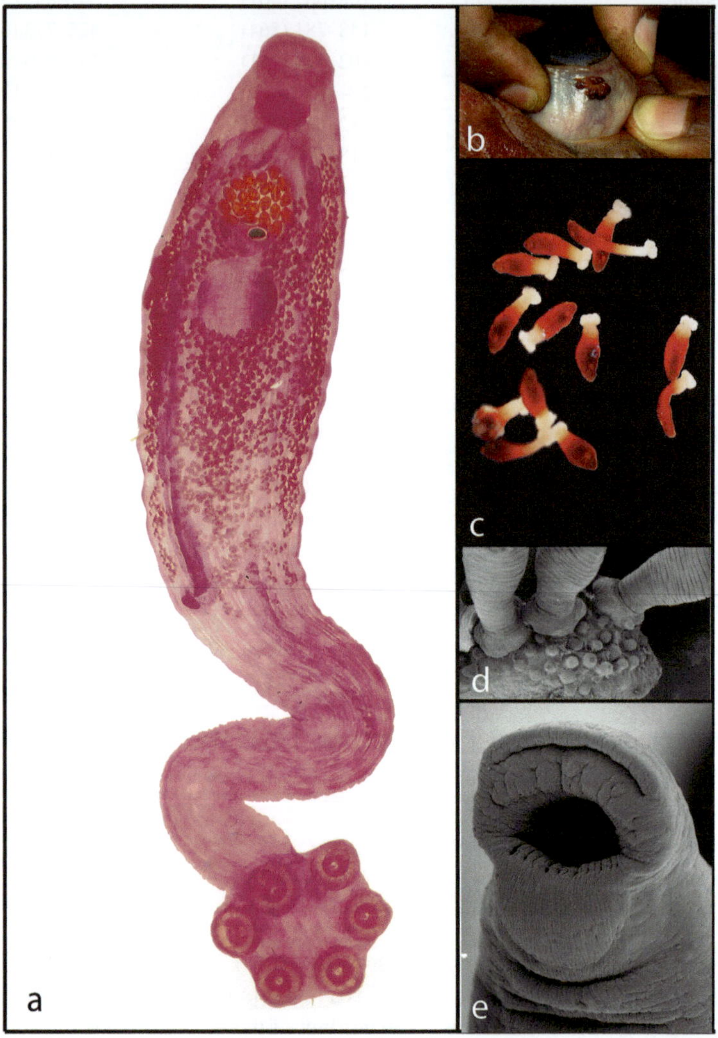

Fig. 5.69 Micrographs of (**a**) *Oculotrema hippopotami*; (**b**) Cluster of parasites on the eye of a hippopotamus; (**c**) Live specimens in a Petri dish; (**d**) SEM micrograph of parasites attached to the eye. Note the buts of tissue where parasites have been removed; (**e**) SEM micrograph of the oral sucker

5.5.1.1 *Oculotrema hippopotami* Stunkard, 1924 (Fig. 5.70)

Collection Specimens deposited in the Nuttall Collection of the Molteno Institute for Research in Parasitology, Cambridge University, Cambridge, England. All specimens has been transferred to the Parasitic Worms Collection in the Natural

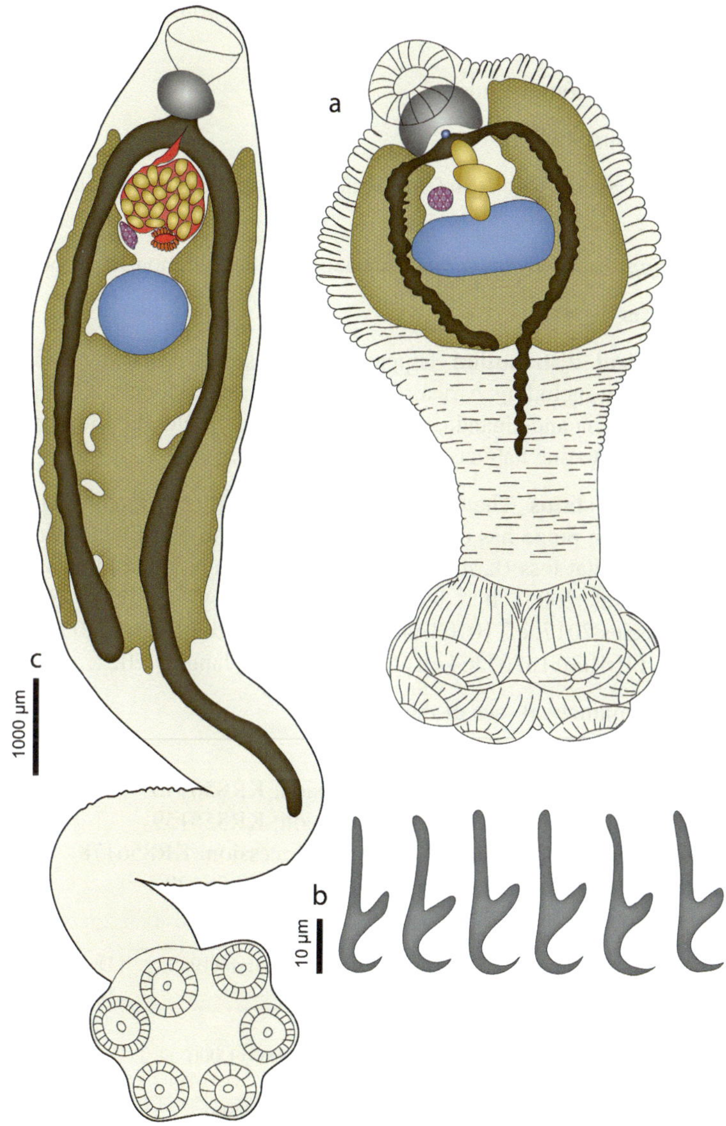

Fig. 5.70 (**a**, **c**) *Oculotrema hippopotami*, (**b**) marginal hooklets. (**a** and **b**) Redrawn from type description (Stunkard, 1924b); (**c**) Redrawn from Du Preez and Moeng (2003)

History Museum, London (NHML); Paratypes also in American Museum of Natural History (AMNH).

Holotype NHML1999.8.26.1.

Paratypes AMNH 142, AMNH 1470.1-4; NHML 1925.1.26.1-5.

Original description Stunkard (1924b).

Other taxonomic contributions Thurston and Laws (1965); Thurston (1968); Du Preez and Moeng (2004).

Etymology Named after the host.

Type locality Cairo, Egypt.

Other localities South Africa (McCully et al. 1967) and Western Uganda (Thurston 1968).

Type host *Hippopotamus amphibius* Linnaeus, 1758, Common Hippopotamus.

Additional hosts None.

Host geographical distribution Hippos are widespread in many countries of sub-Saharan Africa including Africa, Angola, Botswana, Ethiopia, Ghana, Sudan, Mozambique, Nigeria and Uganda, among others. They prefer the lower reaches of rivers as well as estuarine habitats (Okello et al. 2005).

Host conservation status The Common Hippopotamus is a "threatened" species, since its distribution range has declined due to exploitation (hunted for their meat and ivory) and habitat loss (farmers pressure them as they are seen as agricultural pests) (Okello et al. 2005). It is listed as Vulnerable by The IUCN Red List of Threatened Species 2017: e.T10103A18567364. https://doi.org/10.2305/IUCN.UK.2017-2.RLTS.T10103A18567364.en. Accessed on 21 January 2022.

Site on host Conjunctival sacs of the eye.

DNA sequence
- 18S rRNA gene, 1989 bp. GenBank Accession: KR856140
- 28S rRNA gene, 1366 bp. GenBank Accession: KR856159
- Cytochrome c Oxidase I, 341 bp. GenBank Accession: KR856178
- 12S rRNA gene, 483 bp. GenBank Accession: KR856120

Morphology and morphometrics (μm):

Source	Stunkard (1924b)	Thurston and Laws (1965)	Thurston (1968)	Du Preez and Moeng (2004)
Sample size	5			31
Body length	3000–5200	12,000	4000–11,000	(12,310–32,510) 17,280
Greatest width	1100–2200		1000–2500	(2050–3100) 2600
Width at vagina				

(continued)

Morphology and morphometrics (μm):				
Source	Stunkard (1924b)	Thurston and Laws (1965)	Thurston (1968)	Du Preez and Moeng (2004)
Haptor length		1700		(1.590–2.560) 2050
Haptor width		2300		(1.900–2970) 2270
False oral sucker width	360–580			(690–920) 820
Pharynx length				(410–560) 490
Pharynx width	360–580			(620–820) 700
Uterus length				(640–1540) 1050
Uterus width				(540–1670) 1070
Testis length	200–370	300–850		(690–1130) 900
Testis width	700–900	600–1100		(590–1210) 940
Ovary length	300–500			(280–870) 540
Ovary width	140–200			(140–260) 200
Egg length	230–270	235–265		(210–240) 230
Egg width	120–140	135–160		(130–140) 140
Egg incubation time			30 days	
Intra–uterine eggs (n)	4–12	40	1–2 (mature) 40–62 (fully grown worms)	(1–58) 26
Haptoral sucker diameter	500			(400–640) 510
Hamulus hook length				
Marginal hooklet length			15	(14.7–15.2) 14.9
Haptoral L/Body L		0.14		
No of anastomoses	0			0

Remarks None.

References

Arthington AH (2009) Australian lungfish, *Neoceratodus forsteri,* threatened by a new dam. Env Biol Fishes 84:211–221

Baruš V, Moravec F (1967) A survey of helminths from the Cuban turtle *Pseudemys decussata* Gray (Emydidae). Vt Isl Spol Zool 31:313–324

Caballero E (1938) Algunos trematodos de reptiles de Mexico. Anales Inst Biol Univ Nac Autón Mexico 9:103–120

Caballero E, Zerecero MC, Grocott RG (1956) Helmintos de la Republica de Panama. XIX. Algunos trematodos de quelonios de agua dulce. Anales Inst Biol Univ Nac Autón Mexico 27:415–430

Chaabane A, Du Preez L, Johnston G, Verneau O (2022) Revision of the systematics of the Polystomoidinae (Platyhelminthes, Monogenea, Polystomatidae) with redefinition of *Polystomoides* Ward, 1917 and *Uteropolystomoides* Tinsley, 2017. Parasite 29:56. https://doi.org/10.1051/parasite/2022056

Combes C, Ktari MH (1976) *Neopolystoma euzeti* n. sp. (Monogenea: Polystomatidae) premier représentant du genre *Neopolystoma* Price, 1939 en Afrique. Ann Parasitol Hum Comp 51:221–225

Combes C, Kulo SD (1978) *Polystomoides bourgati* n. sp. (Monogenea: Polystomatidae), premier représentant du genre *Polytomoides* Ward, 1917 en Afrique Occidentale. Rev Zool Afr 92:622–626

Combes C, Rohde K (1979) *Polystomoides platynotae* n. sp. (Monogenea: Polystomatidae), parasite du Chélonien d'eau douce *Notochelys platynota* (Gray, 1834) en Malaisie. Vie Milieu, Série C 28–29:69–75

Du Preez LH, Lim LHS (2000) *Neopolystoma liewi* sp. n. (Monogenea: Polystomatidae) from the eye of the Malayan box turtle (*Cuora amboinensis*). Folia Parasitol 47:11–16

Du Preez LH, Moeng IA (2004) Additional morphological information on *Oculotrema hippopotami* Stunkard, 1924 (Monogenea: Polystomatidae) parasitic on the African hippopotamus. Afr Zool 39:225–233

Du Preez LH, Morrison C (2012) Two new polystomes (Monogenea: Polystomatidae) from the eyes of North American freshwater turtles. Zootaxa 3392:47–59

Du Preez LH, Van Rooyen M (2015) A new polystomatid (Monogenea: Polystomatidae) from the mouth of the North American freshwater turtle *Pseudemys nelsoni*. ZooKeys 539:1–9

Du Preez LH, Badets M, Héritier L, Verneau O (2017) Tracking platyhelminth parasite diversity from freshwater turtles in French Guiana: first report of *Neopolystoma* Price, 1939 (Monogenea: Polystomatidae) with the description of three new species. Parasit Vectors 10:53. https://doi.org/10.1186/s13071-017-1986-y

Du Preez L, Domingues MV, Verneau O (2022) Classification of pleurodire polystomes (Platyhelminthes, Monogenea, Polystomatidae) revisited with the description of two new genera from the Australian and neotropical realms. Int J Parasitol Parasites Wildl 19:180–186

Dutton HR, Du Preez LH, Verneau O, Whelan NV, Bullard SA (2021) First record of a polystome from alligator snapping turtle, *Macrochelys temminckii* (Cryptodira: Chelydridae) or Mississippi; with comments on "*Neopolystoma orbiculare* (Stunkard, 1916)" and its junior subjective synonyms. J Parasitol 107:74–88

Euzet L, Combes C (1965) Parasites des chéloniens malgaches. *Polystomoides chabaudi* n. sp. (Monogenea) chez la tortue d'eau douce *Pelomedusa subrufa* Lacépède 1788. Ann Parasitol Hum Comp 40:455–450

Fairfax RA (1990) A new species of *Neopolystoma* (Monogenea) and the occurrence of *Polystomoides* sp. in New Guinea, with notes on some polystomes from North-East Australia. Sci New Guinea 16:109–114

Fischthal JH, Kuntz RE (1964) A monogenetic and seven digenetic trematodes of amphibians and reptiles from Palawan Island, Philippines. Proc Helminthol Soc Wash 31:230–240

Fukui T, Ogata T (1936) Sur deux espèces nouvelles de trématode provenant de l'*Ocadia sinensis*. Zool Mag (Tokyo) 48:765–770

Gonzales JP, Mishra GS (1977) *Polystomoides tunisiensis* n. sp. (Monogenea: Polystomatidae) et *Telorchis temimi* n. sp. (Digenea, Telorchiidae). Deux nouvelles espèces de Trématodes de tortues paludines de Tunisie. Arch Inst Pasteur Tunis 54:29–38

Gupta NK, Randev R (1974) On the histomorphology of *Polystomoides ludhianae*, n. sp. (Monogenea) recovered from the urinary bladder of *Kachuga tectum* and *Kachuga smithi* in North India. Parassitologia 16:225–229

Harwood PD (1932) The helminths parasitic in the Amphibia and Reptilia of Houston, Texas and vicinity. Proc Nat Mus 81:1–71

Héritier L, Badets M, Du Preez LH, Aisien MS, Lixian F, Combes C, Verneau O (2015) Evolutionary processes involved in the diversification of chelonian and mammal polystomatid parasites (Platyhelminthes, Monogenea, Polystomatidae) revealed by palaeoecology of their hosts. Mol Phylogenet Evol 92:1–10

Héritier L, Verneau O, Smith KG, Coetzer C, Du Preez LH (2018) Demonstrating the value and importance of combining DNA barcodes and discriminant morphological characters for polystome taxonomy (Platyhelminthes, Monogenea). Parasitol Int 67:38–46

Knoepffler L-P, Combes C (1977) Présence en Corse de *Polystomoides ocellatum* (Rudolphi, 1819) chez *Emys orbicularis* (L. 1758) (Chelonia, Emydidae). Considérations sur la répartition mondiale du genre *Polystomoides*. Vie Milieu, Série C 27:221–230

Kulo SD (1980) Parasites de Chéloniens en Afrique Intertropicale. *Polystomoides nabedei* n. sp. (Monogenea: Polystomatidae), parasite de la vessie urinaire de la tortue d'eau douce *Pelomedusa subrufa* Lacépède, 1788 (Chélonien, Pelomedusidae). Ann Parasitol Hum Comp 55:367–377

Lamothe-Argumedo R (1972) Monogeneos de reptiles I. Redescripcion de cuatro especies de Monogenea (Polystomatidae) parasitos de la vejiga urinaria de tortugas de Mexico. Anales Inst Biol Univ Nac Autón Mexico 43:1–15

Leidy J (1888) Entozoa of the terrapin. Proc Acad Nat Sci Phila 18:127–128

Lenis C, García-Prieto L (2009) *Polystomoides magdalenensis* n. sp. (Monogenoidea: Polystomatidae), a parasite of buccal cavity of *Trachemys callirostris callirostris* (Testudinata: Emydidae) from Colombia. J Parasitol 95:850–854

MacCallum GA (1918) Studies on the Polystomatidae. Zoopathologica 1:103–120

Mañé-Garzón F (1958) Sobre el hallazgo de *Polystomoides coronatus* (Leidy 1888) en la boca de una Tortuga de Sudamerica. Rev Med Vet Parasitol Maracay 18:35–41

Mañé-Garzón F, Gil O (1961) Trematodos de las tortugas del Urugay. V. Sobre un nuevo Polystomatidae de la faringe de *Phrynops geoffroana hillarii* (D. and B.). Com Zool Mus Hist Nat Montevideo 7:16

Mañé-Garzón F, Gil B (1962) Trematodos de las tortugas del Uruguay, V. Sobre un nuevo Polystomatidae de la faringe de *Phrynops geoffroana hillarii* (D. and B.). Comun Zool Mus Hist Nat Mont 7:1–6

Mañé-Garzón F, Holcman-Spector B (1968) Trematodos de las tortugas del Uruguay, VII. *Polystomoides rohdei* n. sp. de la boca de *Pseudemys dorbigni* (Dum. and Bib.). Comun Zool Mus Hist Nat Mont 9:1–3

McCully RM, Van Niekerk JW, Kruger SP, Jansen BC (1967) Observations on the pathology of bilharziasis and other parasitic infestations of *Hippopotamus amphibius* Linnaeus, 1758, from the Kruger National Park. Onderstepoort J Vet Res 34(2):563–617

Okello JB, Nyakaana S, Masembe C, Siegismund HR, Arctander P (2005) Mitochondrial DNA variation of the common hippopotamus: evidence for a recent population expansion. Heredity 95:206–215

Ozaki Y (1935) Studies on the frog – trematode *Diplorchis ranae*. I. Morphology of the adult form with review of the family Polystomatidae. J Sci Hiroshima Univ 3:193–223

Ozaki Y (1936) Two new trematodes from tortoise *Geoemyda spengleri* (Gmelin). J Sci Hiroshima Univ Ser B (Zool) 4:81–90

Pandey KC (1973) Studies on monogenetic trematodes of India. II. On a new species of the rare genus *Polystomoides* Ward, 1917. Indian J Zootomy 14:143–145

Pandey KC, Agarwal M (1978) A new monogenean, *Polystomoides chauhani* n. sp., from *Hardella thurjii* Gray. Indian J Helminthol 30:126–128

Paul AA (1938) Life history studies of North American freshwater polystomes. J Parasitol 24:489–510

Petzold A, Vargas-Ramirez M, Kehlmaier C, Vamberger M, Branch WR, Du Preez L, Hofmeyr MD, Meyer L, Schleicher A, Široký P, Fritz U (2014) A revision of African helmeted terrapins

(Testudines: Pelomedusidae: *Pelomedusa*), with descriptions of six new species. Zootaxa 3795: 523–548

Pichelin S (1995) The taxonomy and biology of the Polystomatidae (Monogenea) in Australian freshwater turtles (Chelidae, Pleurodira). J Nat Hist 29:1345–1381

Pichelin S, Whittington I, Pearson J (1991) *Concinnocotyla* (Monogenea: Polystomatidae), a new genus for the polystome from the Australian lungfish *Neoceratodus forsteri*. Syst Parasitol 18: 81–93

Platt TR (2000a) Helminth parasites of the western painted turtle, *Chrysemys picta belli* (Gray), including *Neopolystoma elizabethae* n. sp. (Monogenea: Polystomatidae), a parasite of the conjunctival sac. J Parasitol 86:815–818

Platt TR (2000b) *Neopolystoma fentoni* n. sp. (Monogenea: Polystomatidae) a parasite of the conjunctival sac of freshwater turtles in Costa Rica. Mem Inst Ozwaldo Cruz Rio J 95:833–837

Platt TR, Firth A, Sharma RS (2011) Redescription of *Neopolystoma liewi* Du Preez and Lim, 2000 (Monogenea: Polystomatidae), from *Cuora amboinensis* (Testudines: Geomydidae) with notes on specimen preparation. Comp Parasitol 78:286–290

Price EW (1939) North American monogenetic trematodes. IV. The family Polystomatidae (Polystomatoidea). Proc Helminthol Soc Wash 6:80–92

Reichenbach-Klinke H (1966) Eine neue art der Polystomatidengattung *Eupolystoma* Kaw, 1950 (Monogenea, Polystomatidae) von den Kiemen des australischen lungenfisches *Neoceratodus forsteri* Krefft. Zool Anz 176:142–146

Richardson JPM, Brooks DR (1987) *Polystomoidella mayesi* n. sp. (Monogenea: Polystomatidae) from the urinary bladder of a Malaysian box turtle, *Cuora amboinensis*. Can J Zool 65:1567–1569

Rohde K (1963) *Polystomoides malayi* n. sp. (Monogenea: Polystomatidae) aus der Harnblase von *Cyclemys amboinensis* in Malaya. Z Parasitenkd 22:278–282

Rohde K (1965) Studies on the genus *Polystomoides* Ward, 1917 (Monogenea). I. Description of 4 Malayan species, a key to the known species, and a comparison of the subcuticular layers in *Polystomoides* and some digenetic trematodes. Zool Jahrb Abt Syst Oekol Geogr Tiere 92:345–368

Rohde K (1984) Three new species of the genus *Neopolystoma* (Monogenea) from river tortoises in Australia. Syst Parasitol 6:99–105

Rohde K, Pearson JC (1980) Two polystomes (Monogenea) from Australian river tortoises (Pleurodira, Chelidae), *Polystomoides australiensis* sp. nov. from *Emydura krefftii*, and *Neopolystoma chelodinae* (MacCallum, 1919) from *Chelodina longicollis*. Zool Anz 204: 191–208

Rudolphi CA (1819) *Entozoorum zynopsis* cui accedunt mantissa duplex et indices locupletissimi. Sumtibus A. Rücker, Berolini. 811 p

Santana DO, Iwama RE, Teixeira AA, Moura GJ, Faria RG, Mesquita DO (2019) Spatio–temporal variation and the use of host body surface by ectoparasites of the chelonians *Phrynops geoffroanus* and *Mesoclemmys tuberculata* in areas of the Caatinga and Atlantic Forest in Northeast Brazil. Parasitol Res 118:913–926

Sproston NG (1946) A synopsis of the monogenetic trematodes. Trans Zool Soc Lond 25:185–600

Stewart FH (1914) The anatomy of *Polystomum kachugae*, sp. nov, with notes on the genus *Polystomum*. Rec Indian Mus 10:195–205

Strelkov YA (1950) New species of monogenetic trematode of the far-east tortoise *Amyda sinensis*. Docklladi Akademii Nauk SSSR 74:159–162

Stunkard HW (1916) On the anatomy and relationships of some North American trematodes. J Parasitol 3:21–27

Stunkard HW (1924a) On some trematodes from Florida turtles. Trans Am Microsc Soc 43:97–117

Stunkard HW (1924b) A new trematode, *Oculotrema hippopotami* n. g., n. sp., from the eye of the hippopotamus. Parasitology 16:436–440

Thurston JP (1968) The larva of *Oculotrema hippopotami* (Monogenea: Polystomatidae). J Zool 154:475–480

Thurston JP, Laws RM (1965) *Oculotrema hippopotami* (Trematoda:Monogenea) in Uganda. Nature 4976:1127

Timmers SF, Lewis PD (1979) Helminths of *Chrysemys picta belli* in Manitoba including *Polystomoides pauli* sp. n. (Monogenea: Polystomatidae). Can J Zool 57:1046–1051

Tinsley RC (1973) Observations on Polystomatidae (Monogenoidea) from East Africa with a description of *Polystoma makereri* n. sp. Z Parasitenkde 42:251–263

Uetz P, Freed P, Aguilar R, Reyes F, Hošek J (2022) The reptile database. http://www.reptile-database.org. Accessed 23 Aug 2022

Vieira FM, Novelli IA, Sousa BM, De Souzalima S (2008) A new species of *Polystomoides* Ward, 1917 (Monogenea: Polystomatidae) from freshwater chelonians (Testudines: Chelidae) in Brazil. J Parasitol 94:626–630

Whittington ID, Pichelin S, Pearson J (1991) Attachment of eggs by *Concinnocotyla australensis* (Monogenea: Polystomatidae) to the tooth plates of the Australian lungfish, *Neoceratodus forsteri* (Dipnoi). Int J Parasitol 21:341–346

Wright RR (1879) Contributions to American helminthology. Proc Can Inst 1:54–75

Chapter 6
Polystomatidae: Life-History Strategies, the Key to Success

Contents

Abstract Polystomatid flatworms radiated and co-evolved with their sarcopterygian hosts, namely the Australian lungfish, amphibians, freshwater turtles and the common hippopotamus. They are generally strictly host and site-specific with a close synchronization of their life cycle with that of their host, which is the key to their diversification success. This section reviews the main characteristics of their life stages, from eggs to mature worms, and brings together all the known life

L. H. du Preez et al., *Polystomatid Flatworms*, Zoological Monographs 9,
https://doi.org/10.1007/978-3-031-35887-6_6

cycles of polystomes. Regardless of their infection site, gills, urinary bladder, pharyngeal cavity or conjunctival sacs, the variations observed in the distinct life cycles mainly result in egg production, size of the uterus and egg development (external versus *intra-uterine*).

6.1 Introduction

Monogeneans belong to the phylum Platyhelminthes and comprise roughly 20,000 species. According to Odhner (1912), they were classified into two subclasses: the mucophagous Monopisthocotylea and the sanguinivorous Polyopisthocotylea (see Halton and Jennings 1965). They display specialized attachment organs and means of attachment (Chisholm and Whittington 1998), predominantly related to the morphological specialization of the posterior haptor (Cribb et al. 2002) that seem well adapted to the host and environment. Changes of host by monogeneans from fish to rhipidistians revealed a tendency to abandon the exposed ectoparasitic mode of life for an enclosed meso- or endo-parasitic lifestyle. Poulin (2011) stated that an internal environment was generally more predictable than an external one since all conspecific hosts are virtually identical in construction and function, with organs performing the same function or secreting the same chemicals. The nature of the substrate to which these parasites attach, along with water currents, thus may have played an important role in the evolution of attachment organs and associated structures. Chisholm and Whittington (1998) hypothesized that the complexity of the haptor could be related to the parasite's habitat. Parasites that live in habitats exposed to strong water currents, such as the gills and dorsal skin surface, generally have more complex haptors than those living in environments exposed to weaker currents or without currents, such as the nasal fossae or the urogenital system.

Monogeneans are typically soft-bodied organisms and are, therefore, very plastic in body shape. Their hard sclerotized sclerites, however, which consist of scleroproteins (Lyons 1966), are considered to have important functional, taxonomic and evolutionary significance (Ramasamy and Brennan 2000). They therefore probably represent one of the most informative structures for species delimitation. In polystomatids sclerotized structures include marginal hooklets of the oncomiracidium, hamuli, sclerotized support structures in the suckers and genital spines. Marginal hooklets, which aid in the attachment of oncomiracidia to the host, lose their function as the parasites mature. However, they are retained in the soft tissue of adult parasites and, since their morphology is stable within a species, making them important taxonomic characters for polystomes (Du Preez and Maritz 2006).

Firm attachment is critical for parasites to maintain a close relationship with their host, which is key to the infection's success. Monogeneans attach to their host either through an anterior attaching organ known as the prohaptor, or, more often, by the posterior opisthaptor. While the prohaptor is mainly associated with feeding activities, the opisthaptor is primarily associated with attachment to the host (Wright and Dechtiar 1974) thanks to hooks, suckers, clamps and/or chemical bonding (using bioadhesives) (Buchmann and Lindenstrøm 2002). According to Kearn (1994), the secretion of adhesives might have served as an early attachment of monogeneans. These are indeed commonly found in free-living platyhelminthes for temporary attachment to substrates. It is believed that the first monogeneans were small skin parasites that only used small marginal hooks to attach themselves to the host's epidermis (Kearn 1994). Hooks aid in the attachment to host epidermal cells, limiting the force and size of each to avoid separation of the pierced epidermal cell membranes, restricting not only the size of the hooks but also the overall size of the parasite. The incorporation of suckers into the haptor was considered as a significant advance in the development of the monogeneans (Kearn 1994). Judging by the fact that it is present among several monogeneans, along with other skin parasitic invertebrates such as leeches and crustaceans, suction, as a means of attachment to the host's skin, is efficient. Kearn (1994) also made the statement that the hooks originally might have served as internal attachment sites within parasites, only later obtaining a secondary function for attachment to the host. The attachments of monogeneans, such as head organs, anchors, suckers and clamps provided with muscle fibres, help them to function efficiently and effectively.

Depending on the site in the host where polystomes live, they face varying degrees of being dislodged. When oncomiracidia of *Polystoma* Zeder, 1800 or *Metapolystoma* Combes, 1976 enter the spiracle, which is sinistral in most anurans, they attach to the first gill arches using their marginal hooklets. Within 10 days, they develop suckers and migrate out of the water current to the right-hand gill arches (Kok and Du Preez 1989). Neotenic forms do not have any hamuli and attach only using suckers (Kok and Du Preez 1989). Amphibian polystomes infecting the branchial chamber are more or less exposed to the air–water interface and, thus do not experience major flushing forces. Should a parasite accidentally detach, the risk of being flushed out before it can re-attach is small. In these circumstances, where parasites attach to a rigid substrate such as a gill filament, attachment by suckers is sufficient to ensure a firm grip. Inside the urinary bladder, parasites attach to a thin, highly contractile, epithelial membrane. Here, flash urination or rehydration can cause sudden changes to surface area, posing a serious threat of being dislodged and flushed out. These risks of being flushed out are counteracted by large hamuli's gaffing the epithelial membrane (Tinsley and Tinsley 2016).

6.2 Life Stages of Polystomes

6.2.1 The Egg

Polystome eggs (Fig. 6.1a) are generally small, i.e. roughly from 150 to 250 μm, operculated, yellow-tan, smooth surfaced, ovoid to fusiform and without

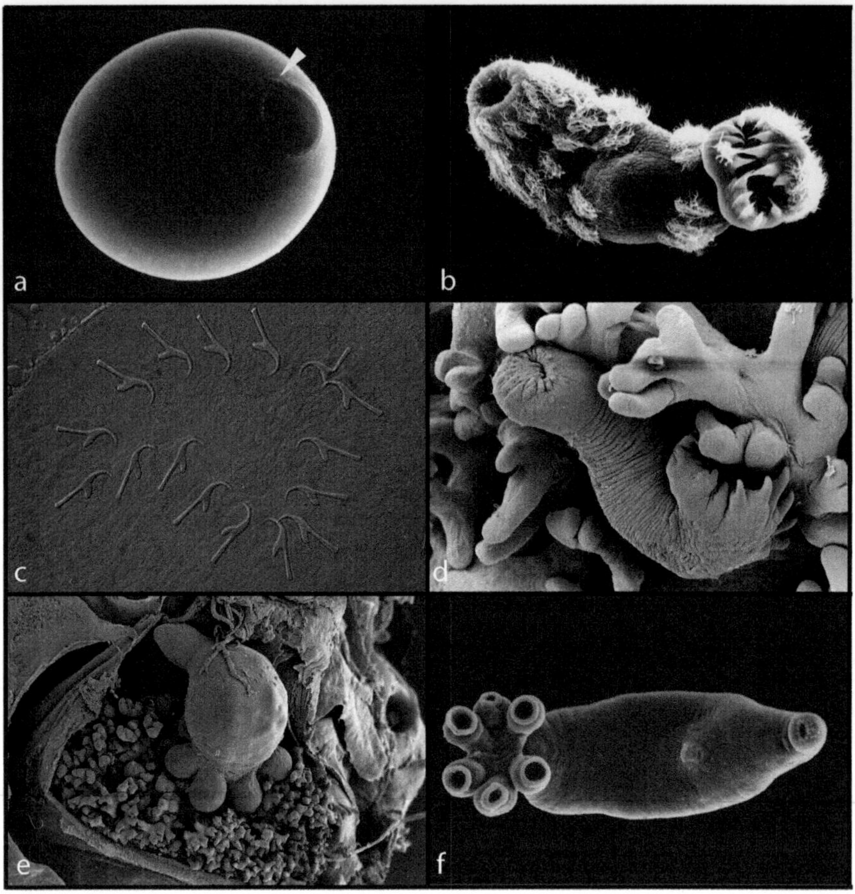

Fig. 6.1 (**a**) Egg of *Polystoma australe* Kok and Van Wyk, 1986 with the hinged operculum (arrow); (**b**) ciliated oncomiracidium of *P. australe*; (**c**) marginal hooklets of *P. australe*; (**d**) juvenile of *P. australe* attached to gill filaments; (**e**) neotenic form of *P. australe* on gills of a tadpole; (**f**) mature bladder parasite of *P. australe*

appendages. Although most eggs are ovoid to pyriform, eggs of *Apaloneotrema* Du Preez and Verneau, 2020 are long ovoid and those of *Fornixtrema* Du Preez and Verneau, 2020 and *Protopolystoma* Bychowsky, 1857 are fusiform or spindle-shaped (Du Preez and Verneau 2020). A rudimentary flagellum, which appears to be non-functional, may sporadically be observed in *Polystoma* (see Du Preez 1996), however, egg filaments are known from the eggs of *Concinnocotyla australensis* (Reichenbach-Klinke, 1966) the polystome from the Australian lungfish (Pichelin et al. 1991). The majority of polystomes have a dark tan coloured egg with a reasonably strong shell wall consisting of scleroproteins, however, in some polystomes of the genera *Pseudodiplorchis* Yamaguti, 1963, *Madapolystoma* Du Preez, Verneau, Raharivololoniaina and Vences, 2010 and *Wetapolystoma* Gray,

1993, the shell wall is a soft, pale-yellow membrane (Raharivololoniaina et al. 2011) and as in *Eupolystoma* Kaw, 1950 these eggs are non-operculated (Du Preez et al. 2003). For these genera with non-operculated eggs, the eggshell membrane simply ruptures when eggs hatch. For the remainder, the shell wall is yellow-tan in colour, 1.4 μm thick in *Diplorchis* Ozaki, 1931 (see Ozaki 1935), to 3.3 μm in *Oculotrema* Stunkard, 1924 (see Du Preez and Moeng 2004). The operculum is apical and circular. For *Polystoma* it has been reported that opercula are hinged (Fig. 6.1a) and will stay attached when oncomiracidia leave the eggs (Du Preez 1996).

Concerning egg production, it varies considerably among the distinct genera. However, this has only been documented for a few polystomes. For *Polystoma* spp. the output is in the order of 8–20 eggs/parasite/day (Tinsley and Owen 1975), however, *Polystoma integerrimum* (Fröhlich, 1791) may produce as many as 2500 eggs/parasite/day (Combes 1972). This high output is, however, short-lived, and roughly 90% of the annual egg production takes place in 4 days (Tinsley 1983). Chelonian polystomes produce in the order of 5–10 eggs/parasite/day. While egg production seems well correlated to the reproductive status of the host, external temperature also influences egg production. For *Protopolystoma* spp. infecting the primarily aquatic, clawed frog, *Xenopus* Wagler, 1827, egg production increases by 10% with each degree centigrade rise in temperature (Jackson and Tinsley 1988). The release of polystome eggs can also be synchronized with host behaviour. In polystomes infecting semi-aquatic frogs, eggs accumulate in the uterus or the host's urinary bladder. When the host enters the water to reproduce, water rapidly accumulates in its bladder. This decrease in osmotic pressure stimulates polystomes occurring in the bladder to release eggs and, when the frog urinates, polystome eggs are flushed out.

6.2.2 The Oncomiracidium

An oncomiracidium (Fig. 6.1b) is a ciliated larva that hatches from the egg with four refractive eye spots, tufts of cilia arranged in groups and patterns, surface sensilla and a haptor with 16 marginal hooklets (Fig. 6.1c). These hooklets are placed peripherally along the outer margin of the haptor, with the posterior-most pair being the largest, like in *Polystoma*. In some other genera, like *Eupolystoma*, the marginal hooklets are of equal size.

Cilia are arranged in groups, and their arrangement is of taxonomic importance. Combes (1968) started to document the arrangement of sensilla and ciliated cells on the tegument of the oncomiracidium. He was followed by Combes et al. (1975, 1978), Lambert and Bourgat (1978), Lambert et al. (1978), Lambert and Kulo (1982), Thurston (1968) and Tinsley (2013). It was noted that, once established on a host, oncomiracidia discard ciliated epidermis cells, which are no longer used during further parasite development (Du Preez and Kok 1987). The life span of an oncomiracidium is short and, in general for monogeneans less than 24 h (Llewellyn 1963). However, it may vary between polystome genera and mainly consists of two phases for most species: a free-swimming phase in search of a host and, upon contact with a suitable host, a gliding or looping phase, on / or in the host, in search of a specific attachment site or the opening to enter the body to reach the target site.

6.2.3 The Mature Parasite

Following establishment at the target site, polystomes develop at varying rates depending on the host group. All amphibian polystomes are sanguiferous and feed on blood in the urinary bladder, kidneys, gills or, on rare occasions, lungs. Polystome life strategies are tightly synchronized with the seasonal reproductive mode of their host, which may vary from a few days to several weeks in the year. Newly established parasites in amphibian hosts usually develop slowly and only reach maturity after several months. In contrast, chelonian polystomes feed on epithelial tissue and mucous and are less influenced by host reproductive activity. Apart from *Uteropolystomoides* Tinsley, 2017, chelonian polystomes do not have a uterus and, thus, cannot accumulate eggs. As their host generally keeps to an aquatic environment, eggs are produced over more extended periods and released into the water, where they develop. The parasites' reproduction is, thus, not restricted to a brief window of time when the host reproduces but occurs when the host is active and enters water bodies as opposed to hibernating on land during winter.

6.3 Attachment Structures of Polystomes

All oncomiracidia have 16 marginal hooklets that secure a firm grip onto the host—that being on the gills of tadpoles for *Polystoma* and *Metapolystoma*, in the kidneys for *Protopolystoma*, in the lungs for *Pseudodiplorchis* and on the eye of the host, in the oral region or the urinary bladder for chelonian polystomes. While marginal hooklets ensure an effective attachment for the oncomiracidium (Fig. 6.1d), as the parasite develops, additional attachment organs in the form of large hooks (hamuli) or haptoral suckers arise (Fig. 6.1e, f). These structures will not play an equally important role regarding the site of infection, which explains why polystomes favour hamuli, suckers or a combination. While all polystomes have haptoral suckers, some lack hamuli altogether, some have a single pair and some have two pairs. Whereas

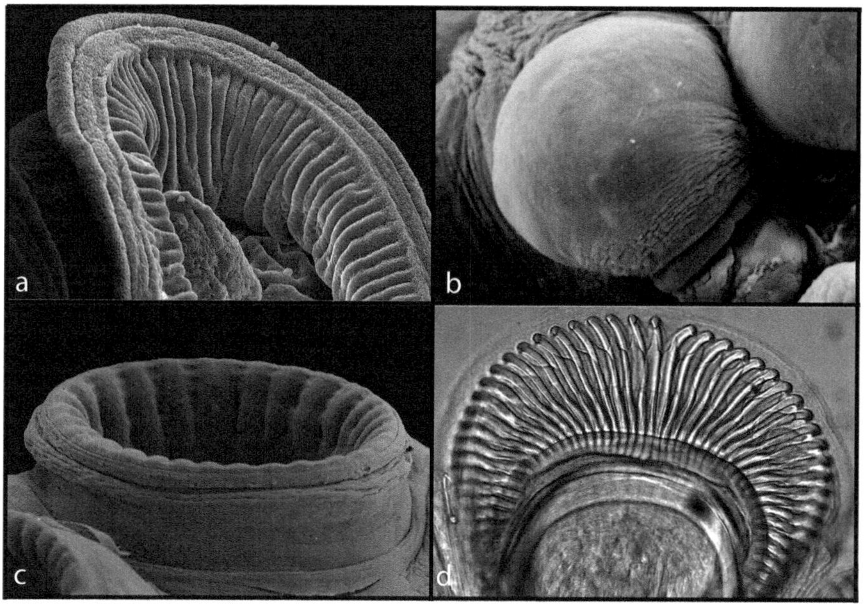

Fig. 6.2 Sucker classification according to Du Preez and Theunissen (2021) (**a**) sucker type 1 in *Protopolystoma*; (**b**) sucker type 2 in *Polystoma*; (**c**) sucker type 3 in *Polystomoidella*; (**d**) sucker type 4 in *Concinnocotyla*

haptoral suckers do not differ dramatically in size, hamuli do. Hamuli could be small, in some genera, to massive in, for example, *Uropolystomoides* Tinsley and Tinsley, 2016.

Based on the early work of Strelkov (1950), Pichelin (1995) classified polystomes into three groups according to the type of suckers. Du Preez and Theunissen (2021), from re-evaluating the sucker structure, proposed dividing the haptoral system into four categories. Type 1 is a soft-bodied, flexible haptor (Fig. 6.2a) that is found in *Protopolystoma*; Type 2 is a firm sucker without any skeletal elements (Fig. 6.2b) and is found in all remaining anuran polystomes; Type 3 is a firm sucker with skeletal elements (Fig. 6.2c) that vary from a skeletal tube in *Nanopolystoma* Du Preez, Wilkinson and Huyse, 2008, to a ring of well-developed skeletal blocks held together by a system of muscles as in chelonian polystomes and *Oculotrema*; Type 4 is an asymmetrical sucker with long skeletal hooks and discs (Fig. 6.2d) which only characterizes *Concinnocotyla* Pichelin, Whittington and Pearson, 1991.

While suckers appear to provide a very firm attachment on harder surfaces, such as the oral region of chelonians and the eye of the hippopotamus, hamuli tend to be the primary form of attachment in the bladder of chelonians (Tinsley and Tinsley 2016). Therefore, for amphibian polystomes, haptoral suckers are used in combination with hamuli when present.

6.4 Life Cycles of Polystome Genera

All polystomes, true to the Monogenea, have direct life cycles and thus no intermediate hosts are involved. Instead, they evolved different infection strategies, aiding in their success of diversification within sarcopterygian host species. Because their life cycle has not been studied and documented for all polystome genera, we present here only the well-documented life cycles.

6.4.1 Concinnocotyla (Fig. 6.3)

Concinnocotyla australensis is found inside the oral cavity, on the gill arches and on the edges of the primary gill lamellae of the Australian lungfish *Neoceratodus forsteri* Krefft, 1870 in Queensland, Australia. The worm does not have a uterus and the long, ellipsoid eggs are attached with an elongated appendage to the walls of the buccal cavity (Whittington and Pichelin 1991). Oncomiracidia are not ciliated and, along with subadult worms, are found on the gill filaments. Transfer of parasites to other host individuals probably occurs during close contact with host individuals during mating.

Fig. 6.3 Life cycle of *Concinnocotyla* (1) mature parasite on gills or in mouth; (2) egg attached to mouth roof or gills; (3) oncomiracidium attaches to gills or (4) leave host for another host

6.4.2 *Protopolystoma* (Fig. 6.4)

Protopolystoma is known only from the primarily aquatic clawed frog, *Xenopus*. Egg production peaks in the warmer months and worms produce about 12 eggs/worm/day at 20 °C (Jackson and Tinsley 1998). Egg production starts 3–4 months post-infection at 22 °C (Tinsley and Owen 1975; Tinsley and Jackson 2002). Since the host is fully aquatic, no need exists to accumulate eggs for polystomes. *Protopolystoma* has no uterus, and eggs are continually expelled directly into the host's urinary bladder from where they are flushed when the frog urinates (Tinsley 2004). Infective oncomiracidium hatches from an operculated egg following incubation of approximately 22 days. It swims around actively searching for a potential host for up to 24 h. Once contact has been made, the oncomiracidium enters the cloaca and migrates to the kidneys of its host (Thurston 1964). The developing parasite larva establishes inside ducts in the kidney where it attaches and feeds on blood for approximately 2–3 months. It subsequently migrates to the urinary

Fig. 6.4 Life cycle of *Protopolystoma* (1) mature parasite in the urinary bladder; (2) eggs expelled; (3) oncomiracidium hatching and finds host; (4) oncomiracidium enters cloaca, migrates to kidney where it feeds and develops and then migrates to the bladder

bladder via the urinary duct, where it continues to develop. It reaches maturity after 3–4 months post-infection (Tinsley and Owen 1975).

6.4.3 *Polystoma* and *Metapolystoma* (and Possibly *Indopolystoma* Chaabane, Verneau and Du Preez, 2019) (Fig. 6.5)

Polystoma and *Metapolystoma* spp. differ from other polystomes by having a neotenic or branchial phase in the life cycle (Murith et al. 1977; Du Preez and Kok 1998). Following spawning, frog eggs usually develop rapidly and, after hatching, tadpoles remain in the area for the first few days. Polystome eggs, which are released at the same time and place, hatch after a period of about 10–16 days. It thus brings the oncomiracidium in close proximity to potential tadpole hosts. Once contact has been made with the tadpole of the correct host species, the oncomiracidium attaches to the surface for a moment and then glides or crawls with a leech-like looping motion over the surface of its host until it locates the sinistrally positioned spiracle, where it enters and establishes on the internal gills (Du Preez and Kok 1997). If an oncomiracidium makes contact with the non-suitable host, it breaks contact and continues to swim in search of another tadpole (Du Preez and Kok 1997). In the correct tadpole the oncomiracidium attaches to the gills, where it starts feeding on blood. If the tadpole happens to be a young tadpole in pre-metamorphosis, the oncomiracidium rapidly develops over approximately 16 days into a neotenic parasite that will boost the parasite population. Neotenic parasites have a short life span and drop out and die as soon as the front legs of the developing frog metamorph break through. If an oncomiracidium attaches to an older tadpole in pro-metamorphosis, it remains on the gills, undergoing very little development. The hamulus primordial starts to develop into hamuli and, when the front legs of the developing frogs break through, the parasite migrates to the outside of it host and enters the cloaca. It then crawls to the urinary bladder where it attaches and starts to feed on blood (Williams 1961; Combes 1968). Within the urinary bladder it slowly develops and matures into an adult parasite that starts producing eggs during the next breeding season (Kok and Du Preez 1987). Neotenic forms differ from parasites found in the urinary bladder in that the haptor and suckers are far more flexible, the ovary is very big and the uterus and hamuli are absent.

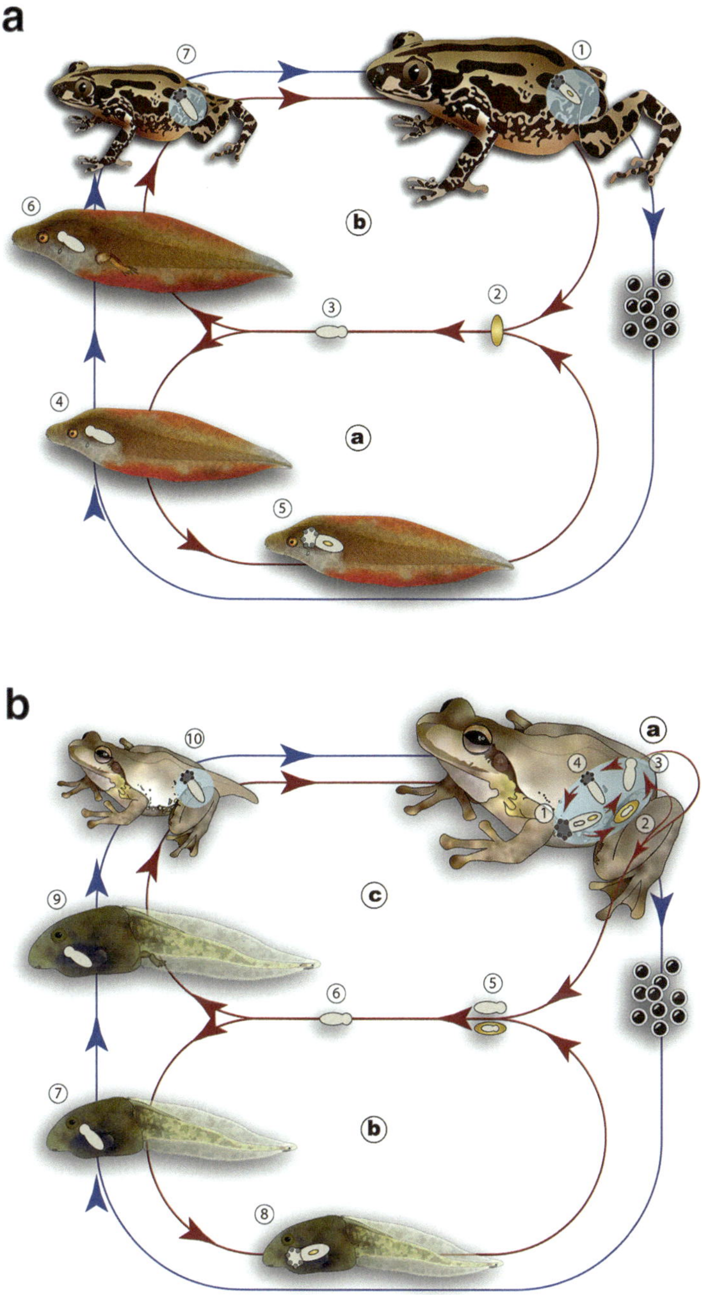

Fig. 6.5 (**a**) Life cycle of *Polystoma*. Blue arrows indicate the reproductive cycle of the host and red the cycles of the parasite (1) mature parasite in the urinary bladder; (2) polystome eggs expelled; (3) ciliated oncomiracidium hatching; (4) oncomiracidium attached to gills of a young tadpole in pro-metamorphosis with (a) = neotenic life cycle; (5) neotenic parasite egg-producing within 16 days; (6) oncomiracidium attached to gills of old tadpole in pre-metamorphosis with (b) = bladder life cycle; (7) host metamorphosis during which migration to the urinary bladder and establishment

6.4.4 *Pseudodiplorchis* and *Neodiplorchis* Yamaguti, 1963 (Fig. 6.6)

Pseudodiplorchis and *Neodiplorchis* spp. infect arid-adapted, spadefoot toads that breed for 1–3 days following the first spring rains. These polystomes accumulate

Fig. 6.6 Life cycle of *Pseudodiplorchis americanus* (Rodgers and Kuntz, 1940). (1) Mature parasite in bladder accumulate embryonated eggs in utero; (2) when frog enters water to breed eggs are released; (3) oncomiracidia hatch and may establish along parent or (4) are expelled; (5) oncomiracidium searches for potential host and enters nostril and (6) migrates to mouth or (7) vocal sac; (8) developing parasites migrate to lungs; (9) then they migrate to mouth and (10) then via gut to cloaca and to the bladder

---

Fig. 6.5 (continued) of the parasite. (**b**) Life cycle of *Metapolystoma*. Blue arrows indicate the reproductive cycle of the host and red the cycles of the parasite (1) mature parasite with eggs in utero; (2) embryonated eggs expelled in the bladder; (3) oncomiracidium hatching; (4) oncomiracidium developing next to its parent in the urinary bladder with (a) = internal life cycle; (5) embryonated eggs or oncomiracidium expelled outside the urinary bladder; (6) free-swimming oncomiracidium; (7) oncomiracidium attached to gills of a young tadpole in pro-metamorphosis with (b) = neotenic life cycle; (8) neotenic parasite egg-producing within 16 days; (9) oncomiracidium attached to gills of old tadpole in pre-metamorphosis with (c) = bladder life cycle; (10) host metamorphosis during which migration to the urinary bladder and establishment of the parasite

large numbers of eggs *intra-uterine*. As very large numbers of frogs enter the newly formed pools polystome eggs are released *en masse* and oncomiracidia hatch in large numbers. The window for parasite transmission is actually much smaller than 3 days of frog activity at the pond as the frogs are nocturnal and only enter the water around 21h00 and leave the water by 04h00, resulting in the total period of possible transmission being less than 24 h a year (Tinsley 1990a). In spite of the very short period of parasite transmission, infection is still highly effective as more than 50% of the frogs are infected with *Pseudodiplorchis* (see Tinsley 1989, 1990b). When frogs enter the water body they hydrate stimulating the worms to release the oncomiracidia *en masse* that are washed out with the frog's urine. Oncomiracidia immediately infect another host by passing through the nostrils, where they accumulate. During the following week they migrate to the mouth and may even enter the vocal sac. From here they migrate through the glottis to the lungs where they establish and feed on blood. By day 21 they return to the mouth and, from day 28, they migrate within a period of 5–20 min through the alimentary canal to the urinary bladder where they attach and continue to grow, becoming sexually mature within 3–4 weeks (Tinsley 1990a). During the migration the worms are protected by tegumental vesicles discharged all along the alimentary canal (Cable and Tinsley 1992).

6.4.5 *Madapolystoma* (Fig. 6.7)

Madapolystoma spp. display advanced *intra-uterine* development with no apparent free-living oncomiracidia. Larvae develop *intra-uterine* to the extent that they have hamuli and two pairs of suckers (Du Preez et al. 2010). Within the developed F1 larva

Fig. 6.7 Life cycle of *Madapolystoma* (1) mature parasite in the urinary bladder; with advanced *intra-uterine* development; (2) release of subadult parasites; (3) infection or re-infection via the cloaca during frogs mating or possibly when revisiting egg clutches

a F2 embryo already starts developing. Because the anuran hosts of *Madapolystoma* breed on land, or in very shallow water, it was hypothesized that larvae leave the parent and simply crawl out and transfer to another frog or to a batch of terrestrial eggs, from where it may infect another frog visiting the eggs (Landman et al. 2018).

6.4.6 *Eupolystoma* (Fig. 6.8)

Eupolystoma shows two distinct life cycles, internal and/or external. Parasites accumulate eggs in their uterus where they develop. Ciliated and unciliated oncomiracidia hatch immediately upon releasing of eggs within the bladder and following a change in osmotic pressure, due to the influx of water (Combes et al. 1973; Fournier and Combes 1979). While ciliated oncomiracidia leave the host and swim into the external environment where they actively search for another, or possibly the same, host to infect (external cycle as in Fig. 6.8a), unciliated oncomiracidia attach directly to the host's bladder wall alongside its parents (internal cycle—auto-infection, as in Fig. 6.8b) (Combes et al. 1973; Tinsley 1990a). In the

Fig. 6.8 Life cycle of *Eupolystoma* (1) mature parasite in the urinary bladder with *intra-uterine* development; (2) embryonated eggs expelled inside the urinary bladder; (3) re-infection with oncomiracidium without cilia corresponding to an internal life cycle; (4) embryonated eggs expelled outside (5) ciliated oncomiracidium hatching expelled outside; (6) free larva; (7) oncomiracidium migration to the urinary bladder of the same or another host; (8) establishment of the oncomiracidium in the urinary bladder

case of external cycle, oncomiracidia infect new host via the cloaca. In some *Eupolystoma* species oncomiracidia migrate via the Mullerian ducts and the kidneys to the bladder (Du Preez, personal observations), while in other species complete larval development takes place solely in the bladder, where young parasites establish and reach maturity (Tinsley 1978). In the case of internal cycle, unciliated oncomiracidia do not migrate and reach maturity within the same host. The occurrence of auto-infection was confirmed by Du Preez et al. (2003), discovering both immature and mature forms of *Eupolystoma vanasi* Du Preez, Tinsley and De Sá, 2003 within the same host individuals. Auto-infection usually results in large numbers of specimens within the bladder of adult frogs. As many as 2000 individuals of *Eupolystoma anterorchis* Tinsley, 1978 were found in a single specimen of *Sclerophrys pantherina* (Smith 1828), the host species (Tinsley 1973). Large numbers of polystomes are accommodated by the huge and vascularized urinary bladders of toads. Species of *Diplorchis, Kankana* Raharivololoniaina, Verneau, Berthier, Vences and Du Preez, 2011, *Mesopolystoma* Vaucher, 1981, *Neodiplorchis, Parapolystoma* Ozaki, 1935, *Sundapolystoma* Lim and Du Preez, 2001 and *Wetapolystoma* that also have enlarged uteri, most likely also display *intra-uterine* development of eggs. However, their life cycle remains to be studied.

6.4.7 *Nanopolystoma* (Fig. 6.9)

Adult worms in the urinary bladder lay unembryonated eggs. Though it is not certain whether eggs are released from the host and whether oncomiracidia are ciliated or not, parasites of all age classes were observed in the urinary bladder of infected hosts (Du Preez et al. 2014) indicating that an internal cycle may occur.

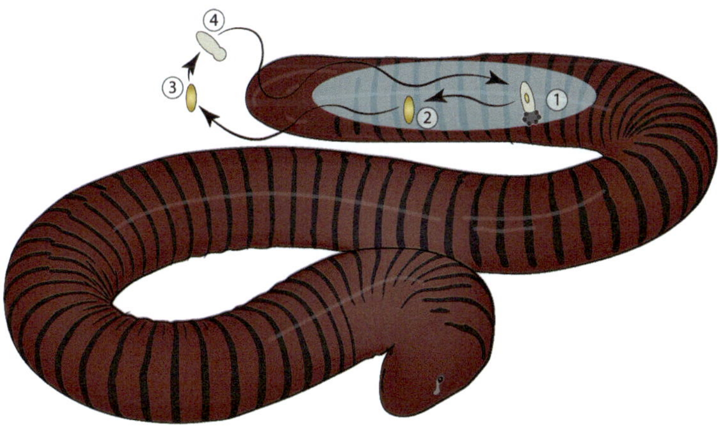

Fig. 6.9 Life cycle of *Nanopolystoma* (1) mature parasite in the urinary bladder or phallodeum; (2) eggs expelled; (3) oncomiracidium hatching; (4) oncomiracidium migration to the urinary bladder of another host

6.4.8 *Sphyranura* Wright, 1879 (Fig. 6.10)

Sphyranura spp. attach to the external gills of aquatic hosts, around the front limbs or inside the mouth of aquatic salamanders. Eggs are expelled as they are formed and sink to the bottom of the water body. Following embryonation of 28 days (Alvey 1936) oncomiracidia hatch from the operculated eggs. They have a life span of a few hours in which to locate a suitable host. When an oncomiracidium makes contact with a salamander it migrates to the gills where it attaches and develops to reach maturity in 2 months (Alvey 1936).

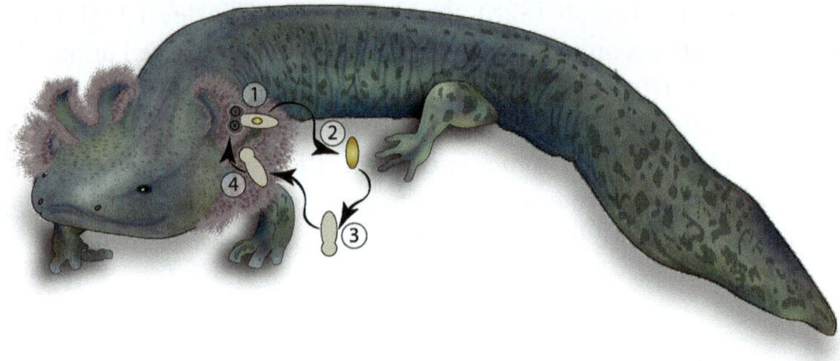

Fig. 6.10 Life cycle of *Sphyranura* (1) mature parasite on the gills; (2) eggs expelled; (3) oncomiracidium hatching; (4) oncomiracidium establishing on the gills of another host

6.4.9 Chelonian Polystomes of the Conjunctival Sacs: *Apaloneotrema, Aussietrema* Du Preez and Verneau, 2020 and *Fornixtrema* (Fig. 6.11a)

Polystomes, in the conjunctival sac of their host, release eggs at a slow rate, but, over a long period of time. These eggs wash off the eyes into the water body where they develop over a period of 20+ days. Oncomiracidia hatch from operculated eggs and actively search for a potential host within the water body. Once a suitable host is found, oncomiracidia establish on the eye and settle under eyelids or nictitating membranes in the fornix. Oncomiracidia develop into mature parasites and start producing eggs. Eggs are expelled almost continuously, due to the host's association with water.

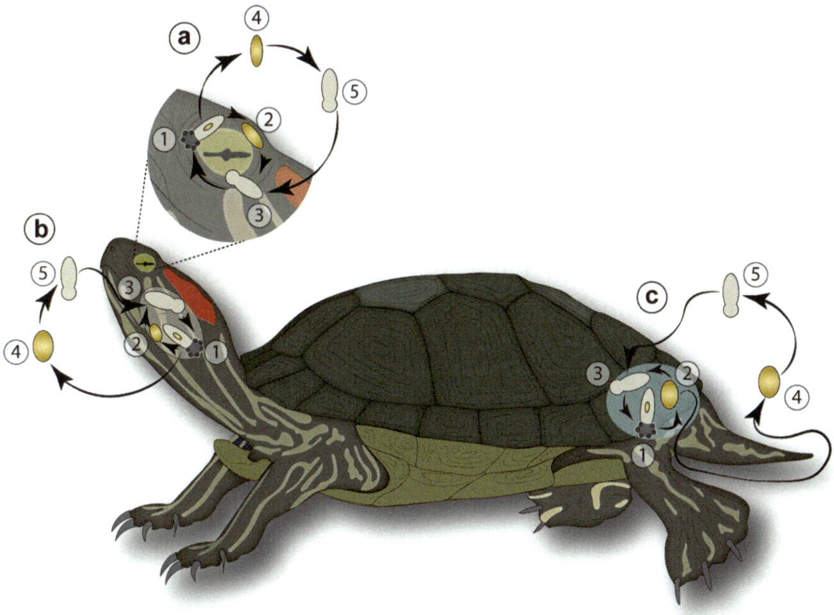

Fig. 6.11 Life cycles of chelonian polystomes (**a**) polystomes of the conjunctival sacs (1) mature parasites of *Apalonotrema, Aussietrema* and *Fornixtrema*; (2) eggs expelled and trapped on eye of the same host; (3) oncomiracidium hatching and developing; (4) eggs washed out of eyes; (5) oncomiracidium hatching and locating another host. (**b**) Polystomes of the oral region (1) mature parasites of certain species of *Pleurodirotrema*, certain species of *Polystomoides* and *Uteropolystomoides*; (2) eggs expelled and trapped in mouth of the same host; (3) oncomiracidium hatching and developing; (4) eggs washed out of mouth; (5) oncomiracidium hatching and locating another host. (**c**) Polystomes of the urinary bladder (1) mature parasites of *Manotrema,* certain species of *Pleurodirotrema, Polystomoidella,* certain species of *Polystomoides* and *Uropolystomoides*; (2) eggs expelled and trapped in the urinary bladder as in *Polystomoidella*; (3) oncomiracidium hatching and developing; (4) eggs expelled from the urinary bladder; (5) oncomiracidium hatching and locating another host

6.4.10 Chelonian Polystomes of the Urinary Bladder: Certain Species of *Pleurodirotrema* Du Preez, Domingues and Verneau, 2022, *Polystomoidella,* Certain Species of *Polystomoides* Ward, 1917 sensu Chaabane et al. (2022) and *Uropolystomoides* (Fig. 6.11b)

Oncomiracidia hatch from operculated eggs and actively search for a potential host within the water body. Once a suitable host is located oncomiracidia enter via the cloaca and establish in the urinary bladder. They develop into mature parasites and start producing eggs. Eggs are expelled almost continuously, due to the long host's association with water.

6.4.11 Chelonian Polystomes of the Oral Region: *Manotrema* Du Preez, Domingues and Verneau, 2022, Certain Species of *Pleurodirotrema* Du Preez, Domingues and Verneau, 2022, Certain Species of *Polystomoides* and *Uteropolystomoides* (Fig. 6.11c)

Oncomiracidia hatch from operculated eggs and actively search for a potential host within the water body. Once a suitable host is located oncomiracidia establish in the mouth and pharyngeal pouches. They develop into mature parasites and start producing eggs. Eggs are expelled almost continuously, without long delays, due to the long host's association with water.

Unlike all other chelonian polystomes infecting the oral cavity of their host, *Uteropolystomoides* has a uterus that holds up to 20 eggs. Because hosts of *Uteropolystomoides*, i.e. *Pseudemys* spp., live in fast-flowing rivers, polystome eggs that would be released midstream of the river would be swept away. Similarly, an oncomiracidium hatching from an egg settled on the bottom some 20+ days later would be swept away very quickly. We thus hypothesize that polystome eggs accumulate in the uterus and are released when turtles are basking. Turtles have their preferred basking sites on rocks, or fallen trees, along the banks of rivers. Eggs wash out of the turtle's mouth when it dives back into the water from its basking sites. They subsequently settle at the bottom among vegetation and develop over a period of 20+ days. Oncomiracidia, that hatch from operculated eggs, then infect turtles approaching the basking sites and establish in the buccal region. They develop into mature parasites and start producing eggs.

6.4.12 *Oculotrema* (Fig. 6.12)

Oculotrema hippopotami Stunkard, 1924 is found in clusters on the eye, under the eyelids and nictitating membrane in the fornix of the hippopotamus eye. Eggs expelled into the external environment hatch after approximately 20 days at 30 °C. However, developing eggs have been found within the mucous of their eyes, as well as different parasite life stages ranging from recently established to immature and mature worms. This suggests that auto-infection and re-infection are therefore both likely to occur. Because hippopotami tend to have their preferred locations in a water body where they bask for hours on end during the daytime, on the one hand, and because they are very social mammals with a close physical relationship with offspring on the other, oncomiracidia that hatch from operculated eggs can easily

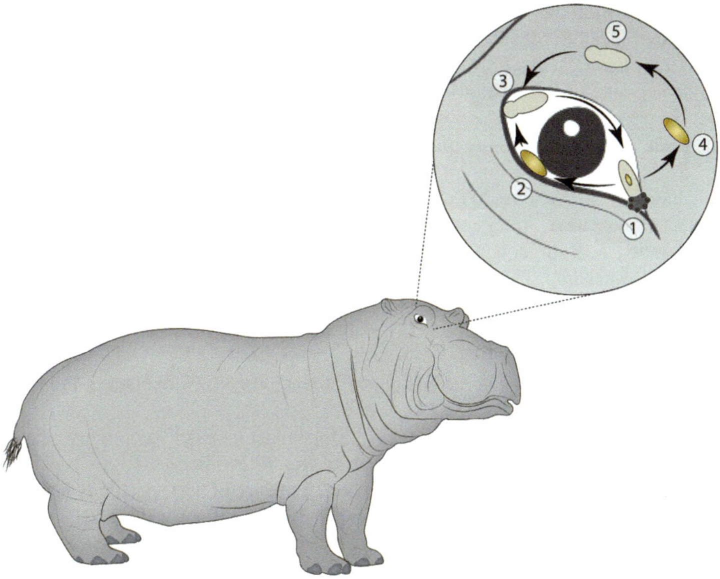

Fig. 6.12 Life cycle of *Oculotrema hippopotami* (1) mature parasites attached on eye or in the conjunctival sacs; (2) eggs expelled and trapped on eye of the same host; (3) oncomiracidium hatching and developing; (4) eggs washed out of; (5) oncomiracidium hatching and locating another host

infect new hosts or possibly the same host enhancing chances of cross infection. These hypotheses are supported by the high levels of *Oculotrema* infections (prevalence of >90%) reported in calves Thurston (1968), and adults (Thurston 1968; Du Preez and Moeng 2004).

References

Alvey CH (1936) The morphology and development of the monogenetic trematode *Sphyranura oligorchis* (Alvey 1933) and the description of *Sphyranura polyorchis* n. sp. Parasitology 28: 229–253

Buchmann K, Lindenstrøm T (2002) Interactions between monogenean parasites and their fish hosts. Int J Parasitol 32:309–319

Cable J, Tinsley RC (1992) Unique ultrastructural adaptations of *Pseudodiplorchis americanus* (Polystomatidae: Monogenea) to a sequence of hostile conditions following host infection. Parasitology 105:229–241

Chaabane A, Du Preez L, Johnston G, Verneau O (2022) Revision of the systematics of the Polystomoidinae (Platyhelminthes, Monogenea, Polystomatidae) with redefinition of *Polystomoides* Ward, 1917 and *Uteropolystomoides* Tinsley, 2017. Parasite 29:56. https://doi.org/10.1051/parasite/2022056

Chisholm LA, Whittington ID (1998) Morphology and development of the haptors among the Monocotylidae (Monogenea). Hydrobiologia 383:251–261

Combes C (1968) Biologie, écologie des cycles et biogéographie de digènes et monogènes d'Amphibiens dans l'est des Pyrénées. Mem Mus Natl Hist Nat. Paris. Nouv Ser A 51:1–195
Combes C (1972) Écologie des Polystomatidae (Monogenea): facteurs influencant le volume et le rythme de la ponte. Int J Parasitol 2:233–238
Combes C, Bourgat R, Salami-Cadoux ML (1973) Biology of the Polystomatidae: the direct internal cycle of *Eupolystoma alluaudi* (de Beauchamp, 1913) (author's transl). Z Parasitenkde 42:69–75
Combes C, Lambert A, Alméras MT (1975) La chétotaxie des larves nageantes de polystomes européens (Monogenea). Ann Parasitol Hum Comp 50:25–37
Combes C, Lambert A, Bourgat R, Salami-Cadoux ML (1978) L'oncomiracidium de *Eupolystoma alluaudi* (de Beauchamp, 1913): Monogène parasite de *Bufo regularis* Reuss en Afrique. Bull Mus Natl Hist Nat Zool 353:211–216
Cribb TH, Chisholm LA, Bray RA (2002) Diversity in the Monogenea and Digenea: does lifestyle matter? Int J Parasitol 32:321–328
Du Preez LH (1996) Morphology of the egg of *Polystoma australis* (Monogenea, Polystomatidae). Mic Soc SA 26:106
Du Preez LH, Kok DJ (1987) *Polystoma australis* (Monogenea): loss of locomotory cilia associated with retarded hatching of oncomiracidia. Parasitol Res 74:50–54
Du Preez LH, Kok DJ (1997) Supporting experimental evidence of host specificity among southern African polystomes (Polystomatidae: Monogenea). Parasitol Res 83:558–562
Du Preez LH, Kok DJ (1998) The relative importance of bladder versus neotenic stages of *Polystoma marmorati* and *P. umthakathi* in natural frog populations in South Africa. J Helminthol 72:117–121
Du Preez LH, Maritz MF (2006) Demonstrating morphometric protocols using polystome marginal hooklet measurements. Syst Parasitol 63:1–5
Du Preez LH, Moeng IA (2004) Additional morphological information on *Oculotrema hippopotami* Stunkard, 1924 (Monogenea: Polystomatidae) parasitic on the African hippopotamus. Afr Zool 39:225–233
Du Preez LH, Theunissen M (2021) A sucker for the job: morphology and functioning of suckers of polystomatid monogeneans. Folia Parasitol 68. https://doi.org/10.14411/fp.2021.006
Du Preez LH, Verneau O (2020) Eye to eye: classification of conjunctival sac polystomes (Monogenea: Polystomatidae) revisited with the description of three new genera *Apaloneotrema* n.g., *Aussietrema* n.g. and *Fornixtrema* n.g. Parasitol Res 119:4017–4031
Du Preez LH, Tinsley RC, De Sá R (2003) Polystomatidae (Monogenea) of Southern African Anura: *Eupolystoma vanasi* n. sp. parasitic in *Schismaderma carens* (Smith). Syst Parasitol 54: 71–79
Du Preez LH, Raharivololoniaina L, Verneau O, Vences M (2010) A new genus of polystomatid parasitic flatworm (Monogenea: Polystomatidae) without free-swimming life stage from the Malagasy poison frogs. Zootaxa 2722:54–68
Du Preez LH, Badets M, Verneau O (2014) Assessment of platyhelminth diversity within amphibians of French Guiana revealed a new species of *Nanopolystoma* (Monogenea: Polystomatidae) in the caecilian *Typhlonectes compressicauda*. Folia Parasitol 61:537–542. https://doi.org/10.14411/fp.2014.065
Fournier A, Combes C (1979) Démonstration d'une dualité évolutive des embryons chez *Eupolystoma alluaudi* [Monogenea, Polystomatidae] et de son rôle dans la genèse du cycle interne. C R Acad Sc (Paris) 289:745–747
Halton DW, Jennings JB (1965) Observations on the nutrition of monogenetic trematodes. Biol Bull 129:257–272
Jackson HC, Tinsley RC (1988) Environmental influences on egg production by the monogenean *Protopolystoma xenopodis*. Parasitology 97:115–128
Jackson JA, Tinsley RC (1998) Effects of temperature on oviposition rate in *Protopolystoma xenopodis* (Monogenea: Polystomatidae). Int J Parasitol 28:309–315
Kearn GC (1994) Evolutionary expansion of the Monogenea. Int J Parasitol 24:1227–1271

Kok DJ, Du Preez LH (1987) *Polystoma australis* (Monogenea): life-cycle studies in experimental and natural infections of normal and substitute hosts. J Zool (Lond) 212:235–243

Kok DJ, Du Preez LH (1989) *Polystoma australis* (Monogenea): development and reproduction in neotenic parasites. Afr Zool 24:225–230

Lambert A, Bourgat R (1978) L'oncomiracidium de *Metapolystoma brygoonis* (Euzet et Combes, 1964) parasite de l'amphibien Malgache *Ptychadena mascareniensis* (Duméril et Bibron). Ann Parasitol Hum Comp 53:547–549

Lambert A, Kulo SD (1982) Existence d'une dualité morphologique chez l'oncomiracidium de *Polystomoides nabedei* Kulo, 1980. Ann Parasitol Hum Comp 57:237–243

Lambert A, Combes C, Ktari MH (1978) Morphologie de l'oncomiracidium de *Polystomoides* Ward, 1917 (Monogenea) et situation du genre parmi les Polystomatidae. Z Parasitenkde 56: 175–181

Landman W, Verneau O, Du Preez LH (2018) First record of viviparity in polystomatid flatworms (Monogenea: Polystomatidae) with the description of two new species of *Madapolystoma* from the Madagascan anuran hosts *Blommersia domerguei* and *Mantella expectata*. Int J Parasitol Parasites Wildl 7:343–354. https://doi.org/10.1016/j.ijppaw.2018.09.004

Llewellyn J (1963) Larvae and larval development of monogeneans. Adv Parasitol 1:287–326

Lyons KM (1966) The chemical nature and evolutionary significance of monogenean attachment sclerites. Parasitology 56:100

Murith D, Vaucher C, Combes C (1977) Parasitologie – coexistence de la néoténie et du cycle interne chez un Polystomatidae (Monogenea). C R Acad Sc (Paris) 284:187–190

Odhner T (1912) Die Homologien der weiblichen Genitalwege bei den Trematoden und Cestoden. Nebst Bemerkungen zum natürlichen System der monogenen Trematoden. Zool Anz 39:337–351

Ozaki Y (1935) Studies on the frog-trematode, *Diplorchis ranae*. II. Morphology and behaviour of the swimming larva. J Sci Hiroshima Univ 4:23–34

Pichelin S (1995) The taxonomy and biology of the Polystomatidae (Monogenea) in Australian freshwater turtles (Chelidae, Pleurodira). J Nat Hist 29:1345–1381

Pichelin S, Whittington I, Pearson J (1991) *Concinnocotyla* (Monogenea: Polystomatidae), a new genus for the polystome from the Australian lungfish *Neoceratodus forsteri*. Syst Parasitol 18: 81–93

Poulin R (2011) Evolutionary ecology of parasites. Princeton University Press, Princeton, NJ

Raharivololoniaina L, Verneau O, Berthier P, Vences M, Du Preez LH (2011) First monogenean flatworm from a microhylid frog host: *Kankana*, a new polystome genus from Madagascar. Parasitol Int 60:465–473

Ramasamy P, Brennan GP (2000) Ultrastructure of the surface structures and haptor of *Empleurosoma pyriforme* (Ancyrocephalinae; Monopisthocotylea: Monogenea) from the gills of the teleost fish *Therapon jarbua*. Parasitol Res 86:129–139

Strelkov YA (1950) New species of monogenetic trematode of the far-east tortoise *Amyda sinensis*. Docklladi Akademii Nauk SSSR 74:159–162. (in Russian)

Thurston JP (1964) The morphology and life cycle of *Protopolystoma xenopi* (Price) Bychovsky in Uganda. Parasitology 54:441–450

Thurston JP (1968) The larva of *Oculotrema hippopotami* (Monogenea: Polystomatidae). J Zool (Lond) 154:475–480

Tinsley RC (1973) Observations on Polystomatidae (Monogenoidea) from East Africa with a description of *Polystoma makereri* n.sp. Z Parasitenkde 42:251–263

Tinsley RC (1978) The morphology and distribution of *Eupolystoma* species (Monogenoidea) in Africa, with a description of *E. anterorchis* sp. n. from *Bufo pardalis* at the Cape. J Helminthol 52:291–302

Tinsley RC (1983) Ovoviviparity in platyhelminth life cycles. Parasitology 86:161–196

Tinsley RC (1989) The effect of host sex on transmission success. Parasitology 5:190–195

Tinsley RC (1990a) Host behaviour and opportunism in parasite life cycles. In: Barnard CJ, Behmke JM (eds) Parasitism and host behaviour. Taylor and Francis, London, pp 158–192

Tinsley RC (1990b) The influence of parasite infection on mating success in spadefoot toads, *Scaphiopus couchii*. Am Zool 30:313–324

Tinsley RC (2004) Platyhelminth parasite reproduction: some general principles derived from monogeneans. Can J Zool 82:270–291

Tinsley RC (2013) The oncomiracidium of *Oculotrema hippopotami* Stunkard, 1924 and relationships within the Polystomatidae (Monogenea). Syst Parasitol 84:123–135

Tinsley RC, Jackson JA (2002) Host factors limiting monogenean infections: a case study. Int J Parasitol 32:353–366

Tinsley RC, Owen RW (1975) Studies on the biology of *Protopolystoma xenopodis* (Monogenoidea): the oncomiracidium and life cycle. Parasitology 71:445–463

Tinsley RC, Tinsley MC (2016) Tracing ancient evolutionary divergence in parasites. Parasitology 143:1902–1916

Whittington ID, Pichelin S (1991) Attachment of eggs by *Concinnocotyla australensis* (Monogenea: Polystomatidae) to the tooth plates of the Australian lungfish, *Neoceratodus forsteri* (Dipnoi). Int J Parasitol 21:341–346

Williams JB (1961) The dimorphism of *Polystoma integerrimum* (Frölich) Rudolphi and its bearing on relationships within the Polystomatidae: part III. J Helminthol 35:181–202

Wright K, Dechtiar A (1974) Light and scanning electron microscopy of attachment organs of three monogeneans (Monogenoidea: Polyopisthocotylea). Can J Zool 52:183–187

Chapter 7
Insights into the Origin and Evolution of the Polystomatidae

Contents

Abstract Phylogenetic relationships within polystomatids have been extensively studied for the past 20 years by means of several molecular markers, including 18S and 28S rRNA slow-evolving genes to explore deepest evolutionary nodes, and 12S rRNA, COI and ITS1 fast-evolving genes and intergene to investigate recent ones. In this chapter, we underline all the main results that were published on the evolution of these platyhelminthes, among others those which concern the origin and evolution of the earliest polystomatids, the chelonian polystomes and *Oculotrema hippopotami* Stunkard, 1924 which is the single polystome species infecting a mammal. Emphasis is also placed on the modes of diversification of amphibian polystomes, on the phylogeography of *Polystoma* Zeder, 1800, which is the most widespread genus and on the origin and evolution of polystomes infecting Madagascan frogs. Finally, we discussed very recent polystome dispersals following host switching resulting from the international turtle pet trade.

7.1 Origin of the Earliest Polystomatids

In line with all other monogeneans, polystomatids display a direct life cycle which facilitates parasite transmission from host to host in aquatic environments. They are distributed worldwide and are usually host and site-specific. Amphibian polystomes

L. H. du Preez et al., *Polystomatid Flatworms*, Zoological Monographs 9,
https://doi.org/10.1007/978-3-031-35887-6_7

are mainly found in the urinary bladder of their host, whereas chelonian polystomes are located either in the urinary bladder and cloaca, the oral and pharyngeal cavity or the conjunctival sacs. Because of all these peculiarities and the tight relationship which exists between the Polystomatidae and their sarcopterygian hosts, namely lungfishes and semi- or fully-aquatic tetrapods adapted to freshwater environments, it was first suggested by Claude Combes in the early 1990s that the family could have evolved in ancient geological times during the ecological transition from aquatic to terrestrial environments. This could have happened when actinopterygians and sarcopterygians diverged at about 425 Million years (My) based on their reported palaeontological evidence, as suggested by Janvier (1998) and Ahlberg (1999). Therefore, the phylogenetic relationships between and within the main genera of polystomatids were investigated from the analysis of several molecular markers, i.e. partial 18S rRNA gene sequences (Sinnappah et al. 2001; Verneau et al. 2002), ITS1 sequences (Bentz et al. 2001, 2006), combined complete 18S and partial 28S rRNA gene sequences (Verneau et al. 2009) and partial mitochondrial 12S and COI sequences combined with nuclear complete 18S and partial 28S rRNA gene sequences (Héritier et al. 2015), in order to explore patterns and processes of polystomatids' evolution.

In the most comprehensive phylogeny, which included 55 polystomatids belonging to 19 genera, Héritier et al. (2015) showed, from the analysis of a molecular data set covering sequences of four nuclear and mitochondrial genes, a basal position of *Concinnocotyla australensis* (Reichenbach-Klinke, 1966), which also occupied an intermediate position between monogeneans of fishes and tetrapods. This result indicated an ancient origin of the Polystomatidae by the Late Silurian–Early Devonian periods, as previously suggested from the analysis of partial 18S rRNA gene sequences (Verneau et al. 2002). Two major monophyletic groups were also evidenced. The first includes all polystomes infesting hosts of the Batrachia. The second includes parasites infesting mainly hosts of the Testudines, *Nanopolystoma tinsleyi* Du Preez, Badets and Verneau, 2014 infesting a caecilian host, i.e. *Typhlonectes compressicauda* (Duméril and Bibron, 1841) and *Oculotrema hippopotami* Stunkard, 1924 infesting the common hippopotamus *Hippopotamus amphibius* Linnaeus, 1758. The overall topology of Héritier et al. (2015), which was consistent with the phylogenetic tree depicted by Verneau et al. (2002), suggested co-evolution between polystomatids and Lissamphibian hosts since their origin. *Sphyranura* Wright, 1849 and *Protopolystoma* Bychowsky, 1957, which infect urodeles and pipids, respectively, were actually the most basal genera within the clade, including strictly amphibian parasites, while *N. tinsleyi* was the most basal taxon within the second clade. Furthermore, divergence-time estimates, inferred from molecular calibrations, indicated a split between these two lineages at about 334 My, suggesting that polystomes infecting tetrapods may have originated in the ancestor of the Lissamphibia. The age estimate for this node was indeed in line with palaeontological evidence of the hosts (Anderson et al. 2008) and to molecular time estimates of the basal divergence in the Lissamphibia. According to Héritier et al. (2015), polystomatids would have, subsequently, co-evolved within anurans on one side—assuming host switching from anurans to urodeles along the evolution of amphibians—and within caecilians on the other, considering that primitive amniotes were commonly adapted to terrestrial environments.

7.2 Evolution of Chelonian Polystomatids

Verneau et al. (2002) postulated that chelonian polystomes would have originated following a switch from, presumably extinct, primitive amniotes to freshwater chelonians. This hypothesis held until the recent discovery of a new genus *Nanopolystoma* Du Preez, Wilkinson and Huyse, 2008 within caecilians (Du Preez et al. 2008) and the placement of *N. tinsleyi* in a phylogenetic tree that encompassed a great diversity of polystomatids (Héritier et al. 2015). Considering that primitive polystomes may have co-evolved within amphibians, the most plausible scenario was to consider a switch from caecilians to the ancestor of pleurodires and cryptodires at about 178 My (Héritier et al. 2015). Assuming that polystomes infecting hippopotami would have arisen from caecilian hosts (see below), another possibility was to consider a switch from caecilians to freshwater turtles more recently, i.e. at about 152 My according to Héritier et al. (2015). Regardless of the time window that should be considered for the origin of chelonian polystomatids, it is still, however, very difficult to explain the discordance between parasite phylogenetic patterns and host relationships and *in fine* evolution of chelonian polystomes. The three main lineages that were evidenced within chelonian polystomes might have arisen following a combination of numerous non-exclusive factors such as the evolution of plate tectonics, processes of co-divergences, dispersals through host switching and intra-host duplications. Though Du Preez and Verneau (2020) extended the phylogeny of chelonian polystomes from 20 to 28 taxa, it is still extremely difficult to explain which factors were involved in the speciation and diversification processes of the six apparent and main lineages.

7.3 Origin and Evolution of *Oculotrema hippopotami*

The origin of *O. hippopotami* has been for a long time a mystery as this species is the sole polystome known from a mammalian host (Stunkard 1924). Because all polystomatids have an obligatory infective aquatic larval stage along their direct life cycle, the occurrence of this enigmatic species on *H. amphibius* was probably due to the ecological preferences of its host for freshwater aquatic environments. Regarding morphological and biological similarities between *O. hippopotami* and chelonian polystomes (Du Preez and Moeng 2004; Tinsley 2013), it was hypothesized that *O. hippopotami* may have diverged from representatives of chelonian polystomes (Du Preez and Moeng 2004), more specifically from *Polystomoides* species infecting pharyngeal cavities (Tinsley 2013). According to the phylogenetic tree of Héritier et al. (2015), *O. hippopotami* showed a sister group relationship to a lineage associating all chelonian polystomes with *N. tinsleyi* being rejected at the base of that clade. Considering molecular calibrations and the phylogenetic position of *O. hippopotami*, several hypotheses were put forward by Héritier et al. (2015) as its origin: (1) it would have diverged from the stem branch of chelonian polystomatids, at about 152 My, following a switch from ancestral turtles to either hippopotami or early mammals adapted to aquatic environments. These two possibilities were, however, ruled out regarding the Miocene origin of hippopotamids (see Orliac et al. 2010; Pickford 2011), on the one hand, and the absence of polystomes within extant mammals with the exception of *H. amphibius*, on the other; (2) it would have isolated from not yet identified or extinct chelonian polystomes. If the first of these two possibilities was dismissed owing to the diversity of polystomes used to reconstruct the phylogeny, the second hypothesis was difficult to validate due to the scarcity of the parasite fossil records (see De Baets et al. 2015); (3) it would have diverged from a polystome lineage occurring in another extant African caecilian lineage following host switching. Because all polystome species that are currently recorded within *Nanopolystoma* were described exclusively from South American hosts in the late 2000s and early 2010s (Du Preez et al. 2008, 2014), many more *Nanopolystoma* species may be expected to be discovered within caecilian hosts. This is why Héritier et al. (2015) considered that the last hypothesis was the most plausible. They, therefore, suggested that those polystomes may be found in African caecilian families in Central and East Africa—firstly because they have never been surveyed and secondly because hippopotami also occur in these areas. Following an extensive polystome survey that was conducted in 2019 across freshwater chelonians in the USA, Verneau et al. (2023) showed a close phylogenetic relationship between *O. hippopotami* and *Apaloneotrema moleri* (Du Preez and Morrison, 2012), which was recovered from the Florida Softshell turtle *Apalone ferox* (Schneider, 1783). This association suggested that the plesiomorphic host for this parasite would have actually been a turtle rather than a caecilian. Because *A. ferox* only occurs in Alabama, Florida, Georgia and South Carolina in the USA (Rhodin et al. 2021), assuming a switch from that turtle to *H. amphibius* is impossible. According to the distribution and evolution of trionychid turtles, Li et al. (2017) showed that softshell turtles would have originated in Asia at about 108 My and then dispersed to Africa, Insular South-East Asia and North America in the Late Cretaceous–Early Eocene. Therefore, the ancestral turtle from which polystomes could have switched to *H. amphibius* may be found among the five extant species of the African trionychid, namely *Cyclanorbis elegans* (Gray, 1869), *Cyclanorbis senegalensis* (Duméril and Bibron, 1835), *Cycloderma aubryi* (Duméril, 1856), *Cycloderma frenatum* Peters, 1854 and/or *Trionyx triunguis* (Forskål, 1775) (see Verneau et al. 2023). Regardless of who the plesiomorphic host for *O. hippopotami* is, host switching would have occurred in recent times since it is no more than 21 My hippopotamids may have been adapted to semi-aquatic habitats (see Boisserie et al. 2011). The proximity, between hippopotami and freshwater turtles, which both occur in the same aquatic environments, could have thus enhanced the chances of parasite transmission.

7.4 Diversification Patterns Within Amphibian Polystomatids

Even though numerous questions concerning the evolutionary processes driving the evolution of chelonian polystomes still remain unanswered, the modes of diversification within amphibian polystomes are currently better understood than they previously were. Geographical distributions and species evolution provide a critical framework to infer the historical biogeography of organisms. Besides these, the host phylogeny actually offers a powerful timeframe to inspect parasite evolution through geological times, on the one hand, and co-evolution patterns which give valuable information to assess past climatic and environmental factors that have shaped host–parasite associations, on the other. Given the worldwide distribution of polystomes of the Neobatrachia, Badets et al. (2011) aimed to investigate whether evolution of this unique group of polystomes was correlated to the early evolution of their host species as well as to plate tectonics in the Early Jurassic. According to the phylogeny of Badets et al. (2011), which was inferred from the analysis of 18S and partial 28S rRNA gene sequences, four polystome lineages were revealed whose distributions were limited to specific geographical areas. The first lineage encompassed species of two genera, i.e., *Diplorchis* Ozaki, 1931 and *Parapolystoma* Ozaki, 1935, which were sampled from Japanese ranids and Australian hylids, respectively; the second lineage clustered only polystomes recovered from Asian *Rhacophorus* Kuhl and Van Hasselt, 1822, frog species of India, China and Japan. If these polystomes were initially considered as part of *Polystoma* Zeder, 1800, they are now recognized as representatives of *Indopolystoma* Chaabane, Verneau and Du Preez, 2019 (Chaabane et al. 2019); the third lineage was also composed of species of a single genus, i.e., *Eupolystoma* (de Beauchamp, 1913) that were sampled from African bufonids; the fourth lineage associated *Wetapolystoma* Gray, 1993 and species of the most diversified genus, namely *Polystoma*, all of them being collected from bufonids and ranids of South America, North America, Europe and Africa. Since the phylogenetic relationships of these four polystome lineages reflected sequential processes of plate tectonics, Badets et al. (2011) explored, in greater depth, the links between evolution of the main lineages and the Earth's history at the time of the Gondwana breakup. They relied on two vicariant biogeographical methodologies, i.e., TreeMap (Charleston and Page 2002) and DIVA for Dispersal-Vicariance Analysis (Ronquist 1997), based, respectively, on a priori and a posteriori assumptions of plate tectonics relationships. Both approaches gave similar results. Indeed, they suggested that polystomes of the Neobatrachia would have originated in the Gondwana, according to plate tectonics and reticulate relationships between areas. Vicariance biogeography would have played a major role in the earliest evolution of this group of parasites and, following subsequent continental drifting and colliding, dispersal between terrestrial blocks could have been possible allowing subsequent speciation and diversification. The most basal polystome lineage could have been isolated in the Australian block at a geological period where India was still bound to the western part of Gondwana. Sooner or later, the second polystome lineage could have been isolated in India, following the separation of the western part of Gondwana to India. Finally, following the divergence of South American and African plates, the last two lineages could have been isolated in Africa and South America, respectively. Whether all these biogeographic vicariance events occurred well within the divergence-time estimates that were proposed for polystome divergences, dispersal events that would also have occurred from Australia to Eurasia, from India to Eurasia and from South America to Eurasia, would have shaped polystome evolution.

Similarly, Badets et al. (2011) conducted co-phylogenetic analyses in this host–parasite association to investigate processes of polystome diversification as well as to identify ancestral host lineages. The parasite phylogeny that was generated from the analysis of 18S and 28S nuclear markers was compared to a gene tree of host species that was inferred from the analysis of combined nuclear and mitochondrial markers. Host and parasite phylogenies were then compared following the TreeMap procedure that was used for investigating the historical biogeography of polystomes. The most significant reconstructions found, i.e., those with the minimal total cost assuming that co-divergence was less costly than any other processes, always implied three to four host-switching events, as well as a great number of duplication and extinction events. Each of the above scenarios was equally unlikely with regard to the biology of polystomes, the direction of host switching and the supposed routes of polystome dispersal from South America to Africa (see Bentz et al. 2006). Duplication events implied the divergence of parasite lineages in the same host species. This had never been illustrated within frogs. Similarly, some of the host switchings that were highlighted implied that extant African *Polystoma* species would have originated in ancestral ranoids on the Gondwana, which was in contradiction to a more recent origin suggested by Bentz et al. (2001). Consequently, the most plausible hypothesis to explain the incongruence between co-phylogenetic and biogeographic patterns, considered that co-divergence events within polystomes may have been hidden by subsequent extinctions of ancestral parasites and/or host lineages. Studying this host–parasite association can therefore provide interesting results about the evolutionary ecology of frogs, as the presence of some parasites in specific clades and in discrete geographic areas, would indicate the occurrence and radiation of amphibians over geological times, especially when the fossil record is missing.

7.5 Phylogeography of *Polystoma*, the Most Diverse Genus

According to the geographical distribution of the most diverse genus in the Polystomatidae, namely *Polystoma,* which has been reported everywhere in the world, with the exception of Australia, it was originally postulated that its present-day distribution could be the consequence of mid-Cretaceous geological events. In order to investigate the processes of *Polystoma* evolution and diversification, a phylogeny was reconstructed from the analysis of complete ITS1 and 18S sequences (Bentz et al. 2006). It showed the paraphyly of American *Polystoma*, on the one hand, and a nested position of the Eurafrican clade within South American *Polystoma* on the other. The results indicated that the genus might have originated in South America and then invaded new areas, following successive waves of colonization. According to the historical biogeography of frogs that was interpreted regarding mid-Cretaceous vicariance events—following the split between South American and African plates during the fragmentation of Gondwana—*Polystoma* would have originated within South American hylids after they had been isolated. Vertebrate exchanges between South America and North America, which would have included bufonids, were considered possible in the Palaeocene. Thus, the present-day distribution of *Polystoma* in North America and Eurasia, could be explained by mid-Cenozoic dispersal of bufonids or hylids from South America. Regardless of the ancestral host that would have carried along polystomes to Eurasia via the land connection Beringia, polystomes would have secondarily switched to European ranoids and pelobatid species. From a phylogeny based on the ITS1 marker, it was also suggested that there was a single event of colonization to Africa regarding the paraphyly of European *Polystoma* species with respect to the African clade (Bentz et al. 2001). Dispersal to that continent would have involved the crossing of the Mediterranean Sea during the late Miocene, when Spain and Morocco were connected for a brief period. If pelobatids were suspected of being the ancestral hosts that would have served as carriers for dispersing European polystomes across African frogs, the African polystome diversification could have been related to the *Ptychadena* Boulenger, 1917 radiation following parallel evolution between hosts and their parasites (Bentz et al. 2001). Compared with the polystome richness of Eurasia, North and South America, Africa has the highest diversity of *Polystoma* species, *Ptychadena* being the most suitable host for polystomes (Du Preez and Kok 1992). Regardless of the complex process that was involved in the polystome diversification within neobatrachian hosts, Africa would have definitely been the last continent to be invaded by *Polystoma* (Bentz et al. 2001).

7.6 Origin and Evolution of Amphibian Polystomes of Madagascar

Though the first record of polystomes in Madagascar dates back to 1964, with the report of *Metapolystoma brygoonis* (Euzet and Combes, 1964) from the host *Ptychadena mascareniensis*[1] (Duméril and Bibron, 1841), it is only recently that a spectacular polystome diversity has been evidenced following intensive field surveys conducted in suitable habitats for frogs (Verneau et al. 2009; Du Preez et al. 2010; Raharivololoniaina et al. 2011; Berthier et al. 2014; Landman et al. 2018,

[1] *Ptychadena mascareniensis* is considered today a complex species including two species and ten operational taxonomic units (Zimkus et al. 2017).

2021, 2023). Because Madagascar is characterized by a very high species diversity, pronounced endemism and threats to the fauna and flora, it was identified as a hotspot of biodiversity (Myers et al. 2000). Among the most diverse groups of animals on the island, frogs have an exceptional richness with just over 400 species (Frost 2023). In spite of the taxonomy and classification of this group being extensively studied during the past 20 years (Bossuyt and Milinkovitch 2001; Vences et al. 2003; Andreone et al. 2004; Van der Meijden et al. 2004, 2005, 2007; Glaw et al. 2006; Wollenberg et al. 2007, 2008; Kurabayashi et al. 2008), several issues were still in question—among others, the biogeographic origins of some families like the Mantellidae. It was, therefore, extremely interesting to investigate the phylogenetic relationships of their polystomes in order to track both host and parasite occurrences over geological times. Interestingly, the phylogeny that was reconstructed from 18S and 28S rRNA gene sequences indicated the presence of two independent polystome clades, *Metapolystoma* Combes, 1976 and *Madapolystoma* Du Preez, Raharivololoniaina, Verneau and Vences, 2010, suggesting a double origin for Madagascan polystomes (Verneau et al. 2009). According to the sister group relationship between *Madapolystoma* and African *Eupolystoma*, on the one hand, and molecular dating estimates for the crown group of *Madapolystoma* on the other (which ranged from 135 to 40 My), several scenarios were outlined: (1) the ancestor of *Madapolystoma* would have colonized the island with ancestral mantellids—however, this scenario required that *Madapolystoma* be related to rhacophorid polystomes, which was not the case; (2) ancestors of *Madapolystoma* would have been carried to Madagascar along with hyloid frogs and would have then switched to mantellids—however, despite the occurrence of hyloid fossils in Madagascar, they had no affinities with bufonid frogs which might have been the ancestral host for *Madapolystoma*; (3) ancestors of *Madapolystoma* would have been introduced with microhylids that colonized Madagascar with mantellids. This hypothesis was also discarded by Verneau et al. (2009) owing to the absence of polystomes within microhylids at the time of publication, but it is felt that renewed interest should be shown in the investigation of the historical biogeography of *Kankana* Verneau, Berthier, Vences and Du Preez, 2011 following the discovery of *Kankana manampoka* Raharivololoniaina, Verneau, Berthier, Vences and Du Preez, 2011, a species that was further described from *Cophyla pollicaris* (Boulenger, 1888) in Madagascar. Indeed, this species is more closely related to *Madapolystoma* than *Madapolystoma* to *Eupolystoma* (Raharivololoniaina et al. 2011).

The second clade, including Madagascan polystomes, i.e., *Metapolystoma*, was found nested within the genus *Polystoma*, with strong affinities to African polystomes. It was therefore suggested, according to the time window for the colonization of Madagascar by *Metapolystoma*, i.e., 14.2–2.3 My, that *Ptychadena* host species would have carried polystomes to the island via oceanic dispersals. Those parasites would have secondarily switched to mantellid frogs that had already diversified. Though it is generally impossible to interpret the evolutionary history of parasites without a thorough investigation of the evolution of their hosts, parasites, like polystomes, may provide additional data to support key events in the host evolution, for instance transoceanic colonization of Madagascar by some frogs of the *Ptychadena* complex (Vences et al. 2004; Measey et al. 2007).

7.7 Polystome Host Switching Following Turtle Pet Trade

For a long time, polystomes have been considered host- and site-specific, however, it can be seen in the literature that some species were reported to occur in multiple host species. *Polystomoides orbicularis* (Stunkard, 1916), for instance, was originally described from the urinary bladder of *Chrysemys picta* (Schneider, 1783) and *Trachemys scripta* (Thunberg, 1792) and later on reported from *T. decussata* (Gray, 1831). *Polystomoidella whartoni* (Wright, 1879) was reported from the urinary bladder of several hosts like *Sternotherus odoratus* (Latreille, 1802), *Sternotherus carinatus* (Gray, 1856), *C. picta*, *Chelydra serpentina* (Linnaeus, 1758) and *Kinosternon subrubrum* Bonnaterre, 1789. *Polystomoides coronatus* (Leidy, 1788) was described from the oral cavity of an unknown host species but later reported from the oral cavity of *Actinemys marmorata* (Baird and Girad, 1852), *C. picta*, *T. s. elegans* (Wied, 1838) and *Apalone spinifera* (Lesueur, 1827), from the urinary bladder of *T. s. elegans* and from the conjunctival sacs of *A. ferox* and *Graptemys pseudogeographica* (Gray, 1831).

Even if one questions the validity of certain descriptions and, therefore, assumes misidentifications of species in some cases, one can also consider that polystome species are able to adapt to new environments. As there are more and more turtle farms, for the pet trade and for turtles exhibition all over the world, one wonders whether polystomes are able to switch from host to host in artificial environments. We, therefore, investigated the polystome diversity within several introduced American turtle species—*A. spinifera*, *C. picta*, *G. pseudogeographica* and *T. scripta,* as well as in two native species—the European pond turtle *Emys orbicularis* (Linnaeus, 1758) and the Mediterranean pond turtle *Mauremys leprosa* (Schweigger, 1812) (Verneau et al. 2011) at a turtle farm in France. After collecting polystome eggs from infected turtles, we used a DNA barcoding approach to identify the polystome species diversity. COI distance analyses combined with phylogenetic relationships of the polystomes sampled in the turtle farm and from several wild turtles, collected long before in their natural environments in Africa, Australia, Asia and the Americas, showed the occurrence of several polystome species in the two endemic turtle species. Using molecular species delimitation inferred from polystomes (Verneau et al. 2011), we showed that most of them were previously identified from American turtles, thus illustrating multiple cases of host switching between American turtles to *E. orbicularis* and *M. leprosa* in captivity, but also between captive American turtles. In the Verneau et al. (2011) study it was demonstrated that at least some species of American polystomes exhibited low host-specificity when offered new opportunities for host switching, particularly when the turtles were sharing the same aquatic habitat.

The American red-eared slider, *T. s. elegans*, was the most popular turtle pet in the second half of the twentieth century. As a consequence, in the 1980s and 1990s, this species was exported worldwide, especially to Asian and European markets. This has resulted in it being extensively released in suitable environments well outside its natural range, where it has established feral populations and has become the most widely invasive reptile species in the world (Kraus 2009). As a result of Verneau et al. (2011) showing that polystomes of *T. s. elegans* were able to switch to *E. orbicularis* and *M. leprosa* in artificial ponds, one may wonder to what extent host switching could have take place between the invasive freshwater turtle species and the two native species, i.e. *M. leprosa* and *E. orbicularis*, in wild habitats of northern Spain and southern France (Meyer et al. 2015; Héritier et al. 2017). The polystome diversity was, therefore, assessed within both turtle species based on phylogenetic analyses of all samples collected in the wild and on the COI threshold for species delimitation determined in Verneau et al. (2011) and refined later in Héritier et al. (2017). Besides the two native polystome species that were recovered from *M. leprosa*, namely *Polystomoides tunisiensis* Gonzales and Mishra, 1977 in the oral cavity and *Polystomoides euzeti* (Combes and Ktari, 1976) in the urinary bladder, six non-native polystome species were also recorded from *M. leprosa*, among which were the two American polystome *P. orbicularis* in the urinary bladder and *Polystomoides oris* Paul, 1938 in the oral cavity of *C. picta* (Meyer et al. 2015). Because *T. scripta* and *M. leprosa* were reported to occur in the same aquatic habitats as *P. orbicularis*, it clearly indicated host switching in the wild. *Trachemys scripta elegans* would have served as a carrier for *P. orbicularis* and would have secondarily transmitted that polystome species to native turtles in natural environments. On the contrary, because *T. scripta* was not reported in the natural environments where *P. oris* was found in *M. leprosa*, it was assumed that already infected specimens of *M. leprosa* could have been released into the wild. The study that was carried out with *E. orbicularis* in wild environments gave very similar results. Besides the native polystome species that was recovered from *E. orbicularis*, namely *Polystomoides ocellatus* (Rudolphi, 1819) in the oral cavity, at least eight non-native polystome species were also recorded from *E. orbicularis* (Héritier et al. 2017), among which were *P. orbicularis* and *P. oris* that were also reported by Meyer et al. (2015) in *M. leprosa*. Regarding the polystome diversity of *E. orbicularis*, on the one hand, and the distribution of *T. scripta* in the wild on the other, the same conclusions were reached. If some host switching could have occurred directly in the wild from *T. scripta* to *E. orbicularisin* some places, the release of some already infected turtles in the wild could not be excluded in some other places.

References

Ahlberg PE (1999) Palaeontology: something fishy in the family tree. Nature 397:564–565. https://doi.org/10.1038/17484

Anderson JS, Reisz RR, Scott D, Fröbisch NB, Sumida SS (2008) A stem batrachian from the early Permian of Texas and the origin of frogs and salamanders. Nature 453:515–518. https://doi.org/10.1038/nature06865

Andreone F, Vences M, Vieites DR, Glaw F, Meyer A (2004) Recurrent ecological adaptations revealed through a molecular analysis of the secretive cophyline frogs of Madagascar. Mol Phylogenet Evol 34:315–322. https://doi.org/10.1016/j.ympev.2004.10.013

Badets M, Whittington I, Lalubin F, Allienne JF, Maspimby JL, Bentz S, Du Preez LH, Barton D, Hasegawa H, Tandon V, Imkongwapang R, Ohler A, Combes C, Verneau O (2011) Correlating early evolution of parasitic platyhelminths to Gondwana breakup. Syst Biol 60:762–781. https://doi.org/10.1093/sysbio/syr078

Bentz S, Leroy S, Du Preez L, Mariaux J, Vaucher C, Verneau O (2001) Origin and evolution of African *Polystoma* (Monogenea: Polystomatidae) assessed by molecular methods. Int J Parasitol 31:697–705. https://doi.org/10.1016/S0020-7519(01)00179-5

Bentz S, Sinnappah-Kang ND, Lim L-HS, Lebedev B, Combes C, Verneau O (2006) Historical biogeography of amphibian parasites, genus *Polystoma* (Monogenea: Polystomatidae). J Biogeogr 33:742–749. https://doi.org/10.1111/j.1365-2699.2005.01402.x

Berthier P, Du Preez LH, Raharivololoniaina L, Vences M, Verneau O (2014) Two new species of polystomes (Monogenea: Polystomatidae) from the anuran host *Guibemantis liber*. Parasitology 63:108–119. https://doi.org/10.1016/j.parint.2013.09.014

Boisserie J-R, Fisher RE, Lihoreau F, Weston EM (2011) Evolving between land and water: key questions on the emergence and history of the Hippopotamidae (Hippopotamoidea, Cetancodonta, Cetartiodactyla). Biol Rev 86:601–625. https://doi.org/10.1111/j.1469-185X.2010.00162.x

Bossuyt F, Milinkovitch MC (2001) Amphibians as indicators of early tertiary 'out-of-India' dispersal of vertebrates. Science 292:93–95. https://doi.org/10.1126/science.1058875

Chaabane A, Verneau O, Du Preez L (2019) *Indopolystoma* n. gen. (Monogenea, Polystomatidae) with the description of three new species and reassignment of eight known *Polystoma* species from Asian frogs (Anura, Rhacophoridae). Parasite 26:67. https://doi.org/10.1051/parasite/2019067

Charleston MA, Page RDM (2002) TreeMap 2.0β. A Macintosh R program for the analysis of how dependent phylogenies are related, by cophylogeny mapping. University of Sydney, Faculty of Engineering and IT. http://sydney.edu.au/engineering/it/_mcharles/software/treemap/treemap.html

De Baets K, Dentzien-Dias P, Upeniece I, Verneau O, Donoghue PCJ (2015) Constraining the deep origin of parasitic flatworms and host-interactions with fossil evidence. In: De Baets K, Littlewood DTJ (eds) Fossil parasites, vol 90. Elsevier, pp 93–135. ISBN: 9780128040010

Du Preez LH, Kok DJ (1992) The frog genus *Ptychadena* as host for polystomatid (Monogenea) parasites in Africa. Afr J Herpetol 40:47–49

Du Preez LH, Moeng IA (2004) Additional morphological information on *Oculotrema hippopotami* Stunkard, 1924 (Monogenea: Polystomatidae) parasitic on the African hippopotamus. Afr Zool 39:225–233. https://doi.org/10.1080/15627020.2004.11657219

Du Preez LH, Verneau O (2020) Eye to eye: classification of conjunctival sac polystomes (Monogenea: Polystomatidae) revisited with the description of three new genera *Apaloneotrema* n. g., *Aussietrema* n. g. and *Fornixtrema* n. g. Parasitol Res 119:4017–4031. https://doi.org/10.1007/s00436-020-06888-w

Du Preez LH, Wilkinson M, Huyse T (2008) The first record of polystomes (Monogenea: Polystomatidae) from caecilian hosts (Amphibia: Gymnophiona), with the description of a new genus and two new species. Syst Parasitol 69:201–209. https://doi.org/10.1007/s11230-007-9120-8

Du Preez LH, Raharivololoniaina L, Verneau O, Vences M (2010) A new genus of polystomatid parasitic flatworm (Monogenea: Polystomatidae) without free-swimming life stage from the Malagasy poison frogs. Zootaxa 2722:54–68

Du Preez LH, Badets M, Verneau O (2014) Assessment of platyhelminth diversity within amphibians of French Guiana revealed a new species of *Nanopolystoma* (Monogenea: Polystomatidae) in the caecilian *Typhlonectes compressicauda*. Folia Parasitol 61:537–542. https://doi.org/10.14411/fp.2014.065

Frost DR (2023) Amphibian species of the world: an online reference. Version 6.1 (22 March 2023). American Museum of Natural History, New York, USA. https://amphibiansoftheworld.amnh.org/index.php. https://doi.org/10.5531/db.vz.0001

Glaw F, Hoegg S, Vences M (2006) Discovery of a new basal relict lineage of Madagascan frogs and its implications for mantellid evolution. Zootaxa 1334:27–43

Héritier L, Badets M, Du Preez LH, Aisien MS, Lixian F, Combes C, Verneau O (2015) Evolutionary processes involved in the diversification of chelonian and mammal polystomatid parasites (Platyhelminthes, Monogenea, Polystomatidae) revealed by palaeoecology of their hosts. Mol Phylogenet Evol 92:1–10. https://doi.org/10.1016/j.ympev.2015.05.026

Héritier L, Valdeón A, Sadaoui A, Gendre T, Ficheux S, Bouamer S, Kechemir-Issad N, Palacios C, Du Preez LH, Verneau O (2017) Introduction and invasion of the red-eared slider and its parasites in freshwater ecosystems of Southern Europe: risk assessment for the European pond turtle in wild environments. Biodivers Conserv 26:817–1843. https://doi.org/10.1007/s10531-017-1331-y

Janvier P (1998) Forerunners of four legs. Nature 395:748–749. https://doi.org/10.1038/27332

Kraus F (2009) Alien reptiles and amphibians: a scientific compendium and analysis. Springer, Dordrecht. 576 pp

Kurabayashi A, Sumida M, Yonekawa H (2008) Phylogeny, recombination, and mechanisms of stepwise mitochondrial genome reorganization in mantellid frogs from Madagascar. Mol Biol Evol 25:874–891. https://doi.org/10.1093/molbev/msn031

Landman W, Verneau O, Du Preez L (2018) Two new species of *Madapolystoma* (Monogenea: Polystomatidae) from the Madagascan anuran hosts *Blommersia domerguei* and *Mantella expectata*. Int J Parasitol: Parasites Wildl 7:343–354. https://doi.org/10.1016/j.ijppaw.2018.09.004

Landman W, Verneau O, Raharivololoniaina L, Du Preez LH (2021) First record of *Metapolystoma* (Monogenea: Polystomatidae) from *Boophis* tree frogs in Madagascar, with the description of five new species. Int J Parasitol Parasites Wildl 14:161–178. https://doi.org/10.1016/j.ijppaw.2021.01.012

Landman W, Verneau O, Vences M, Du Preez L (2023) *Metapolystoma ohlerianum* n. sp. (Monogenea: Polystomatidae) from *Aglyptodactylus madagascariensis* (Anura: Mantellidae). Acta Parasitol. https://doi.org/10.1007/s11686-023-00668-z

Li H, Liu J, Xiong L, Zhang H, Zhou H, Yin H, Jing W, Li J, Shi Q, Wang Y, Liu J, Nie L (2017) Phylogenetic relationships and divergence dates of softshell turtles (Testudines: Trionychidae) inferred from complete mitochondrial genomes. J Evol Biol 30:1011–1023. https://doi.org/10.1111/jeb.13070

Measey GJ, Vences M, Drewes RC, Chiari Y, Melo M, Bourlès B (2007) Freshwater paths across the ocean: molecular phylogeny of the frog *Ptychadena newtoni* gives insights into amphibian colonization of oceanic islands. J Biogeogr 34:7–20. https://doi.org/10.1111/j.1365-2699.2006.01589.x

Meyer L, Du Preez LH, Bonneau E, Héritier L, Quintana M-F, Valdeón A, Sadaoui A, Kechemir-Issad N, Palacios C, Verneau O (2015) Parasite host-switching from the invasive American red-eared slider, *Trachemys scripta elegans*, to the native Mediterranean pond turtle, *Mauremys leprosa*, in natural environments. Aquat Invasions 10:79–91. https://doi.org/10.3391/ai.2015.10.1.08

Myers N, Mittermeier RA, Mittermeier CG, da Fonseca GAB, Kent J (2000) Biodiversity hotspots for conservation priorities. Nature 403:853–858. https://doi.org/10.1038/35002501

Orliac M, Boisserie J-R, MacLatchy L, Lihoreau F (2010) Early Miocene hippopotamids (Cetartiodactyla) constrain the phylogenetic and spatiotemporal settings of hippopotamid origin. Proc Natl Acad Sci U S A 107:11871–11876. https://doi.org/10.1073/pnas.1001373107

Pickford M (2011) *Morotochoerus* from Uganda (17.5 Ma) and *Kenyapotamus* from Kenya (13–11 Ma): implications for hippopotamid origins. Estud Geol Madrid 67:523–540. https://doi.org/10.3989/egeol.40393.205

Raharivololoniaina L, Verneau O, Berthier P, Vences M, Du Preez LH (2011) First monogenean flatworm from a microhylid frog host: *Kankana*, a new polystome genus from Madagascar. Parasitol Int 60:465–473. https://doi.org/10.1016/j.parint.2011.08.001

Rhodin AGJ, Iverson JB, Bour R, Fritz U, Georges A, Shaffer HB, van Dijk PP (2021) Turtles of the world: annotated checklist and atlas of taxonomy, synonymy, distribution, and conservation status. 9th edition. Chelonian Res Monogr 8:1–472

Ronquist F (1997) Dispersal-vicariance analysis: a new approach to the quantification of historical biogeography. Syst Biol 46:195–203. https://doi.org/10.1093/sysbio/46.1.195

Sinnappah ND, Lim LHS, Rohde K, Tinsley R, Combes C, Verneau O (2001) A paedamorphic parasite associated with a neotenic amphibian host: phylogenetic evidence suggests a revised systematic position for Sphyranuridae within anuran and turtle Polystomatoineans. Mol Phylogenet Evol 18:189–201. https://doi.org/10.1006/mpev.2000.0877

Stunkard HW (1924) A new trematode, *Oculotrema hippopotami* n. g., n. sp., from the eye of the hippopotamus. Parasitology 16:436–440. https://doi.org/10.1017/S0031182000020333

Tinsley RC (2013) The oncomiracidium of *Oculotrema hippopotami* Stunkard, 1924 and relationships within the Polystomatidae (Monogenea). Syst Parasitol 84:123–135. https://doi.org/10.1007/s11230-012-9392-5

Van der Meijden A, Vences M, Meyer A (2004) Novel phylogenetic relationships of the enigmatic brevicipitine and scaphiophrynine toads as revealed by sequences from the nuclear Rag-1 gene. Proc R Soc B 271:S378–S381. https://doi.org/10.1098/rsbl.2004.0196

Van der Meijden A, Vences M, Hoegg S, Meyer A (2005) A previously unrecognized radiation of ranid frogs in southern Africa revealed by nuclear and mitochondrial DNA sequences. Mol Phylogenet Evol 37:674–685. https://doi.org/10.1016/j.ympev.2005.05.001

Van der Meijden A, Vences M, Hoegg S, Boistel R, Channing A, Meyer A (2007) Nuclear gene phylogeny of narrow-mouthed toads (family: Microhylidae) and a discussion of competing hypotheses concerning their biogeographical origins. Mol Phylogenet Evol 44:1017–1030. https://doi.org/10.1016/j.ympev.2007.02.008

Vences M, Vieites DR, Glaw F, Brinkmann H, Kosuch J, Veith M, Meyer A (2003) Multiple overseas dispersal in amphibians. Proc R Soc B 270:2435–2442. https://doi.org/10.1098/rspb.2003.2516

Vences M, Kosuch J, Rödel MO, Lötters S, Channing A, Glaw F, Böhme W (2004) Phylogeography of *Ptychadena mascareniensis* suggests transoceanic dispersal in a widespread African-Malagasy frog lineage. J Biogeogr 31:593–601. https://doi.org/10.1046/j.1365-2699.2003.01031.x

Verneau O, Bentz S, Sinnappah ND, Du Preez LH, Whittington I, Combes C (2002) A view of early vertebrate evolution inferred from the phylogeny of polystome parasites (Monogenea: Polystomatidae). Proc R Soc Lond Ser B Biol Sci 269:535–543. https://doi.org/10.1098/rspb.2001.1899

Verneau O, Du Preez LH, Laurent V, Raharivololoniaina L, Glaw F, Vences M (2009) The double odyssey of Madagascan polystome flatworms leads to new insights on the origins of their amphibian hosts. Proc R Soc B Biol Sci 276:1575–1583. https://doi.org/10.1098/rspb.2008.1530

Verneau O, Palacios C, Platt T, Alday M, Billard E, Allienne JF, Basso C, Du Preez LH (2011) Invasive species threat: parasite phylogenetics reveals patterns and processes of host-switching between non-native and native freshwater turtles. Parasitology 138:1778–1792. https://doi.org/10.1017/S0031182011000333

Verneau O, Johnston GR, Du Preez L (2023) A quantum leap in the evolution of platyhelminths: host-switching from turtles to hippos illustrated from a phylogenetic meta-analysis of polystomes (Monogenea, Polystomatidae). Int J Parasitol 53:317–325. https://doi.org/10.1016/j.ijpara.2023.03.0010020-7519/2023

Wollenberg KC, Glaw F, Meyer A, Vences M (2007) Molecular phylogeny of Malagasy reed frogs *Heterixalus*, and the relative performance of bioacoustics and color patterns for resolving their systematics. Mol Phylogenet Evol 45:14–22. https://doi.org/10.1016/j.ympev.2007.06.024

Wollenberg KC, Vieites DR, Van der Meijden A, Glaw F, Cannatella DC, Vences M (2008) Patterns of endemism and species richness in Malagasy cophyline frogs support a key role of mountainous areas for speciation. Evolution 62:1890–1907. https://doi.org/10.1111/j.1558-5646.2008.00420.x

Zimkus BM, Lawson LP, Barej MF, Barratt CD, Channing A, Dash KM, Dehling JM, Du Preez L, Gehring P-S, Gvoždík V, Harvey J, Kielgast J, Kusamba C, Nagy ZT, Pabijan M, Penner J, Rödel M-O, Vences M, Lötters S (2017) Leapfrogging into new territory: how Mascarene ridged frogs diversified across Africa and Madagascar to maintain their ecological niche. Mol Phylogenet Evol 106:254–269. https://doi.org/10.1016/j.ympev.2016.09.018

Chapter 8
Polystomatid Studies: Future Prospects

Contents

Abstract Regarding the diversity of polystomatids (Chaps. 4 and 5), along with their unique life-history strategies (Chap. 6) and evolution (Chap. 7), it is evident that much remains to be studied and endless species remain undiscovered. Here we provide a list of topics that, in our opinion, require further attention. Working more closely with herpetologists should be beneficial for parasitologists to extend the polystome species diversity within amphibians and turtles and, to some extent, revise classifications. Some chelonian groups should be investigated in greater depth for their polystomes. This is, for example, the case in the African trionychids and Australian pleurodires of the Testudines in order to highlight the origin of *Oculotrema hippopotami* Stunkard, 1924, on the one hand, and to test biogeographic scenarios that could be correlated to plate tectonics on the other. Furthermore, we recommend an expansion of phylogenetic studies that will contribute to a better understanding of taxonomy, systematics and evolution of polystomatids.

L. H. du Preez et al., *Polystomatid Flatworms*, Zoological Monographs 9,
https://doi.org/10.1007/978-3-031-35887-6_8

8.1 Expanding the Polystome Species Diversity

Since the description of the first polystomes in the late 1780s, with *Polystoma integerrimum* (Fröhlich, 1791) within frogs and *Polystomoides coronatus* (Leidy, 1888) within freshwater turtles, 135 polystome species have been described from amphibians (Chap. 4) and 65 from 46 distinct freshwater turtle species (Chap. 5). With more than 8687 amphibian species currently known (Frost 2023) and the moderate estimate of turtle and tortoise diversity of 357 species (Rhodin et al. 2021), it implies that globally 1 in every 63 amphibians, and roughly one in eight turtle species, harbour a polystome. Du Preez, Kok and co-workers reported seven polystomes with a further two yet undescribed species (Du Preez 1996, 2011; Du Preez and Kok 1993, 1995; Du Preez et al. 2002; Kok and Seaman 1987; Van Niekerk et al. 1993) from the Vernon Crookes Nature Reserve (South Africa) comprising an area of only 2189 ha and with a frog diversity of 27 species (Du Preez 1996). Thus, a ratio of one polystome for every three frog species. With many amphibian species globally yet undescribed and regarding all amphibian and chelonian species that have never been surveyed for polystomes, it is evident that a vast number of polystomes remain unknown. One may expect to discover and describe many more polystome species all over the world, particularly in areas like the Australian, Neotropical, Oriental and Palaearctic Realms. Since fieldwork sampling may be time-consuming and expensive, one way to facilitate more studies is to encourage herpetologists to communicate about their field activities and share their materials with parasitologists. This may contribute to expantion of parasitological collections and will, thereby increase the number of parasite descriptions, regardless of the types of host specimens collected.

8.2 Revising Polystome Classifications

Further investigation of the list of polystomes described from amphibians revealed that some polystome species were reported in several distinct host species and that certain host species were apparently infected by several polystome species. Because there have been strong arguments for strict host- and site-specificity within polystomes, on the one hand, and because, sometimes, identification of both hosts and their parasites can be extremely difficult in the absence of relevant morphological characters on the other, we may expect some confusion when delimitating parasite and host species. That could be the case for numerous parasite species like *Eupolystoma alluaudi* (de Beauchamp, 1913), *Polystoma africanum* (Szidat, 1932), *Polystoma aethiopiense* Meskal, 1970, *Polystoma gabonense* Euzet, Combes and Knoepffler, 1966, *Polystoma australe* Kok and van Wyk, 1986, *Polystoma ebriense* Maeder, 1973, *Polystoma grassei* Euzet, Combes and Knoepffler, 1966, *Polystoma ragnari* Maeder, Euzet and Combes, 1970, *Polystoma togoense* Bourgat, 1977, *Protopolystoma fissile* Tinsley and Jackson, 1998, *Protopolystoma simplicis* Tinsley and Jackson, 1998 and *Protopolystoma xenopodis* (Price, 1943)—each of which was reported from at least two distinct African anuran species. It was also the case for *Neodiplorchis scaphiopodis* (Rodgers, 1941), *Polystoma naevius* Caballero and Cerecero, 1941, *Polystoma napoense* Vaucher, 1987, *Polystoma nearcticum* (Paul, 1935), *Riojatrema bravoae* Lamothe, 1963 and *Sphyranura euryceae* Hughes and Moore, 1943 described from American amphibian species; for *Polystoma ezoense* Uchida, Machida, Uchida and Itagaki, 1988 and *P. integerrimum* described from European frog species and for *Parapolystoma bulliense* (Johnston, 1912) described from Australian frogs. For all of these parasites, a re-description of some specimens collected from the distinct host species, complemented by molecular investigation following COI sequencing, should help to re-evaluate the specific status of those polystomes. Finally, we should be able to conclude whether a single specific parasite species may occur in distinct host species or if only one parasite species, as was initially reported by authors of original descriptions, may infect several host species. The occurrence of several parasite species within the same host species was reported from African host species such as *Sclerophrys regularis* (Reuss, 1833), *Bufo gutturalis* Power, 1927, *Ptychadena bibroni* (Hallowell, 1845), *Ptychadena mascareniensis* (Duméril and Bibron, 1841), *Ptychadena porosissima* (Steindachner, 1867), *Ptychadena pumilio* (Boulenger, 1920), *Ptychadena superciliaris* (Günther, 1858), *Afrana angolensis* (Bocage, 1866), *Amnirana albolabris* (Hallowell, 1856), *Xenopus fraseri* Boulenger, 1905, *Xenopus wittei* Tinsley, Kobel and Fischberg, 1979, *Xenopus muelleri* (Peters, 1844), *Xenopus l. victorianus* Ahl, 1924 and *Xenopus l. poweri* Hewitt, 1927. Several parasite species were also reported from North American host species including *Necturus maculosus* (Rafinesque, 1818), from South American host species including *Osteocephalus taurinus* Steindachner, 1862 and *Rhinella margaritifera* (Laurenti, 1768), and from European host species such as *Hyla arborea* (Linnaeus, 1758) and *Rana temporaria* Linnaeus, 1758.

While parasite misidentification or description could also explain the presence of two or more polystome species within certain host species, misidentification of host species is also, to some extent, quite common, particularly for complexes of cryptic species. This is the case for the Mascarene ridged frog, *P. mascareniensis*, which occurs in humid savannas and open forests of mainland Africa, but also in Madagascar, the Seychelles and the Mascarene Islands. It is known today as a species complex including *Ptychadena nilotica* (Seetzen, 1855), *Ptychadena newtoni* (Bocage, 1886) and ten operational taxonomic units which deserve further description (Zimkus et al. 2017). Because *P. mascareniensis* has been reported as the host species for three polystome species, *Metapolystoma brygoonis* (Euzet and Combes, 1964) in Madagascar, *P. africanum* in Uganda and *P. aethiopiense* in Ethiopia, it is likely that all these polystome species may infect distinct host species. That's why a thorough revision of the taxonomy of hosts and their parasites, based on morphological and molecular grounds, is required.

The investigation of the polystome species list, described from freshwater turtles, led us to the same observations as were drawn from the amphibian polystome list. Chelonian polystomes can be found in three distinct ecological niches, namely, the conjunctival sacs, the oral cavity and the urinary bladder and associated cloaca of their host, usually with very strict host- and site-specificity. As a result, no more than three distinct polystome species is expected per host, which is not strictly the case regarding the list of already described species. Concerning polystome misidentification, one may wonder whether all of the four polystome species reported from the oral cavity of *Chrysemys picta* (Schneider, 1783) of the USA, namely *Polystomoides microcotyle* (Stunkard, 1916), *Polystomoides megacotyle* (Stunkard, 1916), *Polystomoides oris* Paul, 1938 and *Polystomoides pauli* Timmers and Lewis, 1979, are all valid species; the taxonomy of the polystomes infecting *C. picta* in the wild was indeed questioned with regard to COI sequences obtained from several specimens collected in allopatry (unpublished results). Besides polystome species delimitations, though problems of misidentification of turtle species is certainly less than for frog species, it has been reported in species complexes, as in the case of the African helmeted terrapin, *Pelomedusa subrufa* (Bonnaterre, 1789). It is known today as a complex of ten species in sub-Saharan Africa, Arabian Peninsula and Madagascar (Vargas-Ramírez et al. 2010; Petzold et al. 2014). Two polystome species have been reported from the urinary bladder of *P. subrufa*, namely *Uropolystomoides nabedei* (Kulo, 1980) in Togo and *Uropolystomoides chabaudi* (Euzet and Combes, 1965) in Uganda and Madagascar, which leads one to suspect that each of the *Pelomedusa* spp. reported in Africa may be hosting a distinct polystome species. Nevertheless, the occurrence of several parasite species within the same host species was also documented from COI DNA barcoding within the red-eared slider turtle, *Trachemys scripta elegans* (Thunberg, 1792) (Héritier et al. 2018). Therefore, as was also documented for some of the amphibian polystomes, some polystome species need further investigation for validation.

8.3 Documenting Species Diversity Within Particular Polystome Genera

Polystomoidella Price, 1939 is a genus which differentiates from all other polystome genera infecting freshwater turtles by the presence of a single pair of hamuli. *Polystomoidella oblongum* (Wright, 1879) and *Polystomoidella whartoni* (Wright, 1879) were described more than one century ago from the urinary bladder of distinct American host species of the genera *Kinosternon* Spix, 1824, *Chrysemys* Gray, 1844, *Chelydra* Schweigger, 1812 and *Sternotherus* Bell, 1825. Along with *Polystomoidella mayesi* Richardson and Brooks, 1987 from the urinary bladder of *Cuora amboinensis* (Daudin, 1801) in Malaysia, *Polystomoidella* today comprises only three species. Because *P. oblongum* and *P. whartoni* were each recorded in several divergent host species of distinct genera, one may wonder about the accuracy of these reports. A revision of the taxonomy of *Polystomoidella* from COI mitochondrial markers is therefore required. Similarly, because *Kinosternon* seems to be the preferential host for *Polystomoidella*, investigating *Kinosternon* spp. across their distribution range, from the USA to Peru, should provide a clear picture of the *Polystomoidella* diversity. *Kinosternon* comprises 22 species, 7 in the USA—among which 3 are endemic and 16 in Mexico—among which 10 are endemic (Rhodin et al. 2021). As *Polystomoidella* was already reported from *Kinosternon integrum* Le Conte, 1854 and *Kinosternon hirtipes* (Wagler, 1830) in Mexico, we may expect that many more species of *Kinosternon* are infected. Furthermore, as *Sternotherus* is also a good candidate for *Polystomoidella*, investigating its polystome diversity across its home range, from Ontario and Quebec to the USA (Rhodin et al. 2021), should also provide interesting information. Finally, as *Polystomoidella mayesi* Richardson and Brooks, 1987 is the single polystome species of that genus recorded outside North America, we may question its validity. It could have been introduced following the pet trade of *T. scripta* in Asia, as was illustrated for some polystomes in Europe (see Meyer et al. 2015; Héritier et al. 2017).

Nanopolystoma Du Preez, Wilkinson and Huyse, 2008 is the single genus recorded from caecilians, with only three species described so far in South America. Caecilians are poorly diversified with 221 species (Frost 2023) distributed in tropical regions of Central and South America, Africa, Seychelles islands and South-east Asia (see Zhang and Wake 2009). It may be expected that polystomes of African caecilians also exist and still await discovery. Because some of the caecilian families like the paraphyletic Caeciliidae are distributed in Africa and South America (Zhang and Wake 2009), it would require testing the historical biogeographic hypotheses following plate tectonics during the breakup of Gondwana.

8.4 Looking for the Plesiomorphic Host of *Oculotrema hippopotami*

Because *Oculotrema hippopotami* Stunkard, 1924 is the single polystome species infecting a mammal, namely the common hippopotamus *Hippopotamus amphibius* Linnaeus, 1758, one may have expected it to have originated from either amphibian or chelonian polystomes. Though Héritier et al. (2015) showed it was more closely related to chelonian polystomes than to amphibian polystomes, they hypothesized, with regard to molecular dating, that it would have originated from caecilians. It is only recently that Verneau et al. (2023) showed it would have been derived from freshwater turtles, most likely from African trionychids. Therefore, investigating the diversity of polystomes within trionychid turtles from Africa, namely *Cyclanorbis elegans* (Gray, 1869), *Cyclanorbis senegalensis* (Duméril and Bibron, 1835), *Cycloderma aubryi* (Duméril, 1856), *Cycloderma frenatum* Peters, 1854 and *Trionyx triunguis* (Forskål, 1775) could shed light on the plesiomorphic host for this enigmatic parasite.

8.5 Investigating Polystome Species from New Host Species Groups

The biological life cycle of polystomes involves an aquatic larval stage during which infective oncomiracidia meet new host specimens for transmission. This prerequisite in the parasite transmission implies similar aquatic environments for infection success, which explains how *H. amphibius* may have been infected by polystomes during evolution. Considering that hippopotami are good hosts for polystomes, we expect that other groups of vertebrates like, among others, water snakes, crocodilians, sirenians and even marine turtles, can also be infected.

8.6 Amending the Classification Within Anuran Polystomes

According to the phylogenetic studies of Verneau et al. (2009), anuran polystomes form a clade, which associates several monophyletic genera as *Indopolystoma* Chaabane, Verneau and Du Preez, 2019 (a genus composed of rhacophorid polystome species, in Chaabane et al. 2019), *Eupolystoma* (de Beauchamp, 1913), *Madapolystoma* Du Preez, Raharivololoniaina, Verneau and Vences, 2010 and *Metapolystoma* Combes, 1976. However, if *Metapolystoma* seems a valid genus nested within *Polystoma* Ozaki, 1935 (Verneau et al. 2009; Landman et al. 2021), then *Polystoma* is paraphyletic with respect to *Metapolystoma*. With regard to the most comprehensive phylogeny of anuran polystomes (Verneau et al. 2009), five lineages are well differentiated, including, respectively, an African polystome lineage with *Polystoma dawiekoki* du Preez, Vaucher and Mariaux, 2002 and *Polystoma occipitalis* Maeder, 1973, an Eurafrican polystome lineage with *Polystoma gallieni* Price, 1939, *Polystoma claudecombesi* du Preez and Kok, 1995 and *Polystoma marmorati* van Niekerk, Kok and Seaman, 1993, an European polystome lineage with *P. integerrimum* and two American lineages. Therefore, one may expect that specific morphological characters characterize each of these five lineages. Thus, we propose revising the classification of *Polystoma* by looking for new interesting morphological characters within *Polystoma* collected from all over the world, except in Australia where it has never been reported.

8.7 Expanding the Phylogeny of Chelonian Polystomes of the Suborder Pleurodira

Until Tinsley and Tinsley (2016) erected a new genus within the Polystomoidinae of the Polystomatidae, chelonian polystomes were subdivided into three genera based on the presence of hamuli, namely two pairs of hamuli for *Polystomoides* Ward, 1917, one pair of hamuli for *Polystomoidella* Price, 1939 and none for

Neopolystoma Price, 1939. Considering phylogenetic studies by Héritier et al. (2015), Tinsley and Tinsley (2016) created a new genus *Uropolystomoides* Tinsley and Tinsley, 2016 for the clade which includes all species with two pairs of hamuli infecting the urinary bladder of their host. This genus is well characterized by the size of hamulus 1 which is always greater than the sucker diameter. Then Tinsley (2017) proposed *Uteropolystomoides* Tinsley, 2017 for *Polystomoides nelsoni* Du Preez and Van Rooyen, 2015, which is characterized by an uterus containing numerous eggs but also by a massive genital bulb with a great number of genital spines.

Du Preez and Verneau (2020) created three new genera for all polystomes occurring in the conjunctival sacs of their hosts. *Aussietrema* Du Preez and Verneau, 2020 is mainly characterized by a spherical ovary and egg, *Fornixtrema* Du Preez and Verneau, 2020 by an egg-cell-maturation-chamber and fusiform to diamond-shaped egg with acute tips and *Apaloneotrema* Du Preez and Verneau, 2020 by a large fusiform egg with rounded tips. Du Preez et al. (2022) also created *Pleurodirotrema* Du Preez, Domingues and Verneau, 2022 for Australian chelonian polystomes with latero-ventral vaginae and without hamuli, infecting the oral region or the bladder of pleurodires. Likewise, *Manotrema* Du Preez, Domingues and Verneau, 2022 was created for chelonian polystomes infecting pleurodires from the Neotropical Realm (Du Preez et al. 2022). This genus shows deep cuts between the suckers, latero-ventral vaginae and two pairs of very small hamuli with deep cuts between the handle and the guard of the bigger hamulus.

Finally, with regard to a clade associating species of the two paraphyletic *Neopolystoma* and *Polystomoides* genera (see Héritier et al. 2015), Chaabane et al. (2022) considered only a single genus, *Polystomoides*. This genus, which associates all species infecting the urinary bladder and the oral cavity of turtles of the suborder Cryptodira, with the exception of *Uropolystomoides* spp., shows peripheral vaginae and displays either none or two pairs of small hamuli. In their revision of the classification of chelonian polystomes, Chaabane et al. (2022) still considered *Uteropolystomoides* as a valid taxon but considered *Uteropolystomoides nelsoni* (Du Preez and Van Rooyen, 2015), the unique species originally described from *Pseudemys nelsoni* Carr, 1938 as *Uteropolystomoides multifalx* (Stunkard, 1924) which infects three distinct *Pseudemys* Gray, 1856 species of North America.

As *Uropolystomoides* spp. from the Australian and African Realms differ morphologically from those of the Oriental Realm (see Du Preez et al. 2022), an in-depth investigation of types could be very helpful to conclude and go further in the revision of the classification. Furthermore, a more thorough analysis of the phylogenetic relationships of the Polystomatidae, including a few representatives of the genera *Aussietrema*, *Pleurodirotrema* and *Uropolystomoides* from the Australian Realm as well as species of *Manotrema* will permit the testing of biogeographic scenarios possibly related to plate tectonics.

8.8 Deciphering Patterns and Processes of Chelonian Polystome Evolution

Chelonian polystomes are mainly host- and site-specific with the exception of a few species (see Verneau et al. 2011; Meyer et al. 2015; Héritier et al. 2017). They infect either the conjunctival sacs, the oral cavity and/or the urinary bladder and cloaca of their host, which means that up to three parasite species can be found in the same host species. One therefore wonders, if speciation occurred within or between host species. Littlewood et al. (1997), based on a phylogeny of six species infecting distinct habitats of four chelonian species of Australia and Malaysia, did not find any correlation between parasite speciation and host microhabitat, nor with host speciation and geographical distribution. Héritier et al. (2015), from a phylogeny based on 20 polystome species sampled all over the world from distinct ecological niches of 15 distinct host species, concluded that the evolution of chelonian polystomes could be the consequence of multiple events such as plate tectonics, co-divergences, host dispersal followed by host switching and intra-host speciation. Still later, Verneau et al. (2023) concluded, from a larger sample including 28 species, that host switching across the same ecological habitat would have occurred frequently during the evolution of freshwater turtles. Host switching would be most likely be between sympatric host species of similar ecology rather than between closely related host species. In order to clarify the evolutionary processes involved in the diversification of chelonian polystomes, we now recommend expanding the polystome sampling, especially from pleurodires and to conduct co-phylogenetic studies in a biogeographic context as illustrated for polystomes of the Neobatrachia (see Badets et al. 2011). Because polytomies, or weakly supported nodes within the tree of chelonian polystomes (see Du Preez and Verneau 2020), may blur the co-evolutionary history of hosts and their parasites, we also recommend adding more molecular characters, as, for instance, characters inferred from the sequencing of ITS1, which look promising (Verneau, unpublished observations).

8.9 Determining the Role and Importance of *Trachemys scripta elegans* in the Dissemination of Polystome Parasites

Using DNA barcoding to identify polystome species from adult worms and/or eggs, several cases of host switching from exotic to indigenous turtles were illustrated in turtle farms, demonstrating that captive turtles could act as reservoirs of exotic parasites (Verneau et al. 2011). *Trachemys s. elegans*, which is considered to be among the world's worst invasive species, was introduced almost everywhere as a result of the pet trade of exotic animals. Because feral populations of this species have now become established in natural continental aquatic ecosystems of Southern Europe, the question was whether its parasites may have spread to indigenous turtles like the Mediterranean pond turtle *Mauremys leprosa* (Schweigger, 1812) and the European pond turtle *Emys orbicularis* (Linnaeus, 1758). Polystome surveys that were conducted within these two native turtles showed a polystome diversity much higher than expected (Meyer et al. 2015; Héritier et al. 2017). Besides the native polystome species that were retrieved in both indigenous turtles species, numerous other polystome species were identified—among which were species that were also documented in wild *T. s. elegans*. Because some of these parasites were originally described from American turtles in their home range, like *C. picta*, it was assumed that *T. s. elegans* could act as a taxi for polystome dispersal from American to European wetlands (Meyer et al. 2015). As *T. s. elegans* is also considered a pest in the USA, one may wonder about the polystome diversity of this species inside, but also outside, its home range of the USA, as well as about the host-specificity of its parasites. Since two polystomes were recently described from the pharyngeal cavity of *T. s. elegans*, namely *Polystomoides soredensis* Héritier, Verneau, Smith, Coetzer and Du Preez, 2018 and *Polystomoides scriptanus* Héritier, Verneau, Smith, Coetzer and Du Preez, 2018 and because both polystome species were retrieved recently from *T. s. elegans* and other American turtles of the USA after fieldwork activities in North Carolina and Florida in 2015, 2018 and 2019 (Verneau and Du Preez unpublished observations), an extensive polystome survey across American turtles of the USA should give new insights about the role of *T. s. elegans* in parasite dispersal.

References

Badets M, Whittington I, Lalubin F, Allienne JF, Maspimby JL, Bentz S, Du Preez LH, Barton D, Hasegawa H, Tandon V, Imkongwapang R, Ohler A, Combes C, Verneau O (2011) Correlating early evolution of parasitic platyhelminths to Gondwana breakup. Syst Biol 60:762–781. https://doi.org/10.1093/sysbio/syr078

Chaabane A, Verneau O, Du Preez L (2019) *Indopolystoma* n. gen. (Monogenea, Polystomatidae) with the description of three new species and reassignment of eight known *Polystoma* species from Asian frogs (Anura, Rhacophoridae). Parasite 26:67. https://doi.org/10.1051/parasite/2019067

Chaabane A, Du Preez LH, Johnston GR, Verneau O (2022) Revision of the systematics of the Polystomoidinae (Platyhelminthes, Monogenea, Polystomatidae) with redefinition of *Polystomoides* Ward, 1917 and *Uteropolystomoides* Tinsley, 2017. Parasite 29:56. https://doi.org/10.1051/parasite/2022056

Du Preez LH (1996) Anuran community structure in the Vernon Crookes Nature Reserve. Modoqua 19:25–29

Du Preez LH (2011) *Polystoma vernoni* n. sp. (Monogenea: Polystomatidae) from the sharp nosed grass frog *Ptychadena oxyrhynchus* (Smith, 1849) in South Africa. J Helminthol 85:294–299

Du Preez LH, Kok DJ (1993) Polystomatidae (Monogenea) of Anura in southern Africa: *Polystoma testimagna* n. sp. parasitic in *Strongylopus f. fasciatus* (Smith, 1849). Syst Parasitol 25:213–219

Du Preez LH, Kok DJ (1995) Polystomatidae (Monogenea) of southern African Anura: *Polystoma claudecombesi* n. sp. parasitic in *Rana angolensis* Bocage, 1866. Syst Parasitol 30:223–231

Du Preez LH, Verneau O (2020) Eye to eye: classification of conjunctival sac polystomes (Monogenea: Polystomatidae) revisited with the description of three new genera *Apaloneotrema*

n. g., *Aussietrema* n. g. and *Fornixtrema* n. g. Parasitol Res 119:4017–4031. https://doi.org/10.1007/s00436-020-06888-w

Du Preez LH, Vaucher C, Mariaux J (2002) Polystomatidae (Monogenea) of African Anura. *Polystoma dawiekoki* n. sp. parasitic in *Ptychadena anchietae* (Bocage). Syst Parasitol 52:35–41

Du Preez L, Domingues MV, Verneau O (2022) Classification of pleurodire polystomes (Platyhelminthes, Monogenea, Polystomatidae) revisited with the description of two new genera from the Australian and Neotropical Realms. Int J Parasitol: Parasites Wildl 19:180–186. https://doi.org/10.1016/j.ijppaw.2022.09.004

Frost DR (2023) Amphibian species of the world: an online reference. Version 6.1 (13 October 2023). American Museum of Natural History, New York, USA. https://amphibiansoftheworld.amnh.org/index.php. https://doi.org/10.5531/db.vz.0001

Héritier L, Badets M, Du Preez LH, Aisien MS, Lixian F, Combes C, Verneau O (2015) Evolutionary processes involved in the diversification of chelonian and mammal polystomatid parasites (Platyhelminthes, Monogenea, Polystomatidae) revealed by palaeoecology of their hosts. Mol Phylogenet Evol 92:1–10. https://doi.org/10.1016/j.ympev.2015.05.026

Héritier L, Valdeón A, Sadaoui A, Gendre T, Ficheux S, Bouamer S, Kechemir-Issad N, Palacios C, Du Preez LH, Verneau O (2017) Introduction and invasion of the red-eared slider and its parasites in freshwater ecosystems of Southern Europe: risk assessment for the European pond turtle in wild environments. Biodivers Conserv 26:817–1843. https://doi.org/10.1007/s10531-017-1331-y

Héritier L, Verneau O, Smith KG, Coetzer C, Du Preez LH (2018) Demonstrating the value and importance of combining DNA barcodes and discriminant morphological characters for polystome taxonomy (Platyhelminthes, Monogenea). Parasitol Int 67:38–46. https://doi.org/10.1016/j.parint.2017.03.002

Kok DJ, Seaman MT (1987) Polystomatidae (Monogenea) parasitic in the anuran genus *Natalobatrachus* in South Africa. S Afr J Zool 22:258–263

Landman W, Verneau O, Raharivololoniaina L, Du Preez L (2021) First record of *Metapolystoma* (Monogenea: Polystomatidae) from *Boophis* tree frogs in Madagascar, with the description of five new species. Int J Parasitol: Parasites Wildl 14:161–178. https://doi.org/10.1016/j.ijppaw.2021.01.012

Littlewood DTJ, Rohde K, Clough KA (1997) Parasite speciation within or between host species? – Phylogenetic evidence from site-specific polystome monogeneans. Int J Parasitol 27:1289–1297

Meyer L, Du Preez LH, Bonneau E, Héritier L, Quintana M-F, Valdeón A, Sadaoui A, Kechemir-Issad N, Palacios C, Verneau O (2015) Parasite host-switching from the invasive American red-eared slider, *Trachemys scripta elegans*, to the native Mediterranean pond turtle, *Mauremys leprosa*, in natural environments. Aquat Invasions 10:79–91. https://doi.org/10.3391/ai.2015.10.1.08

Petzold A, Vargas-Ramirez M, Kehlmaier C, Vamberger M, Branch WR, Du Preez L, Hofmeyr MD, Meyer L, Schleicher A, Siroky P, Fritz U (2014) A revision of African helmeted terrapins (Testudines: Pelomedusidae: *Pelomedusa*), with descriptions of six new species. Zootaxa 3795:523–548. https://doi.org/10.11646/zootaxa.3795.5.2

Rhodin AGJ, Iverson JB, Bour R, Fritz U, Georges A, Shaffer HB, van Dijk PP (2021) Turtles of the world: annotated checklist and atlas of taxonomy, synonymy, distribution, and conservation status. 9th edition. Chelonian Res Monogr 8:1–472

Tinsley R (2017) Reproductive innovation and the recognition of a new genus within the Polystomatidae (Monogenea) infecting chelonian vertebrates. Folia Parasitol 64:017. https://doi.org/10.14411/fp.2017.017

Tinsley RC, Tinsley MC (2016) Tracing ancient evolutionary divergence in parasites. Parasitology 143:1902–1916. https://doi.org/10.1017/S0031182016001347

Van Niekerk S, Kok DJ, Seaman MT (1993) A new species of *Polystoma* (Monogenea: Polystomatidae) parasitic in *Hyperolius marmoratus* (Anura: Hyperoliidae) in South Africa. Syst Parasitol 25:73–80

Vargas-Ramírez M, Vences M, Branch WR, Daniels SR, Glaw F, Hofmeyr MD, Kuchling G, Maran J, Papenfuss TJ, Siroky P, Vieites DR, Fritz U (2010) Deep genealogical lineages in the widely distributed African helmeted terrapin: evidence from mitochondrial and nuclear DNA (Testudines: Pelomedusidae: *Pelomedusa subrufa*). Mol Phylogenet Evol 56:428–440. https://doi.org/10.1016/j.ympev.2010.03.019

Verneau O, Du Preez LH, Laurent V, Raharivololoniaina L, Glaw F, Vences M (2009) The double odyssey of Madagascan polystome flatworms leads to new insights on the origins of their amphibian hosts. Proc R Soc B Biol Sci 276:1575–1583. https://doi.org/10.1098/rspb.2008.1530

Verneau O, Palacios C, Platt T, Alday M, Billard E, Allienne JF, Basso C, Du Preez LH (2011) Invasive species threat: parasite phylogenetics reveals patterns and processes of host-switching between non-native and native freshwater turtles. Parasitology 138:1778–1792. https://doi.org/10.1017/S0031182011000333

Verneau O, Johnston GR, Du Preez L (2023) A quantum leap in the evolution of platyhelminths: host-switching from turtles to hippos illustrated from a phylogenetic meta-analysis of polystomes (Monogenea, Polystomatidae). Int J Parasitol 53:317–325. https://doi.org/10.1016/j.ijpara.2023.03.0010020-7519/2023

Zhang P, Wake MH (2009) A mitogenomic perspective on the phylogeny and biogeography of living caecilians (Amphibia: Gymnophiona). Mol Phylogenet Evol 53:479–491. https://doi.org/10.1016/j.ympev.2009.06.018

Zimkus BM, Lawson LP, Barej MF, Barratt CD, Channing A, Dash KM, Dehling JM, Du Preez L, Gehring P-S, Gvoždík V, Harvey J, Kielgast J, Kusamba C, Nagy ZT, Pabijan M, Penner J, Rödel M-O, Vences M, Lötters S (2017) Leapfrogging into new territory: how Mascarene ridged frogs diversified across Africa and Madagascar to maintain their ecological niche. Mol Phylogenet Evol 106:254–269. https://doi.org/10.1016/j.ympev.2016.09.018

GPSR Compliance

The European Union's (EU) General Product Safety Regulation (GPSR) is a set of rules that requires consumer products to be safe and our obligations to ensure this.

If you have any concerns about our products, you can contact us on ProductSafety@springernature.com

In case Publisher is established outside the EU, the EU authorized representative is:

Springer Nature Customer Service Center GmbH
Europaplatz 3
69115 Heidelberg, Germany

Batch number: 09223966

Printed by Printforce, the Netherlands